Lecture Notes in Computer Science   11811

Nieves R. Brisaboa · Simon J. Puglisi (Eds.)

String Processing
and Information Retrieval

26th International Symposium, SPIRE 2019
Segovia, Spain, October 7–9, 2019
Proceedings

 Springer

Editors
Nieves R. Brisaboa ⓘ
University of A Coruña
A Coruña, Spain

Simon J. Puglisi ⓘ
University of Helsinki
Helsinki, Finland

ISSN 0302-9743   ISSN 1611-3349   (electronic)
Lecture Notes in Computer Science
ISBN 978-3-030-32685-2   ISBN 978-3-030-32686-9   (eBook)
https://doi.org/10.1007/978-3-030-32686-9

LNCS Sublibrary: SL1 – Theoretical Computer Science and General Issues

This Springer imprint is published by the registered company Springer Nature Switzerland AG
The registered company address is: Gewerbestrasse 11, 6330 Cham, Switzerland

Preface

SPIRE 2019, held October 7–9, 2019, in Segovia, Spain, was the 26th International Symposium on String Processing and Information Retrieval. SPIRE started in 1993 as the South American Workshop on String Processing, therefore it was held in Latin America until 2000, when SPIRE traveled to Europe. From then on, SPIRE meetings have been held in Australia, Japan, UK, Spain, Italy, Finland, Portugal, Israel, Brazil, Chile, Colombia, Mexico, Argentina, Bolivia, and Peru.

In this edition, again in Spain, we continued the long and well-established tradition of encouraging high-quality research at the broad nexus of algorithms and data structures for sequences and graphs, data compression, databases, data mining, information retrieval, and computational biology. As usual, SPIRE 2019 continues to provide an opportunity to bring together specialists and young researchers working in these areas.

This volume contains the 36 papers, out of a total of 59 submissions accepted to be presented in SPIRE 2019. Each submission received at least three reviews. Authors of accepted papers come from 17 countries, across five continents (Africa, Asia, Europe, North America, South America). We thank all authors who submitted their work for consideration to SPIRE 2019 and we especially thank the Program Committee and the external reviewers, whose many thorough reviews helped us select the papers presented. The success of the scientific program is due to their hard work.

Besides the 36 accepted papers, the scientific program included three invited lectures, given by:

- Veli Mäkinen on "When Stringology Meets Graphs"
- Alistair Moffat on "User-Based Evaluation in Information Retrieval"
- Gonzalo Navarro on "Repetitiveness and Indexability"

We thank the invited speakers for accepting our invitation and for their excellent presentations at the conference.

To complete the event, this year for the fourth year running, SPIRE 2019 had a Best Paper Award, sponsored by Springer that was announced during the conference. Besides Springer, we thank the EU project BIRDS (H2020-MSCA-RISE-2015 GA No 690941) for its financial support and the ICT Research Center CITIC at the University of A Coruña and the Segovia Campus of the University of Valladolid whose administrative and financial support we gratefully acknowledge.

October 2019

Nieves R. Brisaboa
Simon J. Puglisi

Organization

Program Committee

Amihood Amir	Bar-Ilan University, Israel and Johns Hopkins University, USA
Ricardo Baeza-Yates	Northeastern University, USA and Pompeu Fabra University, Spain
Hideo Bannai	Kyushu University, Japan
Christina Boucher	University of Florida, USA
Nieves Brisaboa	University of A Coruña, Spain
Antonio Fariña	University of A Coruña, Spain
Johannes Fischer	TU Dortmund, Germany
Jose Fuentes	University of Chile, Chile
Travis Gagie	Dalhousie University, Canada
Pawel Gawrychowski	University of Wroclaw, Poland
Simon Gog	eBay Inc., USA
Inge Li Gørtz	Technical University of Denmark, Denmark
Susana Ladra	University of A Coruña, Spain
Zsuzsanna Liptak	University of Verona, Italy
Miguel A. Martinez-Prieto	University of Valladolid, Spain
Jose R. Parama	University of A Coruña, Spain
Kunsoo Park	Seoul National University, Republic of Korea
Matthias Petri	The University of Melbourne, Australia
Solon Pissis	CWI, The Netherlands
Simon Puglisi	University of Helsinki
Marinella Sciortino	University of Palermo, Italy
Diego Seco	University of Concepción, Chile
Jouni Sirén	University of California at Santa Cruz, USA
Yasuo Tabei	RIKEN Center for Advanced Intelligence Project, Japan
Rossano Venturini	University of Pisa, Italy
Nivio Ziviani	Federal University of Minas Gerais, Brazil

Additional Reviewers

Abedin, Paniz
Ahanonu, Eze
Arroyuelo, Diego
Ayad, Lorraine
Bernardini, Giulia
Bille, Philip
Boneh, Itai
Brandão, Wladmir
Calvo-Zaragoza, Jorge
Cazaux, Bastien
Charalampopoulos, Panagiotis
Clifford, Raphael
De Sensi, Daniele
Epifanio, Chiara
Fici, Gabriele
Freire Castro, Borja
Fujishige, Yuta
Galaktionov, Daniil
Gańczorz, Michał
Glowacka, Dorota
Goto, Keisuke
Gómez-Brandón, Adrián
Holland, William
Inenaga, Shunsuke
Janczewski, Wojciech
Kanda, Shunsuke
Kida, Takuya
Klein, Shmuel Tomi
Kolesnikov, Vladimir
Komusiewicz, Christian

Kondratovsky, Eitan
Köppl, Dominik
Levy, Avivit
MacKenzie, Joel
Mieno, Takuya
Moffat, Alistair
Mukherjee, Kingshuk
Nakashima, Yuto
Nekrich, Yakov
Nishimoto, Takaaki
Ochoa, Carlos
Pibiri, Giulio Ermanno
Piatkowski, Marcin
Pokorski, Karol
Prezza, Nicola
Radoszewski, Jakub
Raman, Rajeev
Rossi, Massimiliano
Salmela, Leena
Schmid, Markus L.
Shalom, Riva
Silvestre Vilches, Jorge
Thankachan, Sharma V.
Trotman, Andrew
Turpin, Andrew
Valle, Dan
Zhukova, Bella
Zobel, Justin
Zuin, Gianlucca

Abstracts of Invited Talks

Repetitiveness and Indexability

Gonzalo Navarro

CeBiB—Center for Biotechnology and Bioengineering,
IMFD—Millennium Institute for Foundational Research on Data,
Department of Computer Science, University of Chile, Chile
gnavarro@dcc.uchile.cl

Abstract. Compressed indexes for highly repetitive text collections can reduce the data size by orders of magnitude while still supporting efficient searches. Compression of this kind of data requires dictionary-based methods, because statistical compression fails to capture repetitiveness. Unlike statistical compression, where the state of the art is mature and indexes reaching entropy size are already several years old, there is not even a clear concept of entropy for highly repetitive collections. There is a wealth of measures, some more ad-hoc and some more principled. Some relations are known between them, other relations are unknown. It is known that no compressor can reach some measures, it is known how to reach others, and for some it is unknown whether this is possible. From the reachable ones, some allow random access to the compressed text, for others it is unknown how to do it. Finally, some admit indexed searches, for others we do not know if this is possible. In this talk I will survey this zoo of measures, show their properties and known relations, show what is known and unknown about them, and point out several open questions that relate repetitiveness with indexability.

Keywords: Repetitive text collections · Compressed text indexing · Entropy

Partially supported by Basal Funds B0001, Conicyt, Chile and by the Millennium Institute for Foundational Research on Data, Mideplan, Chile.

C/W/L Spells "Cool": User-Based Evaluation in Information Retrieval

Alistair Moffat

The University of Melbourne, Australia

Abstract. The Information Retrieval community pride themselves on the strength of their evaluation protocols: working with large test collections; executing dozens or hundreds of queries taken to be representative of typical information requirements; and, in many cases, employing expert assessors to form relevance judgments. System scores using these resources are then computed using an effectiveness metric such as precision at depth k, expected reciprocal rank, or average precision; and champion-versus-challenger evaluations are carried out by considering the two system means through the lens of a statistical significance test.

This presentation focuses on the effectiveness metrics that are at the heart of this batch evaluation pipeline. After describing a range of traditional approaches to measuring effectiveness, the "C/W/L" framework [2, 3] is motivated and defined, and a range of implications of this approach to IR evaluation then explored. Notable in the C/W/L structure is the explicit correspondence between metrics and user models. This relationship makes it possible for metrics to be evaluated and compared in terms of their suitability for different types of search task, based on the extent to which the user model associated with each candidate metric correlates with observed user behavior when performing that task [1, 4, 5]. Measurement accuracy is also considered for C/W/L metrics, together with the implications that certain types of user behavior then have on experimental design.

Keywords: Information retrieval evaluation · Web search · User model · Effectiveness metric

Acknowledgment. The work presented in this talk was carried out in collaboration with Peter Bailey, Falk Scholer, Paul Thomas, Alfan Wicaksono, and Justin Zobel. Their various contributions are gratefully acknowledged.

References

1. Azzopardi, L., Thomas, P., Craswell, N.: Measuring the utility of search engine result pages. In: Proceedings of SIGIR, pp. 605–614 (2018)
2. Moffat, A., Bailey, P., Scholer, F., Thomas, P.: Incorporating user expectations and behavior into the measurement of search effectiveness. ACM Trans. Inf. Sys. **35**(3), 24:1–24:38 (2017)

3. Moffat, A., Thomas, P., Scholer, F.: Users versus models: what observation tells us about effectiveness metrics. In: Proceedings of CIKM, pp. 659–668 (2013)
4. Wicaksono, A.F., Moffat, A.: Empirical evidence for search effectiveness models. In: Proceedings of CIKM, pp. 1571–1574 (2018)
5. Wicaksono, A.F., Moffat, A., Zobel, J.: Modeling user actions in job search. In: Azzopardi, L., Stein, B., Fuhr, N., Mayr, P., Hauff, C., Hiemstra, D. (eds.) Advances in Information Retrieval. ECIR 2019. LNCS, vol. 11437, pp. 652–664. Springer, Cham (2019). https://doi.org/10.1007/978-3-030-15712-8_42

When Stringology Meets Graphs

Veli Mäkinen ⓘ

Department of Computer Science, University of Helsinki, Finland
veli.makinen@helsinki.fi

Abstract. Consider a directed acyclic graph (DAG) G with nodes labelled with characters. We say that a string pattern P occurs in G if there is a path spelling P. When G is deterministic, that is, no node has two edges leading to nodes with the same character label, there is a trivial algorithm to locate P in G: Start at all places and check if path spelling P exists. This trivial algorithm turns out to be optimal under the Strong Exponential Time Hypothesis (SETH). The talk starts by explaining this result by Equi, Grossi, Mäkinen, and Tomescu (*ICALP* 2019).

Quadratic running time for matching pattern P against graph G can slightly be improved without violating SETH, by using bit-parallelism. The talk discusses extensions of Shift-And and Myers' algorithms for exact and approximate pattern matching on graphs as studied by Rautiainen, Mäkinen, and Marschall (*Bioinformatics*, to appear).

Sparse dynamic programming is another technique that can evade the quadratic bound, assuming a sub-quadratic size set of anchors is given as input to limit the alignment options. An anchor defines a plausible alignment of a substring against a subpath. An ordered subset of anchors forms a co-linear chain if the corresponding substrings are in linear order in P and the corresponding subpaths are in linear order in some path of G. Consider the problem of finding a co-linear chain that maximally covers P. This problem is studied by Mäkinen, Tomescu, Kuosmanen, Paavilainen, Gagie, and Chikhi (*ACM Transactions on Algorithms*, 2019), who give an algorithm whose running time depends on the number of paths needed to cover G; the algorithm is optimal once G is just a string. The talk covers the main insights of this algorithm.

The talk concludes with another alignment problem related to path covers. Consider two DAGs G_1 and G_2 each of which is coverable by at most two paths. Such DAGs can be seen as simplest extension of strings into graphs and are also representing diploid genomes. A covering alignment asks for path covers (A, B) and (C, D) of G_1 and G_2, respectively, that minimize the sum of edit distance between A and C and between B and D. Covering alignment turns out to be NP-hard as shown by Rizzi, Cairo, Mäkinen, Tomescu, and Valenzuela (*IEEE/ACM Transactions on Computational Biology and Bioinformatics*). The talk gives an overview of the reduction.

Keywords: String matching · Graphs · SETH · Bit-parallelism · Sparse dynamic programming · Covering alignment

Partially supported by the Academy of Finland (grant 309048).

Contents

Algorithms

Computational Biology

Data Compression

Approximation Ratios of **RePair**, **LongestMatch** and **Greedy** on Unary Strings

Danny Hucke[(⊠)]

University of Siegen, Siegen, Germany
hucke@eti.uni-siegen.de

Abstract. A grammar-based compressor computes for a given input w a context-free grammar that produces only w. So-called global grammar-based compressors (**RePair**, **LongestMatch** and **Greedy**) achieve impressive practical compression results, but the recursive character of those algorithms makes it hard to achieve strong theoretical results. To this end, this paper studies the approximation ratio of those algorithms for unary input strings, which is strongly related to the field of addition chains. We show that in this setting, **RePair** and **LongestMatch** produce equal size grammars that are by a factor of at most $\log_2(3)$ larger than a smallest grammar. We also provide a matching lower bound. The main result of this paper is a new lower bound for **Greedy** of 1.348..., which improves the best known lower bound for arbitrary (not necessarily unary) input strings.

Keywords: Data compression · Grammar-based compression · Approximation algorithm · Addition chain

1 Introduction

The goal of grammar-based compression is to represent a word w by a small context-free grammar that produces exactly $\{w\}$. Such a grammar is called a straight-line program (SLP) for w. In the best case, one gets an SLP of size $\Theta(\log n)$ for a word of length n, where the size of an SLP is the total length of all right-hand sides of the rules of the grammar. A grammar-based compressor is an algorithm that produces an SLP for a given word w. There are various grammar-based compressors that can be found at many places in the literature. Well-known examples are the classical **LZ78**-compressor[1] of Lempel and Ziv [21], **BISECTION** [12] and **SEQUITUR** [17], just to mention a few. In this paper, we study the class of global grammar-based compressors which are also called global algorithms. A key concept of those algorithms are maximal strings. A maximal string of an SLP $\mathbb{A}$ is a word that has length at least two and occurs at least twice

---

[1] While **LZ78** was not introduced as a grammar-based compressor, it is straightforward to compute from the **LZ78**-factorization of w an SLP for w of roughly the same size.

© Springer Nature Switzerland AG 2019
N. R. Brisaboa and S. J. Puglisi (Eds.): SPIRE 2019, LNCS 11811, pp. 3–15, 2019.
https://doi.org/10.1007/978-3-030-32686-9_1

without overlap as a factor of the right-hand sides of the rules of $\mathbb{A}$. Further, no strictly longer word appears at least as many times without overlap as a factor of the right-hand sides of the rules of $\mathbb{A}$. For an input word w, a global grammar-based compressor starts with the SLP that has a single rule $S \to w$, where S is the start nonterminal of the grammar. The SLP is then recursively updated by choosing a maximal string γ of the current SLP and replacing a maximal set of pairwise non-overlapping occurrences of γ by a fresh nonterminal X. Additionally, a new rule $X \to \gamma$ is introduced. The algorithm stops when the obtained SLP has no maximal string. The probably best known example for a global algorithm is RePair [13], which selects in each round a most frequent maximal string. Note that the RePair algorithm as it is proposed in [13] always selects a word of length 2, but in this paper we follow the definition of [8], where the algorithm possibly selects longer words. However, both definitions coincide for unary input strings as considered in this work. Other global algorithms are LongestMatch [11], which chooses a longest maximal string in each round, and Greedy [2–4], which selects a maximal string that minimizes the size of the SLP obtained in the current round. It is again worth mentioning that the Greedy algorithm as originally presented in [2–4] is different from the version studied in this work as well as in [8]: The original Greedy algorithm only considers the right-hand side of the start rule for the choice and the replacement of the maximal string. In particular, all other rules do not change after they are introduced.

In the seminal work of Charikar et al. [8], the worst case approximation ratio of grammar-based compressors is studied. For a grammar-based compressor $\mathcal{C}$ that computes an SLP $\mathcal{C}(w)$ for a given word w, one defines the approximation ratio of $\mathcal{C}$ on w as the quotient of the size of $\mathcal{C}(w)$ and the size $g(w)$ of a smallest SLP for w. The approximation ratio $\alpha_{\mathcal{C}}(n)$ is the maximal approximation ratio of $\mathcal{C}$ among all words of length n. In [8] the authors provide upper and lower bounds for the approximation ratios of several grammar-based compressors (among them are all compressors mentioned so far), but for none of the compressors the lower and upper bounds match. For LZ78 and BISECTION those gaps were closed in [15]. For all global algorithms, the best upper bound on the approximation ratio is $\mathcal{O}((n/\log n)^{2/3})$ [8], while the best known lower bounds are $\Omega(\log n/\log\log n)$ for RePair [14], $\Omega(\log\log n)$ for LongestMatch and $5/(3\log_3(5)) = 1.137...$ for Greedy [8]. In the context of our work, it is worth mentioning that the lower bound for Greedy uses words over a unary alphabet. In general, the achieved bounds "leave a large gap of understanding surrounding the global algorithms" as the authors in [8] conclude.

We aim to strengthen the understanding of global grammar-based compressors in this paper by studying the behavior of these algorithms on unary inputs, i.e., words of the form a^n for some symbol a. Grammar-based compression on unary words is strongly related to the field of addition chains, which has been studied for decades (see [16, Chapter 4.6.3] for a survey) and still is an active topic due to the strong connection to public key cryptosystems (see [18] for a review from that point of view). An addition chain for an integer n of size m is a sequence of integers $1 = k_1, k_2, \ldots, k_m = n$ such that for each d ($2 \leq d \leq m$),

there exists i, j $(1 \leq i, j < d)$ such that $k_i + k_j = k_d$. It is straightforward to compute from an addition chain for an integer n of size m an SLP for a^n of size $2m - 2$. Vice versa, an SLP for a^n of size m yields an addition chain for n of size m. So this paper can also be interpreted as a study of global algorithms as addition chain solvers. For RePair and LongestMatch, the restriction to unary inputs allows a full understanding of the produced SLPs and it turns out that for all unary inputs the SLP produced by RePair has the same size as the SLP produced by LongestMatch. In fact, both algorithms are basically identical to the binary method that produces an addition chain for n by creating powers of two using repeated squaring, and then the integer n is represented as the sum of those powers of two that correspond to a one in the binary representation of n. Based on that information, we show that for any unary input w the produced SLPs of RePair and LongestMatch have size at most $\log_2(3) \cdot g(w)$, and we provide a matching lower bound.

Unfortunately, even for unary inputs it is hard to analyze the general behavior of Greedy due to the discrete optimization problem in each round of the algorithm. The (probably weak) upper bound that we achieve for the approximation ratio of Greedy on unary inputs is $\mathcal{O}(n^{1/4}/\log n)$. We derive this bound by estimating the size of the SLP obtained by Greedy after three rounds, which already indicates space for improvement. For the original Greedy algorithm where only the start rule is compressed, it is a direct consequence of our analysis that the approximation ratio on unary input strings is $\Theta(\sqrt{n}/\log n)$. On the positive side, we provide a new lower bound of 1.348... for the approximation ratio of Greedy (in the variant where all right-hand sides are considered) that improves the best known lower bound for inputs over arbitrary alphabets. The key to achieve the new bound is the sequence $y_k = y_{k-1}^2 + 1$ with $y_0 = 2$, which has been studied in [1] (among other sequences), where it is shown that $y_k = \lfloor \gamma^{2^k} \rfloor$ for $\gamma = 2.258...$. In order to prove the lower bound, we show that the SLP produced by Greedy on input a^{y_k} has size $3 \cdot 2^k - 1$, while a smallest SLP for a^{y_k} has size $3 \cdot \log_3(\gamma) \cdot 2^k + o(2^k)$ (this follows from a construction used to prove the lower bound for Greedy in [8]).

Related Work. One of the first appearances of straight-line programs in the literature are [6,9], where they are called word chains (since they generalize addition chains from numbers to words). In [6], Berstel and Brlek prove that the function $g(k, n) = \max\{g(w) \mid w \in \{1, \ldots, k\}^n\}$ is in $\Theta(n/\log_k n)$. Recall that $g(w)$ is the size of a smallest SLP for the word w and thus $g(k, n)$ measures the worst case SLP-compression over all words of length n over a k-letter alphabet.

The smallest grammar problem is the problem of computing a smallest SLP for a given input word. It is known from [8,20] that in general no grammar-based compressor can solve the smallest grammar problem in polynomial time unless P = NP. Even worse, unless P = NP one cannot compute in polynomial time for a given word w an SLP of size at most $\frac{8569}{8568} \cdot g(w)$ [8]. One should mention that the constructions to prove those hardness results use alphabets of unbounded size. While in [8] it is remarked that the construction in [20] works for words over a ternary alphabet, Casel et al. [7] argue that this is not clear at all and provide

a construction for fixed alphabets of size at least 24. However, for grammar-based compression on unary strings as it is studied in this work (as well as for the problem of computing a smallest addition chain), there is no NP-hardness result, so there might be an optimal polynomial-time algorithm even though it is widely believed that there is none.

Other notable systematic investigations of grammar-based compression are provided in [11,19]. Whereas in [11], grammar-based compressors are used for universal lossless compression (in the information-theoretical sense), it is shown in [19] that the size of so-called irreducible SLPs (that include SLPs produced by global algorithms) can be upper bounded by the (unnormalized) k-th order empirical entropy of the produced string plus some lower order terms.

2   Preliminaries

For $i, j \in \mathbb{N}$, let $[i, j] = \{i, i + 1, \ldots, j\}$ for $i \leq j$ and $[i, j] = \emptyset$ otherwise. For integers m, n, we denote by m div n the integer division of m and n. We denote by $m \bmod n$ the modulo of m and n, i.e., $m \bmod n \in [0, n-1]$ and

$$m = (m \text{ div } n) \cdot n + (m \bmod n).$$

If m/n or $\frac{m}{n}$ is used, then this refers to the standard division over $\mathbb{R}$. Note that m div $n = \lfloor m/n \rfloor$ and $(m \text{ div } n) + (m \bmod n) \geq m/n$.

For an *alphabet* Σ, let $w = a_1 \cdots a_n$ $(a_1, \ldots, a_n \in \Sigma)$ be a *word* or *string* over Σ. The length $|w|$ of w is n and we denote by ε the word of length 0. A *unary word* is a word of the form a^n for $a \in \Sigma$. Let $\Sigma^+ = \Sigma^* \setminus \{\varepsilon\}$ be the set of all nonempty words. For $w \in \Sigma^+$, we call $v \in \Sigma^+$ a *factor* of w if there exist $x, y \in \Sigma^*$ such that $w = xvy$.

2.1   Straight-Line Programs

A *straight-line program*, briefly SLP, is a context-free grammar that produces a single word $w \in \Sigma^+$. Formally, it is a tuple $\mathbb{A} = (N, \Sigma, P, S)$, where N is a finite set of nonterminals with $N \cap \Sigma = \emptyset$, $S \in N$ is the start nonterminal, and P is a finite set of productions (or rules) of the form $A \rightarrow w$ for $A \in N$, $w \in (N \cup \Sigma)^+$ such that:

- For every $A \in N$, there exists exactly one production of the form $A \rightarrow w$, and
- the binary relation $\{(A, B) \in N \times N \mid (A \rightarrow w) \in P, B \text{ occurs in } w\}$ is acyclic.

Every nonterminal $A \in N$ produces a unique, nonempty word. The word defined by the SLP $\mathbb{A}$ is the word produced by the start nonterminal S. The *size* of the SLP $\mathbb{A}$ is $|\mathbb{A}| = \sum_{(A \rightarrow w) \in P} |w|$. We denote by $g(w)$ the size of a smallest SLP producing the word $w \in \Sigma^+$. We will use the following inequalities that can be found in [8]:

Lemma 1 ([8]). *For all unary words w of length n, we have*

$$3\log_3(n) - 3 \le g(w) \le 3\log_3(n) + o(\log n).$$

Note that the first inequality also holds when w is a word over an arbitrary alphabet. The proof of the first inequality can be found in Lemma 1 of [8] and the second inequality is shown in the proof of Theorem 11 of [8].

Approximation Ratio. A *grammar-based compressor* $\mathcal{C}$ is an algorithm that computes for a nonempty word w an SLP $\mathcal{C}(w)$ that produces the word w. The *approximation ratio* $\alpha_{\mathcal{C}}(w)$ of $\mathcal{C}$ for an input w is defined as $|\mathcal{C}(w)|/g(w)$. The worst-case approximation ratio $\alpha_{\mathcal{C}}(k, n)$ of $\mathcal{C}$ is the maximal approximation ratio over all words of length n over an alphabet of size k:

$$\alpha_{\mathcal{C}}(k, n) = \max\{\alpha_{\mathcal{C}}(w) \mid w \in [1, k]^n\} = \max\{|\mathcal{C}(w)|/g(w) \mid w \in [1, k]^n\}$$

In this paper we are mainly interested in the case $k = 1$, i.e., we are interested in grammar-based compression on unary words.

3 Global Algorithms

For a given SLP $\mathbb{A} = (N, \Sigma, P, S)$, a word $\gamma \in (N \cup \Sigma)^+$ is called a *maximal string* of $\mathbb{A}$ if

- $|\gamma| \ge 2$,
- γ appears at least twice without overlap as a factor of the right-hand sides of $\mathbb{A}$,
- and no strictly longer word appears at least as many times as a factor of the right-hand sides of $\mathbb{A}$ without overlap.

A *global grammar-based compressor* starts on input w with the SLP $\mathbb{A}_0 = (\{S\}, \Sigma, \{S \to w\}, S)$. In each round $i \ge 1$, the algorithm selects a maximal string γ of $\mathbb{A}_{i-1}$ and updates $\mathbb{A}_{i-1}$ to $\mathbb{A}_i$ by replacing a largest set of pairwise non-overlapping occurrences of γ in $\mathbb{A}_{i-1}$ by a fresh nonterminal X. Additionally, the algorithm introduces the rule $X \to \gamma$ in $\mathbb{A}_i$. The algorithm stops when no maximal string occurs. Note that the replacement is not unique, e.g. the word a^5 has a unique maximal string $\gamma = aa$, which yields SLPs with rules $S \to XXa, X \to aa$ or $S \to XaX, X \to aa$ or $S \to aXX, X \to aa$. We assume the first variant in this paper, i.e., maximal strings are replaced from left to right.

3.1 Greedy

The global grammar-based compressor Greedy selects in each round $i \ge 1$ a maximal string of $\mathbb{A}_{i-1}$ such that $\mathbb{A}_i$ has minimal size among all possible choices of maximal strings of $\mathbb{A}_{i-1}$.

We start with the main result of this paper, which is a new lower bound for the approximation ratio of Greedy. The best known lower bound [8] so far is

$$\alpha_{\mathsf{Greedy}}(k,n) \geq \frac{5}{3\log_3(5)} = 1.13767699...$$

for all $k \geq 1$ and infinitely many n. This bound is achieved using unary input strings. A key concept to prove a better lower bound is the sequence x_n described in the following lemma by [1]:

Lemma 2 ([1, **Example 2.2**]). *Let* $x_{n+1} = x_n^2 + 1$ *with* $x_0 = 1$ *and*

$$\beta = \exp\left(\sum_{i=1}^{\infty} \frac{1}{2^i} \log\left(1 + \frac{1}{x_i^2}\right)\right).$$

We have $x_n = \lfloor \beta^{2^n} \rfloor$.

In this work, we use the shifted sequence $y_n = x_{n+1}$, i.e., we start with $y_0 = 2$. It follows that $y_n = \lfloor \gamma^{2^n} \rfloor$, where $\gamma = \beta^2 = 2.25851845....$ Additionally, we need the following lemma:

Lemma 3. *Let* $m \geq 1$ *be an integer. Let* $f_m : \mathbb{R}_{>0} \to \mathbb{R}$ *with*

$$f_m(x) = x + \frac{m^2 + 1}{x}.$$

We have $f_m(x) > 2m$ *for all* $x > 0$.

Proof. The unique minimum of $f_m(x)$ is $2\sqrt{m^2 + 1}$ for $x = \sqrt{m^2 + 1}$. It follows that $f_m(x) \geq 2\sqrt{m^2 + 1} > 2\sqrt{m^2} = 2m$.

Now we are able to prove the new lower bound for Greedy:

Theorem 1. *For all* $k \geq 1$ *and infinitely many* n, *we have*

$$\alpha_{\mathsf{Greedy}}(k,n) \geq \frac{1}{\log_3(\gamma)} = 1.34847194... \ .$$

Proof. Let $\Sigma = \{a\}$ be a unary alphabet. We define $w_k = a^{y_k}$. By Lemma 2, we have $|w_k| \leq \gamma^{2^k}$. Applying Lemma 1 yields

$$g(w_k) \leq 3 \cdot \log_3(\gamma) \cdot 2^k + o(2^k).$$

In the remaining proof we show that on input w_k, Greedy produces an SLP of size $3 \cdot 2^k - 1$, which directly implies $\alpha_{\mathsf{Greedy}}(1,n) \geq 3/(3\log_3(\gamma))$. We start with the SLP $\mathbb{A}_0$ which has the single rule $S \to a^{y_k}$. Consider now the first round of the algorithm, i.e., we need to find a maximal string a^x of $\mathbb{A}_0$ such that the grammar $\mathbb{A}_1$ with rules

$$X_1 \to a^x, \quad S \to X_1^{y_k \text{ div } x} a^{y_k \bmod x}$$

has minimal size. We have $|\mathbb{A}_1| = x + (y_k \text{ div } x) + (y_k \text{ mod } x) \geq x + y_k/x$. By the definition of y_k we have $|\mathbb{A}_1| \geq x + (y_{k-1}^2 + 1)/x$. Applying Lemma 3 yields $|\mathbb{A}_1| \geq 2y_{k-1} + 1$. Note that for $x = y_{k-1}$ this minimum is achieved, i.e., we can assume that Greedy selects the maximal string $a^{y_{k-1}}$ and $\mathbb{A}_1$ is

$$X_1 \to a^{y_{k-1}}, \; S \to X_1^{y_{k-1}}a.$$

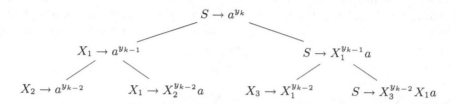

Fig. 1. Three rounds of Greedy on input a^{y_k}.

Each maximal string of $\mathbb{A}_1$ is either a unary word over X or a unary word over a, i.e., we can analyze the behavior of Greedy on both rules independently. The rule $X_1 \to a^{y_{k-1}}$ is obviously treated similarly as the initial SLP $\mathbb{A}_0$, so we continue with analyzing $S \to X_1^{y_{k-1}}a$. But again, the same arguments as above show that Greedy introduces a rule $X_3 \to X_1^{y_{k-2}}$ which yields $S \to X_3^{y_{k-2}}X_1a$ as the new start rule. This process can be iterated using the same arguments for the leading unary strings of length y_i for some $i \in [1, k]$.

The reader might think of this process as a binary tree, where each node is labelled with a rule (the root is labelled with $S \to a^{y_k}$) and the children of a node are the two rules obtained by Greedy when the rule has been processed. We assume that the left child represents the rule for the chosen maximal string and the right child represents the parent rule where all occurrences of the maximal string are replaced by the fresh nonterminal. In Fig. 1 this binary tree is depicted for the steps we discussed above. Note that when a rule is processed, the longest common factor of the two new rules has length 1 (the remainder). More generally, after each round there is no word of length at least two that occurs as a factor in two different rules, since a possibly shared remainder has length 1 and otherwise only fresh nonterminals are introduced. It follows that we can iterate this process independently for each rule until no maximal string occurs. This is the case when each rule starts with a unary string of length $y_0 = 2$ or, in terms of the interpretation as a binary tree, when a full binary tree of height k is produced. Each right branch occurring in this tree adds a new remainder to those remainders that already occur in the parent rule and a left branch introduces a new (smaller) instance of the start problem. We show by induction that on level $i \in [0, k]$ of this full binary tree of height k, there is one rule of size $y_{k-i} + i$ and 2^{i-j-1} many rules of size $y_{k-i} + j$ for $j \in [0, i-1]$. On level 0, this is true since there is only a single rule of size $y_k + 0$. Assuming that our claim is true on level $i < k$, we derive from each rule on level i two new rules on level $i+1$: A right branch yields a rule that starts with a leading unary string of size y_{k-i-1}

and adds a new remainder to the parent rule. A left branch yields a rule that contains only a unary string of size y_{k-i-1}. If we first consider the left branches, we derive that each of the 2^i many rules on level i adds a rule of size y_{k-i-1} on level $i+1$. For the right branches, the single rule of size $y_{k-i}+i$ on level i yields a rule of size $y_{k-i-1}+i+1$ on level $i+1$. Further, each of the 2^{i-j-1} many rules of size $y_{k-i}+j$ ($j \in [0, i-1]$) yields a rule of size $y_{k-i-1}+j+1$. When we put everything together, we get that on level $i+1$ there is a single rule of size $y_{k-i-1}+i+1$ and 2^{i-j} many rules of size $y_{k-i-1}+j$ for $j \in [0, i]$. That finishes the induction. It follows that the final SLP (which consists of the rules on level k) has a single rule of size $y_0+k = 2+k$ and 2^{k-j-1} many rules of size $2+j$ for $j = 0, \ldots, k-1$. This gives a total size of

$$
\begin{aligned}
2 + k + \sum_{j=0}^{k-1} 2^{k-j-1}(2+j) &= 2 + k + 2^k \sum_{j=0}^{k-1} 2^{-j} + 2^k \sum_{j=0}^{k-1} 2^{-j-1} j \\
&= 2 + k + 2^k(2 - 2^{-k+1}) + 2^k(-2^{-k}k - 2^{-k} + 1) \\
&= 2 + k + 2^{k+1} - 2 - k - 1 + 2^k \\
&= 2^{k+1} + 2^k - 1 \\
&= 3 \cdot 2^k - 1.
\end{aligned}
$$

In the remaining part of this section, we prove an upper bound on the size of the SLP produced by Greedy on input a^n:

Theorem 2. *The SLP produced by* Greedy *(after three rounds) on input a^n has size $\mathcal{O}(n^{1/4})$.*

Proof. Consider an input a^n with $n \geq 4$ (otherwise $S \to a^n$ is the final SLP since there is no maximal string). The SLP $\mathbb{A}_1$ obtained by Greedy after the first round has the form

$$
X \to a^x, \quad S \to X^{n \text{ div } x} a^{n \text{ mod } x}, \tag{1}
$$

where a^x is the selected maximal string. We first show

$$
\frac{1}{3}\sqrt{n} \leq x \leq 3\sqrt{n}, \quad \frac{1}{3}\sqrt{n} \leq n \text{ div } x \leq 3\sqrt{n}, \quad n \text{ mod } x < 3\sqrt{n}.
$$

Assume $x = \lceil \sqrt{n} \rceil$ in Eq. (1). In this case, the size of the SLP is

$$
\lceil \sqrt{n} \rceil + \left\lfloor \frac{n}{\lceil \sqrt{n} \rceil} \right\rfloor + n \text{ mod } \lceil \sqrt{n} \rceil \leq 3\sqrt{n} + 1.
$$

Since the maximal string a^x is selected greedily such that $\mathbb{A}_1$ has minimal size, we have $|\mathbb{A}_1| \leq 3\sqrt{n} + 1$. It follows that $x \leq 3\sqrt{n}$, because otherwise the size of $\mathbb{A}_1$ would be at least $3\sqrt{n} + 2$ due to the fact that a maximal string (represented by the nonterminal X) occurs at least twice. It follows that $n \text{ mod } x < 3\sqrt{n}$ and

n div $x \geq 1/3\sqrt{n}$. Further, we have n div $x \leq 3\sqrt{n}$ because otherwise the size of $\mathbb{A}_1$ would be at least $3\sqrt{n} + 2$ due to the fact that $x \geq 2$ (a maximal string has length at least two). It follows that $x \geq 1/3\sqrt{n}$. Actually, a slightly more careful analysis allows sharper bounds for x, n div x and n mod x, but for the matter of this proof it is easier to work with the constants 3 and $1/3$.

Now the only maximal strings occurring in $\mathbb{A}_1$ are of the form X^y or a^z (for integers $y, z \geq 2$) since no other factor of the right-hand sides of $\mathbb{A}_1$ occurs at least twice. Note that both optimization problems (for X^y and a^z) are independent, so we assume the chosen maximal string in the second round has the form X^z, afterwards we proceed with a^z. Let $d = n$ div x, where a^x is again the maximal string that has been selected in the first round. Then the SLP $\mathbb{A}_2$ obtained after the second round of Greedy has the form

$$X \to a^x, \; Y \to X^y, \; S \to Y^{d \text{ div } y} X^{d \bmod y} a^{n \bmod x}. \tag{2}$$

Let $g(x) = x + (n \bmod x)$, which is the size of those parts of $\mathbb{A}_2$ that are independent of the choice of the maximal string X^y. Assume $y = \lceil n^{1/4} \rceil$ in (2) and let n be large enough such that Y occurs at least twice in the start rule. This yields an SLP of size

$$\lceil n^{\frac{1}{4}} \rceil + \left\lfloor \frac{d}{\lceil n^{\frac{1}{4}} \rceil} \right\rfloor + d \bmod \lceil n^{\frac{1}{4}} \rceil + g(x) \leq 5n^{\frac{1}{4}} + 1 + g(x).$$

The inequality is achieved by using $d = n$ div $x \leq 3\sqrt{n}$ as argued above. It follows again from the greedy nature of the algorithm that $|\mathbb{A}_2| \leq 5n^{1/4} + 1 + g(x)$ and similar arguments as above show that the exponents y, d div y and d mod y can be upper bounded by $5n^{1/4}$.

All maximal strings occurring in $\mathbb{A}_2$ are again unary words, but since a^x has length at least $1/3\sqrt{n}$ and the lengths of all other unary factors over X or Y are bounded by $5n^{1/4}$, we can assume that (for n large enough) Greedy selects a^z for some integer $z \geq 2$ as the maximal string in order to achieve a minimal size SLP $\mathbb{A}_3$. Note that if we would have assumed that the chosen maximal string in round two is a^z instead of X^y, then similar arguments would show that X^y is selected in round three if n is large enough. Now, let $e = n$ mod x, then the SLP $\mathbb{A}_3$ obtained after the third round has the form

$$Z \to a^z, \; X \to Z^{x \text{ div } z} a^{x \bmod z}, \; Y \to X^y,$$
$$S \to Y^{d \text{ div } y} X^{d \bmod y} Z^{e \text{ div } z} a^{e \bmod z}. \tag{3}$$

Let $h(y) = y + (d \text{ div } y) + (d \bmod y)$, which is the size of those parts of $\mathbb{A}_3$ that are independent of the choice of the maximal string a^z. Assume now $z = \lceil n^{1/4} \rceil$ and let n be large enough such that Z occurs at least twice in the right-hand sides of (3). The obtained SLP has size

$$\lceil n^{1/4} \rceil + \left\lfloor \frac{x}{\lceil n^{1/4} \rceil} \right\rfloor + (x \bmod \lceil n^{1/4} \rceil) + \left\lfloor \frac{e}{\lceil n^{1/4} \rceil} \right\rfloor + (e \bmod \lceil n^{1/4} \rceil) + h(y).$$

Using $x \le 3\sqrt{n}$ and $e = n \bmod x < 3\sqrt{n}$ we can upper bound this size by $9n^{1/4}+1+h(y)$. Note that we have already bounded the size of $h(y)$ by $5n^{1/4}+1$ in the previous step, so the SLP $\mathbb{A}_3$ obtained by Greedy after three rounds has size at most $14n^{1/4} + 2$.

The bound on the approximation ratio is now achieved using Lemma 1, which shows that a smallest grammar for a^n has size $\Omega(\log n)$.

Corollary 1. *We have* $\alpha_{\mathsf{Greedy}}(1, n) \in \mathcal{O}(n^{1/4}/\log n)$.

The reader might wonder why our estimation stops after three rounds of Greedy. The most important reason is that a precise invariant is missing in order to iterate our arguments for a non constant number of rounds. On the other hand, it seems likely that similar arguments as provided in the proof of Theorem 2 can be used to show that the SLP produced by Greedy after some more rounds has size $\mathcal{O}(n^{1/8})$ and maybe again after some rounds has size $\mathcal{O}(n^{1/16})$. However, further analysis would require more and more case distinctions since it is not clear anymore that the selected maximal string is always unary as the reader can see in (3), where factors of the form Z^*a^* can occur more than once on the right-hand sides. It seems therefore necessary to apply some new information in order to improve the upper bound beyond $\mathcal{O}(n^{1/2^k})$ for some fixed k.

An interesting consequence of the proof of Proposition 2 applies to the originally proposed Greedy variant [2–4]. Recall from the introduction that in this setting, the algorithm recursively chooses the maximal string only in dependence on the right-hand side of the start rule and replaces the occurrences of the chosen string only there. In other words, the right-hand side of the start rule is compressed in a greedy way and all other rules do not change after they are introduced. Note that in the first round both variants of Greedy (the one studied here and the original one) are identical, because in this case the only rule of the SLP is the start rule $S \to a^n$. Hence, our analysis of the first step in the proof of Proposition 2 applies to the original variant as well. We have shown that the selected maximal string in the first round (and thus the right-hand side of the introduced rule) has length $\Theta(\sqrt{n})$ and since the original variant does not modify the corresponding rule any further, it follows directly that the SLP produced by the original algorithm has size $\Omega(\sqrt{n})$. But since the modified start rule has also size $\Theta(\sqrt{n})$ after the first step, it follows that the SLP produced by the original algorithm has size $\mathcal{O}(\sqrt{n})$ as well (the size of the SLP does not increase later). Together with Lemma 1, it follows that this variant of the Greedy algorithm has approximation ratio $\Theta(\sqrt{n}/\log n)$ on unary inputs of length n.

3.2   RePair and LongestMatch

In this section we analyze the global grammar-based compressors RePair and LongestMatch. In each round i, RePair selects a most frequent maximal string of $\mathbb{A}_{i-1}$ and LongestMatch selects a longest maximal string of $\mathbb{A}_{i-1}$.

We will abbreviate the approximation ratio $\alpha_{\mathsf{LongestMatch}}$ by α_{LM} for better readability. We will first show that RePair and LongestMatch produce SLPs of

equal size for unary inputs a^n and we prove the exact size of those SLPs in dependency on n. In a second step, we use this information to obtain our result for $\alpha_{\mathsf{RePair}}(1, n)$, respectively $\alpha_{\mathsf{LM}}(1, n)$. Fix an integer $n \geq 2$ and consider the binary representation

$$n = \sum_{i=0}^{\lfloor \log_2 n \rfloor} b_i \cdot 2^i \tag{4}$$

of n, where $b_i \in \{0, 1\}$ for $i \in [0, \lfloor \log_2 n \rfloor]$. We denote by $\nu(n)$ the number of 1's in the binary representation of n, i.e.,

$$\nu(n) = \sum_{i=0}^{\lfloor \log_2 n \rfloor} b_i.$$

For example, we have $11 = 1 \cdot 2^3 + 0 \cdot 2^2 + 1 \cdot 2^1 + 1 \cdot 2^0$ and thus $b_0 = b_1 = b_3 = 1$, $b_2 = 0$ and $\nu(11) = 3$.

Proposition 1. *For $n \geq 2$, let $\mathbb{A}$ be the SLP produced by* RePair *on input a^n and $\mathbb{B}$ be the SLP produced by* LongestMatch *on input a^n. We have*

$$|\mathbb{A}| = |\mathbb{B}| = 2\lfloor \log_2 n \rfloor + \nu(n) - 1.$$

Proof. If $n = 2$ or $n = 3$ (we only consider $n \geq 2$), then a^n has no maximal string and thus the final SLP of any global algorithm has a single rule $S \to a^n$. The reader can easily verify the claimed result for those cases.

We assume $n \geq 4$ in the following. Let $m = \lfloor \log_2 n \rfloor - 1$. We prove the claim for RePair first, afterwards we proceed with LongestMatch. On input a^n, RePair runs for exactly m rounds and creates rules $X_1 \to aa$ and $X_i \to X_{i-1}X_{i-1}$ for $i \in [2, m]$, i.e., the nonterminal X_i produces the string a^{2^i}. This rules have total size $2m$. After this steps, the start rule is

$$S \to X_m X_m X_m^{b_m} X_{m-1}^{b_{m-1}} \cdots X_1^{b_1} a^{b_0},$$

where the b_i's are the coefficients occurring in the binary representation of n, see Eq. (4). In other words, the symbol a only occurs in the start rule if the least significant bit $b_0 = 1$, and the nonterminal X_i ($i \in [1, m-1]$) occurs in the start rule if and only if $b_i = 1$. Since RePair only replaces words with at least two occurrences, the most significant bit $b_{m+1} = 1$ is represented by $X_m X_m$. A third X_m occurs in the start rule if and only if $b_m = 1$. The size of the start rule is $2 + \sum_{i=0}^{m} b_i$. It follows that the total size of the SLP produced by RePair on input a^n is $2m + 2 + \sum_{i=0}^{m} b_i$, which together with $m = \lfloor \log n \rfloor - 1$ and $b_{\lfloor \log n \rfloor} = 1$ (the most significant bit is always 1) yields the claimed size.

Now we prove the same result for LongestMatch. In the first round, the chosen maximal string is $a^{\lfloor n/2 \rfloor}$, which yields rules $X_1 \to a^{\lfloor n/2 \rfloor}$ and $S \to X_1 X_1 a^{b_0}$, i.e., the symbol a occurs in the start rule if and only if n is odd and thus the least significant bit $b_0 = 1$. Assuming $n \geq 8$, this procedure is now repeated for the rule $X_1 \to a^{\lfloor n/2 \rfloor}$ (for $n < 8$ there is no maximal string and the algorithm stops after

the first round). This yields $X_2 \to a^{\lfloor n/4 \rfloor}$, $X_1 \to X_2 X_2 a^{b_1}$ and $S \to X_1 X_1 a^{b_0}$ (note that $\lfloor (\lfloor n/2 \rfloor)/2 \rfloor = \lfloor n/4 \rfloor$). After $m = \lfloor \log n \rfloor - 1$ steps, the iteration of that process results in the final SLP with rules $S \to X_1 X_1 a^{b_0}$, $X_i \to X_{i+1} X_{i+1} a^{b_i}$ for $i \in [1, m-1]$ and $X_m \to aaa^{b_m}$. The size of this SLP is $2 \cdot (m+1) + \sum_{i=0}^{m} b_i$, which directly implies the claimed result for LongestMatch.

Using Proposition 1, we prove the matching bounds for $\alpha_{\mathsf{RePair}}(1, n)$, $\alpha_{\mathsf{LM}}(1, n)$:

Theorem 3. *For all n, we have $\alpha_{\mathsf{RePair}}(1, n) = \alpha_{\mathsf{LM}}(1, n) \leq \log_2(3)$.*

Proof. As a consequence of Proposition 1, RePair and LongestMatch produce on input a^n SLPs of size at most $3 \log_2 n$, since $\nu(n) - 1 \leq \log_2 n$. By Lemma 1, we have $g(a^n) \geq 3 \log_3 n - 3$. The equality $\log_2 n / \log_3 n = \log_2(3)$ finishes the proof.

Theorem 4. *For infinitely many n, we have $\alpha_{\mathsf{RePair}}(1, n) = \alpha_{\mathsf{LM}}(1, n) \geq \log_2(3)$.*

Proof. Let $w_k = a^{2^k - 1}$. We have $2^k - 1 = \sum_{i=0}^{k-1} 2^i$ and thus $\nu(2^k - 1) = k$. By Proposition 1, the size of the SLPs produced by RePair and LongestMatch is $3k - 3$. By Lemma 1, we have

$$g(w_k) \leq 3 \log_3(2^k - 1) + o(\log(2^k - 1) \leq 3 \log_3(2) \cdot k + o(k).$$

The equality $1 / \log_3(2) = \log_2(3)$ finishes the proof.

4   Future Work

The obvious question concerns the gap between the lower and upper bound for Greedy. First of all, it might be possible to improve our lower bound by finding a similar sequence such that Greedy produces larger remainders in each round, but care has to be taken since for larger remainders it is not true anymore that the rules can be analyzed independently because the rules could share factors of length greater 1. Concerning the upper bound, we conjecture that Greedy achieves logarithmic compression for all unary inputs and thus the approximation ratio is constant, but the direct analysis of the algorithm as we tried in Theorem 2 misses a clear invariant for a non constant number of rounds. For arbitrary alphabets, a non-constant lower bound for Greedy as well as an improvement of the upper bound of $\mathcal{O}((n/\log n)^{2/3})$ for any global algorithm seems to be natural starting points for future work.

References

1. Aho, A.V., Sloane, N.J.A.: Some doubly exponential sequences. Fib. Quart. **11**, 429–437 (1973)
2. Apostolico, A., Lonardi, S.: Some theory and practice of greedy off-line textual substitution. In: Proceeding of the DCC 1998, pp. 119–128. IEEE Computer Society (1998)

3. Apostolico, A., Lonardi, S.: Compression of biological sequences by greedy off-line textual substitution. In: Proceedings of the DCC 2000, pp. 143–152. IEEE Computer Society (2000)
4. Apostolico, A., Lonardi, S.: Off-line compression by greedy textual substitution. Proc. IEEE **88**(11), 1733–1744 (2000)
5. Arpe, J., Reischuk, R.: On the complexity of optimal grammar-based compression. In: Proceedings of the DCC 2006, pp. 173–182. IEEE Computer Society (2006)
6. Berstel, J., Brlek, S.: On the length of word chains. Inf. Process. Lett. **26**(1), 23–28 (1987)
7. Casel, K., Fernau, H., Gaspers, S., Gras, B., Schmid, M.L.: On the complexity of grammar-based compression over fixed alphabets. In: Proceedings ICALP 2016, Lecture Notes in Computer Science. Springer (1996)
8. Charikar, M., et al.: The smallest grammar problem. IEEE Trans. Inf. Theory **51**(7), 2554–2576 (2005)
9. Diwan, A.A.: A New Combinatorial Complexity Measure for Languages. Tata Institute, Bombay (1986)
10. Jeż, A.: Approximation of grammar-based compression via recompression. Theor. Comput. Sci. **592**, 115–134 (2015)
11. Kieffer, J.C., Yang, E.-H.: Grammar-based codes: a new class of universal lossless source codes. IEEE Trans. Inf. Theory **46**(3), 737–754 (2000)
12. Kieffer, J.C., Yang, E.-H., Nelson, G.J., Cosman, P.C.: Universal lossless compression via multilevel pattern matching. IEEE Trans. Inf. Theory **46**(4), 1227–1245 (2000)
13. Larsson, N.J., Moffat, A.: Offline dictionary-based compression. In: Proceedings of the DCC 1999, pp. 296–305. IEEE Computer Society (1999)
14. D. Hucke, A. Jeż, and M. Lohrey. Approximation ratio of RePair. Technical report, arxiv. org (2017). https://arxiv.org/abs/1703.06061
15. Hucke, D., Lohrey, M., Reh, C.P.: The smallest grammar problem revisited. In: Inenaga, S., Sadakane, K., Sakai, T. (eds.) SPIRE 2016. LNCS, vol. 9954, pp. 35–49. Springer, Cham (2016). https://doi.org/10.1007/978-3-319-46049-9_4
16. Knuth, D.E.: The Art of Computer Programming Volume II. Seminumerical Algorithms, 3rd edn. Addison Wesley, Reading (1998)
17. Nevill-Manning, C.G., Witten, I.H.: Identifying hierarchical structure in sequences: a linear-time algorithm. J. Artif. Intell. Res. **7**, 67–82 (1997)
18. Noma, A.M., Muhammed, A., Mohamed, M.A., Zulkarnain, Z.A.: A review on heuristics for addition chain problem: towards efficient public key cryptosystems. J. Comput. Sci. **13**(8), 275–289 (2017)
19. Ochoa, C., Navarro, G.: RePair and all irreducible grammars are upper bounded by high-order empirical entropy. IEEE Trans. Inf. Theory **65**(5), 3160–3164 (2019)
20. Storer, J.A., Szymanski, T.G.: Data compression via textual substitution. J. ACM **29**(4), 928–951 (1982)
21. Ziv, J., Lempel, A.: Compression of individual sequences via variable-rate coding. IEEE Trans. Inf. Theory **24**(5), 530–536 (1977)

Lossless Image Compression Using List Update Algorithms

Arezoo Abdollahi[1], Neil Bruce[2], Shahin Kamali[1(✉)], and Rezaul Karim[1]

[1] Department of Computer Science, University of Manitoba, Winnipeg, Canada
abdollaa@myumanitoba.ca, {shahin.kamali,karimr}@cs.umanitoba.ca
[2] Department of Computer Science, Ryerson University, Toronto, Canada
bruce@ryerson.ca

Abstract. We consider lossless image compression using a technique similar to bZip2 for sequential data. Given an image represented with a matrix of pixel values, we consider different approaches for linearising the image into a sequence and then encoding the sequence using the Move-To-Front list update algorithm. In both linearisation and encoding stages, we exploit the locality present in the images to achieve encodings that are as compressed as possible. We consider a few approaches, and in particular Hilbert space-filling curves, for linearising the image. Using a natural model of locality for images introduced by Albers et al. [J. Comput. Syst. Sci. 2015], we establish the advantage of Hilbert space-filling curves over other linearisation techniques such as row-major or column-major curves for preserving the locality during the linearisation. We also use a result by Angelopoulos and Schweitzer [J. ACM 2013] to select Move-To-Front as the best list update algorithm for encoding the linearised sequence. In summary, our theoretical results show that a combination of Hilbert space-filling curves and Move-To-Front encoding has advantage over other approaches. We verify this with experiments on a dataset consisting of different categories of images.

Keywords: Lossless image compression · Move-To-Front encoding · Hilbert space-filling curve · List update

1 Introduction

Lossless compression is a data encoding mechanism that allows reconstruction of the original data from the code without having it altered. For digital images, commonly used encoding schemes such as JPEG offer good compression via a *lossy* encoding of images. In these schemes, the original image is different from the encoded one. Such difference, however, is often not visible to human eyes and hence lossy compression remains popular for everyday use. Nevertheless, lossless image compression is important in applications where there is a need to save bandwidth and storage through compression while the actual image content is also required. Notable applications of lossless image compression include medical and scientific imaging [1–3].

© Springer Nature Switzerland AG 2019
N. R. Brisaboa and S. J. Puglisi (Eds.): SPIRE 2019, LNCS 11811, pp. 16–34, 2019.
https://doi.org/10.1007/978-3-030-32686-9_2

In the past few decades, various algorithms and industry standards have been proposed for lossless image compression (see Sect. 4 for a summary of previous work). Generally speaking, a compression algorithm exploits the redundancy and locality of information that is present in images to encode them in a compressed manner. A good compression algorithm is desired to output encodings that are as compressed as possible. This is measured with *compression ratio*, which is the ratio between the size of the encoded data and the original data. In addition to achieving good compression ratio, a compression scheme should be efficient in terms of using computational resources. Naturally, using more resources, in particular spending more time, can help in getting further compression to a certain limit.

An image is often represented with three "channels" for three primary colors. For each channel, a pixel receives an integer value (typically in the range $[0,256)$). In order to compress an image, the three channels that form the image are encoded separately. Consequently, encoding an image translates to encoding three matrices formed by non-negative integer values. In most images, pixel values are similar to those nearby. That is, images have a high *locality* of pixel values in the sense that the number of distinct pixel values in any sub-image is relatively small compared to the size of the sub-image. In order to exploit the locality present in images, various entropy-reduction techniques are embedded in compression schemes. These include prediction-based (e.g., [4,5]), wavelet-transform-based (e.g., [6]), and deep learning (e.g., [7,8]) approaches. The trade off between the achieved compression ratio and and the amount of computational resources that is used (particularly the time complexity) implies that most existing compression schemes are not directly comparable, and depending on the application, one might have an advantage over the other.

When it comes to sequential (one-dimensional) data, a promising family of algorithms use *list update* encodings. Given a sequence σ formed by characters from a universe U, these algorithms store members of U in a list that is maintained by a list update algorithm, typically the Move-To-Front (MTF) algorithm. When encoding σ, the index of the next character x in σ is encoded in the compressed file using a self-delimiting code (e.g., Elias gamma code [9]), and the list is subsequently updated by the list update algorithm. When using the MTF rule, x is simply moved to the front of the list. In the likely case that x appears again in σ (as implied by locality), the index encoded for the next appearances of x will be smaller and hence will be encoded using a shorter code. The MTF encoding lies at the heart of practical data compression schemes such as bZip2 [10], where it is a part of a pipeline that includes the Burrows-Wheeler Transform (BWT) [11] at the beginning and run-length encoding at the end.

Image compression schemes often work with implicit or explicit *linearising* of a 2-dimensional image into a 1-dimensional sequence and then encoding this sequence using various compression techniques. While some compression schemes use simple linearisation techniques such as row-major (row by row visit of pixels), others use more involved linearisation methods such as Hilbert space-filling curves. Using Hilbert curves for image compression was first suggested by Lempel

and Ziv [12] who showed applying Lempel-Ziv dictionary-based encodings (e.g., LZ encoding of [13]) in conjunction with Hilbert curve results in asymptotically optimal compression schemes for large images in terms of information entropy. These dictionary-based approaches, however, are very sensitive to dictionary size [14] and are generally slow compared to other schemes such as bZip2 [15,16].

Contribution

In this paper, we study the application of list update algorithms, and in particular the Move-To-Front algorithm, for lossless image compression. Different linearisation methods paired with different list update algorithms result in various compression schemes that can be enhanced with some pre-processing and post-processing stages. The main contribution of the paper is to show Hilbert curve linearisation followed by Move-To-Front encoding has a strict advantage over schemes that use other practical linearisation and list-update algorithms. In our analysis, we extend the concept "Max-Model" model of locality introduced by Albers [17] for sequential data into 2-dimensional images. An image has locality under the Max-Model iff the number of distinct pixel values in any square sub-image formed by k^2 pixels is at most $f(k^2)$ where f is a concave function. We prove that an image has locality under the Max-Model if and only if the sequence formed by the Hilbert curve of the image also has locality under Max-Model, that is, the number of different values in any window of size x in the sequence is upper bounded by a concave function of x. This result certifies that Hilbert curve linearisation "preserves" the locality of images. Meanwhile, we show that some of the basic linearisation techniques do not provide such guarantee and hence Hilbert curve traversal has a strict advantage over these linearisation methods. The sequence formed by the Hilbert curve can be encoded using any list update algorithm. We use a result of Angelopoulos and Schweitzer [18] to show that Move-To-Front is strictly better than any other list update algorithm for sequences that have locality under the Max-Model. In summary, our theoretical results show the advantage of Hilbert curve in conjunction with the MTF encoding over other linearisation and list update algorithms. We verify these results using an experimental study of different linearisation techniques combined with the MTF encoding and observe the advantage of Hilbert curve over other linearisation methods as well as the effectiveness of the MTF encoding for improving the compression ratio.

2   Proposed Approach

Our proposed approach consists of two main stages in a compression pipeline, namely, linearising an image into a sequence using a space-filling curve such as a Hilbert curve, and then using a list update algorithm such as Move-To-Front to encode such sequence. In what follows, we discuss the details of each step with the motivation and theoretical grounds for them.

2.1   Linearisation of an Image into a Sequence

Linearisation involves defining a sequential order for the pixels in the image. The simplest way to linearise an image is to visit pixels row by row, pretty much in the same way that pixels are sequentially stored in a 2-dimensional array. Column-major traversal is defined similarly by visiting columns one by one. Another way to define ordering of pixels is to use space-filling curves, which are ways to continuously "traverse" all pixels in the image. While the continuity is not necessary for image compression, it often helps in preserving locality. The simplest space filling curve is a snake-move curve that visits pixels row by row from top to bottom such that pixels in the odd rows are visited from left to right and pixels in the even rows are visited from right to left. Another space filling curve is a spiral curve which visits pixels in clockwise order and in the non-increasing order of their distance to the closest image boundary.

Hilbert space filling curves, also known as Hilbert curves, are relatively simple space filling curves with properties that makes them suitable for linearisation purposes (see, e.g., [19–21]). A Hilbert curve of order 1 is simply a (rotated) u-shape curve covering a square of size 2×2. A Hilbert curve of order k covers a square of size $2^k \times 2^k$ and is defined by connecting four copies of Hilbert curve of order $k - 1$ which are placed in the four quadrants of the square. The copies in the lower-left and lower-right are rotated 90 degrees in respectively clockwise and counter-clockwise directions, while the copies in the upper quadrants appear without rotation. Given an image of size $2^k \times 2^k$, a Hilbert curve of order k visits every pixel once. The Hilbert curves of the first four orders are shown in Fig. 1. We note that Hilbert curves can be computed efficiently using bit operations [22] and lookup tables [23]. In general cases where the images are not squared or have a length that is not a power of 2, we assume the image is placed at the bottom left of a larger squared image whose length is the power of 2 that is next to the larger length of the image, as shown in Fig. 2.

A good linearisation method should preserve the locality present in images. To capture this locality, we consider the "Max-Model" model of locality introduced by Albers [17] for sequential data. Given a sequence σ, a *window* of size w in σ is defined as a subsequence of w consecutive requests in σ. Now, σ is said to be *consistent* with some increasing concave function f if the number of distinct requests in any window of size w is at most $f(w)$, for any $w \in \mathbb{N}$. We extend the

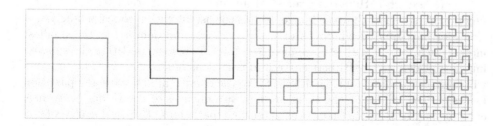

Fig. 1. Recursive construction of the Hilbert curves of the first four orders.

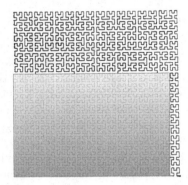

Fig. 2. Linearisation of a non-square image using the Hilbert curve of a larger square whose side-lengths are a power of 2.

Max-Model from sequential data to 2-dimensional images by defining a window to be a square sub-image of size $y = x \times x$. As before, an image is said to be consistent with a function g if the number of distinct pixel values in any window of size y is at most $g(y)$. For example, an image is consistent with $g(y) = \log(y)$ if there are at most $\log(y)$ distinct requests in any square with y pixels.

Theorem 1. *An image is consistent with a concave function g if and only if its Hilbert curve is consistent with some concave function f.*

Proof. First, assume an image is consistent with a concave function g. We need to present a concave function f such that the Hilbert sequence is consistent with f. Consider a window of length w in the Hilbert sequence. Define j as the smallest integer such that $2^j \times 2^j \geq w$; we have $2^{2j} < 4w$. Partition the (possibly larger) squared image which defines the Hilbert curve into squares of length 2^j. The recursive definition of Hilbert curve implies that it never leaves a partition before visiting all its pixels. Consequently, the window w includes pixels from at most two partitions. Since each partition is a square of 2^{2j} pixels, by the locality of image, there are at most $g(2^{2j})$ different pixel values in each partition. Hence, there are at most $2 \times g(2^{2j}) < 2g(4w)$ different pixel values in the windows w. So, if we define $f(w) = 2g(4w)$, the Hilbert curve sequence will be consistent with f. Note that since g is concave, f is also concave.

Next, assume a Hilbert traversal of an image is consistent with a concave function f. We show that the image itself is consistent with a concave function g. Consider any squared window s of size $y = x \times x$, and define j as the smallest integer such that $2^j \times 2^j \geq y$; we have $2^{2j} < 4y$. As before, partition the squared image that defines the Hilbert curve into partitions of size $2^j \times 2^j$. The square s will intersect at most four partitions. Since the curve will not leave any partition before visiting all its pixels, there are four subsequences of the Hilbert sequence, each of length $2^j \times 2^j$, that cover these four partitions. By the locality of the the Hilbert sequence, we know there are at most $f(2^{2j})$ different pixel values in each of these four subsequences, that is, there are at most $4f(2^{2j})$ different

pixel values in the four partitions. Since these four partitions cover the square s, there are at most $4f(2^{2j}) < 4f(4y)$ different pixel values in s. So, if we define $g(y) = 4f(4y)$, the Hilbert curve sequence will be consistent with g. Note that since f is concave, g will be concave as well. $\square$

While the above theorem implies that Hilbert curve traversal preserves the locality of images, other simple linearisation methods do not have this property. Consider images of size $n \times n$ in which pixels take values from 1 to n in the same order that they are arranged by the linearisation method (see Fig. 3). Clearly, any window of size n in the linearised sequence includes n distinct pixel values. On the other hand, one can check that any square s of $x \times x$ pixels includes $O(x)$ different pixel values. For row-major and column-major methods, s includes one pixel per column and per row, respectively; so there are x different pixel values in s. For the snake method, each column takes at most two distinct values, which gives at most $2x$ different pixel values. For the spiral move, each diagonal ray in s takes at most 3 values, which gives a total of at most $3x$ different pixel values. In summary, the number of different pixel values in each square of $y = x^2$ pixels is at most $3 \times \sqrt{y}$, which is a concave function and ensures locality under the Max-Model, while the number of different values in every window in the linearised sequence is y. Hence, these linearisation methods do not preserve locality.

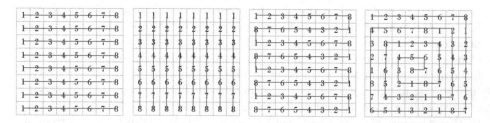

Fig. 3. Examples that show simple linearisation methods do not preserve locality. From left to right, figures correspond to row-major, column-major, snake-move, and spiral linearisation methods. In all cases, the number of distinct pixel values in any square of size x^2 is upper bounded by $3x$ while the linearised sequence includes x^2 different pixel values in a window of size $x^2 \leq n$, where n is the side-length of the image (here $n = 8$).

Proposition 1. *The Hilbert curve preserves the locality of images under the Max-Model while simple linearisation methods like row-major, column-major, snake curve, and spiral curve fail to preserve locality.*

2.2   List Update Encoding

The second step in our image compression scheme involves applying a list update algorithm to encode the linearised sequence. An instance of the list update problem involves a sequence formed by requests to items in a universe U that are

stored in a linked list. To answer each request, a list update algorithm performs a linear scan of the list to access the item. Accessing an item at position i has an *access cost* of i. After the access, the item can be moved closer to the front of the list at no additional cost using a *free exchange*. In addition, the algorithm can re-arrange the list using *paid exchanges* each swapping two consecutive items in the list at a cost of 1. Move-To-Front is a simple list update algorithm that moves an accessed item to the front of the list using a free exchange. Transpose is another simple algorithm that moves the accessed item one unit closer to the front using a free exchange. List update is an important problem in the context of online algorithms where it has contributed a lot to the concept of *competitive analysis*, which is a worst-case measure for comparing online algorithms. We refer the reader to [24] for a review of the list update problem.

List update is widely used in compression algorithms. Consider each value in a (linearised) sequence as an item in the list. A list update encoding writes an arbitrary initial configuration in the compressed file, as well as the access costs of a list update algorithm $\mathcal{A}$ for accessing each character in the list. For decoding, the algorithm initiates the list using the stored configuration and follows the same steps by reading the access costs written in the encoded text.

Any list update algorithm can be used to encode an input sequence (see, e.g., [25–28]). Different algorithms perform differently from one sequence to another. We note that the results for competitive analysis are not useful for comparing list update algorithms in the context of data compression as competitive analysis only compares algorithms over their worst-case sequences. A more practical measure for comparing list update algorithms is *bijective analysis* [29]. Consider all sequences of a given length n for a given list. An algorithm A is said to be no worse than algorithm B under bijective analysis iff there is a bijection (one to one mapping) b between sequences such that the cost of A for serving any sequence σ is no more than that of B for serving $b(\sigma)$. If A is no worse than B, then the two algorithms are said to be *equal* if B is also no worse than A; otherwise, A is *strictly better* than B. Competitive analysis involves comparing algorithms over all sequences in a given universe and hence provides theoretical guarantees for *typical* performance of algorithms, which is more relevant for data compression.

Theorem 2 [30]. For sequences with locality under the Max-Model, Move-To-Front is strictly better than any other algorithm under bijective analysis.

The above theorem can be used to conclude that MTF encodings are strictly better for encoding sequences formed by Hilbert curve traversal of images. We have to make two points, however, before drawing such a conclusion. First, while Theorem 2 concerns the "total cost" of list update algorithm (the cost over all requests), the bijections in its proof are defined in a way that the cost of MTF for each individual access in σ is no more the cost of the other algorithm for the request at the same index in $b(\sigma)$. This implies that if we change the cost model such that accessing an item at position i has access cost $\lceil \log i \rceil + O(\log \log i)$ instead of i, Theorem 2 still holds. This particular cost model is more relevant

for compression [26,28] as it is consistent with the length of a self-delimiting code for storing an index i. Second, Theorem 1 implies that any sequence which has locality under the Max-Model can be looked at as the Hilbert curve traversal of some image which has locality under the Max-Model (this is because the two directions hold in the theorem). Consequently, Theorem 2 implies that the MTF encoding has advantage over other list update encodings for images with locality under Max-Model.

Proposition 2. *Under bijective analysis, Move-To-Front is strictly better than any other list update algorithm for encoding Hilbert sequence of images which have locality under Max-Model.*

3   Experimental Results

In this section, we provide an experimental evaluation of different linearisation algorithms when paired with list update encoding for lossless image compression. Our focus in our experiments is comparing the effect of different linearisation methods, and therefore we have not applied other optimization stages of the bzip2 scheme, assuming these optimizations are orthogonal generally benefit all schemes equally [26].

In the light of Proposition 2 and previous work that show advantage of MTF over other list update algorithms (e.g., [26,28]), we fix the Move-To-Front as the list update encoding and experiment on different linearisation methods. We assume that the MTF encoding is followed by the Huffman entropy encoding to capitalized the reduction in the entropy resulted from the MTF encoding

We use *entropy ratio* and compression ratio to compare different compression schemes. The entropy ratio is defined as the ratio between the entropy of the linearised sequences after and before applying the MTF encoding. Given the fact that the goal of the linearisation and MTF encoding is mainly reducing the entropy, the entropy ratio provides a good measure to compare different linearisation methods separately from other stages of a compression pipeline. We also consider compression ratio as the ratio between the the original file size and compressed file size. Since we are aimed at reducing the entropy and eventually the size of compressed file, smaller entropy and higher compression ratios imply better results.

Dataset. Similarly to previous works, (e.g., [31]) we use four publicly available datasets named Classic, Medical, Kodak, and Digital-Camera images. The Classic dataset is composed of four images widely used by image processing community (see Fig. 4a. The set of Medical images consists of several Positron Emission Tomography (PET) images of human brain, digital camera images for eyes and eye-grounds, and endoscope images of the human intestine (see Fig. 4b for some of these images). The Kodak set contains 24 scene images from "Kodak Photo CD Photo Sampler", which are photographic quality images of a variety of subjects in many locations and under a variety of lighting conditions (see Fig. 4c for

some of the Kodak images). We note that Kodak dataset has been widely used in testing previous lossless image compression schemes [32,33]. Finally, Digital-Camera images are images from commercial digital cameras including Nikon D90 and Olympus E-P1 (see Fig. 4d for sample images from this data set). We refer the reader to [31] for details about the datasets used in our experiments.

Entropy Ratio. The linearisation stage, followed by the MTF encoding, are aimed at producing a sequence with small information entropy. The entropy ratio of different linearisation techniques for the Classic and Medical datasets are respectively presented in Tables 1 and 2. Similar results for the Kodak and Digital-Camera datasets are provided in Tables 3 and 4. In all cases, we achieve better entropy reduction using Hilbert space-filling curve linearisation. These results are consistent with Proposition 1 and confirm that the Hilbert curve traversal preserves the locality in practice (as it theoretically does under Max-Model) and consequently there is a bigger reduction in entropy when using Hilbert curve linearisation.

Table 1. Entropy ratio of different linearisation methods for images in the Classic dataset.

Image	Column major	Row major	Snake	Spiral	Hilbert curve
Mandrill	0.9656	0.9519	0.9480	0.9537	**0.9094**
Peppers	0.8904	0.8718	0.8664	0.8735	**0.7746**
Barbara	0.9054	0.8837	0.8791	0.8962	**0.8102**
Average	0.9112	0.8810	0.8755	0.8922	**0.8110**

Compression Ratio. Tables 5, 6, 7, and 8 show the compression ratio of different linearisation methods when paired with the MTF encoding[1]. As expected, the Hilbert linearisation method results in the highest compression. These results are in line with our theoretical results and show the advantage of using Hilbert curve linearisation when paired with the Move-To-Front coding before applying the entropy coding (e.g., Huffman coding).

As reported in [14], applying Huffman encoding just after the linearisation step (without the MTF encoding) does not give a promising compression ratio. The results in Table 9 show the compression ratio of different images in the absence and presence of MTF encoding before applying Huffman coding when images are linearised by the Hilbert curve. In all cases, applying the MTF encoding results in significant improvement in compression ratio. This highlights the importance of an entropy reduction step like the MTF encoding in the compression pipeline for image compression. We note that the MTF encoding can be preceded and succeeded with other stages of a compression pipeline

---

[1] Note that each image has three color channels and hence the reported file sizes are three times the size of mono-chrome images associated with each color.

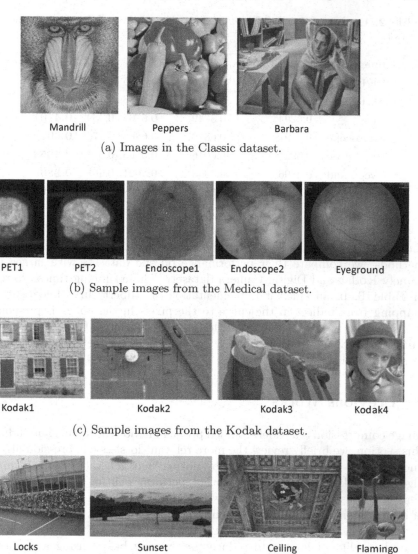

(a) Images in the Classic dataset.

Mandrill   Peppers   Barbara

PET1   PET2   Endoscope1   Endoscope2   Eyeground

(b) Sample images from the Medical dataset.

Kodak1   Kodak2   Kodak3   Kodak4

(c) Sample images from the Kodak dataset.

Locks   Sunset   Ceiling   Flamingo

(d) Sample images from the Digital-Camera dataset.

Fig. 4. Some of the images from the four datasets used in our experiments. (Color figure online)

(e.g., Burrows-Wheeler transform at the beginning and run-length encoding at the end). Assuming that all linearisation methods benefit from these extra stages, we excluded these steps in our experiments. The reported compression ratios are expected to improve when adding more stages (this comes at the price of spending more computational resources).

Table 2. Entropy ratio of different linearisation methods for images in the Medical dataset.

Image	Column major	Row major	Snake	Spiral	Hilbert curve
PET1	0.5309	0.5740	0.5686	0.4745	**0.4722**
PET2	0.5428	0.6024	0.5953	0.5193	**0.4901**
PET3	0.5294	0.5749	0.5701	0.4892	**0.4701**
Endoscope1	0.7197	0.6876	0.6803	0.6763	**0.6070**
Endoscope2	0.6021	0.6016	0.5905	0.5903	**0.4454**
Eyeground	0.4896	0.4963	0.4921	0.4632	**0.3866**
Average	0.5690	0.5894	0.5828	0.5354	**0.4786**

Running Time. In order to compare the running time of different linearisation methods, we compare their performance on the datasets with larger images, namely Kodak and Digital-Camera datasets. These running times are reported in Table 10. In an efficient implementation of Hilbert curve linearisation, the mapping from indices in the curve to the pixels in the image is pre-computed and stored in a lookup table. As such, the computation of Hilbert curve does not cause an overhead, and the running times of different methods are comparable as reflected in the table.

4   Previous Works

Image compression is a widely studied problem in both academia and industry. In this section, we briefly review the more relevant lossless compression algorithms and industry standards. We start by noting that general-purpose compression algorithms like Huffman coding do not perform well for images on their own. This is because these encodings rely on the redundancy of codewords or entropy of information and do not exploit the spatial locality that is inherent in images. As such, image compression algorithms often have a mechanism to utilize the spatial locality of images to reduce the entropy before using any entropy or arithmetic coding. Accordingly, image compression algorithms fall broadly into two categories: prediction based and transformation based.

In the prediction based approaches, a prediction of the next pixel is generated from previous pixels and the error of prediction is compressed with entropy coding or arithmetic coding. Intuitively, the error codes utilize the locality of pixel intensities to reduce entropy. The first remarkable prediction based approach was the median-edge predictor used in the JPEG-LS [34]. This method generates a prediction of each pixel based on three previous pixels (left, top, top-left) and encodes the prediction error. The context-based, adaptive, lossless image codec (CALIC) family of schemes [5] generate a prediction that is based on the local gradient information. Several other prediction based models [35,36] have been

Table 3. Entropy ratio of different linearisation methods for the Kodak dataset.

Image	Column major	Row major	Snake-like	Spiral	Hilbert curve
Kodak01	0.9212	0.9378	0.9332	0.9237	**0.888**
Kodak02	0.8022	0.8228	0.8145	0.8041	**0.7523**
Kodak03	0.6992	0.7214	0.7106	0.7097	**0.6074**
Kodak04	0.8181	0.8486	0.8375	0.8242	**0.7219**
Kodak05	0.9612	0.9748	0.9701	0.9601	**0.8993**
Kodak06	0.8104	0.8813	0.8699	0.8466	**0.7869**
Kodak07	0.7802	0.8157	0.8085	0.7849	**0.7101**
Kodak08	0.9469	0.9296	0.9242	0.9389	**0.8856**
Kodak09	0.8128	0.7775	0.7687	0.7824	**0.7065**
Kodak10	0.8313	0.8266	0.817	0.8133	**0.7212**
Kodak11	0.8366	0.8853	0.8793	0.8584	**0.8008**
Kodak12	0.7295	0.7902	0.7738	0.7515	**0.6707**
Kodak13	0.9534	0.9846	0.9782	0.965	**0.9162**
Kodak14	0.8966	0.9416	0.934	0.9044	**0.8317**
Kodak15	0.7886	0.7433	0.7315	0.7734	**0.6625**
Kodak16	0.7474	0.8749	0.8598	0.808	**0.7268**
Kodak17	0.8623	0.8441	0.8364	0.8572	**0.7439**
Kodak18	0.949	0.9606	0.9535	0.9524	**0.8847**
Kodak19	0.867	0.8415	0.8308	0.8544	**0.77**
Kodak20	0.785	0.8544	0.8488	0.8174	**0.7389**
Kodak21	0.8251	0.8902	0.8827	0.8506	**0.7982**
Kodak22	0.8591	0.8821	0.8701	0.8713	**0.7872**
Kodak23	0.7915	0.7511	0.7392	0.7781	**0.6392**
Kodak24	0.9033	0.886	0.8806	0.8978	**0.8157**
Average	0.8412	0.8604	0.8516	0.8458	**0.77**

Table 4. Entropy ratio of different linearisation methods for the Digital-Camera dataset.

Image	Column major	Row major	Snake-like	Spiral	Hilbert curve
Locks	0.7076	0.7301	0.7252	0.7315	**0.6393**
Sunset	0.5034	0.591	0.5841	0.5289	**0.4899**
Ceiling	0.8043	0.8565	0.8524	0.8077	**0.6969**
Flamingo	0.529	0.5551	0.5504	0.5508	**0.4421**
Berry	0.7178	0.7343	0.7314	0.73	**0.6165**
Fireworks	0.648	0.6594	0.6583	0.6557	**0.553**
Flower	0.7615	0.7676	0.7637	0.7658	**0.6042**
Park	0.566	0.6095	0.6051	0.5795	**0.545**
Average	0.655	0.6888	0.6837	0.671	**0.5725**

Table 5. Compression ratio of different linearisation methods for images in the Classic dataset.

Image	Size	Pixels	Column major	Row major	Snake	Spiral	Hilbert curve
Mandrill	3 × 512 × 512	786,432	1.06	1.08	1.09	1.08	**1.12**
Peppers	3 × 512 × 512	786,432	1.16	1.19	1.19	1.19	**1.33**
Barbara	3 × 640 × 512	983,040	1.15	1.18	1.18	1.12	**1.28**
Average	-	-	1.12	1.15	1.15	1.13	**1.24**

Table 6. Compression ratio of different linearisation methods for images in the Medical dataset.

Image	Size	Pixels	Column major	Row major	Snake	Spiral	Hilbert curve
PET1	3 × 256 × 256	196,608	4.0	3.85	3.85	4.35	**4.55**
PET2	3 × 256 × 256	196,608	3.85	3.57	3.57	4.0	**4.17**
PET3	3 × 256 × 256	196608	3.85	3.7	3.7	4.17	**4.35**
Endoscope1	3 × 603 × 552	998,568	1.54	1.59	1.61	1.61	**1.82**
Endoscope2	3 × 568 × 506	862,224	1.79	1.79	1.82	1.79	**2.38**
Eyeground	3 × 1600 × 1216	5,836,800	2.63	2.56	2.56	2.63	**3.23**
Average	-	-	2.94	2.84	2.85	3.09	**3.42**

proposed which mainly differ in the method of generating prediction. Differential encoding schemes [37] are also used as a prediction method for lossless image compression.

Variants of wavelet transforms [6] are widely used in both lossy and lossless image compression schemes. Most of the wavelet-based lossless image compression methods (e.g., [38,39]) use discrete, or integer variants of wavelet transform that are reversible, and encode the transformed coefficients. Integer to integer wavelet transforms [40] and binary wavelet transform [41] are also proposed for lossless image compression. Variants of cosine transforms [42,43] are also proposed for lossless image compression.

Entropy coding is frequently used as the final step of most of the lossless image compression. Run length Coding, Huffman Coding, Golomb Rice coding [44] and Arithmetic Coding are the most widely used as entropy coding scheme. Adaptive run-length coding with Move-To-Front transform [45] has been used for lossless image compression. Huffman coding and Golomb Rice coding use variable length codewords. Arithmetic coding is a variant of entropy coding where a set of codewords are encoded as a whole. Several adaptive arithmetic codings [46] have been proposed for lossless image compression. We note that these methods can be used in conjunction with the MTF encoding studied in this paper.

More recently, approaches that are based on deep-learning [7] or a combination of deep-learning with a transform based approach [8] are also proposed. Medical images generally have more locality of pixels. consequently, there have been a range of solutions especially designed for medical images (e.g., [1,2]). Other popular industry standards for lossless image compression schemes include

Table 7. Compression ratio of different linearisation methods for the Kodak dataset.

Image	Size	Pixels	Column major	Row major	Snake-like	Spiral	Hilbert curve
kodak01.ppm	(3 × 768 × 512)	1,179,648	1.18	1.16	1.16	1.09	**1.22**
kodak02.ppm	(3 × 768 × 512)	1,179,648	1.45	1.43	1.43	1.33	**1.56**
kodak03.ppm	(3 × 768 × 512)	1,179,648	1.54	1.47	1.52	1.39	**1.75**
kodak05.ppm	(3 × 768 × 512)	1,179,648	1.11	1.1	1.1	1.02	**1.19**
kodak04.ppm	(3 × 768 × 512)	1,179,648	1.3	1.25	1.27	1.19	**1.47**
kodak06.ppm	(3 × 768 × 512)	1,179,648	1.3	1.2	1.22	1.15	**1.35**
kodak07.ppm	(3 × 768 × 512)	1,179,648	1.39	1.33	1.35	1.27	**1.54**
kodak08.ppm	(3 × 768 × 512)	1,179,648	1.1	1.11	1.12	1.02	**1.18**
kodak09.ppm	(3 × 768 × 512)	1,179,648	1.37	1.43	1.45	1.3	**1.56**
kodak10.ppm	(3 × 768 × 512)	1,179,648	1.33	1.33	1.35	1.27	**1.54**
kodak11.ppm	(3 × 768 × 512)	1,179,648	1.33	1.25	1.27	1.19	**1.39**
kodak12.ppm	(3 × 768 × 512)	1,179,648	1.49	1.37	1.41	1.33	**1.61**
kodak13.ppm	(3 × 768 × 512)	1,179,648	1.11	1.08	1.08	1.01	**1.15**
kodak14.ppm	(3 × 768 × 512)	1,179,648	1.18	1.11	1.12	1.06	**1.27**
kodak15.ppm	(3 × 768 × 512)	1,179,648	1.33	1.43	1.45	1.27	**1.59**
kodak16.ppm	(3 × 768 × 512)	1,179,648	1.45	1.25	1.27	1.23	**1.49**
kodak17.ppm	(3 × 768 × 512)	1,179,648	1.25	1.28	1.28	1.16	**1.45**
kodak18.ppm	(3 × 768 × 512)	1,179,648	1.18	1.16	1.18	1.09	**1.27**
kodak19.ppm	(3 × 768 × 512)	1,179,648	1.22	1.25	1.27	1.14	**1.37**
kodak20.ppm	(3 × 768 × 512)	1,179,648	1.67	1.54	1.54	1.47	**1.79**
kodak21.ppm	(3 × 768 × 512)	1,179,648	1.33	1.23	1.25	1.19	**1.37**
kodak22.ppm	(3 × 768 × 512)	1,179,648	1.25	1.22	1.23	1.14	**1.37**
kodak23.ppm	(3 × 768 × 512)	1,179,648	1.32	1.39	1.41	1.23	**1.64**
kodak24.ppm	(3 × 768 × 512)	1,179,648	1.23	1.25	1.27	1.15	**1.37**
Average	-	-	1.31	1.28	1.29	1.2	**1.44**

Table 8. Compression ratio of different linearisation methods for the Digital-Camera dataset.

Image	Size	Pixels	Column major	Row major	Snake-like	Spiral	Hilbert curve
locks.ppm	(3 × 4288 × 2848)	36636672	1.47	1.43	1.43	1.32	**1.64**
fireworks.ppm	(3 × 3024 × 4032)	36578304	2.13	2.08	2.08	1.96	**2.5**
sunset.ppm	(3 × 4288 × 2848)	36636672	2.08	1.75	1.79	1.82	**2.13**
flower.ppm	(3 × 4032 × 3024)	36578304	1.33	1.32	1.32	1.27	**1.67**
park.ppm	(3 × 4032 × 3024)	36578304	2.0	1.85	1.85	1.89	**2.08**
flamingo.ppm	(3 × 2848 × 4288)	36636672	2.04	1.96	1.96	1.82	**2.44**
ceiling.ppm	(3 × 4288 × 2848)	36636672	1.39	1.3	1.32	1.27	**1.61**
berry.ppm	(3 × 2848 × 4288)	36636672	1.41	1.37	1.39	1.28	**1.64**
Average	-	-	1.73	1.63	1.64	1.58	**1.96**

CALIC [47], Lossless JPEG and it's derivatives JPEG-LS and JPEG 2000 (see, e.g., [48,49]), LZ77, LZW [50], and Free Lossless Image Format (FLIF) [51]. These methods generally use a combination of methods in a pipeline for compressing image file.

In this work, we studied the Hilbert space-filling curve and MTF encodings for reducing entropy. Our solution, when boosted with other stages of compression pipeline (e.g., Burrows-Wheeler transform and run-length encoding) could provide a simple and efficient compression scheme. Such a scheme would not necessarily give a compression ratio better than more complicated methods discussed

Table 9. The impact of the MTF encoding in improving compression ratio for some of the images is from the four datasets when linearised using Hilbert curve.

Image	Compression Ratio	
	Huffman	MTF+Huffman
Mandrill	0.9751	0.886
Peppers	0.9618	0.7467
Barbara	0.9665	0.7828
PET1	0.4072	0.2226
PET2	0.4325	0.2373
PET3	0.4254	0.2287
Endoscope1	0.9054	0.5522
Endoscope2	0.9324	0.4175

Image	Compression Ratio	
	Huffman	MTF+Huffman
Kodak01	0.9188	0.8173
Kodak02	0.854	0.6436
Kodak03	0.9363	0.5685
Kodak04	0.94	0.6797
Locks	0.9618	0.6148
Fireworks	0.7204	0.4009
Sunset	0.9484	0.4715
Flower	0.9881	0.5981
Park	0.8828	0.4838

Table 10. Running time of different linearisation methods (in seconds).

Image	Pixels	Column major	Row major	Snake	Spiral	Hilbert curve
kodak01	1,179,648	3.2	3.5	3.3	3.5	3.4
kodak02	1,179,648	2.6	2.7	2.6	2.8	2.8
kodak03	1,179,648	2.7	2.8	2.7	3.0	2.7
kodak04	1,179,648	3.0	3.2	3.0	3.3	2.9
kodak05	1,179,648	3.6	3.7	3.7	4.1	3.6
kodak06	1,179,648	2.9	3.5	3.2	3.4	3.1
kodak07	1,179,648	2.9	3.0	2.9	3.3	3.0
kodak08	1,179,648	3.8	3.7	3.6	4.1	3.6
kodak09	1,179,648	3.0	2.8	2.7	3.1	2.9
kodak10	1,179,648	3.0	2.9	2.8	3.1	2.9
kodak11	1,179,648	2.9	3.1	3.1	3.3	3.0
kodak12	1,179,648	2.7	2.8	2.7	3.1	2.8
kodak13	1,179,648	3.5	3.8	3.7	4.0	3.7
kodak14	1,179,648	3.4	3.6	3.5	3.6	3.2
kodak15	1,179,648	3.1	2.9	2.7	3.2	2.9
kodak16	1,179,648	2.6	3.2	3.0	3.2	2.9
kodak17	1,179,648	3.2	3.0	3.0	3.4	2.9
kodak18	1,179,648	3.3	3.2	3.2	3.5	3.3
kodak19	1,179,648	3.2	3.0	3.0	3.4	3.2
kodak20	1,179,648	2.6	3.1	2.9	3.1	2.8
kodak21	1,179,648	2.9	3.2	3.2	3.3	3.2
kodak22	1,179,648	3.0	3.1	3.0	3.3	3.1
kodak23	1,179,648	3.1	2.9	2.8	3.5	2.8
kodak24	1,179,648	3.1	3.2	3.1	3.4	3.1
Average (Kodak dataset)	-	3.1	3.2	3.1	3.4	3.1
locks	36,636,672	88.0	92.0	101.8	112.6	92.4
fireworks	36,578,304	82.5	84.1	83.4	89.3	77.7
sunset	36,636,672	71.5	79.6	76.9	79.8	84.2
flower	36,578,304	99.7	102.8	96.8	101.9	84.2
park	36,578,304	75.3	78.7	77.7	78.5	78.8
flamingo	36,636,672	71.9	74.8	72.8	80.3	81.5
ceiling	36,636,672	89.7	91.5	91.5	93.3	89.0
berry	36,636,672	104.7	108.0	107.1	114.5	94.4
Average (Digital-Camera dataset)	-	85.4	88.9	88.5	93.8	85.3

above (e.g., deep learning methods). It could, however, be used in applications where simplicity and time complexity are more important.

5 Concluding Remarks

In this paper, we presented a simple and efficient lossless image compression method that is based on the Hilbert space-filling curve and Move-To-Front encoding. Our theoretical and experimental results show the advantage of Hilbert curves over other simple linearisation techniques and also effectiveness of MTF encodings for image compression. These results, paired with the simplicity and efficiency of computing Hilbert curves and MTF codes, the benefit of using them in certain practical applications. From a theoretical point of view, it is interesting to study other (possibly more complex) space-filling curves for image linearisation. A natural question is whether there is a space-filling curve that preserves locality "better" than Hilbert curves do. To answer this question, it is necessary to quantify the locality in the Max-Model or use another measure of locality. We conjecture that Hilbert curves are as good as any other space-filling curve such as Morton curves [52] for image compression schemes that use MTF codes, and leave the study of other curves in this context as a topic for future work. The compression scheme proposed in this paper might be useful for encoding multi-spectral images. Finally, the techniques in this paper might be paired with other techniques that enable querying a compressed image without decompressing it (e.g., [53,54]); investigating such hybrid compression schemes is another topic for future work.

References

1. Malathkara, N.V., Soni, S.K.: Low-complexity and lossless image compression algorithm for capsule endoscopy. In: Proceedings of the 3rd International Conference on Internet of Things and Connected Technologies (ICIoTCT) (2018)
2. Reddy, P., Reddy, V.R., Bindu, S.: The lossless medical image compression for telemedicine applications with delimiter. J. Adv. Res. Dyn. Control. Syst. 10(3), 74–79 (2018)
3. Murtagh, F., Louys, M., Starck, J.-L., Bonnarel, F.: Compression of grayscale scientific and medical image data. Data Sci. J. 1(1), 111–127 (2002)
4. Martucci, S.A.: Reversible compression of HDTV images using median adaptive prediction and arithmetic coding. In: IEEE International Symposium on Circuits and Systems, pp. 1310–1313 (1990)
5. Xiaolin, W., Memon, N., Sayood, K.: A context-based, adaptive, lossless/nearly-lossless coding scheme for continuous-tone images. ISO/IEC JTC 1, 12 (1995)
6. Antonini, M., Barlaud, M., Mathieu, P., Daubechies, I.: Image coding using wavelet transform. IEEE Trans. Image Process. 1(2), 205–220 (1992)
7. Schiopu, I., Munteanu, A.: Residual-error prediction based on deep learning for lossless image compression. Electron. Lett. 54(17), 1032–1034 (2018)
8. Ahanonu, E., Marcellin, M., Bilgin, A.: Lossless image compression using reversible integer wavelet transforms and convolutional neural networks. In: 2018 Data Compression Conference, p. 395 (2018)

9. Elias, P.: Universal codeword sets and the representation of the integers. IEEE Trans. Inf. Theory **21**, 194–203 (1975)
10. Seward, J.: bZip2 compression program. http://www.bzip.org/
11. Burrows, M., Wheeler, D.J.: A block-sorting lossless data compression algorithm. Technical report 124, DEC SRC (1994)
12. Lempel, A., Ziv, J.: Compression of two-dimensional data. IEEE Trans. Inf. Theory **32**(1), 2–8 (1986)
13. Ziv, J., Lempel, A.: Compression of individual sequences via variable-rate coding. IEEE Trans. Inf. Theory **24**(5), 530–536 (1978)
14. Liang, J.-Y., Chen, C.-S., Huang, C.-H., Liu, L.: Lossless compression of medical images using Hilbert space-filling curves. Comput. Med. Imaging Graph. **32**(3), 174–182 (2008)
15. Collin, L.: A Quick Benchmark: Gzip vs. Bzip2 vs. LZMA (2005)
16. Klausmann, T.: Gzip, Bzip2 and LZMA compared (2008). Blog Archives. https://web.archive.org/web/20130106193958/http://blog.i-no.de/archives/2008/05/08/index.html
17. Albers, S., Favrholdt, L.M., Giel, O.: On paging with locality of reference. In: Proceedings of the Thiry-Fourth Annual ACM Symposium on Theory of Computing, pp. 258–267. ACM (2002)
18. Angelopoulos, S., Schweitzer, P.: Paging and list update under bijective analysis. In: Proceedings of the Twentieth Annual ACM-SIAM Symposium on Discrete Algorithms, pp. 1136–1145. SIAM (2009)
19. Sagan, H.: Space-Filling Curves. Springer, New York (2012)
20. Moon, B., Jagadish, H.V., Faloutsos, C., Saltz, J.H.: Analysis of the clustering properties of the Hilbert space-filling curve. IEEE Trans. Knowl. Data Eng. **13**(1), 124–141 (2001)
21. Mokbel, M.F., Aref, W.G., Kamel, I.: Analysis of multi-dimensional space-filling curves. GeoInformatica **7**(3), 179–209 (2003)
22. Butz, A.R.: Convergence with Hilbert's space filling curve. J. Comput. Syst. Sci. **3**(2), 128–146 (1969)
23. Kamata, S., Eason, R.O., Bandou, Y.: A new algorithm for N-dimensional Hilbert scanning. IEEE Trans. Image Process. **8**(7), 964–973 (1999)
24. Kamali, S., López-Ortiz, A.: A survey of algorithms and models for list update. In: Brodnik, A., López-Ortiz, A., Raman, V., Viola, A. (eds.) Space-Efficient Data Structures, Streams, and Algorithms - Papers in Honor of J. Ian Munro on the Occasion of His 66th Birthday. LNCS, vol. 8066, pp. 251–266. Springer, Heidelberg (2013). https://doi.org/10.1007/978-3-642-40273-9_17
25. Bentley, J.L., Sleator, D., Tarjan, R.E., Wei, V.K.: A locally adaptive data compression scheme. Commun. ACM **29**, 320–330 (1986)
26. Dorrigiv, R., López-Ortiz, A., Ian Munro, J.: An application of self-organizing data structures to compression. In: Proceedings of the 8th International Symposium on Experimental Algorithms (SEA), vol. 5526, pp. 137–148 (2009)
27. Kamali, S., Ladra, S., López-Ortiz, A., Seco, D.: Context-based algorithms for the list-update problem under alternative cost models. In: Data Compression Conference (DCC), pp. 361–370 (2013)
28. Kamali, S., López-Ortiz, A.: Better compression through better list update algorithms. In: Data Compression Conference (DCC), pp. 372–381 (2014)
29. Angelopoulos, S., Dorrigiv, R., López-Ortiz, A.: On the separation and equivalence of paging strategies and other online algorithms. Algorithmica **81**(3), 1152–1179 (2019)

30. Angelopoulos, S., Schweitzer, P.: Paging and list update under bijective analysis. J. ACM **60**(2), 7:1–7:18 (2013)
31. Kim, S., Cho, N.I.: Hierarchical prediction and context adaptive coding for lossless color image compression. IEEE Trans. Image Process. **23**(1), 445–449 (2014)
32. Malvar, H.S., Sullivan, G.J.: Progressive-to-lossless compression of color-filter-array images using macropixel spectral-spatial transformation. In: Data Compression Conference (DCC 2012), pp. 3–12 (2012)
33. Zhang, N., Xiaolin, W.: Lossless compression of color mosaic images. IEEE Trans. Image Process. **15**(6), 1379–1388 (2006)
34. Weinberger, M.J., Seroussi, G., Sapiro, G.: From LOCO-I to the JPEG-LS standard. Hewlett Packard Laboratories (1999)
35. Meyer, B., Tischer, P.: TMW-a new method for lossless image compression. ITG-Fachbericht, pp. 533–540 (1997)
36. Li, X., Orchard, M.T.: Edge-directed prediction for lossless compression of natural images. IEEE Trans. Image Process. **10**(6), 813–817 (2001)
37. Chianphatthanakit, C., Boonsongsrikul, A., Suppharangsan, S.: A lossless image compression algorithm using differential subtraction chain. In: International Conference on Knowledge and Smart Technology (KST), pp. 84–89 (2018)
38. Shapiro, J.M.: Embedded image coding using zerotrees of wavelet coefficients. IEEE Trans. Image Process. **41**(12), 3445–3462 (1993)
39. Taubman, D.: High performance scalable image compression with EBCOT. IEEE Trans. Image Process. **9**(7), 1158–1170 (2000)
40. Robert Calderbank, A., Daubechies, I., Sweldens, W., Yeo, B.-L.: Lossless image compression using integer to integer wavelet transforms. In: International Conference on Image Processing, vol. 1, pp. 596–599 (1997)
41. Pan, H., Siu, W.C., Law, N.F.: Lossless image compression using binary wavelet transform. IET Image Process. **1**(4), 353 (2007)
42. Ahmed, N., Natarajan, T., Rao, K.R.: Discrete cosine transform. IEEE Trans. Comput. **100**(1), 90–93 (1974)
43. Mandyam, G., Ahmed, N., Magotra, N.: Lossless image compression using the discrete cosine transform. J. Vis. Commun. Image Represent. **8**(1), 21–26 (1997)
44. Starosolski, R.: Simple fast and adaptive lossless image compression algorithm. Softw. Pract. Exp. **37**(1), 65–91 (2007)
45. Zalik, B., Lukac, N.: Chain code lossless compression using move-to-front transform and adaptive run-length encoding. Signal Process. Image Commun. **29**(1), 96–106 (2014)
46. Triantafyllidis, G.A., Strintzis, M.G.: A context based adaptive arithmetic coding technique for lossless image compression. IEEE Signal Process. Lett. **6**(7), 168–170 (1999)
47. Wu, X., Memon, N.: CALIC-a context based adaptive lossless image codec. In: IEEE International Conference on Acoustics, Speech, and Signal Processing, ICASSP 1996, vol. 4, pp. 1890–1893 (1996)
48. Pennebaker, W.B., Mitchell, J.L.: JPEG: Still Image Data Compression Standard. Springer, New York (1992)
49. Christopoulos, C., Skodras, A., Ebrahimi, T.: The JPEG2000 still image coding system: an overview. IEEE Trans. Consum. Electron. **46**(4), 1103–1127 (2000)
50. Abu Taleb, S.A., Musafa, H.M.J., Khtoom, A.M., Gharaybih, I.K.: Improving LZW image compression. Eur. J. Sci. Res **44**(3), 502–509 (2010)
51. Sneyers, J., Wuille, P.: FLIF: free lossless image format based on MANIAC compression. In: IEEE International Conference on Image Processing (ICIP), pp. 66–70 (2016)

52. Morton, G.M.: A computer oriented geodetic data base and a new technique in file sequencing. International Business Machines (1966)
53. Pajarola, R., Widmayer, P.: An image compression method for spatial search. IEEE Trans. Image Process. **9**(3), 357–365 (2000)
54. Pinto, A., Seco, D., Gutiérrez, G.: Improved queryable representations of rasters. In: Data Compression Conference (DCC), pp. 320–329 (2017)

Rpair: Rescaling RePair with Rsync

Travis Gagie[1,2], Tomohiro I[3], Giovanni Manzini[4(✉)], Gonzalo Navarro[1,5],
Hiroshi Sakamoto[3], and Yoshimasa Takabatake[3]

[1] CeBiB—Center for Biotechnology and Bioengineering, Santiago, Chile
[2] Faculty of Computer Science, Dalhousie University, Halifax, Canada
[3] Department of Artificial Intelligence, Kyushu Institute of Technology,
Fukuoka, Japan
[4] Department of Science and Technological Innovation,
University of Eastern Piedmont, Alessandria, Italy
giovanni.manzini@uniupo.it
[5] Department of Computer Science, University of Chile, Santiago, Chile

Abstract. Data compression is a powerful tool for managing massive but repetitive datasets, especially schemes such as grammar-based compression that support computation over the data without decompressing it. In the best case such a scheme takes a dataset so big that it must be stored on disk and shrinks it enough that it can be stored and processed in internal memory. Even then, however, the scheme is essentially useless unless it can be built on the original dataset reasonably quickly while keeping the dataset on disk. In this paper we show how we can preprocess such datasets with context-triggered piecewise hashing such that afterwards we can apply RePair and other grammar-based compressors more easily. We first give our algorithm, then show how a variant of it can be used to approximate the LZ77 parse, then leverage that to prove theoretical bounds on compression, and finally give experimental evidence that our approach is competitive in practice.

1 Introduction

Dictionary compression has proved to be an effective tool to exploit the repetitiveness that most of the fastest-growing datasets feature [24]. Lempel-Ziv (LZ77 for short) [23,33] stands out as the most popular and effective compression method for repetitive texts. Further, it can be run in linear time and even in external memory [18]. LZ77 has the important drawback, however, that accessing random positions of the compressed text requires, essentially, to decompress it from the beginning. Therefore, it is not suitable to be used as a *compressed data structure* that represents the text in little space while simulating direct access

Partially funded with Basal Funds FB0001, Conicyt, Chile. Partially funded by JST CREST Grant Number JPMJCR1402 (TI, HS, YT), KAKENHI Grant Numbers 19K20213 (TI), 17H01791 (HS), 18K18111 (YT). Partially funded by PRIN Grant Number 2017WR7SHH and by the LSBC_19-21 Project from the University of Eastern Piedmont (GM).

© Springer Nature Switzerland AG 2019
N. R. Brisaboa and S. J. Puglisi (Eds.): SPIRE 2019, LNCS 11811, pp. 35–44, 2019.
https://doi.org/10.1007/978-3-030-32686-9_3

to it. Grammar compression [19] is an alternative that offers better guarantees in this sense. The aim is to build a small context-free grammar (or Straight-Line Program, SLP) that generates (only) the text. The smallest SLP generating a text is always larger than its LZ77 parse, but only by a logarithmic factor that is rarely reached in practice. With an SLP we can access any text substring with only an additive logarithmic time penalty [3,5], which has led to the development of various self-indexes building on SLPs [4,9,12,13,15,26]. Many other richer queries on sequences have also been supported by associating summary information with the nonterminals of the SLP [1,2,5,7,10,11]. There are applications in which SLPs are preferable to LZ77 for other reasons, as well; see, e.g., [22,25].

Although finding the smallest SLP for a text is NP-complete [8,28], there are several grammar construction algorithms that guarantee at most a logarithmic blowup on the LZ77 parse [8,16,17,28,29]. In practice, however, they are sharply outperformed by RePair [21], a heuristic that runs in linear time and obtains grammars of size very close to that of the LZ77 parse in most cases. This has made RePair the compressor of choice to build grammar-based compressed data structures [1,7,10,11]. A serious problem with RePair, however, is that, despite running in linear time and space, in practice the constant of proportinality is high and it can be built only on inputs that are about one tenth of the available memory. This significantly hampers its applicability on large datasets.

In this paper we introduce a scalable SLP compression algorithm that uses space very close to that of RePair and can be applied on very large inputs. We prove a constant-approximation factor with respect to any SLP construction algorithm to which our technique is applied. Our experimental results show that we can compress a very repetitive 50 GB text in less than an hour, using less than 650MB of RAM and obtaining very competitive compression ratios.

2   Preliminaries

For the sake of brevity, we assume the reader is familiar with SLPs, LZ77, and the links between the two. To prove theoretical bounds for our approach, we consider a variant of LZ77 in which if $S[i..j]$ is a phrase then either $i = j$ and $S[i]$ is the first occurrence of a distinct character, or $S[i..j]$ occurs in $S[1..j-1]$ and $S[i..j+1]$ does not occur in $S[1..j]$. We refer to this variant as LZSS due to its similarity to Storer and Szymanski's version of LZ77 [30], even though they allow substrings to be stored as raw text and we do not.

The best-known algorithm for building SLPs is probably RePair [21], for which there are many implementations (see [14] and references therein). It works by repeatedly finding the most common pair of symbols and replacing them with a new non-terminal. Although it is not known to have a good worst-case approximation ratio with respect to the size of LZ77 parsing, in practice it outperforms other constructions. RePair uses linear time and space but the coefficient in the space bound is quite large and so the standard implementations are practical only on small inputs. A more recent and more space economical alternative to RePair is SOLCA [31] that we will consider in Sect. 5.

Algorithm 1. Rpair: use Rsync parsing to build an SLP for a given string S

1. build an Rsync dictionary and parse for S;
2. generate SLPs for the distinct blocks as follows:
   (a) append a unique separator character to each block in the dictionary and then concatenate the blocks (in the order of their first appearances in S) into a string D;
   (b) build an SLP for D;
   (c) delete from the SLP any non-terminal that occurs only once in the parse tree (and any rule including it);
   (d) delete from the SLP the separator characters (and any rules including them);
   (e) list the non-terminals at the roots of the maximal remaining subtrees of the parse tree;
   (f) divide the list into sublists such that the concatenation of the expansions of the non-terminals in the ith sublist is the i block in D;
   (g) create a set of rules generating the ith sublist from a new non-terminal X_i;
3. build an SLP for the parse P;
4. replace by X_i each occurrence in P of the terminal for the ith block in D;
5. combine the SLP for P with the SLPs for the blocks.

Context-triggered piecewise hashing (CTPH) is a technique for parsing strings into blocks such that long repeated substrings are parsed the same way (except possibly at the beginning or end of the substrings). The name CTPH seems to be due to Kornblum [20] but the ideas go back to Tridgell's Rsync [32] and Spamsum (https://www.samba.org/ftp/unpacked/junkcode/spamsum/README): "The core of the spamsum algorithm is a rolling hash similar to the rolling hash used in 'rsync'. The rolling hash is used to produce a series of 'reset points' in the plaintext that depend only on the immediate context (with a default context width of seven characters) and not on the earlier or later parts of the plaintext."

Specifically, in this paper we choose a rolling hash function and a threshold p, run a sliding window of fixed size w over S and end the current block whenever the window contains a triggering substring, which is a substring of length w whose hash is congruent to 0 modulo p. When we end a block, we shift the window ahead w characters so all the blocks are disjoint and form a parse, which we call the *Rsync parse*. We call the set of distinct blocks the *Rsync dictionary*: if the input text contains many repetitions, we expect the dictionary to be much smaller than the text.

3   Algorithms

Given a string S, we can use Rsync parsing to help build an SLP for S with Algorithm 1 ("Rpair"). The final SLP can be viewed as first generating the parse, then replacing each block ID in the parse by the sublist of non-terminals that generate each block, and finally replacing the sublists by the blocks themselves.

Algorithm 2. Rparse: use Rsync to build an LZSS-like parse for a string S

---

1. build an Rsync dictionary and parse for S;
2. append a unique separator character to each block in the dictionary and concatenate the blocks (in the order of their first appearances in S) into a string D;
3. compute the LZSS parse of D;
4. compute the LZSS parse of the parse P, treating each block as a meta-character;
5. map D's and P's parses onto S:
   (a) discard any separator character $D[j]$ in D;
   (b) turn the first occurrence $D[j]$ of any other character in D into the first occurrence $S[j']$ of that character in S;
   (c) turn each phrase $D[j..j+\ell-1]$ in block B with source $D[i..i+\ell-1]$ in block B', into a phrase $S[j'..j'+\ell-1]$ with source $S[i'..i'+\ell-1]$, where $S[j']$ and $S[i']$ have the same respective offsets from the beginnings of the first occurrences of B and B' in S, as $D[j]$ and $D[i]$ have from the beginnings of B and B' in D;
   (d) discard the first occurrence $P[j]$ of each block in P;
   (e) turn each phrase $P[j..j+\ell-1]$ with source $P[i..i+\ell-1]$, into a phrase $S[j'..j'+\ell'-1]$ with source $S[i'..i'+\ell'-1]$, where $S[j']$ and $S[i']$ are the first characters in the jth and ith blocks, respectively, and ℓ' is the total length of the jth through $(j+\ell-1)$st blocks (and thus also the total length of the ith through $(i+\ell-1)$st blocks).

Since each separator character appears only once in D and its parse tree, any non-terminal whose expansion includes a separator character also appears only once and is deleted. Since the parse tree of an SLP is binary and each non-terminal we delete appears only once, the number of distinct non-terminals we delete is at least the length of the list of non-terminals at the roots of the maximal remaining subtrees of the parse tree, minus one. Therefore, creating rules to generate the sublists does not cause the number of distinct non-terminals to grow to more than the number in the original SLP for D, plus one.

Algorithm 1 works with any algorithm for building SLPs for D and P. In Sect. 4 we show that, if we choose an algorithm that builds SLPs for D and P at most an α-factor larger than their LZ77 parses, then we obtain an SLP an $O(\alpha)$-factor larger than the LZ77 parse of S. In the process we will refer to Algorithm 2 ("Rparse"), which produces an LZSS-like parse of S but is intended only to simplify our analysis of Algorithm 1 (not to compete with cutting-edge LZ-based compressors). By "LZSS-like" we mean a parse in which each phrase is either a single character that has not occurred before, or a copy of an earlier substring. We note in passing that, if the parse in Step 3 is still too big for a normal construction, then we can apply Algorithm 1 to it. We will show in the full version of this paper that, if we recurse only a constant number of times, then we worsen our compression bounds by only a constant factor.

4   Analysis

The main advantage of using Rsync parsing to preprocess S is that Rsync parsing is quite easy to parallelize, apply over streamed data, or apply in external
memory. The resulting dictionary and parse may be significantly smaller than
S, making it easier to apply grammar-based compression. In the full version of
this paper we will analyze how much time and workspace Algorithms 1 and 2
use in terms of the total size of the dictionary and parse, but for now we are
mainly concerned with the quality of the compression.

Let b be the number of distinct blocks in the Rsync parse of S, and let z be
the number of phrases in the LZ77 parse of S. The first block is obviously the
first occurrence of that substring and if $S[i..j]$ is the first occurrence of another
block, then $S[i-w..j]$ (i.e., the block extended backward to include the previous
triggering substring) is the first occurrence of that substring. Since the first
occurrence of any non-empty substring overlaps or ends at a phrase boundary
in the LZ77 parse, we can charge $S[i..j]$ to such a boundary in $S[i-w..j]$. Since
blocks have length at least w and overlap by only w characters when extended
backwards, each boundary has the first occurrences of at most two blocks charged
to it, so $b = O(z)$.

In Step 5 of Algorithm 2, we discard $O(b)$ of the phrases in the LZSS parses
of D and P when mapping to the phrases in the LZSS-like parse of S. Therefore,
by showing that the number of phrases in the LZSS-like parse of S is $O(z)$, we
show that the total number of phrases in the LZSS parses of D and P is also
$O(z + b) = O(z)$, so the total number of phrases in their LZ77 parses is $O(z)$ as
well.

Lemma 1. *If the t-th phrase in the LZSS parse of S is $S[j..j + \ell - 1]$ then the
$5t$-th phrase resulting from Algorithm 2, if it exists, ends at or after $S[j + \ell - 1]$.*

Proof. Our claim is trivially true for $t = 1$, since the first phrases in both parses
are the single character $S[1]$, so let t be greater than 1 and assume our claim
is true for $t - 1$, meaning the $5(t - 1)$st phrase in our parse ends at $S[k - 1]$
with $k \geq j$. If $k \geq j + \ell$ then our claim is also trivially true for t, so assume
$j \leq k < j + \ell$. We must show that our parse divides $S[k..j + \ell - 1]$ into at most
five phrases, in order to prove our claim for t.

First suppose that $S[k..j + \ell - 1]$ does not completely contain a triggering
substring, so it overlaps at most two blocks. (It can overlap two blocks without
containing a triggering substring if and only if a prefix of length less than w lies
in one block and the rest lies in the next block.) Let $S[i..i+\ell-1]$ be $S[j..j+\ell-1]$'s
source and let $k' = i + k - j$, so in the LZSS parse $S[k..j + \ell - 1]$ is copied from
$S[k'..i + \ell - 1]$. Since $S[k'..i + \ell - 1]$ does not completely contain a triggering
substring either, it too overlaps at most two blocks.

Without loss of generality (since the other cases are easier), assume $S[k..j +
\ell - 1]$ and $S[k'..i + \ell - 1]$ each overlap two blocks and they are split differently:
$S[k..k + d - 1]$ lies in one block and $S[k + d..j + \ell - 1]$ lies in the next, and
$S[k'..k' + d' - 1]$ lies in one block and $S[k' + d'..i + \ell - 1]$ in the next, with $d \neq d'$.

Assume also that $d < d'$, since the other case is symmetric. Since $S[k..k+d-1]$ is completely contained in a block and occurs earlier completely contained in a block, as $S[k'..k'+d-1]$, our parse does not divide it. Similarly, since $S[k+d..k+d'-1]$ and $S[k+d'..j+\ell-1]$ are each completely contained in a block and occur earlier each completely contained in a block, as $S[k'+d..k'+d'-1]$ and $S[k'+d'..i+\ell-1]$, respectively, our parse does not divide them. Therefore, our parse divides $S[k..j+\ell-1]$ into at most three phrases.

Now suppose the first and last triggering substrings completely contained in $S[k..j+\ell-1]$ are $S[x..x+w-1]$ and $S[y..y+w-1]$ (possibly with $x = y$). By the arguments above, our parse divides $S[k..x+w-1]$ into at most three phrases. Since $S[x+w..y+w-1]$ is a sequence of complete blocks that have occurred earlier (in $S[k'..i+\ell-1]$), our parse does not divide it unless $S[k..x+w-1]$ is a complete block that has occurred before as a complete block, in which case it may divide $S[k..y+w-1]$ once between $S[x+w]$ and $S[y+w-1]$. Since $S[y+w..j+\ell-1]$ is completely contained in a block and occurs earlier completely contained in a block (in $S[k'..i+\ell-1]$), our parse does not divide it. Therefore, our parse divides $S[k..j+\ell-1]$ into at most five phrases.   □

We note that we can quite easily can reduce the five in Lemma 1, at the cost of complicating our algorithm slightly. We leave a detailed analysis for the full version of this paper.

Corollary 1. *Algorithm 2 yields an LZSS-like parse of S with at most five times as many phrases as its LZSS parse.*

Proof. If the LZSS parse has t phrases then the t-th phrase ends at $S[n]$ so, by Lemma 1, Algorithm 2 yields a parse with at most $5t$ phrases.   □

Theorem 1. *Algorithm 2 yields an LZSS-like parse of S with $O(z)$ phrases.*

Proof. It is well known that the LZSS parse of S has at most twice as many phrases as the its LZ77 parse (since dividing each LZ77 phrase into a prefix with an earlier occurrence and a mismatch character yields an LZSS-like parse with at most twice as many phrases, and the LZSS parse has the fewest phrases of any LZSS-like parse). Therefore, by Corollary 1, Algorithm 2 yields a parse with at most $O(z)$ phrases.   □

Corollary 2. *The LZ77 parses of D and P have $O(z)$ phrases.*

Proof. Immediate, from Theorem 1, the fact that the LZ77 parse is no larger than the LZSS parse, and inspection of Algorithm 1.   □

Let A be any algorithm that builds an SLP at most an α-factor larger than the LZ77 parse of its input. For example, with Rytter's construction [28] we have $\alpha = O(\log(|S|/z))$.

By Corollary 2, applying A to D—Step 2b in Algorithm 1—yields an SLP for D with $O(\alpha z)$ rules. As explained in Sect. 3, Steps 2c to 2g then increase the number of rules by at most one while modifying the SLP such that, for each block in the dictionary, there is a non-terminal whose expansion is that block.

Similarly, applying A to P—Step 3—yields an SLP for P with $O(z)$ rules. Replacing the terminals in the SLP by the non-terminals generating the blocks and then combining the two SLPs—Steps 4 and 5—yields an SLP for S with $O(\alpha z)$ rules. This gives us our main result of this section:

Theorem 2. *Using A in Steps 2b and 3 of Algorithm 1 yields an SLP for S with $O(\alpha z)$ rules.*

5   Experiments

We use two genome collections in our experiments: cN consists of N concatenated variants of the human chromosome chr19, of about 59 MB each; sN consists of N concatenated variants of salmonella genomes, of widely different sizes.

The chr19 collection was downloaded from the 1000 Genomes Project. Each chr19 sequence was derived by using the bcftools consensus tool to combine the haplotype-specific (maternal or paternal) variant calls for an individual with the chr19 sequence in the GRCH37 human reference. The salmonella genomes were downloaded from NCBI (BioProject PRJNA183844) and preprocessed by assembling each individual sample with IDBA-UD [27] setting kMaxShortSequence to 1024 per public advice from the author to accommodate the longer paired end reads that modern sequencers produce. More details of the collections are available in previous work [6, Sec. 4].

We compare two grammar compressors: RePair [21] produces the best known compression ratios but uses a lot of main memory space, whereas SOLCA [31] aims at optimizing main memory usage. Their versions combined with parallelized CTPH parsing are BigRepair and BigSOLCA. RePair could be run only on the smaller collections. Our experiments ran on a Intel(R) I7-4770 @ 3.40 GHz machine with 32 GB memory using 8 threads; currently only the CTPH parsing takes advantage of the multiple threads.

For RePair we use Navarro's implementation for large files, at http://www. dcc.uchile.cl/gnavarro/software/repair.tgz, letting it use 10 GB of main memory, whereas the implementation of SOLCA is at https://github.com/tkbtkysms/ solca. To measure their compression ratios in a uniform way, we consider the following encodings of their output: if RePair produces r (binary) rules and an initial rule of length c, we account $2r$ bits to encode the topology of the pruned parse tree (where the nonterminal ids become the preorder of their internal node in this tree) and $(r+c)\lceil \log_2 r \rceil$ bits to encode the leaves of the tree and the initial rule. SOLCA is similar, with $c = 1$. Our code is available at https://gitlab.com/ manzai/bigrepair.

Table 1 shows the results in terms of compression ratio, time, and space in RAM. On the more repetitive chr19 genomes, BigRePair is clearly the best choice for large files. It loses to RePair in compression ratio, but RePair takes 11 h just to process 5.5 GB, so it is not a choice for larger files. Instead, BigRepair processes 55 GB in about 20 min and 6.5 GB. Similarly, SOLCA obtains better compression but more compression time than BigSOLCA, though the latter uses more space.

Table 1. Performance of the compressors. File sizes are expressed in GB, compression ratios in percentage of compressed file over uncompressed file, compression times in seconds per input GB, and compression main memory usage in MBs per input GB.

File	Size	RePair			BigRePair			SOLCA			BigSOLCA		
		Ratio	Time	Spc	Ratio	Time	Spc	Ratio	Time	Spc	Ratio	Time	Spc
c50	2.75	0.80%	1832	3842	0.91%	29.30	454.7	1.35%	244.1	107.4	1.54%	66.47	183.4
c100	5.51	0.30%	7311	3155	0.48%	25.05	246.4	0.77%	236.4	53.67	0.86%	56.96	130.4
c250	13.8				0.23%	22.10	119.8	0.40%	239.0	29.78	0.44%	48.55	95.00
c500	27.5				0.14%	22.31	118.0	0.28%	237.4	17.05	0.30%	47.46	84.72
c1000	55.1				0.10%	22.61	117.3	0.22%	237.3	13.56	0.23%	47.79	78.82
s815	3.75	1.72%	8478	3726	1.93%	51.70	2254	3.01%	317.7	161.0	3.50%	104.1	291.4
s2073	9.72				2.01%	55.48	1055	3.01%	370.9	153.1	3.53%	116.9	285.9
s4570	22.0				2.61%	201.1	534.2	3.57%	480.6	154.4	4.24%	142.8	335.1
s11264	53.1				1.51%	2560	294.2	2.20%	620.2	92.60	2.61%	113.1	206.7

The comparison between the two compressors shows that BigRepair performs better than both SOLCA and BigSOLCA in both compression ratio (reaching nearly half the compressed size of SOLCA on the largest files) and time (half the time of BigSOLCA). Still SOLCA uses much less space: it compresses 55 GB in 3.6 h, but using less than 750 MB.

The results start similarly on the less compressible salmonella collection, but, as the size of the input grows, there are significant differences. The time of BigRePair on chr19 was stable around 2GBs per minute, but on salmonella it is not: When moving from 10 GB to 20 GB of input data, the time per processed GB of BigRePair jumps by a factor of 3.6, and when moving from 20 GB to 50 GB it jumps by more than 10. To process the largest 53 GB file, BigRePair requires more than 37 h and over 15 GB of RAM. SOLCA, instead, handles this file in nearly 9 h and less than 5 GB, and BigSOLCA in less than 2 h and 11 GB, being the fastest. What happens is that, being less compressible, the output of the CTPH parse is still too large for RePair, and thus it slows down drastically as soon as it cannot fit its structures in main memory. The much lower memory footprint of SOLCA, instead, pays off on these large and less compressible files, though its compression ratio is worse than that of BigRePair. In the full version of this paper we will investigate applying BigRePair and BigSOLCA recursively, following the strategy mentioned at the end of Sect. 3.

References

1. Abeliuk, A., Cánovas, R., Navarro, G.: Practical compressed suffix trees. Algorithms **6**(2), 319–351 (2013)
2. Bannai, H., Gagie, T., I, T.: Online LZ77 parsing and matching statistics with RLBWTs. In: CPM, pp. 7:1–7:12 (2018)
3. Belazzougui, D., Cording, P.H., Puglisi, S.J., Tabei, Y.: Access, rank, and select in grammar-compressed strings. In: ESA, pp. 142–154 (2015)
4. Bille, P., Ettienne, M.B., Gørtz, I.L., Vildhøj, H.W.: Time-space trade-offs for Lempel-Ziv compressed indexing. In: CPM, pp. 16:1–16:17 (2017)

5. Bille, P., Landau, G.M., Raman, R., Sadakane, K., Rao, S.S., Weimann, O.: Random access to grammar-compressed strings and trees. SIAM J. Comput. **44**(3), 513–539 (2015)
6. Boucher, C., Gagie, T., Kuhnle, A., Manzini, G.: Prefix-free parsing for building big BWTs. In: WABI, pp. 2:1–2:16 (2018)
7. Brisaboa, N., Gómez-Brandón, A., Navarro, G., Paramá, J.: Gract: a grammar-based compressed index for trajectory data. Inf. Sci. **483**, 106–135 (2019)
8. Charikar, M., et al.: The smallest grammar problem. IEEE Trans. Inf. Theory **51**(7), 2554–2576 (2005)
9. Christiansen, A.R., Ettienne, M.B.: Compressed indexing with signature grammars. In: Bender, M.A., Farach-Colton, M., Mosteiro, M.A. (eds.) LATIN 2018. LNCS, vol. 10807, pp. 331–345. Springer, Cham (2018). https://doi.org/10.1007/978-3-319-77404-6_25
10. Claude, F., Fariña, A., Martínez-Prieto, M., Navarro, G.: Universal indexes for highly repetitive document collections. Inf. Sys. **61**, 1–23 (2016)
11. Claude, F., Munro, J.I.: Document listing on versioned documents. In: Kurland, O., Lewenstein, M., Porat, E. (eds.) SPIRE 2013. LNCS, vol. 8214, pp. 72–83. Springer, Cham (2013). https://doi.org/10.1007/978-3-319-02432-5_12
12. Claude, F., Navarro, G.: Self-indexed grammar-based compression. Fund. Inf. **111**(3), 313–337 (2010)
13. Claude, F., Navarro, G.: Improved grammar-based compressed indexes. In: Calderón-Benavides, L., González-Caro, C., Chávez, E., Ziviani, N. (eds.) SPIRE 2012. LNCS, vol. 7608, pp. 180–192. Springer, Heidelberg (2012). https://doi.org/10.1007/978-3-642-34109-0_19
14. Furuya, I., Takagi, T., Nakashima, Y., Inenaga, S., Bannai, H., Kida, T.: MR-RePair: grammar compression based on maximal repeats. In: DCC, pp. 508–517 (2019)
15. Gagie, T., Gawrychowski, P., Kärkkäinen, J., Nekrich, Y., Puglisi, S.J.: LZ77-based self-indexing with faster pattern matching. In: Pardo, A., Viola, A. (eds.) LATIN 2014. LNCS, vol. 8392, pp. 731–742. Springer, Heidelberg (2014). https://doi.org/10.1007/978-3-642-54423-1_63
16. Jeż, A.: Approximation of grammar-based compression via recompression. Theor. Comput. Sci. **592**, 115–134 (2015)
17. Jeż, A.: A really simple approximation of smallest grammar. Theor. Comput. Sci. **616**, 141–150 (2016)
18. Kärkkäinen, J., Kempa, D., Puglisi, S.J.: Lempel-Ziv parsing in external memory. In: DCC, pp. 153–162 (2014)
19. Kieffer, J.C., Yang, E.-H.: Grammar-based codes: a new class of universal lossless source codes. IEEE Trans. Inf. Theory **46**(3), 737–754 (2000)
20. Kornblum, J.D.: Identifying almost identical files using context triggered piecewise hashing. Digit. Invest. **3**, 91–97 (2006)
21. Larsson, J., Moffat, A.: Off-line dictionary-based compression. Proc. IEEE **88**(11), 1722–1732 (2000)
22. Lasch, R., Oukid, I., Dementiev, R., May, N., Demirsoy, S.S., Sattler, K.-U.: Fast & strong: The case of compressed string dictionaries on modern CPUs. In: DaMoN, pp. 4:1–4:10 (2019)
23. Lempel, A., Ziv, J.: On the complexity of finite sequences. IEEE Trans. Inf. Theory **22**(1), 75–81 (1976)
24. Navarro, G.: Indexing highly repetitive collections. In: Arumugam, S., Smyth, W.F. (eds.) IWOCA 2012. LNCS, vol. 7643, pp. 274–279. Springer, Heidelberg (2012). https://doi.org/10.1007/978-3-642-35926-2_29

25. Nevill-Manning, C.G., Witten, I.H.: Identifying hierarchical structure in sequences: a linear-time algorithm. J. Artif. Intell. Res. **7**, 67–82 (1997)
26. Nishimoto, T., I, T., Inenaga, S., Bannai, H., Takeda, M.: Dynamic index, LZ factorization, and LCE queries in compressed space. CoRR, abs/1504.06954 (2015)
27. Peng, Y., Leung, H.C.M., Yiu, S.M., Chin, F.Y.L.: IDBA-UD: a de novo assembler for single-cell and metagenomic sequencing data with highly uneven depth. Bioinformatics **28**(11), 1420–1428 (2012)
28. Rytter, W.: Application of Lempel-Ziv factorization to the approximation of grammar-based compression. Theor. Comput. Sci. **302**(1–3), 211–222 (2003)
29. Sakamoto, H.: A fully linear-time approximation algorithm for grammar-based compression. J. Discr. Algorithm **3**(2–4), 416–430 (2005)
30. Storer, J.A., Szymanski, T.G.: Data compression via textual substitution. J. ACM **29**(4), 928–951 (1982)
31. Takabatake, Y., I, T., Sakamoto, H.: A space-optimal grammar compression. In: ESA, pp. 67:1–67:15 (2017)
32. Tridgell, A.: Efficient Algorithms for Sorting and Synchronization. Ph.D. thesis, The Australian National University (1999)
33. Ziv, J., Lempel, A.: A universal algorithm for sequential data compression. IEEE Trans. Inf. Theory **IT–23**(3), 337–349 (1977)

Information Retrieval

Position Bias Estimation for Unbiased Learning-to-Rank in eCommerce Search

Grigor Aslanyan[(⊠)] and Utkarsh Porwal

eBay Inc., 2025 Hamilton Avenue, San Jose, CA 95125, USA
{gaslanyan,uporwal}@ebay.com

Abstract. The Unbiased Learning-to-Rank framework [16] has been recently proposed as a general approach to systematically remove biases, such as position bias, from learning-to-rank models. The method takes two steps - estimating click propensities and using them to train unbiased models. Most common methods proposed in the literature for estimating propensities involve some degree of intervention in the live search engine. An alternative approach proposed recently uses an Expectation Maximization (EM) algorithm to estimate propensities by using ranking features for estimating relevances [21]. In this work we propose a novel method to directly estimate propensities which does not use any intervention in live search or rely on modeling relevance. Rather, we take advantage of the fact that the same query-document pair may naturally change ranks over time. This typically occurs for eCommerce search because of change of popularity of items over time, existence of time dependent ranking features, or addition or removal of items to the index (an item getting sold or a new item being listed). However, our method is general and can be applied to any search engine for which the rank of the same document may naturally change over time for the same query. We derive a simple likelihood function that depends on propensities only, and by maximizing the likelihood we are able to get estimates of the propensities. We apply this method to eBay search data to estimate click propensities for web and mobile search and compare these with estimates using the EM method [21]. We also use simulated data to show that the method gives reliable estimates of the "true" simulated propensities. Finally, we train an unbiased learning-to-rank model for eBay search using the estimated propensities and show that it outperforms both baselines - one without position bias correction and one with position bias correction using the EM method.

1 Introduction

Modern search engines rely on machine learned methods for ranking the matching results for a given query. Training and evaluation of models for ranking is commonly known as **Learning-to-Rank (LTR)** [18]. There are two common approaches for collecting the data for LTR - **human judgements** and **implicit user feedback**. For human judgements samples of documents are gathered

© Springer Nature Switzerland AG 2019
N. R. Brisaboa and S. J. Puglisi (Eds.): SPIRE 2019, LNCS 11811, pp. 47–64, 2019.
https://doi.org/10.1007/978-3-030-32686-9_4

for a sample of queries and sent to human judges who analyze and label each document. The labels can be as simple as *relevant* vs. *not relevant* or can involve more levels of relevance. This labeled data is then used for training and/or evaluation of LTR models. Collecting human judged data can be expensive and time consuming and often infeasible. On the other hand, data from implicit user feedback, such as clicks, is essentially free and abundant. For that reason it is often the preferred method for collecting data for LTR. A major drawback of this method is that the data can be heavily biased. For example, users can only click on documents that have been shown to them (presentation bias) and are more likely to click on higher ranked documents (position bias). A lot of work in the LTR literature has focused on accounting for and removing these biases. In particular, the recent paper by Joachims et al. [16] has proposed a framework for systematically removing the biases from user feedback data. Following the title of the paper we will refer to this framework as **Unbiased Learning-to-Rank**. In particular, the authors have focused on removing the position bias by first estimating the click propensities and then using the inverse propensities as weights in the loss function. They have shown that this method results in an unbiased loss function and hence an unbiased model.

Unbiased Learning-to-Rank is an appealing method for removing the inherent biases. However, to apply it one needs to first get a reliable estimate of click propensities. The method proposed in [16] uses result randomization in the live search engine to estimate propensities. This can negatively impact the quality of the search results, which will in turn result in poor user experience and potential loss of revenue for the company [21]. It also adds bookkeeping overhead. Wang et al. [21] have proposed a regression-based Expectation Maximization (EM) method for estimating click propensities which does not require result randomization. However, this method uses the ranking features to estimate relevances and can result in a biased estimate of propensities unless the relevance estimates are very reliable, which is difficult to achieve in practice.

In this paper we propose a novel method for estimating click propensities without any intervention in the live search results page, such as result randomization. We use query-document pairs that appear more than once at different ranks to estimate click propensities. In comparison to the EM-based algorithm in [21] our method does not rely on modeling the relevance using ranking features. In fact, we completely eliminate the relevances from the likelihood function and directly estimate the propensities by maximizing a simple likelihood function.

Agarwal et al. [1] have proposed a similar approach for estimating propensities without interventions, which has been done in parallel to our work. The approach developed there relies on having multiple different rankers in the system, such as during A/B tests. They also derive a likelihood function to estimate the propensities, called an *AllPairs* estimator, which depends on terms for all combinations of rank pairs. In comparison to the method in [1] our method is more general and does not rely on having multiple rankers in the system. Although requiring multiple rankers is better than intervention it may still have a similar cost. For example, a different ranker could result in a different user experience

and extra book keeping overhead. In contrast, our proposed approach leverages the organic ranking variation because of time dependent features and does not result in extra costs. That said, our method can naturally take advantage of having multiple rankers, if available. More importantly, our likelihood function depends on the propensities only, rather than terms for all combinations of pairs. The number of unknown parameters to estimate for our method is linear, rather than quadratic, in the number of ranks, which is a major advantage. Our method can therefore give reliable estimates for much lower ranks using much less data.

We use simulated data to test our method and get good results. We then apply our method on actual data from eBay search logs to estimate click propensities for both web and mobile platforms and compare them with estimates using the EM method [21]. Finally, we use our estimated propensities to train an unbiased learning-to-rank model for eBay search and compare it with two baseline models - one which does not correct for position bias and one which uses EM-based estimates for bias correction. Our results show that both unbiased models significantly outperform the "biased" baseline on our offline evaluation metrics, with our model also outperforming the EM method [21].

The main novel contributions of this work can be summarized as follows:

- We present a new approach for directly estimating click propensities without any interventions in live search. Compared with other approaches in the literature [1,21], our approach does not require multiple rankers in the system and large amounts of data for each pair of ranks from different rankers. Moreover, our proposal gives direct estimates of the propensity without having to model relevance. This makes our approach more robust and general.
- Under a mild assumption we derive a simple likelihood function that depends on the propensities only. This allows for propensity estimation for much lower ranks. We also prove the validity of the method through simulations.
- We estimate propensities up to rank 500 using our method for a large eCommerce search engine. This is a much lower rank than previous methods in the literature have been able to obtain (around rank 20). This may not be important for some search engines but is especially important in the eCommerce domain where people typically browse and purchase items from much lower ranks than for web search.
- To the best of our knowledge this is the first paper to do a detailed study of the unbiased learning-to-rank approach for eCommerce search.

The rest of the paper is organized as follows. In Sect. 2 we discuss some of the related work in the literature. In Sect. 3 we introduce our method for estimating click propensities. In Sect. 4 we apply our method to eBay search logs and estimate propensities for web and mobile search, and compare them with EM-based estimates. In Sect. 5 we train and evaluate unbiased learning-to-rank models for eBay search using our estimated propensities as well as the propensities estimated with the EM method [21], and show that our model outperforms both baselines - one without position bias correction and one with bias correction using estimates from the EM method. We summarize our work in Sect. 6 and

discuss future directions for this research. The derivation of our likelihood function is presented in Appendix A. Finally, in Appendix B we apply our method to simulated data and show that we are able to obtain reliable estimates of the "true" simulated propensities.

2   Related Work

Implicit feedback such as clicks are commonly used to train user facing machine learned systems such as ranking or recommender systems. Clicks are preferred over human judged labels as they are available plentifully, are available readily and are collected in a natural environment. However, such user behavior data can only be collected over the items shown to the users. This injects a presentation bias in the collected data. This affects the machine learned systems as they are trained on user feedback data as positives and negatives. It is not feasible to present many choices to the user and it affects the performance of these systems as we can not get an accurate estimate of positives and negatives for training with feedback available only on selective samples. This situation is aggravated by the fact that the feedback of the user not only depends on the presentation, it also depends on where the item was presented. This is a subclass of the presentation bias called position bias. Joachims et al. [16] proved that if the collected user behavior data discounts the position bias accurately then the learned system will be the same as the one learned on true relevance signals.

Several approaches have been proposed to de-bias the collected user behavior data. One of the most common approaches is the use of click models. Click models are used to make hypotheses about the user behavior and then the true relevance is estimated by optimizing the likelihood of the collected clicks. There are several types of click models. One such model is a random click model (RCM) [9] where it is assumed that every document has the same probability of getting clicked and that probability is the model parameter. In a rank based click through rate model (RCTR) it is assumed that the probability of every document being clicked depends on its rank. Therefore, the total number of model parameters is the total number of ranks in the ranking system. Another model is the document based CTR model (DCTR) [8] where the click through rates are estimated for each query-document pair. In this model the total number of model parameters is the total number of query-document pairs. This model is prone to overfitting as the number of parameters grows with the training data size. Most commonly used click models are the position based model (PBM) [8,15] and the cascade model (CM) [8]. In PBM the hypothesis is that a document is only clicked if it is observed and the user found it attractive or relevant. In CM the hypothesis is that the user sequentially scans the whole document top to bottom and clicks when the document is found to be relevant. In this model the top document is always observed and consecutive documents are only observed if the previous ones were observed and were not deemed relevant. In our proposed method we make a similar hypothesis such as the position based method where the observation probability depends on the rank and the probability of relevance

only depends on the query-document pair. However, our approach is to learn the click propensities instead of learning the true relevance by optimizing the likelihood of the collected clicks. More advanced click models, such as the user browsing model (UBM) [9], the dependent click model (DCM) [12], the click chain model (CCM) [11], and the dynamic Bayesian network model (DBN) [6] are also proposed. Chuklin et al. [7] provides a comprehensive overview of click models.

Click models are trained on the collected user behavior data. Interleaving is another option that is deployed at the time of data collection. In interleaving different rank lists can be interleaved together and presented to the user. By comparing the clicks on the swapped results one can learn the unbiased user preference. Different methods for interleaving have been proposed. In the balanced interleave method [17] a new interleaved ranked list is generated for every query. The document constraint method [13] accounts for the relation between documents. Hofmann et al. [14] proposed a probabilistic interleaving method that addressed some of the drawbacks of the balanced interleave method and the document constraint method. One limitation of the interleaving method is that often the experimentation platform in eCommerce companies is not tied to just search. It supports A/B testing for all teams, such as checkout and advertisements. Therefore, the interleaving ranked list may not be supported as it is pertinent only for search ranking.

A more recent approach to address presentation bias is the unbiased learning-to-rank approach. In this click propensities are estimated and then the inverse propensities are used as weights in the loss function. Click propensities are estimated by presenting the same items at different ranks to account for click biases without explicitly estimating the query-document relevance. Click propensity estimation can either be done randomly or in a more principled manner. Radlinski et al. [19] presented the FairPairs algorithm that randomly flips pairs of results in the ranking presented to the user. They called it randomization with minimal invasion. Carterette et al. [4] also presented a minimally invasive algorithm for offline evaluation. Joachims et al. [16] proposed randomized intervention to estimate the propensity model. Radlinski et al. [20], on the other hand, proposed alteration in ranking in a more informed manner using Multi-Armed Bandits. The main drawback of randomization for propensity estimation is that it can cause bad user experience, book keeping overhead, and a potential loss in revenue. Wang et al. [21] proposed a method to estimate propensities without randomization using the EM algorithm. In most of the existing methods, propensity estimation is done first. Once the propensities are learned, an unbiased ranker is trained using the learned propensities. Recently Ai et al. [2] proposed a dual learning algorithm that learns an unbiased ranker and the propensities together.

3   Propensity Estimation Method

The method proposed by Joachims et al. [16] for estimating click propensities is running an experimental intervention in the live search engine, where the

documents at two selected ranks are swapped. By comparing the click through rates at these ranks before and after swapping one can easily estimate the ratios of propensities at these ranks (one only needs the ratio of propensities for removing the position bias [16]). Here we propose a novel methodology for estimating click propensities without any intervention. For some search engines, especially in eCommerce, the same query-document pair may naturally appear more than once at different ranks. Using the click data on such documents we can accurately estimate click propensities. It is not required that the same query-document pair should appear at different ranks a large number of times.

We model clicks by the following simple model (also used in [16]) - *The probability of a click on a given document is the product of the probability of observing the document and the probability of clicking on the document for the given query assuming that it has been observed.* We assume that the probability of observing a document depends only on its rank and the probability of clicking on the document for a given query if it is observed depends only on the query and the document. Mathematically:

$$\begin{aligned} p(c = 1|q, y) &= p(o = 1|q, y)p(c = 1|q, y, o = 1) \\ &= p(o = 1|rank(y))p(c = 1|q, y, o = 1) \qquad (1) \\ &= p_{rank(y)}p(c = 1|q, y, o = 1) \end{aligned}$$

where q denotes a query, y denotes a document, c denotes a click (0 or 1), o denotes observation (0 or 1), and p_i denotes the propensity at rank i.

Let us assume that our data D consists of N query-document pairs x_j for $j \in [1, N]$. For a query-document pair x_j we will denote the probability of clicking on the document after observing it by z_j. For each query-document pair x_j we have a set of ranks r_{jk} where the document has appeared for the query, and clicks c_{jk} denoting if the document was clicked or not (1 or 0) when it appeared at rank r_{jk}, for $k \in [1, m_j]$. Here we assume that the query-document pair x_j has appeared m_j separate times. For now we do not assume that m_j must be greater than 1 - it can be any positive integer.

The probability of a click for query-document pair x_j where the document appeared at rank r_{jk} is, according to (1) $p(c = 1) = p_{r_{jk}}z_j$. It follows that $p(c = 0) = 1 - p_{r_{jk}}z_j$. We can now introduce the following likelihood function:

$$\mathcal{L}(p_i, z_j|D) = \prod_{j=1}^{N} \prod_{k=1}^{m_j} \left[c_{jk}p_{r_{jk}}z_j + (1 - c_{jk})(1 - p_{r_{jk}}z_j) \right] . \qquad (2)$$

Here the parameters are the propensities p_i and the "relevances" z_j (relevance here means probability of clicking for a given query-document pair assuming that the document has been observed). Theoretically, the parameters can be estimated by maximizing the likelihood function above. However, this can be challenging due to the large number of parameters z_j. In fact, we are not even interested in estimating the z_j - we only need to estimate the propensities p_i, and the z_j are nuisance parameters.

The likelihood function above can be simplified under mild and generally applicable assumptions. Firstly, only query-document pairs that appeared at multiple different ranks and got at least one click are of interest. This is because we need to compare click activities for the same query-document pair at different ranks to be able to gain some useful information about propensities with the same "relevance". Secondly, we make the assumption that overall click probabilities are not large (i.e. not close to 1). We discuss this assumption in detail in Appendix A. As we will see in Sect. 4 this is a reasonable assumption for eBay search. This assumption is generally valid for lower ranks (below the top few), and in Appendix A we discuss how to make small modifications to the data in case the assumption is violated for topmost ranks. We also discuss alternative approaches for estimating the click propensities for cases when the our assumption might not work very well (our methodology of simulations in Appendix B can be used to verify the validity of the assumption).

The likelihood can then be simplified to take the following form:

$$\log \mathcal{L}(p_i|D) = \sum_{j=1}^{N} \left(\log(p_{r_{jl_j}}) - \log \sum_{k=1}^{m_j} p_{r_{jk}} \right) . \tag{3}$$

The detailed derivation is presented in Appendix A. Note that the simplified likelihood function (3) only depends on the propensities, which is one of the most important contributions of this work. By maximizing the likelihood function above we can get an estimate of the propensities. Because the likelihood function depends on the propensities only we can estimate the propensities up to much lower ranks than previously done in the literature without having to rely on a large amount of data.

4   Click Propensities for eBay Search

In this section we apply the method developed above on eBay search data to estimate propensities. For comparison, we also estimate the propensities using the EM method [21].

We collected a small sample (0.2%) of queries for four months of eBay search traffic. For each query we keep the top 500 items (in this work we use the terms "item" and "document" interchangeably). There are multiple sort types on eBay (such as Best Match, Price Low to High, Time Ending Soonest) and click propensities may differ for different sort types. In this paper we present our results on Best Match sort, and hence we keep only queries for that sort type. Furthermore, there are multiple different platforms for search (such as a web browser or a mobile app) which can have different propensities. We separate our dataset into two platforms - web and mobile, and estimate click propensities for each platform separately. For web queries we estimate the propensities for list view with 50 items per page (the most common option).

Next, we identify same query-document pairs and find cases where the document appeared at multiple different ranks. We apply certain filters to ensure

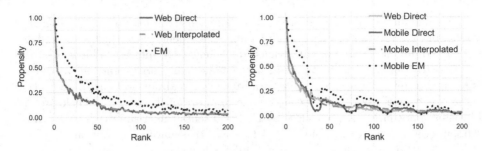

Fig. 1. Click propensity estimated for eBay search for web data (left) and mobile data (right). The solid blue line is the direct estimation of propensities for each rank, the red dashed line is the estimation using interpolation, and the black dotted curve is the estimation using the EM method. For comparison, on the right side we also plot the propensities for web data using interpolation in solid green, which is the same as the red dashed line from the left side. (Color figure online)

that the "relevance" of the document has not changed for the query between multiple appearances, and different click probabilities are only due to different ranks. Namely, we check that the price of the item has not changed and exclude auction items (since their relevance depends strongly on the current bid and the amount of time left). We also keep the same query-document pairs from the same day only to make sure that seasonality effects do not affect the popularity of the item. For the query side we identify two queries to be the same if they have the same keywords, as well as the same category and aspect (such as color, size) constraints. We then keep only those query-document pairs that appeared at two different ranks and got one click in one rank and no click in the other.[1] We have also verified our assumption of not very large click probabilities for our dataset. Note that the validity of the assumption is also verified through simulations in Appendix B where the simulated data has similar click through rates to the actual eBay data.

We first estimate propensities for web queries. Our dataset consists of about 40,000 query-item pairs, each of which appeared at two different ranks and received a click at one of the ranks. We use two methods for estimating propensities - direct and interpolation. In the direct method we treat the propensity at each rank as a separate parameter. We therefore get 500 different parameters to estimate. In the interpolation method we fix a few different ranks and use the propensities at those ranks as our parameters to estimate. The propensities for all the other ranks are computed as a linear interpolation in the log-log space, i.e. we approximate the log of the propensity as a linear function of the log of the rank. This results in the propensity being a power law of the rank. For the

---

[1] Note that keeping only query-document pairs that appeared at two ranks exactly is in no way a requirement of our method. The method is general and can be used for query-document pairs that appeared more than twice. This is just intended to simplify our analysis without a significant loss in data, since it is rare for the same query-document pair to appear at more than two ranks.

interpolation method our fixed ranks are 1, 2, 4, 8, 20, 50, 100, 200, 300, and 500. We choose a denser grid for higher ranks since there is more data and less noise for higher ranks, and the propensities can be estimated more accurately.

Our resulting propensity for web search is shown in Fig. 1 (left). The solid blue line shows the propensities estimated through the direct method, and the red dashed curve shows the propensities estimated through interpolation. Even though we estimate propensities up to rank 500, we plot them only up to rank 200 so that the higher ranks can be seen more clearly. The red dashed curve passes smoothly through the blue solid curve, which is reassuring. Note that the red dashed curve is not a fit to the blue one. The two are estimated directly from the data. For the blue curve the parameters are all of the propensities at each rank, whereas for the red dashed curve we only parametrize the propensities at select ranks and interpolate in between. We then maximize the likelihood for each case to estimate the parameters. The fact that the red dashed line appears to be a smooth fit to the solid blue shows that the interpolation method is useful in obtaining a smooth and less noisy propensity curve which is still very close to the direct estimation.

The propensities estimated from eBay mobile search data are shown in Fig. 1 (right). As in the left plot (web data), the blue solid curve shows direct estimation, and the red dashed curve is estimation using interpolation. For comparison, we plot the propensities from web using interpolation in solid green. The blue solid curve shows a certain periodicity - the propensities seem to drop sharply near rank 25, then go back up at rank 40, drop again around rank 65, then back at rank 80, and so on. In fact, this reflects the way results are loaded in mobile search - 40 at a time. The blue curve seems to indicate that users observe the results at higher ranks with the usual decrease in interest, then they tend to scroll faster to the bottom skipping the results towards the bottom, then as the new batch is loaded they regain interest. The red dashed curve matches the blue one reasonable well, but it fails to capture the periodic dips. This is due to our choice of knots for the linear spline. One can use the blue curve to choose new locations of the knots to be able to get a better interpolation for the propensities. The green solid curve matches fairly well with the blue one except for the dips. This means that the propensities for web and mobile are very similar, except for the periodic dips for mobile. The web results are shown 50 items per page, but we have not found any periodic dips for web search. Perhaps this indicates that for web search users do not tend to scroll quickly towards the end of the page and then regain interest as a new page is loaded. The smooth decline in propensities indicates that for web search users steadily lose interest as they scroll down, but the number of items per page does not affect their behavior.

We have also estimated propensities using the regression-based EM method by Wang et al. [21]. The results are plotted with black dotted lines in Fig. 1. The two methods are very different and use different kinds of data so it is hard to have a fair comparison. However, we have used datasets of similar sizes with similar numbers of queries to make the comparison as fair as possible. For the regression method we have used gradient boosted decision trees [10] using our

top 25 ranking features. The estimates obtained with the EM method are in general higher than the estimates using our method. We have obtained similar periodicity patterns for mobile data from both methods which is reassuring. We do not have the ground truth for comparison since we have not performed any randomization experiments. However, our simulations in the next Section show that our method's predictions are close to the ground truth. We have also used these estimates in Sect. 5 to train unbiased learning-to-rank models and have obtained better offline metrics using our estimates compared to the EM-based estimates.

5   Unbiased Learning-to-Rank Models

In this section we study the improvement in ranking models by using the estimated click propensities for eBay search data. Previous studies have consistently shown that unbiased learning-to-rank models significantly improve ranking metrics compared to their biased counterparts. Specifically, Joachims et al. [16] have shown that an unbiased learning-to-rank model significantly improves the average rank of relevant results for simulated data. Furthermore, they have performed an online interleaving experiment on a live search engine for scientific articles, which resulted in a significant improvement for the unbiased model. Wang et al. [21] have shown an improvement in MRR (Mean Reciprocal Rank) for the unbiased learning-to-rank models for personal search.

We train ranking models to check if unbiased ranking models show improvements over their biased counterparts and to compare our method of propensity estimation to the EM method. For our training data we collect a sample of about 40,000 queries which have received at least one click. The sample is collected from four days of search logs. We train listwise ranking models using the LambdaMART algorithm [3]. We use the DCG metric [18] as our loss function. We define rel_{ij} to be 1 if document j was clicked, and 0 otherwise. We train three models - one without position bias correction (*baseline biased*), one with position bias correction using propensity estimates from the EM method (*baseline EM*), and finally a model with position bias correction using propensity estimates from our method (*proposed method*). All models use DCG as a loss function, with *baseline biased* using no position bias correction and the other models using inverse-propensity weighted relevances [16]. We use the propensities estimated for eBay web search as shown in Fig. 1 (left) - red dashed curve for *proposed method* and black dotted curve for *baseline EM*. Our training and test data are also from web search (i.e. browser) only. We use 25 features for all models, selected from our top ranking features. We use the same hyperparameters for all the models: the number of trees is 100 and the shrinkage is 0.1 (we have fixed the number of trees and tuned the shrinkage for the baseline model, which is then applied to all models).[2]

---

[2] Note that these ranking models are significantly different from the eBay production ranker, the details of which are proprietary.

Table 1. AUC improvement of the proposed method compared to two baselines - *baseline biased* and *baseline EM* [21]. The validation set contains documents from a fixed rank, shown in the first column. The next two columns show the improvements in AUC. Error bars are obtained using 1,000 bootstrap samples of the test data - we show the mean and standard deviation of the improvement over the bootstrap samples.

Rank	Improvement over baseline biased	Improvement over baseline EM
1	$3.4 \pm 1.0\%$	$1.0 \pm 0.4\%$
2	$2.4 \pm 1.1\%$	$0.6 \pm 0.4\%$
4	$4.2 \pm 1.2\%$	$0.7 \pm 0.4\%$
8	$3.3 \pm 1.3\%$	$1.2 \pm 0.5\%$
16	$6.8 \pm 1.7\%$	$1.1 \pm 0.6\%$
32	$0.8 \pm 1.8\%$	$0.8 \pm 0.7\%$

Our test data contains a sample of about 10,000 queries from four days of eBay search logs. Since the test data also has the same position bias as the training data we cannot rely on standard ranking metrics such as DCG, NDCG (Normalized Discounted Cumulative Gain), or MRR (Mean Reciprocal Rank). Another option would be to use inverse-propensity-weighted versions of these metrics to remove the presentation bias. However, the true propensities are unknown to us and we obviously cannot use estimated propensities for evaluation since part of the evaluation is checking if our estimate of propensities is a good one. For that reason we choose a different approach for evaluation. Namely, we fix the rank of items in the test data, i.e. we select items from different queries that appeared at a given fixed rank. By selecting the items from a fixed rank in the evaluation set we effectively eliminate position bias since all of the items will be affected by position bias the same way (the observation probability is the same for all the items since the rank is the same). Then we compare the two ranking models as classifiers for those items, which means that we evaluate how well the models can distinguish items that were clicked from ones that were not. We use AUC (Area Under the Receiver Operating Characteristic Curve) as our evaluation metric.

The results are presented in Table 1, where we show results for fixed ranks 1, 2, 4, 8, 16, and 32. To estimate statistical significance of the improvements we have performed 1,000 bootstrap samples of the test data and computed the improvements on these samples. In Table 1 we show the mean and standard deviation on the bootstrap samples (the distribution of the results on the bootstrap samples is close to Gaussian, as expected, so the mean and standard deviation are enough to describe the full distribution). As we can see, for all ranks the *proposed method* outperforms both baselines. Both unbiased models significantly outperform *baseline biased*. However, our *proposed method* outperforms *baseline EM* as well. The improvements are statistically significant for all ranks, except for rank 32, where the improvements are not as large. For ranks below 32 the improvements become minor.

6 Summary and Future Work

In this work we have introduced a new method for estimating click propensities for eCommerce search without randomizing the results during live search. Our method uses query-document pairs that appear more than once and at different ranks. Although we have used eCommerce search as our main example, the method is general and can be applied to any search engine for which ranking naturally changes over time. The clear advantage of our method over result randomization is that it does not affect live search results, which can have a negative impact on the engine as has been shown in the literature [21]. We have compared our method to the EM (Expectation Maximization) based method proposed in [21] and have shown that our proposed method outperforms the EM based method for eBay data. There is another approach proposed in parallel to our work [1] for direct estimation of propensities. However, our method has a few clear advantages, such as not relying on multiple rankers in the system and not requiring a large amount of data for each pair of ranks. This has allowed us to estimate propensities up to ranks that are much lower than previously computed in the literature. Our proposed approach is robust and we believe that it will find widespread use for unbiased learning-to-rank modeling, especially in the eCommerce domain.

We have used simulated data to show that our method can give accurate estimates of the true propensities. We have applied our method to eBay search results to separately estimate propensities for web and mobile search. We have also trained ranking models and compared the performance of the unbiased model using the estimated propensities to two baselines - one without bias correction and one that corrects position bias using estimates from the EM method. Using a validation dataset of documents from a fixed rank we have shown that our unbiased model outperforms both baselines in terms of the AUC metric.

The focus of this work is propensity estimation from query-document pairs that appear at multiple different ranks. Importantly, we have addressed the case when the same query-document pair appears only a few times at different ranks (can be as few as twice). This method can be generalized to use query-document pairs that appeared at a single rank only by incorporating appropriate priors and using Gibbs sampling to estimate the posterior distribution for propensities. We plan to study this approach in a future work. We are also planning to estimate and compare propensities for different classes of queries (such as queries for electronics versus fashion categories) and user demographics, as well as different sort types, such as sort by price.

A   Likelihood Function Simplification

There are multiple approaches that one can take to estimate the propensities depending on the data itself. Let us first consider the query-document pairs that appeared only at one rank. The parameters p_i and z_j appear only as a product of each other in the likelihood function (2). These query-document pairs

could be helpful in estimating the product of the propensity at the rank that they appeared at and the relevance z_j but not each one individually. With z_j unknown, this would not help to estimate the propensity. We should mention that in the presence of a reliable prior for z_j and/or p_i the likelihood function above can be used even for those query-document pairs that appeared only at one rank. In this case it would be more useful to take a Bayesian approach and estimate the posterior distribution for the propensities, for example using Gibbs sampling [5].

From now on we will assume that the query-document pairs appear at least at two different ranks. Another extreme is the case when each query-document pair appears a large number of times at different ranks. This will mean that we will get a large number of query-document pairs at each rank. In this case the propensity ratios for two ranks can be simply estimated by taking the ratio of click through rates of same query-document pairs at these ranks.

Let us now consider the case when the data consists of a large number of query-document pairs that appeared a few times (can be as few as twice) at different ranks, but the query-document pairs do not appear a large enough number of times to be able to get reliable estimates of propensities from taking the ratio of click through rates. In this case we will actually need to maximize the likelihood above and somehow eliminate the nuisance parameters z_j to get estimates for the p_i. We will focus the rest of this work on this case. Also, the data we have collected from eBay search logs falls in this category, as discussed in Sect. 4.

If a query-document pair appeared only a few times there is a good chance that it did not receive any clicks. These query-document pairs will not help in estimating the propensities by likelihood maximization because of the unknown parameter z_j. Specifically, for such query-document pairs we will have the terms $\prod_{k=1}^{m_j}(1 - p_{r_{jk}}z_j)$. If we use the maximum likelihood approach for estimating the parameters then the maximum will be reached by $z_j = 0$ for which the terms above will be 1. So the query-document pairs without any clicks will not change the maximum likelihood estimate of the propensities. For that reason we will only keep query-document pairs that received at least one click. However, we cannot simply drop the terms from the likelihood function for query-document pairs that did not receive any clicks. Doing so would bias the data towards query-document pairs with a higher likelihood of click. Instead, we will replace the likelihood function above by a conditional probability. Specifically, the likelihood function (2) computes the probability of the click data $\{c_{jk}\}$ obtained for that query-document pair. We need to replace that probability by a conditional probability - the probability of the click data $\{c_{jk}\}$ under the condition that there was at least one click received: $\sum_k c_{jk} > 0$. The likelihood function for the query-document pair x_j will take the form:

$$\mathcal{L}_j(p_i, z_j | D_j) = P\left(D_j \Big| \sum_k c_{jk} > 0\right)$$
$$= \frac{P(D_j \cap \sum_k c_{jk} > 0)}{P(\sum_k c_{jk} > 0)} = \frac{P(D_j)}{P(\sum_k c_{jk} > 0)} \tag{4}$$
$$= \frac{\prod_{k=1}^{m_j} \left[c_{jk} p_{r_{jk}} z_j + (1 - c_{jk})(1 - p_{r_{jk}} z_j)\right]}{1 - \prod_{k=1}^{m_j} (1 - p_{r_{jk}} z_j)}.$$

Here $\mathcal{L}_j$ denotes the likelihood function for the query-document pair x_j, $D_j = \{c_{jk}\}$ denotes the click data for query-document pair j, and P denotes probability. $\sum_k c_{jk} > 0$ simply means that there was at least one click. In the first line above we have replaced the probability of data D_j by a conditional probability. The second line uses the formula for conditional probability. The probability of D_j and at least one click just equals to probability of D_j since we are only keeping query-document pairs that received at least one click. This is how the second equality of the second line is derived. Finally, in the last line we have explicitly written out $P(D_j)$ in the numerator as in (2) and the probability of at least one click in the denominator (the probability of no click is $\prod_{k=1}^{m_j}(1 - p_{r_{jk}} z_j)$ so the probability of at least one click is 1 minus that).

The full likelihood is then the product of $\mathcal{L}_j$ for all query-document pairs:

$$\mathcal{L}(p_i, z_j | D) = \prod_{\substack{j=1 \\ \sum_k c_{jk} > 0}}^{N} \frac{\prod_{k=1}^{m_j} \left[c_{jk} p_{r_{jk}} z_j + (1 - c_{jk})(1 - p_{r_{jk}} z_j)\right]}{1 - \prod_{k=1}^{m_j} (1 - p_{r_{jk}} z_j)}. \tag{5}$$

From now on we will assume by default that our dataset contains only query-document pairs that received at least one click and will omit the subscript $\sum_k c_{jk} > 0$.

Our last step will be to simplify the likelihood function (5). Typically the click probabilities $p_i z_j$ are not very large (i.e. not close to 1). This is the probability that the query-document pair j will get a click when displayed at rank i. To simplify the likelihood for each query-document pair we will only keep terms linear in $p_i z_j$ and drop higher order terms like $p_{i_1} z_{j_1} p_{i_2} z_{j_2}$. We have verified this simplifying assumption for our data in Sect. 4. In general, we expect this assumption to be valid for most search engines. It is certainly a valid assumption for lower ranks since click through rates are typically much smaller for lower ranks. Since we are dropping product terms the largest ones would be between ranks 1 and 2. For most search engines the click through rates at rank 2 are around 10% or below, which we believe is small enough to be able to safely ignore the product terms mentioned above (they would be at least 10 times smaller than linear terms). We empirically show using simulations in Appendix B that this assumption works very well for data similar to eBay data. If for other search engines the click through rates are much larger for topmost ranks we suggest keeping only those query-document pairs that appeared at least once at a lower enough rank. Also, using the methodology of simulations from Appendix B one can verify how well this assumption works for their particular data.

Under the simplifying assumption we get for the denominator in (5):

$$1 - \prod_{k=1}^{m_j}(1 - p_{r_{jk}}z_j) \simeq 1 - \left(1 - \sum_{k=1}^{m_j} p_{r_{jk}}z_j\right) = z_j \sum_{k=1}^{m_j} p_{r_{jk}} . \tag{6}$$

Let us now simplify the numerator of (5). Firstly, since the click probabilities are not large and each query-document pair appears only a few times we can assume there is only one click per query-document pair[3]. We can assume $c_{jl_j} = 1$ and $c_{jk} = 0$ for $k \neq l_j$. The numerator then simplifies to

$$\prod_{k=1}^{m_j} \left[c_{jk}p_{r_{jk}}z_j + (1 - c_{jk})(1 - p_{r_{jk}}z_j)\right] = p_{r_{jl_j}}z_j \prod_{\substack{k=1 \\ k\neq l_j}}^{m_j}(1 - p_{r_{jk}}z_j) \simeq p_{r_{jl_j}}z_j . \tag{7}$$

Using (6) and (7) the likelihood function (5) simplifies to

$$\mathcal{L}(p_i, z_j|D) = \prod_{j=1}^{N} \frac{p_{r_{jl_j}}z_j}{z_j \sum_{k=1}^{m_j} p_{r_{jk}}} = \prod_{j=1}^{N} \frac{p_{r_{jl_j}}}{\sum_{k=1}^{m_j} p_{r_{jk}}} . \tag{8}$$

In the last step z_j cancels out from the numerator and the denominator. Our assumption of small click probabilities, together with keeping only query-document pairs that received at least one click allowed us to simplify the likelihood function to be only a function of propensities. Now we can simply maximize the likelihood (8) to estimate the propensities.

Equation (8) makes it clear why we need to include the requirement that each query-document pair should appear more than once at different ranks. If we have a query-document pair that appeared only once (or multiple times but always at the same rank) then the numerator and the denominator would cancel each other out in (8). For that reason we will keep only query-document pairs that appeared at two different ranks at least.

It is numerically better to maximize the log-likelihood, which takes the form:

$$\log \mathcal{L}(p_i|D) = \sum_{j=1}^{N} \left(\log(p_{r_{jl_j}}) - \log \sum_{k=1}^{m_j} p_{r_{jk}}\right) . \tag{9}$$

B   Results on Simulations

In this Appendix we use simulated data to verify that the method of estimating propensities developed in Sect. 3 works well. For our simulations we choose the following propensity function as truth:

$$p_i^{\text{sim}} = \min\left(\frac{1}{\log i}, 1\right) \tag{10}$$

---

[3] This is true for our data as discussed in Sect. 4. For the cases when most query-document pairs receive multiple clicks we suggest using a different method, such as computing the ratios of propensities by computing the ratios of numbers of clicks.

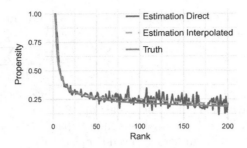

Fig. 2. Propensity estimated from simulated data. The green solid curve shows the "true" propensity (10). The blue solid curve is the estimated propensity using the direct estimation method. The red dashed curve is the estimation using interpolation. (Color figure online)

which assigns propensity of 1 for ranks 1 and 2, and then decreases as the inverse of the log of the rank.

Other than choosing our own version of propensities we simulate the data to be as similar to the eBay dataset as possible. We generate a large number of query-document pairs and randomly choose a mean rank $rank_{mean}$ for each query-document pair uniformly between 1 and 500. We randomly generate a click probability z for that query-document pair depending on the mean rank $rank_{mean}$. We choose the distribution from which the click probabilities are drawn such that the click through rates at each rank match closely with the click through rates for real data, taking into account the "true" propensities (10). We then generate two different ranks drawn from $\mathcal{N}(rank_{mean}, (rank_{mean}/5)^2)$. For each rank i we compute the probability of a click as $z p_i^{\mathrm{sim}}$. Then we keep only those query-document pairs which appeared at two different ranks and got at least one click, in agreement with our method used for real eBay data. Finally, we keep about 40,000 query-document pairs so that the simulated data is similar to the eBay web search data in size. This becomes the simulated data.

The estimated propensities on the simulated dataset are shown in Fig. 2. The green solid curve shows the true propensity (10), the blue solid curve shows the estimated propensity using the direct estimation method, and the red dashed curve is the estimated propensity using interpolation. As we can see, the estimations closely match with the truth. Furthermore, we can see that the interpolation method gives a better result by reducing the noise in the estimate. These results show that the propensity estimation method developed in this paper works well.

References

1. Agarwal, A., Zaitsev, I., Wang, X., Li, C., Najork, M., Joachims, T.: Estimating position bias without intrusive interventions. In: Proceedings of the Twelfth ACM International Conference on Web Search and Data Mining, pp. 474–482. ACM (2019)

2. Ai, Q., Bi, K., Luo, C., Guo, J., Croft, W.B.: Unbiased learning to rank with unbiased propensity estimation. In: The 41st International ACM SIGIR Conference on Research & Development in Information Retrieval, pp. 385–394. ACM (2018)
3. Burges, C.J.: From ranknet to lambdarank to lambdamart: An overview. Technical report, June 2010
4. Carterette, B., Chandar, P.: Offline comparative evaluation with incremental, minimally-invasive online feedback. In: The 41st International ACM SIGIR Conference on Research & #38; Development in Information Retrieval, SIGIR 2018, pp. 705–714. ACM, New York (2018). https://doi.org/10.1145/3209978.3210050
5. Casella, G., George, E.I.: Explaining the gibbs sampler. Am. Stat. **46**(3), 167–174 (1992)
6. Chapelle, O., Zhang, Y.: A dynamic bayesian network click model for web search ranking. In: Proceedings of the 18th International Conference on World Wide Web, pp. 1–10. ACM (2009)
7. Chuklin, A., Markov, I., Rijke, M.D.: Click models for web search. Synth. Lect. Inf. Concepts Retrieval Serv. **7**(3), 1–115 (2015)
8. Craswell, N., Zoeter, O., Taylor, M., Ramsey, B.: An experimental comparison of click position-bias models. In: Proceedings of the 2008 International Conference on Web Search and Data Mining, pp. 87–94. ACM (2008)
9. Dupret, G.E., Piwowarski, B.: A user browsing model to predict search engine click data from past observations. In: Proceedings of the 31st Annual International ACM SIGIR Conference on Research and Development in Information Retrieval, pp. 331–338. ACM (2008)
10. Friedman, J.H.: Greedy function approximation: a gradient boosting machine. Ann. Statist. **29**(5), 1189–1232 (2001). https://doi.org/10.1214/aos/1013203451
11. Guo, F., et al.: Click chain model in web search. In: Proceedings of the 18th International Conference on World Wide Web, pp. 11–20. ACM (2009)
12. Guo, F., Liu, C., Wang, Y.M.: Efficient multiple-click models in web search. In: Proceedings of the Second ACM International Conference on Web Search and Data Mining, pp. 124–131. ACM (2009)
13. He, J., Zhai, C., Li, X.: Evaluation of methods for relative comparison of retrieval systems based on clickthroughs. In: Proceedings of the 18th ACM Conference on Information and Knowledge Management, pp. 2029–2032. ACM (2009)
14. Hofmann, K., Whiteson, S., De Rijke, M.: A probabilistic method for inferring preferences from clicks. In: Proceedings of the 20th ACM International Conference on Information and Knowledge Management, pp. 249–258. ACM (2011)
15. Joachims, T., Granka, L., Pan, B., Hembrooke, H., Gay, G.: Accurately interpreting clickthrough data as implicit feedback. In: Proceedings of the 28th Annual International ACM SIGIR Conference on Research and Development in Information Retrieval, pp. 154–161. SIGIR 2005. ACM, New York (2005). https://doi.org/10.1145/1076034.1076063
16. Joachims, T., Swaminathan, A., Schnabel, T.: Unbiased learning-to-rank with biased feedback. In: Proceedings of the Tenth ACM International Conference on Web Search and Data Mining. WSDM 2017, pp. 781–789. ACM, New York (2017). https://doi.org/10.1145/3018661.3018699
17. Joachims, T., et al.: Evaluating retrieval performance using clickthrough data (2003)
18. Li, H.: A short introduction to learning to rank. IEICE Trans. Inf. Syst. **94**(10), 1854–1862 (2011)
19. Radlinski, F., Joachims, T.: Minimally invasive randomization for collecting unbiased preferences from clickthrough logs (2006)

20. Radlinski, F., Kleinberg, R., Joachims, T.: Learning diverse rankings with multi-armed bandits. In: Proceedings of the 25th International Conference on Machine Learning, pp. 784–791. ACM (2008)
21. Wang, X., Golbandi, N., Bendersky, M., Metzler, D., Najork, M.: Position bias estimation for unbiased learning to rank in personal search. In: Proceedings of the Eleventh ACM International Conference on Web Search and Data Mining, WSDM 2018, pp. 610–618. ACM, New York (2018). https://doi.org/10.1145/3159652.3159732

BM25 Beyond Query-Document Similarity

Billel Aklouche[1,2,4](✉) ⓘ, Ibrahim Bounhas[1,4] ⓘ, and Yahya Slimani[1,3,4] ⓘ

[1] LISI Laboratory of Computer Science for Industrial System,
INSAT, Carthage University, Tunis, Tunisia
bounhas.ibrahim@gmail.com, yahya.slimani@gmail.com
[2] National School of Computer Science (ENSI),
La Manouba University, Manouba, Tunisia
billel.aklouche@ensi-uma.tn
[3] Higher Institute of Multimedia Arts of Manouba (ISAMM),
La Manouba University, Manouba, Tunisia
[4] JARIR: Joint group for Artificial Reasoning and Information Retrieval,
Manouba, Tunisia
http://www.jarir.tn/

Abstract. The massive growth of information produced and shared online has made retrieving relevant documents a difficult task. Query Expansion (QE) based on term co-occurrence statistics has been widely applied in an attempt to improve retrieval effectiveness. However, selecting good expansion terms using co-occurrence graphs is challenging. In this paper, we present an adapted version of the BM25 model, which allows measuring the similarity between terms. First, a context window-based approach is applied over the entire corpus in order to construct the term co-occurrence graph. Afterward, using the proposed adapted version of BM25, candidate expansion terms are selected according to their similarity with the whole query. This measure stands out by its ability to evaluate the discriminative power of terms and select semantically related terms to the query. Experiments on two ad-hoc TREC collections (the standard Robust04 collection and the new TREC Washington Post collection) show that our proposal outperforms the baselines over three state-of-the-art IR models and leads to significant improvements in retrieval effectiveness.

Keywords: Query expansion · Co-occurrence graph · BM25 · Term discriminative power · Ad-hoc IR

1 Introduction

The main purpose of information retrieval (IR) systems is to provide a set of relevant documents according to a user's specified need. A number of ranking models have been proposed in the literature [4,22,32], all of which intend to retrieve the most relevant documents in response to a query. The difference in

© Springer Nature Switzerland AG 2019
N. R. Brisaboa and S. J. Puglisi (Eds.): SPIRE 2019, LNCS 11811, pp. 65–79, 2019.
https://doi.org/10.1007/978-3-030-32686-9_5

retrieval results from one model to another is manifested in the set of returned documents and in the order of their appearance. Among these models, Okapi BM25 [22] is a pre-eminent probabilistic model, which has proven its effectiveness as a state-of-the-art IR model and has been widely used, especially in TREC experiments. The BM25 model incorporates information about both terms and documents, which includes local terms frequencies, global terms frequencies and document length. Since it was introduced, several studies have been presented proposing extensions and improvements [5, 13, 15, 21, 23, 26].

However, despite the improvements that can be made to ranking models, the user's query remains the key factor that controls the relevance of retrieval results. Indeed, it is often too short and insufficient to allow the selection of documents that meet the user needs. In most cases, the latter does not know exactly what he wants or how to express it. Therefore, the returned results are unlikely to be relevant. To overcome this problem, Query Expansion (QE) refers to techniques that reformulate the original query by adding new terms to those entered by the user to better express his need and improve retrieval performance.

A main challenge in QE is the selection of good expansion terms which do not hurt, but improve, retrieval performance. The strength of the BM25 model is that it allows capturing the behavior of terms not only in a document, but also in the entire collection. It assumes that a good document descriptor is a quite frequent term in this document, which is relatively infrequent in the entire document collection [14]. Based on these assumptions, we propose an approach to QE by adapting BM25 to work on term co-occurrence graphs. The main motivation is to model the discriminative power of terms using a measure analogous to the inverse document frequency (IDF) factor of TF-IDF [25]. We define a good expansion term as one that frequently co-occurs with the query terms and has a relatively rare co-occurrence with the rest of the vocabulary.

We evaluate our proposal using two ad-hoc TREC collections: the standard TREC Robust04 collection with 249 queries (TREC 2004 Robust Track) and the newest TREC Washington Post collection with 50 queries (TREC 2018 Common Core Track). Experimental results show that our proposal outperforms the baselines by significant margins in terms of MAP and precision.

The remainder of the paper is organized as follows. In the next section, we discuss some related work on QE. We describe the proposed adapted version of BM25 for QE in Sect. 3. The Experimental setup and the obtained results are presented in Sect. 4. Finally, Sect. 5 concludes the paper and provides insights for future work.

2   Related Work

For several years, great effort has been devoted to the development of new QE approaches [8]. Corpus-based QE approaches are among the most popular techniques that have been widely applied [29]. The corpus itself serves as a source for selecting expansion terms. Indeed, the broad range of corpus-based QE approaches can be divided into two main classes: local approaches and global approaches [28, 29].

Local approaches use the top-ranked documents, retrieved in response to the initial query, in order to select expansion terms, mostly using pseudo-relevance feedback (PRF), where the top k ranked documents in the initial retrieval results are assumed to be relevant. For example, authors in [28] presented a PRF technique called LCA (Local Context Analysis) in which candidate expansion terms are selected on the basis of their co-occurrence relationship with query terms within pseudo-relevant documents. They showed the effectiveness of the proposed PRF technique using different languages. In [12], authors presented a concept-based PRF technique. They built a directed query relations graph to extract concepts that are related to the query. The query relations were mined using association rules. Authors in [27] discussed the contribution of linear methods for PRF. They used an inter-term similarity matrix to get expansion terms. In [31], authors presented a matrix factorization technique using pseudo-relevant documents. They considered PRF as a recommendation task for selecting useful expansion terms. They demonstrated the effectiveness of this technique on two retrieval models: the language model and the vector space model.

Unlike local QE, global approaches allow selecting expansion terms without regard to the initial retrieval results. In this case, expansion terms are selected by analyzing the entire corpus in order to discover term associations and co-occurrence relationships [29]. For example, authors in [33] proposed a technique to expand short queries for microblog retrieval. They explored the use of Wikipedia, DBpedia and association rules mining for selecting semantically related terms to the queries. In [6], authors addressed QE by using co-occurrence relationships and inferential relationships between terms. They proposed to integrate QE into language modeling and demonstrated the feasibility of this integration.

The use of term co-occurrence statistics is one of the earliest QE approaches, in which terms that are statistically related to the query are considered as potential expansion candidates. However, a basic issue in this approach is the selection of discriminative terms using co-occurrence statistics. Usually, the selected terms tend to occur frequently in the entire collection and thus are unlikely to be discriminative. This limitation is mainly due to the way in which the similarity between terms is measured [19].

Several measures have been used to evaluate the similarity between pairs of terms. We may cite Cosine similarity, Jaccard index, Dice coefficient and Mutual Information [8]. Recently, QE based on word embedding [1,3,30] leads to an interesting improvement on retrieval effectiveness by exploring word relationships from embedding vectors. In these methods, term co-occurrence statistics are employed to learn word vector representations using word embedding algorithms such as word2vec [18] and Glove [20]. Indeed, terms co-occurrence within the same context window is used to produce word vectors [30]. We use the same approach, i.e. a context window-based approach applied over the entire corpus, in order to build our term co-occurrence graphs.

3   An Adaptation of BM25 for Query Expansion Based on Term Co-occurrence Graphs

In this section, we describe our QE approach and we present the proposed adaptation of BM25 for term co-occurrence graphs. Figure 1 depicts the general architecture of our QE system. We select semantically related terms to the query following two steps. First, a term co-occurrence graph is constructed over the entire corpus using a context window-based approach. This approach has been used in multiple IR and Natural Language Processing (NLP) tasks such as word embedding [18,20]. Indeed, the co-occurrence of terms within a specified context window is used to capture semantic relations between terms. For instance, given the sentence "The SPIRE conference covers research on string processing and information retrieval." and taking "conference" as the target term with a window-size equal to 2, its context terms will be "The", "SPIRE", "covers" and "research". Second, using an adapted version of the BM25 model to measure the similarity of terms in co-occurrence graphs, candidate expansion terms are scored according to their similarity with the query as a whole.

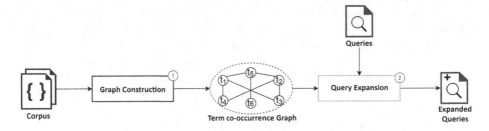

Fig. 1. General architecture of the proposed QE approach.

The Okapi BM25 model calculates the score of a document D given a query Q as follows [22]:

$$BM25(Q, D) = \sum_{t \in Q} IDF(t) \times \frac{(k_1 + 1) \times tf(t, D)}{k_1 \times (1 - b + b \times \frac{dl}{avgdl}) + tf(t, D)} \quad (1)$$

where:

- $IDF(t)$ is the Inverse Term Frequency of t and it is computed as follows:

$$IDF(t) = log\frac{N - df(t) + 0.5}{df(t) + 0.5} \quad (2)$$

- N is the number of documents in the collection.
- $df(t)$ is the number of documents containing term t.

- $tf(t, D)$ is the term frequency of t, i.e., the number of occurrences of term t in the document D.
- dl and $avgdl$ denote the length of document D and the average document length in the collection, respectively.
- k_1 and b are free hyper-parameters.

We propose to contribute in developing one-to-many association measures which are computed on a symmetric co-occurrence graph. This measure is inspired from Okapi BM25 [22]. A symmetric co-occurrence graph is an undirected graph $G = (V, E)$, where V is a set of nodes and E is a set of weighted edges. We also define the symbols cited hereafter as follows:

- n_i: the node number i in the graph.
- $e(n_i, n_c) = e(n_c, n_i)$: the weight of the edge linking n_i and n_c.
- $co\_degree(n_i)$: the number of nodes in G having n_i as destination.
- $sum\_e(n_c)$: the sum of the weights of the edges having n_c as destination.
- $avgsum\_e$: the average of the previous parameter (i.e. $sum\_e(n_c)$) over all the possible destination nodes in G.
- N is the number of all possible destination nodes in G.

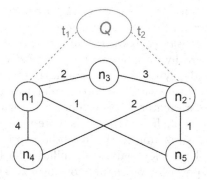

Fig. 2. Example of a query projected on the graph. n_1, n_2 are the query terms and n_3, n_4, n_5 are candidate expansion terms.

We formalize similarity calculus in graphs as follows. Let $C = \{n_1, \ldots, n_m\}$ be the projection of query $Q = \{t_1, \ldots, t_m\}$ on the co-occurrence graph G and n_c be a candidate node in G. An example of a query Q projected on a co-occurrence subgraph is illustrated in Fig. 2. We propose to compute the relevance of a node n_c given C in a co-occurrence graph. C is considered as a query, n_c as a document and the relevance is assessed using an adapted version of BM25. That is, we follow other research which used IR models to compute similarities between queries and terms [9,10]. We define the relevance of a node n_c given another node n_i as a product of a local weight $L_{i,c}$ and a global weight G_i:

$$Sim(n_i, n_c) = L_{i,c} \times G_i \qquad (3)$$

Local Weights. The main hypothesis behind local weights is that terms which co-occur frequently are likely to be similar. Local weights apply to the following constraints:

- $L_{i,c} = 0$ if $e(n_i, n_c) = 0$.
- $L_{i,c}$ increases with $e(n_i, n_c)$.
- $L_{i,c}$ approaches a maximum value of 1.

In classical document indexing with TF-IDF [25], local weights are normalized by document length, which is equivalent in our case to the sum of the weights of the edges linking to n_c (cf. Formula 4). A term which appears n times in a short document is more significant compared to the case in which it appears the same number of times in a longer one. In our case, n_i is more relevant to n_c when the latter has a lower degree.

$$L_{i,c} = \frac{e(n_i, n_c)}{sum\_e(n_c)} \tag{4}$$

In BM25, local weights are computed as parameterized frequencies based on a 2-Poison model:

$$L_{i,c} = \frac{e(n_i, n_c)}{k_1 + e(n_i, n_c)} \tag{5}$$

$L_{i,c}$ is nonlinear to k_1. This is justified by Robertson et al. [24] by the fact that *"the information gained on observing a term the first time is greater than the information gained on subsequently seeing the same term"*. Robertson and Walker [22] considered two hypotheses, namely Verbosity and Scope. The first hypothesis allows to handle synonyms. Let consider two synonyms represented by the nodes n_c^1 and n_c^2 (i.e. both terms are similar to an input node n_i). Let also suppose that n_c^1 has a greater co-occurrence value with n_i (i.e. $e(n_i, n_c^1) > e(n_i, n_c^2)$). We say that one of the two nodes is more verbose (i.e. more likely to be used) than the other. According to this hypothesis, we should normalize the co-occurrence values to obtain close values of similarity for both nodes. However, applying to the scope hypothesis, we should not normalize as we would prefer to return n_c^1. We suppose in this case that the more frequent term is more likely to represent the sense which is shared by both nodes. However, n_c^2 is less frequent and thus unable to add much information to the original query. Both hypotheses are complementary. In real co-occurrence graphs, both scenarios are present, each of them constitutes a partial explanation. To insure that both hypotheses are respected, document lengths are normalized. In our case, we compute the average of the weighted degrees of nodes as follows:

$$avgsum\_e = \frac{\sum_i sum\_e(n_i)}{N} \tag{6}$$

Then weighted degrees are normalized as follows:

$$norm\_sum\_e(n_i) = \frac{sum\_e(n_i)}{avgsum\_e} \tag{7}$$

To allow adjust and tune this score, it is reformulated as follows:

$$norm\_sum\_e(n_i) = 1 - b + b \times \frac{sum\_e(n_i)}{avgsum\_e} \qquad (8)$$

The constant b determines the scaling by the degree of the target node (document length in query-document matching). $b = 1$ means fully scaling the term weight, while $b = 0$ disables normalization. The quantity obtained by Eq. 8 is used to normalize the local weight computed by Eq. 5. Thus we have:

$$L_{i,c} = \frac{e(n_i, n_c) \times (k_1 + 1)}{k_1 \times (1 - b + b \times \frac{sum\_e(n_c)}{avgsum\_e}) + e(n_i, n_c)} \qquad (9)$$

Global Weights. Global weights are defined according to the Probabilistic Model of Robertson and Spärck-Jones [16]. Given a node n_i, we would like to know if a node n_c is relevant (i.e. similar to n_i) based on the probabilistic IR framework [7]. The contingency table (Table 1) defines the main parameters used to estimate the probability of relevance of n_c. This table is defined in a scenario of relevance feedback where a user selects the terms which are relevant to a given query.

Table 1. Contingency table of main parameters

	n_c is relevant	n_c is not relevant	Total
n_c co-occurs with n_i	s	$co\_degree(n_i)$-s	$co\_degree(n_i)$
n_c does not co-occur with n_i	S-s	$(N$- $co\_degree(n_i))$-(S-s)	N-$co\_degree(n_i)$
Total	S	N-S	N

In this table, s is the number of terms which are relevant to the query which co-occur with n_i . The relevance of n_c may be estimated as follows:

$$INF(n_i) = log \frac{\frac{s}{S-s}}{\frac{co\_degree(n_i)-s}{(N-co\_degree(n_i))-(S-s)}} \qquad (10)$$

In this model, it is fairly standard to add 0.5 to the quantities which may be null (i.e. the cells of the second and the third column of the contingency table) [16]. A second variant is thus defined as follows:

$$INF(n_i) = log \frac{\frac{s+0.5}{S-s+0.5}}{\frac{co\_degree(n_i)-s+0.5}{(N-co\_degree(n_i))-(S-s+0.5)}} \qquad (11)$$

Using 0.5 is a kind of smoothing which is justified by the limits of maximum likelihood estimate (or MLE) which penalizes rare events. Smoothing allows handling events which has never been seen nor observed [16]. In the absence of

relevance feedback, we have s=S=0. If we adopt Formula 10 (without smoothing), we obtain:

$$INF(n_i) = log\frac{N - co\_degree(n_i)}{co\_degree(n_i)} \tag{12}$$

With smoothing (Formula 11), we get:

$$INF(n_i) = log\frac{N - co\_degree(n_i) + 0.5}{co\_degree(n_i) + 0.5} \tag{13}$$

In all the cases, $INF(n_i)$ reflects how much a term is distributed over the others. It just checks if it is common or rare across all the other terms. That is, terms which tend to co-occur with many terms (e.g. stop words) will get null or low values. However, it provides an absolute evaluation of the discriminative power of a term which does not depend on the original query.

In Eq. 3, we replace global and local weights computed respectively by Formulas 13 and 9. Besides, we compute the sum of the similarity of n_c and all the terms of C. Thus, using an adapted version of BM25, noted here $BM25_{cog}$ (BM25 for co-occurrence graphs), we calculate the score of each candidate node n_c as follows:

$$BM25_{cog}(C, n_c) = \sum_{n_i \in C} INF(n_i) \times \frac{(k_1 + 1) \times e(n_i, n_c)}{k_1 \times (1 - b + b \times \frac{sum\_e(n_c)}{avgsum\_e}) + e(n_i, n_c)} \tag{14}$$

The constant k_1 determines how relevance changes when the number of co-occurrence $e(n_i, n_c)$ increases. A null value of k_1 means disabling term weight (using only $INF(n_i)$). If k_1 is large, the term weight component would increase nearly linearly with $e(n_i, n_c)$. Using the default value of this parameter means that after three or four co-occurrences, additional co-occurrences will have a little impact [24].

This adapted version of the BM25 model stands out by the following aspects. First, it allows both one-to-one and one-to-many associations. On the other hand, the INF factor allows to evaluate the discriminative power of terms. That is, terms that co-occur with many other terms are penalized. Moreover, it has two hyper-parameters, which may be tuned to enhance results. We used $BM25_{cog}$ in a PRF scenario in [2]. The obtained results showed significant improvements over the state-of-the-art baselines.

4   Experiments

In this section, we first present our test collections and describe the experimental setup. Then we discuss the experimental results.

4.1   Experimental Setup

We used two TREC collections in our experiments. The first is the standard Robust04 collection which is available in TREC disks 4 and 5[1]. It consists of

---

[1] https://trec.nist.gov/data/cd45/.

news articles from different sources. This collection was used in TREC 2004 Robust Track. The second is the newest TREC Washington Post collection[2] provided by TREC 2018 Common Core Track, which consists of news articles and blog posts published by Washington Post from January 2012 through August 2017. Statistics of these collections are presented in Table 2.

Table 2. TREC collections statistics.

Collection	Document set	#docs	Size	#query	#qrels
Robust04	TREC Disks 4 & 5 minus Congressional Record	528k	1.9 GB	249	17,412
WAPOST	TREC Washington Post Corpus	608k	6.9 GB	50	3,948

All experiments were conducted using the Terrier 4.2 IR platform[3]. For both collections, preprocessing involved stopword removal using the Terrier's standard stopword list and stemming using the Porter stemmer. We only considered the title of the TREC topics as queries (i.e., short queries).

We use Mean Average Precision (MAP), precision at top 5 documents (P@5) and precision at top 10 documents (P@10) as evaluation measures. MAP serves as the objective evaluation measure for parameter tuning. Statistically significant differences in terms of retrieval performance are computed using the two-tailed paired t-test at a 95% confidence level.

4.2 Parameter Tuning

In order to construct the term co-occurrence graph, we need to choose the value of the window size parameter. We explored different values of this parameter to see the effect they have on effectiveness. Window size values of 2–10, plus a window size equal to sentence length, were tested. We found that a window size of 7 terms gives the best results on the Robust04 collection, whereas best results on the Washington Post collection are obtained using a dynamic window size equal to sentence length. This is consistent with previous research [11,17], stating that the best choice of context window size is collection-dependent. We should note that we used a symmetric window size, i.e., a window size of n means n terms to the left and n terms to the right of the target term. The optimal parameter value for each collection was used to construct the term co-occurrence graph.

The model hyper-parameters were tuned using 5-fold cross-validation over the queries of each collection, where topics were randomly split into 5 folds. The hyper-parameters were tuned on 4-of-5 folds and tested on the final fold. This process is carried out 5 times, each time using one fold. The results presented are the mean of the 5 runs. We varied the value of b from 0.1 to 0.9 and the value of k_1 from 0.1 to 3.0 in increments of 0.1. The number of expansion terms was empirically set to 10.

---

[2] https://trec.nist.gov/data/wapost/.
[3] http://terrier.org/.

4.3   Results

In this subsection, we evaluate the effectiveness of the proposed approach. We consider three state-of-the-art IR models as baselines, namely: Okapi BM25 model [22], Language Model with Jelinek-Mercer smoothing [32] and Divergence from Randomness (DFR) PL2 model [4]. Besides, we consider the classical PRF approach and embedding-based QE approach by using word2vec[4] (W2V) to train word vectors over the target corpus. The obtained results are reported in Tables 3 and 4. Superscripts 1/2/3 indicate that the improvements over the unexpanded baselines, PRF and W2V, respectively, are statistically significant (t-test with $p\_value < 0.05$).

Table 3. Retrieval results on the Robust04 collection.

Retrieval model	Method	MAP	P@5	P@10
BM25	Baseline	0.2363	0.4691	0.4100
	PRF	0.2537	0.4378	0.3835
	W2V	0.2396	0.4627	0.4133
	$BM25_{cog}$	$\mathbf{0.2589^{1,3}}$	$\mathbf{0.4723^{2}}$	$\mathbf{0.4289^{1,2,3}}$
LM	Baseline	0.2155	0.3952	0.3651
	PRF	0.2437	0.4008	0.3747
	W2V	0.2259	0.4201	0.3743
	$BM25_{cog}$	$\mathbf{0.2469^{1,3}}$	$\mathbf{0.4394^{1,2,3}}$	$\mathbf{0.3940^{1}}$
PL2	Baseline	0.2239	0.4578	0.4032
	PRF	0.2287	0.4185	0.3763
	W2V	0.2303	**0.4683**	0.4092
	$BM25_{cog}$	$\mathbf{0.2464^{1,2,3}}$	0.4667^{2}	$\mathbf{0.4221^{1,2,3}}$

According to these tables, the proposed QE approach outperforms the state-of-the-art baselines in terms of MAP, P@5 and P@10 in all cases. The MAP improvements are always statistically significant in both collections. As for precision, we can see that improvements are also statistically significant in most cases. This shows that our QE approach, which generates semantically related terms to the query as a whole, leads to improvement in retrieval performance of the state-of-the-art models.

Comparing our QE and the classical PRF approach, we observe in Table 4 that the latter outperforms our approach in terms of MAP in the Washington Post collection. This shows in this case the advantage of local analysis over global analysis. Whereas, we see in Table 3 that our approach outperforms PRF in terms of MAP in the Robust04 collection. In terms of precision, we can remark that our approach outperforms PRF by significant margins in both collections. Besides,

---

[4] We used the CBOW implementation of word2vec and we set the vectors dimension to 300.

Table 4. Retrieval results on the Washington post collection.

Retrieval model	Method	MAP	P@5	P@10
BM25	Baseline	0.2385	0.4920	0.4300
	PRF	**0.2865**	0.4640	0.4160
	W2V	0.2436	0.4400	0.4080
	$BM25_{cog}$	$0.2687^{1,3}$	$\mathbf{0.5120^{3}}$	$\mathbf{0.4400^{3}}$
LM	Baseline	0.2065	0.3760	0.3560
	PRF	**0.2612**	0.4000	0.3780
	W2V	0.2119	0.3600	0.3560
	$BM25_{cog}$	$0.2505^{1,3}$	$\mathbf{0.4720^{1,3}}$	$\mathbf{0.4080^{1,3}}$
PL2	Baseline	0.2274	0.4880	0.4100
	PRF	**0.2754**	0.4400	0.4080
	W2V	0.2330	0.4800	0.4080
	$BM25_{cog}$	$0.2599^{1,3}$	$\mathbf{0.5120^{2}}$	$\mathbf{0.4640^{1,2,3}}$

by comparing PRF and the unexpanded baselines, it can be observed that PRF hurts the precision in the majority of cases. This shows that our proposal is able to generate better expansion terms and can filter out non-discriminative ones, which co-occur with too many terms, thus improving the precision at top-ranked documents.

By comparing our results to those obtained using word embedding-based QE, we can remark that our proposal yields better results on both collections with significant margins in the majority of cases. This confirms the effectiveness of the proposed approach for QE.

In this set of experiments, retrieval models were used with their suggested default parameters. These default settings are unlikely to be optimal for different collections and query lengths. Therefore, we next investigate the impact of parameters tuning on retrieval performance in both collections. To this end, the three models were extensively tuned using 5-fold cross-validation over the queries of each collection. Optimal parameter settings are listed in Table 5. We tuned parameters b and λ for the BM25 model and the LM model, respectively, from 0.10 to 0.90 in increments of 0.01. For the PL2 model, parameter c was tuned from 1.0 to 20.0 in increments of 0.1. We tuned the k_1 parameter of the BM25 model but it had little impact on retrieval effectiveness. We therefore used the default value in Terrier (k_1=1.2). Table 6 presents the MAP results achieved by the proposed QE approach and the baselines for each of the three IR models. Superscript 1 indicates that the improvements over the baselines are statistically significant (t-test with $p\_value < 0.05$). We can see that, in both collections, our QE outperforms the unexpanded baselines with statistically significant improvements in all cases. These results confirm the effectiveness of our proposal regardless of the ranking model. Furthermore, it is worth noting that our best result on the Washington Post collection is equal to the result of the best official automatic run in TREC 2018 Common Core Track.

Table 5. Optimal parameter settings.

Query	Original			Expanded		
Retrieval model	BM25	LM	PL2	BM25	LM	PL2
Parameter	b	λ	c	b	λ	c
Robust04	0.322	0.616	8.480	0.404	0.684	7.500
WAPOST	0.418	0.470	5.380	0.545	0.478	4.460

Table 6. Comparison of MAP results between the expanded queries and the baselines with optimal parameter settings for the three retrieval models.

Collection	Retrieval model	Baseline	$BM25_{cog}$
Robust04	BM25	0.2498	0.2643^{1} (+5.80%)
	LM	0.2285	0.2593^{1} (+13.48%)
	PL2	0.2529	$\mathbf{0.2697}^{1}$ (+6.64%)
WAPOST	BM25	0.2506	0.2725^{1} (+8.74%)
	LM	0.2180	0.2642^{1} (+21.19%)
	PL2	0.2481	$\mathbf{0.2761}^{1}$ (+11.29%)

5   Conclusion

In this paper, we proposed an adaptation of the state-of-the-art probabilistic model BM25 to measure the similarity between terms in a co-occurrence graph for QE. The proposed measure allows to evaluate the discriminative power of terms and to obtain semantically related terms to the whole query. Besides, it takes advantage of the BM25's hyper-parameters that can be adjusted to improve retrieval results.

Experiments on the TREC Robust04 and Washington Post collections show significant improvements over the baselines in terms of MAP and precision for three state-of-the-art IR models.

As part of our future work, we plan to investigate the use of external resources (e.g. Wikipedia) to build the term co-occurrence graph. In addition, investigating the use of asymmetric context windows to construct the co-occurrence graph is also an interesting research direction. Another direction for extending this work is to study the use of the new similarity measure for other IR tasks, such as Query Reweighting and Word Sense Disambiguation (WSD).

References

1. Aklouche, B., Bounhas, I., Slimani, Y.: Query expansion based on NLP and word embeddings. In: Proceedings of the The Twenty-Seventh Text Retrieval Conference (TREC 2018), Gaithersburg, Maryland, USA (14–16 November 2018)

2. Aklouche, B., Bounhas, I., Slimani, Y.: Pseudo-relevance feedback based on locally-built co-occurrence graphs. In: Welzer, T., Eder, J., Podgorelec, V., Kamisalic Latific, A. (eds.) Advances in Databases and Information Systems, vol. 11695, pp. 105–119. (2019). https://doi.org/10.1007/978-3-030-28730-6_7
3. ALMasri, M., Berrut, C., Chevallet, J.-P.: A comparison of deep learning based query expansion with pseudo-relevance feedback and mutual information. In: Ferro, N., Crestani, F., Moens, M.-F., Mothe, J., Silvestri, F., Di Nunzio, G.M., Hauff, C., Silvello, G. (eds.) ECIR 2016. LNCS, vol. 9626, pp. 709–715. Springer, Cham (2016). https://doi.org/10.1007/978-3-319-30671-1_57
4. Amati, G.: Probability models for information retrieval based on divergence from randomness. Ph.D. thesis, University of Glasgow, UK (2003)
5. Ariannezhad, M., Montazeralghaem, A., Zamani, H., Shakery, A.: Improving retrieval performance for verbose queries via axiomatic analysis of term discrimination heuristic. In: Proceedings of the 40th International ACM SIGIR Conference on Research and Development in Information Retrieval, Shinjuku, Tokyo, Japan, pp. 1201–1204. ACM, 7–11 August 2017
6. Bai, J., Song, D., Bruza, P., Nie, J.Y., Cao, G.: Query expansion using term relationships in language models for information retrieval. In: Proceedings of the 14th ACM International Conference on Information and Knowledge Management, Bremen, Germany, pp. 688–695. ACM, 31 October–5 November 2005
7. Bounhas, I., Elayeb, B., Evrard, F., Slimani, Y.: ArabOnto: experimenting a new distributional approach for building arabic ontological resources. Int. J. Metadata, Semant. Ontol. 6(2), 81–95 (2011). https://doi.org/10.1504/IJMSO.2011.046578
8. Carpineto, C., Romano, G.: A survey of automatic query expansion in information retrieval. ACM Comput. Surv. (CSUR) 44(1), 11–150 (2012). https://doi.org/10.1145/2071389.2071390
9. Elayeb, B., Bounhas, I., Khiroun, O.B., Evrard, F., Saoud, N.B.B.: A comparative study between possibilistic and probabilistic approaches for monolingual word sense disambiguation. Knowl. Inf. Syst. 44(1), 91–126 (2015). https://doi.org/10.1007/s10115-014-0753-z
10. Elayeb, B., Bounhas, I., Khiroun, O.B., Saoud, N.B.B.: Combining semantic query disambiguation and expansion to improve intelligent information retrieval. In: Duval, B., van den Herik, J., Loiseau, S., Filipe, J. (eds.) ICAART 2014. LNCS (LNAI), vol. 8946, pp. 280–295. Springer, Cham (2015). https://doi.org/10.1007/978-3-319-25210-0_17
11. Fagan, J.: Automatic phrase indexing for document retrieval. In: Proceedings of the 10th Annual International ACM SIGIR Conference on Research and Development in Information Retrieval, New Orleans, Louisiana, USA, pp. 91–101. ACM (3–5 June 1987)
12. Fonseca, B.M., Golgher, P., Pôssas, B., Ribeiro-Neto, B., Ziviani, N.: Concept-based interactive query expansion. In: Proceedings of the 14th ACM International Conference on Information and Knowledge Management, Bremen, Germany, pp. 696–703. ACM (31 October – 05 November 2005)
13. He, B., Huang, J.X., Zhou, X.: Modeling term proximity for probabilistic information retrieval models. Inf. Sci. 181(14), 3017–3031 (2011). https://doi.org/10.1016/j.ins.2011.03.007
14. Jones, K.S., Walker, S., Robertson, S.E.: A probabilistic model of information retrieval: development and comparative experiments: Part 2. Inf. Process. Manag. 36(6), 809840 (2000). https://doi.org/10.1016/S0306-4573(00)00016-9

15. Lv, Y., Zhai, C.: Lower-bounding term frequency normalization. In: Proceedings of the 20th ACM International Conference on Information and Knowledge Management, Glasgow, Scotland, UK, pp. 7–16. ACM, 24–28 October 2011
16. Manning, C.D., Raghavan, P., Schütze, H.: Introduction to Information Retrieval. Cambridge University Press, Cambridge (2008)
17. Metzler, D., Croft, W.B.: A markov random field model for term dependencies. In: Proceedings of the 28th Annual International ACM SIGIR Conference on Research and Development in Information Retrieval, Salvador, Brazil, pp. 472–479. ACM (15–19 August 2005)
18. Mikolov, T., Sutskever, I., Chen, K., Corrado, G.S., Dean, J.: Distributed representations of words and phrases and their compositionality. In: Advances in Neural Information Processing Systems, Proceedings of the 26th International Conference on Neural Information Processing Systems, Lake Tahoe, Nevada, United States, pp. 3111–3119. 5–8 December 2013
19. Peat, H.J., Willett, P.: The limitations of term co-occurrence data for query expansion in document retrieval systems. J. Am. Soc. Inf. Sci. 42(5), 378–383 (1991)
20. Pennington, J., Socher, R., Manning, C.D.: Glove: global vectors for word representation. In: Proceedings of the 2014 Conference on Empirical Methods in Natural Language Processing (EMNLP), Doha, Qatar, pp. 1532–1543. ACL 25–29 October 2014
21. Rasolofo, Y., Savoy, J.: Term proximity scoring for keyword-based retrieval systems. In: Sebastiani, F. (ed.) ECIR 2003. LNCS, vol. 2633, pp. 207–218. Springer, Heidelberg (2003). https://doi.org/10.1007/3-540-36618-0_15
22. Robertson, S.E., Walker, S.: Some simple effective approximations to the 2-poisson model for probabilistic weighted retrieval. In: Croft, B.W., van Rijsbergen, C.J. (eds.) SIGIR 1994, pp. 232–241. Springer, London (1994)
23. Robertson, S.E., Zaragoza, H.: The probabilistic relevance framework: Bm25 and beyond. Found. Trends Inf. Retrieval 3(4), 333–389 (2009). https://doi.org/10.1561/1500000019
24. Robertson, S., Zaragoza, H., Taylor, M.: Simple bm25 extension to multiple weighted fields. In: Proceedings of the Thirteenth ACM International Conference on Information and Knowledge Management, Washington, D.C., USA, pp. 42–49. ACM, 08–13 November 2004
25. Salton, G., McGill, M.: Introduction to Modern Information Retrieval. McGraw-Hill Book Company, USA (1984)
26. Song, R., Taylor, M.J., Wen, J.-R., Hon, H.-W., Yu, Y.: Viewing term proximity from a different perspective. In: Macdonald, C., Ounis, I., Plachouras, V., Ruthven, I., White, R.W. (eds.) ECIR 2008. LNCS, vol. 4956, pp. 346–357. Springer, Heidelberg (2008). https://doi.org/10.1007/978-3-540-78646-7_32
27. Valcarce, D., Parapar, J., Barreiro, A.: Lime: Linear methods for pseudo-relevance feedback. In: Proceedings of the 33rd Annual ACM Symposium on Applied Computing, Pau, France, pp. 678–687. ACM, 09–13 April 2018
28. Xu, J., Croft, W.B.: Improving the effectiveness of information retrieval with local context analysis. ACM Trans. Inf. Syst. (TOIS) 18(1), 79–112 (2000). https://doi.org/10.1145/333135.333138
29. Xu, J., Croft, W.B.: Query expansion using local and global document analysis. In: Proceedings of the 19th Annual International ACM SIGIR Conference on Research and Development in Information Retrieval, Zurich, Switzerland, pp. 4–11. ACM, 18–22 August 1996

30. Zamani, H., Croft, W.B.: Relevance-based word embedding. In: Proceedings of the 40th International ACM SIGIR Conference on Research and Development in Information Retrieval, Shinjuku, Tokyo, Japan, pp. 505–514. ACM, 7–11 August 2017
31. Zamani, H., Dadashkarimi, J., Shakery, A., Croft, W.B.: Pseudo-relevance feedback based on matrix factorization. In: Proceedings of the 25th ACM International on Conference on Information and Knowledge Management, Indianapolis, Indiana, USA, pp. 1483–1492. ACM, 24–28 October 2016
32. Zhai, C., Lafferty, J.: A study of smoothing methods for language models applied to ad hoc information retrieval. In: Proceedings of the 24th Annual International ACM SIGIR Conference on Research and Development in Information Retrieval, New Orleans, Louisiana, USA, pp. 334–342. ACM, 9–13 September 2001
33. Zingla, M.A., Chiraz, L., Slimani, Y.: Short query expansion for microblog retrieval. In: Knowledge-Based and Intelligent Information & Engineering Systems: Proceedings of the 20th International Conference KES-2016, York, UK, pp. 225–234. Elsevier, 5–7 September 2016

Network-Based Pooling for Topic Modeling
on Microblog Content

Anaïs Ollagnier$^{(\boxtimes)}$ (iD) and Hywel Williams (iD)

Computer Science, University of Exeter, Exeter EX4 4QE, UK
{a.ollagnier,h.t.p.williams}@exeter.ac.uk

Abstract. Topic modeling with tweets is difficult due to the short and informal nature of the texts. Tweet-pooling (aggregation of tweets into longer documents prior to training) has been shown to improve model outputs, but performance varies depending on the pooling scheme and data set used. Here we investigate a new tweet-pooling method based on network structures associated with Twitter content. Using a standard formulation of the well-known Latent Dirichlet Allocation (LDA) topic model, we trained various models using different tweet-pooling schemes on three diverse Twitter datasets. Tweet-pooling schemes were created based on mention/reply relationships between tweets and Twitter users, with several (non-networked) established methods also tested as a comparison. Results show that pooling tweets using network information gives better topic coherence and clustering performance than other pooling schemes, on the majority of datasets tested. Our findings contribute to an improved methodology for topic modeling with Twitter content.

Keywords: Microblogs · LDA · Information retrieval · Aggregation · User networks

1 Introduction

Micro-blogging platforms such as Twitter have witnessed a rapid and impressive expansion, creating a popular new mode of public communication. Currently, Twitter has 6000 tweets written every second per day on average[1]. Twitter has become a significant source of information for a broad variety of applications, but the volume of data makes human analysis intractable. There is therefore considerable interest in adaptation of computational techniques for large-scale analyses, such as opinion mining, machine translation, and social information retrieval, among others. Application of topic modeling techniques to Twitter content is non-trivial due to the noisy and short texts associated with individual tweets. In the literature, topic models such as Latent Dirichlet Allocation (LDA) [1] or the Author Topic Model (ATM) [2] have proved their success in several applications (e.g. news articles, academic abstracts). However, results are more mixed when applied on short texts due to the data sparsity in each individual document.

Several approaches have been proposed to design longer pseudo-documents by aggregating multiple short texts (tweets). Each document results from a pooling strategy

---

[1] http://www.internetlivestats.com/twitter-statistics/ Date of access: 28th Jul 2019.

© Springer Nature Switzerland AG 2019
N. R. Brisaboa and S. J. Puglisi (Eds.): SPIRE 2019, LNCS 11811, pp. 80–87, 2019.
https://doi.org/10.1007/978-3-030-32686-9_6

applied in a pre-processing stage. In [3], an author-based tweet pooling scheme is used which builds documents by combining all tweets posted by the same author. A hashtag-based tweet pooling method is proposed by [4], which creates documents consisting of all tweets containing the same hashtag. The main goal behind these approaches is to improve topic model performance by training on the pooled documents, with efficacy measured against similar topic models trained on the unpooled tweets. Empirical studies with these approaches highlight inconsistencies in the homogeneity of generated topics. To overcome this problem, [5] propose a conversation-based pooling technique which aggregates tweets occurring in the same user-to-user conversation. This approach outperforms other pooling methods in terms of clustering quality and document retrieval. More recently, [6] propose to prune irrelevant tweets through a pooling strategy based on information retrieval (IR) in order to place related tweets in the same cluster. This method provides an interesting improvement in a variety of measures for topic coherence, in comparison to unmodified LDA baseline and a variety of other pooling schemes.

Several IR applications in context of microblogs use network representations [7] (e.g. document retrieval, document content). Here, we evaluate a novel network-based tweet pooling method that aggregates tweets based on user interactions around each item of content. Our intuition behind this method is to expose connections between users and their interest in a given topic; by pooling tweets based on relational information (user interactions) we hope to create an improved training corpus. To evaluate this method, we perform a comprehensive empirical comparison against four state-of-the-art pooling techniques chosen after a literature survey. Across three Twitter datasets, we evaluate the pooling techniques in terms of topic coherence and clustering quality. The experimental results show that the proposed technique yields superior performance for all metrics on the majority of datasets and takes considerably less time to train.

2 Tweet-Pooling Methods

Tweet texts are qualitatively different to conventional texts, being typically short (≤ 280 characters[2]) with a messy structure including platform-specific objects (e.g. hashtags, shortened urls, user names, emoticons/emojis). In this context, tweet-pooling has been developed to better capture reliable document-level word co-occurrence patterns. Here, we evaluate four existing unsupervised tweet pooling schemes alongside our proposed network-based scheme:

Unpooled Scheme: The default approach used as a baseline in which each tweet is considered as a single document.

Author Pooling: Each tweet authored by a single user is aggregated as a single document, so the number of documents is the same as the number of unique users. This approach outperforms the unpooled scheme [9].

---

[2] In September 2017, Twitter expanded the original 140-character limit to 280 characters. See: https://blog.twitter.com/official/en_us/topics/product/2017/tweetingmadeeasier.html. Date of access: 11th Feb 2019.

Hashtag Pooling: Tweets using similar hashtags are aggregated as a single document. The number of documents is equal to the number of unique hashtags, but a tweet can appear in several documents if it contains multiple hashtags. Tweets without hashtags are considered as individual documents. This method was shown [5] to outperform unpooled schemes. (Note that [4] showed improved performance by assigning hashtag labels to tweets without hashtags, but this technique adds computational cost and was not used here).

Conversation Pooling: Each document consists of all tweets in the corpus that belong to the conversation tree for a chosen seed tweet. The conversation tree includes tweets written in reply to an original tweet, as well as replies to those replies, and so on. Tweets without replies are considered as individual documents. In [5], conversation pooling outperforms alternative pooling schemes.

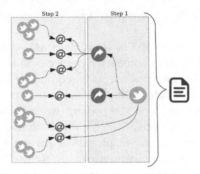

Fig. 1. Network-based tweet pooling. Each document is initialised with a seed tweet. In Step 1, the first layer of direct replies to the seed tweet are added. In Step 2, all tweets by users mentioned in the set of tweets resulting from Step 1 are also added.

Fig. 2. Example content of a document created by network-based tweet pooling.

Network-Based Pooling: In this novel scheme, each document is aggregated from all tweets within the corpus that are associated with the seed tweet by a simple network structure (Figs. 1 and 2). In Step 1, tweets are aggregated that were written in reply to the seed tweet. In Step 2, we identify all mentioned users in the set of tweets from Step 1 (i.e. all users that are referenced in tweet text using the @ symbol). We then aggregate to the document all other tweets in the corpus that are authored by this user set.

This scheme differs from conversation pooling in two aspects. First, only direct replies are aggregated i.e. the first layer of replies from the conversation tree. Manual inspection of full tweet conversation trees showed that the conversation thread can shift in topic as the tree increases in depth. Use of the full tree can thereby capture topics which are not anymore related to those of the seed tweet. To identify reply tweets, we

used the `in_reply_to_status_id` field returned by the Twitter API for each tweet. Second, exploiting tweets of all mentioned users allows the network-based pooling to access additional content from users interested in the topics of the original seed tweet. Leveraging this information, we construct a network based on both interactions and connections between users.

3 Tweet Corpus Building

Table 1. Distribution of latent categories in the datasets (labelled by search theme)

Dataset	No. of tweets	Category / % of Documents
Generic	658,492	Music/24.4 - Business/10.2 - Movie/18.5 - Health/14.7 - Family/7.4 - Sport/24.8
Specific	445,852	Arts&entertainment/9.7 - Business/12.4 - Law Enforcement&Armed Forces/6.2 - Science&technology/36.8 - Healthcare&medicine/25.5 - Service/9.4
Events	188,000	Natural disasters/37.1 - Transport/15.4 - Industrial/10.2 - Health/9.7 - Terrorism/27.6

To evaluate the portability of different pooling schemes we collected three tweet datasets with different levels of underlying thematic/topical heterogeneity. Data was collected using the public Twitter Search API[3] during 2018 and 2019. Each collection was created with a different list of API keywords and included tweets collected on different themes. For each chosen theme a list of terms was manually created. All tweets returned were collated in a single corpus, labelled by the theme. The three datasets collected were:

Generic. A wide range of themes. Tweets from 11 Dec'18 to 30 Jan'19 collected using keywords related to a range of themes ('music', 'business', 'movies', 'health', 'family', 'sports').

Event. Tweets from 23 Mar'18 to 22 Jan'19 associated with various events ('natural disasters', 'transport', 'industrial', 'health', 'terrorism'). Search terms were manually collated based on reading a sample of posts about disaster events.

Specific. Tweets from 21 Feb'18 to 11 Feb'19 associated with job adverts for different industries ('arts & entertainment', 'business', 'law enforcement & armed forces', 'science & technology', 'healthcare & medicine', 'service'). Search terms manually collated based on reading a sample of posts about job advertisements.

For each dataset, tweets retrieved by more than one query have been removed in order to preserve uniqueness of tweet labels. Table 1 illustrates the distribution of latent categories in each dataset. Each retrieved tweet was labeled according to a category corresponding to the query submitted. We leverage these labels to evaluate the topics produced by each model in term of clustering quality.

---

[3] https://dev.twitter.com/rest/public/search. Date of access: 19th Feb 2019.

4   Evaluation Metrics

According to metrics used in previous studies [4–6], we evaluate models both in terms of clustering quality (purity and normalized mutual information (NMI)) and semantic topic coherence (pointwise mutual information (PMI)).

Formally, let T_i be the set of tweets assigned to topic i and let $T = \{T_1, \ldots, T_{|T|}\}$ be the set of topic clusters arising from a LDA model that produces $|T|$ topics. Then let L_j be the set of tweets with ground-truth topic j and let $L = \{L_1, \ldots, L_{|L|}\}$ be the set of of ground-truth topic labels with $|L|$ labels in total. Our clustering-based metrics are defined as follows:

Purity: Purity score is used to measure the fraction of tweets in each assigned LDA topic cluster with the true label for that cluster, where the 'true' label is defined as the most frequent ground-truth label found in that cluster. Formally:

$$Purity(T,L) = \frac{1}{|T|} \sum_{i \in (1,|T|)} \max_{j \in (1,|L|)} |T_i \cap L_j|$$

Higher purity scores indicate better reconstruction of the original 'true' topic assignments by the model.

Normalized Mutual Information (NMI): The NMI score estimates how much information is shared between assigned topics T and the ground-truth labeling L. NMI is defined as follows:

$$NMI(T,L) = \frac{2I(T,L)}{H(T) + H(L)}$$

where respectively, $I(\cdot,\cdot)$ corresponds to mutual information and $H(\cdot)$ is entropy as defined in [8]. NMI is a number between 0 and 1. A score close to 1 means an exact matching of the clustering results.

Pointwise Mutual Information (PMI): The PMI score [10] evaluates the quality of inferred topics based on the top-10 words associated with each modeled topic. This measure is based on PMI which is computed as $PMI(u,v) = log\left(\frac{p(u,v)}{p(u)p(v)}\right)$ where u and v are a given pair of words. The probability $p(x)$ is derived empirically as the frequency of word x in the whole tweet corpus, while probability $p(x,y)$ is the likelihood of observing both x and y in the same tweet. Coherence of a topic k is computed as the average score of PMI for all possible pairs of the ten highest probability words for topic k (i.e. $W_k = \{w_1, \ldots, w_{10}\}$). Formally:

$$PMI - Score(T_k) = \frac{1}{100} \sum_{i=1}^{10} \sum_{j=1}^{10} PMI(w_i, w_j)$$

where $w_i, w_j \in W_k$. Then coherence of a whole topic model is calculated as the average PMI-Score for all topics generated by the model.

5   Results

For each combination of the three datasets (Sect. 3) and five pooling schemes (Sect. 2), we calculated three evaluation metrics (purity scores, NMI scores and PMI scores; Sect. 4) by training LDA models with 10 topics.

Table 2 presents various statistics of the training sets obtained by applying the different pooling schemes. We filtered the datasets to keep only tweets written in English and those with more than three tokens. Tweets were converted to lowercase and all URLs, mentions (except with the network pooling scheme) and stop-words were removed. After the tokenization process, all tokens based only on non-alphanumeric characters (emoticons) and all short tokens (with < 3 characters) were also deleted. Test sets have been randomly extracted (30%) from each dataset preserving the same distribution of tweet categories. For each topic model we conduct five cross-validations.

Table 2. Corpus statistics.

Scheme	No. of documents			No. of tokens		
	general	specific	event	general	specific	event
Unpooled	658492	445852	188000	18991	14794	9454
Author pooling	504253	340826	157377	18339	14091	9222
Conversation Pooling	649389	440682	185737	19301	15061	9668
Hashtag pooling	585171	387522	174501	19868	15185	9348
Network pooling	585171	402687	171266	19868	20065	13051

Table 3. Clustering metrics and coherence scores for different schemes and datasets.

Scheme	Purity			NMI			PMI score		
	general	specific	event	general	specific	event	general	specific	event
Unpooled	0.396	0.316	0.220	0.176	0.108	0.058	−0.131	0.224	0.307
Author pooling	0.377	0.399	0.326	**0.181**	0.176	0.124	0.892	−0.116	0.338
Conversation pooling	0.341	0.359	0.310	0.136	0.141	0.110	−0.131	0.062	−0.131
Hashtag pooling	0.337	0.250	0.245	0.145	0.045	0.071	0.293	0.347	**0.851**
Network Pooling	**0.418**	**0.503**	**0.362**	0.173	**0.228**	**0.155**	**0.912**	**0.582**	0.794

Table 3 summarises the average results obtained with each pooling scheme and dataset. According to the clustering evaluation metrics (purity and NMI), Network Pooling produced the best model performance on all datasets, with the exception of NMI scores on the General dataset, where it was narrowly outperformed by Unpooled and Author Pooling.

Results for other pooling schemes vary by metric and dataset. Author Pooling is the second-ranked scheme for most metrics/datasets, with Conversation Pooling also outperforming the Unpooled scheme in most cases. It is interesting to notify that Hashtag Pooling is mostly ineffective and gives performance worse than the baseline in most

cases. This finding can perhaps be explained by the observation that hashtags are typically present in a minority of tweets (e.g. 19.6% of tweets have hashtags in the Specific dataset). Concerning the measure of the topic interpretability, coherence scores show that the Network Pooling scheme gives better performance on all datasets, with the exception on the Event dataset, where it was narrowly outperformed by Hashtag Pooling.

6 Conclusion

Methods for aggregating tweets to form longer documents more amenable to topic modeling have been shown here and elsewhere to improve model performance. Here we have proposed a new network-based pooling scheme for topic modeling with Twitter data, that takes into account the network of users that engage with a particular tweet. Our approach improves topic extraction despite different levels of underlying thematic/topical heterogeneity of each dataset. While similar to conversation-based pooling in its use of reply tweets, the network approach includes otherwise un-linked content from users who authored replies. Experimental results showed that for the tests performed in this study, the network-based pooling scheme considerably outperformed other methods and was portable between datasets. Model outputs were improved on both clustering metrics (purity and NMI) and topic coherence (PMI).

Although the experiments presented have been conducted on the corpora collected on specific time intervals which reduces the shifting of conversation threads, especially when we collect documents authored by a cited user in response to the seed tweet. On a larger scale, topic shifting might be handled by adding conditions on document timestamps or topic correlation. In addition, the experimental findings suggest that network-based approaches might offer a useful technique for topic modeling with Twitter data, subject to further testing and validation with other datasets.

Acknowledgements. This work was supported by the Institute of Coding which received funding from the Office for Students (OfS) in the United Kingdom.

References

1. Blei, D.M., Ng, A.Y., Jordan, M.I.: Latent dirichlet allocation. J. Mach. Learn. Res. **3**, 993–1022 (2003)
2. Rosen-Zvi, M., Griffiths, T., Steyvers, M., Smyth, P.: The author-topic model for authors and documents. In: Proceedings of the 20th Conference on Uncertainty in Artificial Intelligence, pp. 487–494. AUAI Press (2004)
3. Hong, L., Davison, B.D.: Empirical study of topic modeling in twitter. In: Proceedings of the 1st Workshop on Social Media Analytics, pp. 80–88. ACM (2010)
4. Mehrotra, R., Sanner, S., Buntine, W., Xie, L.: Improving lDA topic models for microblogs via tweet pooling and automatic labeling. In: Proceedings of the 36th International ACM SIGIR Conference on Research and Development in Information Retrieval, pp. 889–892. ACM (2013)
5. Alvarez-Melis, D., Saveski, M.: Topic modeling in twitter: aggregating tweets by conversations. In: Proceedings of the 10th International AAAI Conference on Web and Social Media, pp. 519–522 (2016)

6. Hajjem, M., Latiri, C.: Combining IR and LDA topic modeling for filtering microblogs. Procedia Comput. Sci. **112**, 761–770 (2017)
7. Ahmad, W., Ali, R.: Information retrieval from social networks: a survey. In: Proceedings of the 3rd International Conference on Recent Advances in Information Technology (RAIT), pp. 631–635. IEEE (2016)
8. Manning, C., Raghavan, P., Schütze, H.: Introduction to information retrieval. Natural Lang. Eng. **16**(1), 100–103 (2010)
9. Zhao, W.X., Jiang, J., Weng, J., He, J., Lim, E.-P., Yan, H., Li, X.: Comparing twitter and traditional media using topic models. In: Clough, P., Foley, C., Gurrin, C., Jones, G.J.F., Kraaij, W., Lee, H., Mudoch, V. (eds.) ECIR 2011. LNCS, vol. 6611, pp. 338–349. Springer, Heidelberg (2011). https://doi.org/10.1007/978-3-642-20161-5_34
10. Lau, J.H., Newman, D., Baldwin, T.: Machine reading tea leaves: automatically evaluating topic coherence and topic model quality. In: Proceedings of the 14th Conference of the European Chapter of the Association for Computational Linguistics, pp. 530–539 (2014)

String Algorithms

Bounds and Estimates on the Average Edit Distance

Michele Schimd$^{(\boxtimes)}$ and Gianfranco Bilardi

Department of Information Engineering, University of Padova, Padua, Italy
{schimdmi,bilardi}@dei.unipd.it

Abstract. The edit distance is a metric of dissimilarity between strings, widely applied in computational biology, speech recognition, and machine learning. Let $e_k(n)$ denote the average edit distance between random, independent strings of n characters from an alphabet of a given size k. An open problem is the exact value of $\alpha_k(n) = e_k(n)/n$. While it is known that, for increasing n, $\alpha_k(n)$ approaches a limit α_k, the exact value of this limit is unknown, for any $k \geq 2$. This paper presents an upper bound to α_k based on the exact computation of some $\alpha_k(n)$ and a lower bound to α_k based on combinatorial arguments on edit scripts. Statistical estimates of $\alpha_k(n)$ are also obtained, with analysis of error and of confidence intervals. The techniques are applied to several alphabet sizes k. In particular, for a binary alphabet, the rigorous bounds are $0.1742 \leq \alpha_2 \leq 0.3693$ while the obtained estimate is $\alpha_2 \approx 0.2888$; for a quaternary alphabet, $0.3598 \leq \alpha_4 \leq 0.6318$ and $\alpha_4 \approx 0.5180$. These values are more accurate than those previously published.

Keywords: Edit distance · Average analysis · Upper and lower bounds · Statistical estimates

1 Introduction

Measuring distance between strings is a fundamental problem in computer science, with applications in computational biology, speech recognition, machine learning, and other fields. One commonly used metric is the *edit distance* (or *Levenshtein distance*), defined as the minimum number of substitutions, deletions, and insertions necessary to transform one string into the other.

It is natural to ask what is the average distance between two randomly generated strings, as the string size grows; knowledge of the asymptotic behavior has proved useful in computational biology (Ganguly *et al.* [10]) and in nearest neighbour search (Rubinstein [14]). In computational biology, for example, the edit distance can be used to the test the hypothesis that two subsequences

---

This work was partially supported by University of Padova projects CPDA152255/15 and CPGA3/13; by MIUR, the Italian Ministry of Education, University and Research, under Grant 20174LF3T8 AHeAD: efficient Algorithms for HArnessing networked Data; and by an IBM SUR Grant.

N. R. Brisaboa and S. J. Puglisi (Eds.): SPIRE 2019, LNCS 11811, pp. 91–106, 2019.
https://doi.org/10.1007/978-3-030-32686-9_7

originated from the same portion of DNA. Even for the case of uniform and independent strings, the study of average edit distance appears to be challenging and little work has been reported on the problem. In contrast, the closely related problem of finding the average length of the *longest common subsequence* has been extensively studied, since the seminal work by Chvátal and Sankoff [8].

Using Fekete's lemma, it can be shown that both metrics tend to grow linearly with n (Steele [18]). Let $e_k(n)$ denote the average edit distance between two random, independent strings of length n on a k-ary alphabet; then $\alpha_k(n) = e_k(n)/n$ approaches a limit $\alpha_k \in [0,1]$. Similarly, let $\ell_k(n)$ denote the length of the longest common subsequence; then $\gamma_k(n) = \ell_k(n)/n$ approaches a limit $\gamma_k \in [0,1]$. The γ_k's are known as the Chvátal-Sankoff constants. The determination of the exact values of α_k and γ_k is an open problem. This paper addresses the problem of estimating and bounding α_k, for various alphabet sizes k.

Related Work. There is limited literature directly pursuing bounds and estimates for α_k. It is also interesting to review results on γ_k: on the one hand, bounds to γ_k give bounds to α_k; on the other hand, techniques for analyzing γ_k can be adapted for analyzing α_k.

The only published estimates of α_k can be found in [10] which gives $\alpha_4 \approx 0.518$ for the quaternary alphabet and $\alpha_2 \approx 0.29$ for the binary alphabet. For γ_k, the best available estimates are given by Bundschuh [5], in particular $\gamma_2 \approx 0.8126$ and $\gamma_4 \approx 0.6544$. Estimates of γ_k by sampling are given by Ning and Choi [13], their results, however, appear in contrast with estimates of [5]. In particular they conjectured that $\gamma_2 > 0.82$ contradicting the estimate of γ_2 given in [5].

The best published analytical lower bounds to α_k are $\alpha_4 \geq 0.3383$ for quaternary alphabet and $\alpha_2 \geq 0.1578$ for binary alphabet [10]. To the best of our knowledge, no systematic study of upper bounds to α_k have been published. The best known analytical lower and upper bounds to γ_2 are given by Lueker [11], who obtained $0.7881 \leq \gamma_2 \leq 0.8263$. For larger alphabets, the best results appear in Dancík [9], including $0.5455 \leq \gamma_4 \leq 0.7082$. From known relations between edit distance and the length of the longest common subsequence, it follows that $1 - \gamma_k \leq \alpha_k \leq 2(1 - \gamma_k)$. Thus, upper and lower bounds to α_k can be respectively obtained from lower and upper bounds to γ_k. From $\gamma_2 \leq 0.8263$ [11], we obtain $\alpha_2 \geq 0.1737$, which is tighter than the bound given in [10]. Instead $\gamma_4 \leq 0.7082$ [9] yields $\alpha_4 \geq 0.2918$, which is weaker than the bound $\alpha_4 \geq 0.3383$ [10]. From the weaker relation $(1 - \gamma_4)/2 \leq \alpha_4$, [14] obtained the looser bound $\alpha_2 \geq 0.0869$. Improved bounds, for both α_2 and α_4, are derived in this paper. Some of our techniques resemble those used in Baeza-Yates *et al.* [4] for estimating γ_k.

1.1   Paper Contributions

The contributions of this paper include:

– statistical estimates of $\alpha_k(n)$ with error analysis,
– upper bounds to α_k by exhaustive computation of $\alpha_k(n)$ for small n, and
– lower bounds to α_k through analytical counting arguments.

Our numerical results for $k = 2$ and $k = 4$ are compared with previously known values in Table 1. The methods used to derive such values are presented throughout the paper, which is organized as follows. Section 2 introduces notation and definitions. Section 3 presents statistical estimates, with error analysis. Section 4 describes an algorithm for computing upper bounds. Section 5 develops lower bounds analysis based on counting edit scripts. Section 6 shows and discusses numerical results on bounds and estimates of α_k. Finally, Sect. 7 gives conclusions and future directions of investigation.

Table 1. Our results on α_2 and α_4 compared with previously published ones.

	Lower bound		Estimate		Upper bound	
	Previous	This work	Previous	This work	Previous	This work
α_2	0.1737 [11]	0.1742	0.290 [10]	0.2888	-	0.3693
α_4	0.3383 [10]	0.3598	0.518 [10]	0.5180	-	0.6318

2 Preliminaries

2.1 Notation and Definitions

Let Σ_k be a finite alphabet of size $k \geq 2$ and let $n \geq 1$ be an integer; a *string* x is a sequence of symbols $x[1]x[2] \ldots x[n]$ where $x[i] \in \Sigma_k$; n is called the *length* of x, also denoted by $|x|$. Σ_k^n is the set of all strings of length n.

Edit Distance. We consider the following *edit operations* on a string x: the *match* of $x[i]$, the *substitution* of $x[i]$ with a different symbol $b \in \Sigma_k \setminus \{x[i]\}$, the *deletion* of $x[i]$, and the *insertion* of $b \in \Sigma_k$ in position $j = 0, \ldots, n$ (insertion in j means b goes after $x[j]$ or at the beginning if $j = 0$); an *edit script* is an ordered sequence of edit operations. To each type of edit operation is associated a cost; throughout this paper, matches have cost 0 and other operations have cost 1. The cost of a script is the sum of the costs of its operations. The *edit distance* between x and y, $d_E(x, y)$, is the minimum cost of any script transforming x into y. It is easy to see that $||x| - |y|| \leq d_E(x, y) \leq \max(|x|, |y|)$.

Random Strings and the Limit Constant. A *random string* $X = X[1]X[2] \ldots X[n]$ is a sequence of *random symbols* $X[i]$ each generated according to some distribution. We will assume $X[i]$ uniformly and independently sampled from Σ_k, hence $\mathrm{P}[X = x] = k^{-n}$ for every $x \in \Sigma_k^n$. For a string x we define the *eccentricity*, $\mathrm{ecc}(x)$, as its expected distance from a random string $Y \in \Sigma_k^n$:

$$\mathrm{ecc}(x) = k^{-n} \sum_{y \in \Sigma_k^n} d_E(x, y). \tag{1}$$

The expected edit distance between two random, independent strings of Σ_k^n is:

$$e_k(n) = k^{-2n} \sum_{x \in \Sigma_k^n} \sum_{y \in \Sigma_k^n} d_E(x, y)$$

$$= k^{-n} \sum_{x \in \Sigma_k^n} \mathrm{ecc}(x). \tag{2}$$

Let $\alpha_k(n) = e_k(n)/n$; it can be shown (Fekete's lemma from ergodic theory; see, for example, Lemma 1.2.1 in [18]) that there exists a real number $\alpha_k \in [0, 1]$, depending only on k, such that

$$\lim_{n \to \infty} \alpha_k(n) = \alpha_k.$$

The main objective of this paper is to derive estimates and bounds to α_k.

2.2   Computing the Edit Distance

Edit distance and length of the longest common subsequence (LCS) can be computed using a *dynamic programming* algorithm. For the edit distance, given two strings x and y with length n and m respectively, the algorithm fills an $(n + 1) \times (m + 1)$ matrix $\mathbf{M}$. The values of $\mathbf{M}$ are computed according to the following recurrence:

$$
\begin{aligned}
M_{i,0} &= i & \text{for } i = 0, \ldots, n \\
M_{0,j} &= j & \text{for } j = 0, \ldots, m \quad (3) \\
M_{i,j} &= \min\left\{ M_{i-1,j-1} + \xi_{i,j}; M_{i-1,j} + 1; M_{i,j-1} + 1 \right\} & \text{otherwise}
\end{aligned}
$$

where $\xi_{i,j} = 0$ if $x[i] = y[j]$ and $\xi_{i,j} = 1$ otherwise.[1] The edit distance between x and y is the value computed in the entry $M_{n,m}$. This algorithm takes $\mathcal{O}(nm)$ time and space. From the matrix $\mathbf{M}$, an edit script realizing the transformation of x into y can be obtained using *backtracking* which produces a *path* from cell (n, m) to cell $(0, 0)$ of $\mathbf{M}$. For both problems, the approach by Masek and Paterson [12] (using the method of the Four Russians) reduces the time to $\mathcal{O}(\frac{n^2}{\log n})$ (assuming $n \geq m$). It is known that both the edit distance and the length of the LCS cannot be computed in time $\mathcal{O}(n^{2-\epsilon})$, unless the *Strong Exponential Time Hypothesis (SETH)* is false (Abboud *et al.* [1], Backurs and Indyk [3]). For the edit distance, a $(\log n)^{\mathcal{O}(1/\epsilon)}$ approximation, computable in time $\mathcal{O}(n^{1+\epsilon})$, is given by Andoni *et al.* [2], while a constant approximation algorithm with running time $\mathcal{O}(n^{1+5/7})$ is given by Chakraborty *et al.* [7]. A very recent work by Rubinstein and Song [15], gives a reduction from approximate length of the longest common subsequence to approximate edit distance, proving that the algorithm in [7] can also be used to approximate the length of the LCS.

---

[1] A similar algorithm computes the length of the LCS. The recurrence (3) becomes $M_{i,0} = 0$, $M_{0,j} = 0$, and $M_{i,j} = \max\left\{ M_{i-1,j-1} + (1 - \xi_{i,j}); M_{i-1,j}; M_{i,j-1} \right\}$.

In order to compute upper bounds to α_k, we propose an algorithm related to the approaches developed by Calvo-Zaragoza et al. [6] and [11]. In these works, portions of the dynamic programming matrix are associated to states of a finite state machine. Our algorithm conceptually simulates all possible execution of a machine similar to the one defined in [6].

3   Statistical Estimates of α_k

In this section, we discuss how to develop statistical estimates of $\alpha_k(n)$, by sampling. In Table 2, we show results for $k = 4$, a case of special interest in DNA analysis [10]. Although $\alpha_k < \alpha_k(n)$ for every finite n, when n is sufficiently large, estimates of $\alpha_k(n)$ provide approximations to α_k.

3.1   Estimates of $\alpha_k(n)$ and Confidence Interval

Let $(x_1, y_1), \ldots, (x_N, y_N)$ be N random and independent pairs of strings from Σ_k^n. The sample mean

$$\tilde{e}_k(n) = \frac{1}{N} \sum_{i=1}^{N} d_E(x_i, y_i) \tag{4}$$

provides an estimate of $e_k(n)$. We determine confidence intervals of such an estimate, using the sample variance

$$S_k^2(n) = \frac{1}{N-1} \sum_{i=1}^{N} (d_E(x_i, y_i) - \tilde{e}_k(n))^2. \tag{5}$$

Table 2 presents values of $\tilde{e}_4(n)$ and $S_4^2(n)$ for $N = 5000$ and $n = 2^4, 2^5, \ldots, 2^{14}$. Our analysis on confidence intervals is based on the work by Saw et al. [16], which extends Chebyshev's inequality to the cases where both mean and variance are not known, but rather estimated with (4) and (5), respectively. For this analysis to apply, it is sufficient for a random variable to have finite first and second order moments, a condition certainly satisfied by the edit distance of a random pair of strings of a given length.

Proposition 1 (Eq. (2.2) in [16]). *Let $\tilde{e}_k(n)$ and $S_k^2(n)$ be given by (4) and (5), respectively. For any $t \geq 1$,*

$$P[|e_k(n) - \tilde{e}_k(n)| \leq \sqrt{(N+1)/N} t S_k(n)] \geq 1 - \left(\frac{N-1}{N} \frac{1}{t^2} + \frac{1}{N} \right). \tag{6}$$

From (6) we get the confidence interval on $\alpha_k(n)$

$$P[\alpha_k(n) \in [\tilde{e}_k(n)/n \pm \sqrt{(N+1)/N} t S_k(n)]] \geq 1 - \left(\frac{N-1}{N} \frac{1}{t^2} + \frac{1}{N} \right).$$

Table 2. Results of statistical estimates of $\alpha_4(n)$ for $n = 2^4, 2^5, \ldots, 2^{14}$ obtained from $N = 5000$ samples. The table shows: n, sample mean $\tilde{e}_4(n)$, sample variance $S_4^2(n)$, estimate $\tilde{\alpha}_4(n)$, and the value $S_4(n)/n$ used to the compute confidence intervals.

n	$\tilde{e}_4(n)$	$S_4^2(n)$	$\tilde{\alpha}_4(n)$	$S_4(n)/n$
16	10.0164	2.0814	0.6260	0.0902
32	18.9460	3.3306	0.5920	0.0570
64	36.4370	5.4487	0.5693	0.0365
128	70.5634	9.1274	0.5513	0.0236
256	138.0370	14.4977	0.5392	0.0149
512	272.1636	24.3117	0.5315	0.0096
1024	538.7120	39.6606	0.5261	0.0062
2048	1070.2178	65.2186	0.5226	0.0039
4096	2131.4744	111.4540	0.5204	0.0026
8192	4251.1936	178.6490	0.5189	0.0016
16384	8486.4712	323.7883	0.5180	0.0011

The values $S_4(n)/n$ are shown in Table 2 for each n. For example with $n = 2^{14}$, $N = 5000$, and $t = 5$, we get

$$P[\alpha_4(2^{14}) \in [0.518 \pm 0.0055]] \geq 1 - \left(\frac{4999}{5000} \frac{1}{5^2} + \frac{1}{5000} \right) = 0.9598.$$

Since $\alpha_k < \alpha_k(n)$, we conclude that

$$P[\alpha_4 < 0.5235] \geq 0.9598. \tag{7}$$

3.2   The $(2w + 1)$-bandwidth Algorithm for Approximate Distance

Although better approximations of α_k can, in principle, be obtained using larger values of n, the quadratic complexity of the dynamic programming algorithm limits the values of n we can practically test. To partially circumvent this obstacle, for $n > 2^{14}$, we have estimated $\alpha_k(n)$ resorting to an algorithm that computes an approximation (from above) to $d_E(x, y)$. The algorithm is parameterized by an integer $w \geq 0$ called *bandwidth*. It computes the portion of a matrix $\mathbf{Q}$ corresponding to the $2w + 1$ central diagonals. The algorithm performs the following steps.

1. For $h = 0, \ldots, w + 1$, set $Q_{h,0} = h$ and $Q_{0,h} = h$.
2. For $h = w + 2, \ldots, n$, set $Q_{h-w-1,h}$ and $Q_{h,h-w-1}$ to h.
3. For $j = 1, \ldots, n$ and $i = \max(1, j - w), \ldots, \min(j + w, n)$, apply (3) to $Q_{i,j}$.

We observe that the values of the entries of $\mathbf{Q}$ on the boundary of the region where such matrix is computed (steps 1 and 2) are set to upper bounds to the

corresponding entries of the matrix $\mathbf{M}$, reviewed in Sect. 2.2. Since the function that updates an entry in terms of its neighbours (step 3) is monotone non-decreasing, we have that $Q_{i,j} \geq M_{i,j}$, whence the output $Q_{n,n}$ of the bandwidth algorithm computes an upper bound to $d_E(x, y)$. It can be shown that $Q_{n,n} = d_E(x, y)$ whenever there exists an optimal path in $\mathbf{M}$ confined to the $2w + 1$ central diagonals, a condition that is always met if $d_E(x, y) \leq w$.

We carried out simulations setting $w = \sqrt{n}$, so that the bandwidth algorithm runs in time is $\mathcal{O}(n^{3/2})$. Application of Eq. (6) to $n = 2^{20}$ with $t = 5$, gives for $\hat{e}_4(2^{20})/2^{20}$, the confidence interval $[0.5162, 0.5174]$ with probability at least 0.9598. Since $\hat{e}_4(n)/n$ is an estimate of an upper bound to $e_4(n)/n$, we obtain

$$P[\alpha_4 < 0.5174] \geq 0.9598. \tag{8}$$

We observe that bound (8) improves on bound (7), obtained via the exact distance algorithm. This indicates that the loss of precision due to the use of an approximate algorithm is more than compensated by the ability to process larger string sizes.

4   Upper Bounds for α_k

This section presents methods to derive upper bounds to α_k based on the exact computation of $\alpha_k(n) = e_k(n)/n$ for some n, and on the relation $\alpha_k \leq \alpha_k(n)$, valid for all $n \geq 1$. The computation of $e_k(n)$ can be reduced to that of the eccentricity, as in Eq. (2) repeated here for convenience:

$$e_k(n) = k^{-n} \sum_{x \in \Sigma_k^n} \mathrm{ecc}(x). \tag{9}$$

If $\mathrm{ecc}(x)$ is computed according to Eq. (2) and the distance $d_E(x, y)$ is computed by the $\mathcal{O}(n^2)$-time dynamic programming algorithm for each of the k^n strings $y \in \Sigma_k^n$, then the overall computation time is $\mathcal{O}(n^2 k^n)$ for $\mathrm{ecc}(x)$ and $\mathcal{O}(n^2 k^{2n})$ for $e_k(n)$, since the eccentricity of each of the k^n strings $x \in \Sigma_k^n$ is needed in Eq. (9). Below, we propose a more efficient algorithm to speed up the computation of $\mathrm{ecc}(x)$ and, in turn, that of $e_k(n)$, achieving time $\mathcal{O}(n^2 \min(k, 3)^n k^n) = \mathcal{O}(n^2 3^n k^n)$. We also show how to exploit some symmetries of $\mathrm{ecc}(x)$ in order to limit its evaluation to a suitable subset of Σ_k^n.

4.1   The Coalesced Dynamic Programming Algorithm for Eccentricity

Let $\mathbf{M}(x, y)$ be the matrix produced by the dynamic programming algorithm (reviewed in Sect. 2.2) to compute $d_E(x, y)$, with $x, y \in \Sigma_k^n$. We develop a strategy to coalesce the computations of $\mathbf{M}(x, y)$ for different $y \in \Sigma_k^n$, while keeping x fixed. To this end, we choose to generate the entries of $\mathbf{M}(x, y)$, according to Eq. (3), in column-major order. Clearly, the j-th column is fully determined by x and by the prefix of y of length j. Define now the *column multiset* $\mathcal{C}_j$

Algorithm 1. Coalesced dynamic programming algorithm to compute $ecc(x)$

1: **procedure** ECCENTRICITY(x)
2:     $n \leftarrow |x|$
3:     $\mathcal{C}_0 \leftarrow \{((0, 1, \ldots, n), 1)\}$
4:     **for** $j \leftarrow 1$ **to** n **do**
5:         $\mathcal{C}_j \leftarrow \emptyset$
6:         **for** $\mathbf{c} \in \mathcal{C}_{j-1}$ **do**
7:             **for** $b \in \Sigma_k$ **do**
8:                 $\mathbf{c}' \leftarrow$ NEXTCOLUMN$(x, \mathbf{c}, j, b)$
9:                 INSERT$(\mathcal{C}_j, (\mathbf{c}', \mu(\mathbf{c})))$
10:            **end for**
11:        **end for**
12:    **end for**
13:    $e \leftarrow 0$
14:    **for** $\mathbf{c} \in \mathcal{C}_n$ **do**
15:        $e \leftarrow e + \mu(\mathbf{c}) * \mathbf{c}[n]$
16:    **end for**
17:    **return** e/k^n
18: **end procedure**

---

containing the j-th (i.e., the last) column of $\mathbf{M}(x, y[1] \ldots y[j])$ for each string $y[1] \ldots y[j] \in \Sigma_k^j$. Multiset $\mathcal{C}_j$ is a function of (just) x, although, for simplicity, the dependence upon x is not reflected in our notation.

The *Coalesced Dynamic Programming* (CDP) algorithm described below (referring also to the line numbers of Algorithm 1), constructs the sequence of multisets $\mathcal{C}_0, \mathcal{C}_1, \ldots, \mathcal{C}_n$. A column multiset $\mathcal{C}$ will be represented as a set of pairs $(\mathbf{c}, \mu(\mathbf{c}))$, one for each distinct member $\mathbf{c}$, with $\mu(\mathbf{c})$ being the multiplicity of $\mathbf{c}$ in $\mathcal{C}$. The eccentricity of x is obtained (lines 13–17) as the weighted average of the n-th element of all columns in $\mathcal{C}_n$:

$$ecc(x) = k^{-n} \sum_{\mathbf{c} \in \mathcal{C}_n} \mu(\mathbf{c})\mathbf{c}[n]. \tag{10}$$

As can be seen from Eq. (3), multiset $\mathcal{C}_0$ contains just column $(0, 1, \ldots, n)$, with multiplicity 1 (line 3). For $j = 1, \ldots, n$, $\mathcal{C}_j$ is obtained by scanning all $\mathbf{c} \in \mathcal{C}_{j-1}$ (line 6) and all $b \in \Sigma_k$ (line 7), and by

- computing the j-th column $\mathbf{c}'$ resulting from Eq. (3) when the $(j-1)$-st column is $\mathbf{c}$ and $\xi_{i,j} = 0$ if $x[i] = b$ or else $\xi_{i,j} = 1$ (call to NEXTCOL-UMN$(x, \mathbf{c}, j, b)$, line 8);
- inserting $\mu(\mathbf{c})$ copies of $\mathbf{c}'$ in $\mathcal{C}_j$, by either creating a new pair $(\mathbf{c}', \mu(\mathbf{c}))$ when $\mathbf{c}'$ is not present in the multiset or by incrementing its multiplicity by $\mu(\mathbf{c})$ otherwise (call to INSERT$(\mathcal{C}_j, (\mathbf{c}', \mu(\mathbf{c})))$, line 9).

The correctness of the CDP algorithm is pretty straightforward to establish. A few observations are however necessary in order to describe and analyze a data structure that can efficiently implement, in our specific context, multisets with

the insertion operation. The key property is that, for $j = 0, 1 \ldots, n$, the column of $\mathbf{M}(x, y)$ with index j satisfies the conditions (a) $M_{0,j} = j$ and (b) $(M_{i,j} - M_{i-1,j}) \in \{-1, 0, 1\}$, for $i = 1, \ldots, n$. Using this property, the set of distinct columns that belong to the multiset $\mathcal{C}_j$ can be represented as a ternary tree where each arc has a label from the set $\{-1, 0, 1\}$ and a column $(M_{0,j}, M_{1,j}, \ldots, M_{n,j})$ is mapped to a leaf v such that the n arcs in the path from the root to v have labels $(M_{1,j} - M_{0,j}), \ldots, (M_{n,j} - M_{n-1,j})$. Each leaf stores the multiplicity of the corresponding column. The size of the tree for $\mathcal{C}_j$ is $\mathcal{O}(\min(3^n, k^j))$, since there are at most 3^n columns satisfying the constrains and k^j k-ary strings that contribute (not necessarily distinct) columns. Hence, the body of the loop whose iteration space is defined in lines 4, 6, and 7 is executed $nk\mathcal{O}(\min(3^n, k^j))$ times. Considering that one call to NextColumn() as well as one call to Insert() can be easily performed in $\mathcal{O}(n)$ time, we can summarize the previous discussion as follows, where we also consider that, at any given time, the algorithm only needs to store two consecutive column multisets.

Proposition 2. *The* CPD *algorithm computes the eccentricity* ecc(x) *of a string x of length n over a k-ary alphabet in time $T = \mathcal{O}(n^2 k \min(3^n, k^n))$ and space $S = \mathcal{O}(\min(3^n, k^n))$. Correspondingly, the average distance $e_k(n)$ can be computed in time $T = \mathcal{O}(n^2 k^{n+1} \min(3^n, k^n))$ and space $S = \mathcal{O}(\min(3^n, k^n))$.*

4.2 Exploiting Symmetries of ecc(x) in the Computation of $e_k(n)$

The edit distance enjoys some useful symmetries, which can be easily derived from the definition. One is that, if we let $x^R = x[n] \ldots x[1]$ denote the *reverse* of string $x = x[1] \ldots x[n]$, then $d_E(x, y) = d_E(x^R, y^R)$. Another one is that if $\pi : \Sigma_k \to \Sigma_k$ is a permutation of the alphabet and $\pi(x)$ denotes the string $\pi(x[1]) \ldots \pi(x[n])$, then $d_E(x, y) = d_E(\pi(x), \pi(y))$. The following is a simple, but useful corollary of these properties.

Proposition 3. *For any $x \in \Sigma_k^n$, we have* ecc(x^R) = ecc(x). *Furthermore, for any permutation π of Σ_k, we have* ecc($\pi(x)$) = ecc(x).

It is useful to define the equivalence class of x as the set of strings that have the same eccentricity as x, due to Proposition 3, and denote by $\nu(x)$ the cardinality of such set. If $\mathcal{R}_{k,n} \subseteq \Sigma_k^n$ contains exactly one (representative) member for each equivalence class, then Eq. (9) can be rewritten as

$$e_k(n) = k^{-n} \sum_{x \in \mathcal{R}_{k,n}} \nu(x)\text{ecc}(x). \tag{11}$$

Computing $e_k(n)$ according to Eq. (11) enables one to reduce the number of strings for which the eccentricity has to be computed (via the CDP algorithm) by a factor slightly smaller than $(2k!)$, with a practically appreciable reduction in computation time.

The strategy outlined in this section has been implemented in C++ and run on a 32 core IBM Power7 server. For several alphabet sizes k, we have considered

values of n up to a maximum value n_k^{ub}, under the constraint that the running time would not exceed one week. The resulting values $e_k(n_k^{ub})$ are presented and discussed in Sect. 6.

5   Lower Bounds for α_k

In this section, we prove the theoretical results that are used to obtain the lower bounds α_k^{lb} shown in Sect. 6. To obtain such lower bounds, we will derive lower bounds to $\mathrm{ecc}(x)$ by ignoring the contribution of the strings inside the ball of radius r centered at x and by setting to $r + 1$ the contribution of the string outside the same ball. The objective is to determine the largest value r^* of r for which (it can be shown that) the ball of radius r contains a fraction of Σ_k^n that vanishes with n; then r^* will effectively represent a lower bound to $\alpha_k n$. Below, we formalize this idea and show that we can choose $r^* = \beta n$ for suitable values of β independent on n; this establishes that $\alpha_k \geq \beta$.

5.1   Lower Bounds to ecc(x) Using Upper Bounds to Ball Size

In this subsection, we show how to derive lower bounds to $\mathrm{ecc}(x)$ starting from upper bounds to the size of the ball of radius r centered at x. We also show that, when such bounds are valid for every x, they can be used to compute lower bounds to α_k.

Definition 1. *For a string* $x \in \Sigma_k^n$*, the* ball *of radius* r *centered at* x *is defined as the set of strings having distance at most* r *from* x*:*

$$B_r(x) = \{y \in \Sigma_k^n : d_E(x, y) \leq r\}.$$

Similarly, the shell *of radius* r *centered at* x *is defined as the set of strings having distance exactly* r *from* x*:*

$$S_r(x) = \{y \in \Sigma_k^n : d_E(x, y) = r\}.$$

For a given r, each string in $\Sigma_k^n \setminus B_r(x)$ has distance at least $r + 1$ from x; therefore its contribution to $\mathrm{ecc}(x)$ is at least $r + 1$. Given an upper bound $u_r(x) \geq |B_r(x)|$, the consequent lower bound $|\Sigma_k^n \setminus B_r(x)| \geq (k^n - u_r(x))$ yields the following lower bound to $\mathrm{ecc}(x)$.

Lemma 1. *Let* $u_r(x) \geq |B_r(x)|$*, then for every* $r^* = 0, 1, \ldots, n$*:*

$$\mathrm{ecc}(x) \geq r^* \left(1 - k^{-n} u_{r^*}(x)\right) \tag{12}$$

Proof. We can rewrite (1) as

$$\mathrm{ecc}(x) = k^{-n} \sum_{r=0}^{r^*} r|S_r(x)| + k^{-n} \sum_{r=r^*+1}^{n} r|S_r(x)|$$

$$\geq k^{-n}(r^* + 1) \sum_{r=r^*+1}^{n} |S_r(x)|$$

$$= k^{-n}(r^* + 1) \left(|B_n(x)| - |B_{r^*}(x)|\right)$$

$$> r^* \left(1 - k^{-n}|B_{r^*}(x)|\right)$$

$$\geq r^* \left(1 - k^{-n}u_{r^*}(x)\right).$$

$\square$

In particular when $u_{r^*} \geq |B_{r^*}(x)|$ for every $x \in \Sigma_k^n$, simple manipulations of Eq. (2), recalling that $\alpha_k = \lim_{n \to \infty} \frac{e_k(n)}{n}$, yield

$$\alpha_k \geq \lim_{n \to \infty} \frac{r^*}{n} \left(1 - k^{-n}u_{r^*}\right). \tag{13}$$

The bounds presented next are based on (13). We will show that, for suitable values of β, the quantity $(k^{-n}u_{\beta n})$ converges to 0, so that $\alpha_k \geq \beta$.

5.2   Upper Bounds on Ball Size

To use Lemma 1 we need an upper bound to $|B_r(x)|$. The next proposition develops such an upper bound by (i) associating every string in $B_r(x)$ to a script of certain type with cost r or $r-1$ and (ii) counting such scripts.

Proposition 4. *For any $x \in \Sigma_k^n$ and for any $r = 1, \ldots, n$*

$$|B_r(x)| \leq (k-1)^r \sum_{d=0}^{\lfloor r/2 \rfloor} \binom{n}{d}^2 \binom{n-d+1}{r-2d} \left(\frac{k}{(k-1)^2}\right)^d. \tag{14}$$

Proof. We introduce the notion of *simple script* of cost $r \in \{0, 1, \ldots, n\}$, constructed by the following sequence of choices (shown within square brackets is the number of possible choices):

- $d \in \{0, 1, \ldots, \lfloor r/2 \rfloor\}$
- d positions to delete from x $[\binom{n}{d}]$
- $(r - 2d)$ of the remaining $(n - d)$ positions to be substituted $[\binom{n-d}{r-2d}]$
- d positions to insert in y $[\binom{n}{d}]$
- the symbols in the substitutions $[(k-1)^{r-2d}]$
- the symbols in the insertions $[k^d]$

Straightforwardly, the number of simple scripts of cost r is

$$s_r = \sum_{d=0}^{\lfloor r/2 \rfloor} \binom{n}{d}^2 \binom{n-d}{r-2d} (k-1)^{r-2d} k^d. \tag{15}$$

It is easy to see that optimal scripts are simple.

Next, we prove that any $y \in B_r(x)$ can be obtained from x via a simple script of cost $r-1$ or r. Let $r' = d_E(x, y) \leq r$, we will focus to the case where $r' < r - 1$, since in the complementary case the argument is trivial. Consider an optimal, simple script of cost r' that transforms x into y. By augmenting this script with $\lfloor (r - r')/2 \rfloor$ pairs of deletions and insertions, each pair acting on a matched position, we obtain a simple script of cost r, if $r - r'$ is even, or of cost $r - 1$ if $r - r'$ is odd. The prescribed augmentation is always possible since the number of matches is at least $n - r' \geq r - r' \geq (r - r')/2$.

The thesis is then established by the following sequence of inequalities

$$|B_r(x)| \leq s_r + s_{r-1}$$

$$\leq \sum_{d=0}^{\lfloor r/2 \rfloor} \binom{n}{d}^2 \binom{n-d}{r-2d} (k-1)^{r-2d} k^d + \sum_{d=0}^{\lfloor (r-1)/2 \rfloor} \binom{n}{d}^2 \binom{n-d}{r-1-2d} (k-1)^{r-1-2d} k^d$$

$$\leq \sum_{d=0}^{\lfloor r/2 \rfloor} \binom{n}{d}^2 \binom{n-d}{r-2d} (k-1)^{r-2d} k^d + \sum_{d=0}^{\lfloor r/2 \rfloor} \binom{n}{d}^2 \binom{n-d}{r-1-2d} (k-1)^{r-2d} k^d$$

$$\leq \sum_{d=0}^{\lfloor r/2 \rfloor} \binom{n}{d}^2 \binom{n-d+1}{r-2d} (k-1)^{r-2d} k^d,$$

where we have made use of the identity

$$\binom{n-d}{r-2d} + \binom{n-d}{r-1-2d} = \binom{n-d+1}{r-2d}.$$

$\square$

5.3   Asymptotic Behavior of Ball Size and Bounds for α_k

The next results show that (14), divided by k^n, is bounded by an exponential function where we can choose the exponent in such a way that this function vanishes with n. This can then be used in (13) to obtain lower bounds to α_k.

Definition 2. *Let $H(\beta)$ denote the* binary entropy function

$$H(\beta) = -\beta \log_2 \beta - (1-\beta) \log_2 (1-\beta).$$

Definition 3. *For $\beta \in [0, 1]$ and $\delta \in [0, \beta/2]$ we define the function*

$$g_k(\beta, \delta) = (\beta - 2\delta) \log_2 (k-1) - (1-\delta) \log_2 k$$

$$+ 2H(\delta) + (1-\delta) H\left(\frac{\beta - 2\delta}{1-\delta} \right). \tag{16}$$

Lemma 2. *Let u_r be given by the right hand side of (14) and $g_k(\beta,\delta)$ given by (16). For every $\beta \in [0,1]$*

$$k^{-n}u_{\beta n} \leq (n+1) \sum_{d=0}^{\lfloor \beta n/2 \rfloor} 2^{ng_k\left(\beta,\frac{d}{n}\right)}. \tag{17}$$

Proof. Using the relation

$$\binom{n-d+1}{r-2d} = \frac{n-d+1}{n-r+d+1}\binom{n-d}{r-2d} \leq (n+1)\binom{n-d}{r-2d},$$

the bound $\binom{n}{k} \leq 2^{nH(k/n)}$ (see, e.g., Eq. (5.31) in Spencer [17]), and defining $\beta = r/n$ we get

$$k^{-n}u_r \leq k^{-n}(k-1)^r \sum_{d=0}^{\lfloor r/2 \rfloor} \binom{n}{d}^2 \binom{n-d+1}{r-2d} \left(\frac{k}{(k-1)^2}\right)^d$$

$$\leq (n+1) \sum_{d=0}^{\lfloor r/2 \rfloor} 2^{2nH\left(\frac{d}{n}\right)+(n-d)H\left(\frac{r-2d}{n-d}\right)+(r-2d)\log_2(k-1)^2+(d-n)\log_2 k}$$

$$= (n+1) \sum_{d=0}^{\lfloor \beta n/2 \rfloor} 2^{ng_k\left(\beta,\frac{d}{n}\right)}.$$

$\square$

Theorem 1. *Letting $A_k = \{\beta \in [0,1] : \forall \delta \in [0,\beta/2]\ (g_k(\beta,\delta) < 0)\}$, we have:*

$$\alpha_k \geq \sup A_k.$$

Proof. We begin by observing that A_k is not empty, since $0 \in A_k$. In fact, when $\beta = 0$, the condition $\delta \in [0,\beta/2]$ is satisfied only by $\delta = 0$, and $g_k(0,0) = -\log_2 k < 0$. Since, by definition, $A_k \subseteq [0,1]$, we conclude that $\sup A_k$ is finite. Next, we define the function

$$G_k(\beta) = \max_{0 \leq \delta \leq \beta/2} g_k(\beta,\delta).$$

This definition of $G_k(\beta)$ is well posed, since $g_k(\beta,\delta)$ is a continuous function, hence it does have a maximum in the compact set $0 \leq \delta \leq \beta/2$. Furthermore, it follows from the definitions of $G_k(\beta)$ and A_k that $G_k(\beta) < 0$, for any $\beta \in A_k$.

For any $\beta \in A_k$, we see from Lemma 2 that

$$k^{-n}u_{\beta n} \leq (n+1) \sum_{d=0}^{\lfloor \beta n/2 \rfloor} 2^{ng_k\left(\beta,\frac{d}{n}\right)} \leq f(n)2^{nG_k(\beta)},$$

where we have used the relation $g_k\left(\beta,\frac{d}{n}\right) \leq G_k(\beta)$ (which follows from the definition of $G_k(\beta)$ and the fact that for, any d in the summation range,

$0 \leq \frac{d}{n} \leq \beta/2$) and we have let $f(n) = (n+1)\left(\left\lfloor \frac{\beta n}{2} \right\rfloor + 1\right)$. Taking now the limit in (13) with $r^* = \beta n$ yields:

$$\alpha_k \geq \lim_{n \to \infty} \beta \left(1 - f(n)2^{nG_k(\beta)}\right) = \beta,$$

as $f(n) = \mathcal{O}(n^2)$ and $2^{nG_k(\beta)}$ is a negative exponential. In conclusion, since α_k is no smaller than any member of A_k, it is also no smaller than $\sup A_k$. $\square$

Theorem 1 gives a criterion to find lower bounds to α_k: choose β such that $g(\beta, \delta) < 0$ for every δ. The lower bounds presented in Sect. 6 are computed using a numerical evaluation of $\sup A_k$.

6   Numerical Results and Discussion

In this section, we present and discuss some numerical results obtained by applying the methodologies developed in previous sections, to alphabets of various size k. These values are reported in the Table 3 along with the indication of the string size n_k^{ub} used in the computation of α_k^{ub}.

Table 3. Results on α_k for several alphabet sizes k. The table shows lower bounds α_k^{lb}, statistical estimates $\widetilde{\alpha}_k$, upper bounds α_k^{ub}. The values of α_k^{lb} are obtained by numerically evaluating $\sup A_k$ (Sect. 5, Theorem 1). Each statistical estimate $\widetilde{\alpha}_k$ is based on $N = 5000$ sample pairs of strings of length $n = 2^{14}$ (Sect. 3.1). The values of α_k^{ub} are based on the exact determination of $\alpha_k(n_k^{ub})$ (Sect. 4). We can observe that $\alpha_k^{lb} < \widetilde{\alpha}_k < \alpha_k^{ub}$.

k	α_k^{lb}	$\widetilde{\alpha}_k$	α_k^{ub}	n_k^{ub}
2	0.1742	0.2888	0.3693	24
3	0.2837	0.4292	0.5343	17
4	0.3598	0.5180	0.6318	15
5	0.4152	0.5806	0.7020	13
6	0.4578	0.6277	0.7515	12
7	0.4918	0.6645	0.7903	11
8	0.5199	0.6946	0.8122	12
16	0.6648	0.8196	0.8955	10
32	0.7387	0.8999	0.9659	6

As already mentioned in the introduction, $1 - \gamma_k \leq \alpha_k$. Since γ_k vanishes with k (Theorem 1 in [8]), we have that $\lim_{k \to \infty} \alpha_k = 1$. The data in Table 3 show a trend consistent with this asymptotic behavior of α_k.

The gap between lower and upper bound indicates that there is room for improving both. Improving the current upper bounds requires substantial

improvements in the way we compute exact values of $\alpha_k(n)$. Improving the lower bounds appears viable by tightening the upper bound to the volume of $B_r(x)$, in Proposition 4. To this end, an avenue to be explored is a refinement of script counting that exploits properties of optimal scripts, not considered in the present arguments.

We observe that, for fixed k and variable n, the values $\alpha_k(n)$ form a sequence of computable upper bounds to α_k that converges to α_k. It would be interesting to find a sequence of computable lower bounds to α_k that converges to α_k. Together, these two sequences would establish the computability of α_k, providing an algorithm that, given as input any $\epsilon > 0$, would output a rational number η such that $|\alpha_k - \eta| < \epsilon$. To the best of our knowledge, whether the α_k's are computable is an open question. In contrast, it is a simple corollary of known results that the γ_k's are computable. On the one hand, a sequence of computable lower bounds converging to γ_k is straightforwardly provided by the values $\gamma_k(n)$. On the other hand, a sequence of computable upper bounds converging to γ_k has been established, by a rather sophisticated approach, in [11] (see, in particular, Theorem 3.13).

7   Conclusions

In this paper, we have explored approaches to obtain statistical estimates, upper bounds, and lower bounds to the asymptotic constant characterizing the average edit distance between random, independent strings. We used such approaches to obtain results for some alphabet sizes k. These numerical results (Table 3) improve over previously known values [10]. There is still a gap between upper and lower bounds which deserves further investigation. The approaches proposed here can be extended to the study of other statistical properties of the edit distance (e.g., the standard deviation, widely studied in the context of the longest common subsequence).

It is interesting to explore the role of statistical properties of the edit distance in string alignment and other key problems in DNA processing. One motivation is provided by the increasing availability of reads coming from third generation sequencers (e.g., PacBio) where sequencing errors can be modelled as edit operations. In this case it will be necessary to study the behaviour of the average edit distance, when strings are generated from non-uniform distributions or from empirical distributions (e.g., the distribution of substrings from the human DNA).

References

1. Abboud, A., Backurs, A., Williams, V.V.: Tight hardness results for LCS and other sequence similarity measures. In: 2015 IEEE 56th Annual Symposium on Foundations of Computer Science, pp. 59–78 (2015). https://doi.org/10.1109/FOCS.2015.14

2. Andoni, A., Krauthgamer, R., Onak, K.: Polylogarithmic approximation for edit distance and the asymmetric query complexity. In: 2010 IEEE 51st Annual Symposium on Foundations of Computer Science, pp. 377–386 (2010). https://doi.org/10.1109/FOCS.2010.43

3. Backurs, A., Indyk, P.: Edit distance cannot be computed in strongly subquadratic time (unless seth is false). In: Proceedings of the Forty-seventh Annual ACM Symposium on Theory of Computing, pp. 51–58. STOC 2015, ACM, New York, NY, USA (2015). https://doi.org/10.1145/2746539.2746612

4. Baeza-Yates, R.A., Gavaldà, R., Navarro, G., Scheihing, R.: Bounding the expected length of longest common subsequences and forests. Theor. Comput. Syst. $32(4)$, 435–452 (1999). https://doi.org/10.1007/s002240000125

5. Bundschuh, R.: High precision simulations of the longest common subsequence problem. Eur. Phys. J. B - Condens. Matter Complex Syst. $22(4)$, 533–541 (2001). https://doi.org/10.1007/s100510170102

6. Calvo-Zaragoza, J., Oncina, J., de la Higuera, C.: Computing the expected edit distance from a string to a probabilistic finite-state automaton. Int. J. Found. Comput. Sci. $28(05)$, 603–621 (2017). https://doi.org/10.1142/S0129054117400093

7. Chakraborty, D., Das, D., Goldenberg, E., Koucky, M., Saks, M.: Approximating edit distance within constant factor in truly sub-quadratic time. In: 2018 IEEE 59th Annual Symposium on Foundations of Computer Science, pp. 979–990 (2018). https://doi.org/10.1109/FOCS.2018.00096

8. Chvátal, V., Sankoff, D.: Longest common subsequences of two random sequences. J. Appl. Probab. $12(2)$, 306–315 (1975). https://doi.org/10.2307/3212444

9. Dancík, V.: Expected length of longest common subsequences. Ph.D. thesis, University of Warwick (1994)

10. Ganguly, S., Mossel, E., Racz, M.Z.: Sequence assembly from corrupted shotgun reads. arXiv preprint arXiv:1601.07086 (2016)

11. Lueker, G.S.: Improved bounds on the average length of longest common subsequences. J. ACM $56(3)$, 17:1–17:38 (2009). https://doi.org/10.1145/1516512.1516519

12. Masek, W.J., Paterson, M.S.: A faster algorithm computing string edit distances. J. Comput. Syst. Sci. $20(1)$, 18–31 (1980). https://doi.org/10.1016/0022-0000(80)90002-1

13. Ning, K., Choi, K.P.: Systematic assessment of the expected length, variance and distribution of longest common subsequences. arXiv preprint arXiv:1306.4253 (2013)

14. Rubinstein, A.: Hardness of approximate nearest neighbor search. In: Proceedings of the 50th Annual ACM SIGACT Symposium on Theory of Computing, pp. 1260–1268. STOC 2018, ACM, New York, NY, USA (2018). https://doi.org/10.1145/3188745.3188916

15. Rubinstein, A., Song, Z.: Reducing approximate longest common subsequence to approximate edit distance. arXiv preprint arXiv:1904.05451 (2019)

16. Saw, J.G., Yang, M.C.K., Mo, T.C.: Chebyshev inequality with estimated mean and variance. Am. Stat. $38(2)$, 130–132 (1984). https://doi.org/10.1080/00031305.1984.10483182

17. Spencer, J.: Asymptopia. Am. Math. Soc., 71 (2014)

18. Steele, J.M.: Probability Theory and Combinatorial Optimization. SIAM, Philadelphia (1997)

Compact Data Structures for Shortest Unique Substring Queries

Takuya Mieno[1](✉), Dominik Köppl[1,2]🆔, Yuto Nakashima[1], Shunsuke Inenaga[1], Hideo Bannai[1]🆔, and Masayuki Takeda[1]

[1] Department of Informatics, Kyushu University, Fukuoka, Japan
{takuya.mieno,dominik.koeppl,yuto.nakashima,inenaga,
bannai,takeda}@inf.kyushu-u.ac.jp
[2] Japan Society for Promotion of Science, Tokyo, Japan

Abstract. Given a string T of length n, a substring $u = T[i..j]$ of T is called a shortest unique substring (SUS) for an interval $[s, t]$ if (a) u occurs exactly once in T, (b) u contains the interval $[s, t]$ (i.e. $i \leq s \leq t \leq j$), and (c) every substring v of T with $|v| < |u|$ containing $[s, t]$ occurs at least twice in T. Given a query interval $[s, t] \subset [1, n]$, the *interval SUS problem* is to output all the SUSs for the interval $[s, t]$. In this article, we propose a $4n + o(n)$ bits data structure answering an interval SUS query in output-sensitive $O(occ)$ time, where occ is the number of returned SUSs. Additionally, we focus on the *point SUS problem*, which is the interval SUS problem for $s = t$. Here, we propose a $\lceil (\log_2 3 + 1)n \rceil + o(n)$ bits data structure answering a point SUS query in the same output-sensitive time.

Keywords: String processing algorithm · Shortest unique substring · Compact data structure

1 Introduction

A substring $u = T[i..j]$ of a string T is called a *shortest unique substring (SUS)* for an interval $[s, t]$ if (a) u occurs exactly once in T, (b) u contains the interval $[s, t]$ (i.e., $i \leq s \leq t \leq j$), and (c) every substring v of T with $|v| < |u|$ containing $[s, t]$ occurs at least twice in T. Given a query interval $[s, t] \subset [1, n]$, the *interval SUS problem* is to output all the SUSs for $[s, t]$. When a query interval consists of a single position (i.e., $s = t$), the SUS problem becomes a so-called *point* SUS problem.

Point SUS Problem. The point SUS problem was introduced by Pei et al. [12]. This problem is motivated by applications in bioinformatics like genome comparisons [4] or PCR primer design [12]. Pei et al. tackled this problem with an $O(n)$ words data structure that can return one SUS for a given query position in constant time. They can compute this data structure in $O(n^2)$ time with $O(n)$ space. Based on that result, Tsuruta et al. [13] provided an $O(n)$ words data

© Springer Nature Switzerland AG 2019
N. R. Brisaboa and S. J. Puglisi (Eds.): SPIRE 2019, LNCS 11811, pp. 107–123, 2019.
https://doi.org/10.1007/978-3-030-32686-9_8

structure answering the same query (returning one SUS) in constant time. Their data structure can be constructed in $O(n)$ time. İleri et al. [7] independently showed another data structure with the same time complexities. For the general point SUS problem, Tsuruta et al. [13] can also resort to their proposed data structure returning all SUSs for a query position in optimal $O(occ)$ time, where occ is the number of returned SUSs.

The aforementioned data structures all take $\Theta(n)$ words. This space can become problematic for large n. This problem was perceived by Hon et al. [5], who proposed a data structure consisting of the input string T and two integer arrays, each of length n. Both arrays store, respectively, the beginning and the ending position of a SUS for each position i with $1 \le i \le n$. Hon et al. provided an algorithm that can construct these two arrays in linear time with $O(\log n)$ bits of additional working space, given that both arrays are stored in $2n \log n$ bits and that $\sigma \le n$. Instead of building a data structure, Ganguly et al. [3] proposed a time-space trade-off algorithm using $O(n/\tau)$ words of additional working space, answering a given query in $O(n\tau^2 \log \frac{n}{\tau})$ time directly, for a trade-off parameter $\tau \ge 1$. They also proposed the first *compact* data structure of size $4n + o(n)$ bits that can answer a query in constant time. They can construct this data structure in $O(n \log n)$ time using $O(n \log \sigma)$ bits of additional working space.

Interval SUS Problem. Hu et al. [6] were the first to consider the interval SUS problem. They proposed a data structure answering a query returning all SUSs for the respective query interval in $O(occ)$ optimal time after $O(n)$ time preprocessing. In the compressed setting, Mieno et al. [11] considered the interval SUS problem when the input string T is given *run-length encoded (RLE)*, and proposed a data structure of size $O(r)$ words answering a query by returning all SUSs for the respective query interval in $O(\sqrt{\log r / \log \log r} + occ)$ time, where r is the number of single character runs in T.

Our Contribution. In this paper, we propose the following two data structures:

(A) A data structure of size $2n + 2m + o(n)$ bits answering an interval SUS query in $O(occ)$ time, where m is the number of *minimal unique substrings* of the input string[1], and occ is the number of SUSs of T for the respective query interval (Theorem 1).

(B) A data structure of size $\lceil (\log_2 3 + 1)n \rceil + o(n)$ bits answering a point SUS query in $O(occ)$ time, where occ is the number of SUSs of T for the respective query point (Theorem 2).

Instead of outputting the answer as a list of substrings of T, it is sometimes sufficient to output only the intervals corresponding to the respective substrings. In such a case, both data structures can answer a query *without* the need of the input string. The data structure (A) is the first data structure of size $O(n)$ bits for the interval SUS problem. Also, the data structure (B) is the first data

---

[1] We show later in Lemma 1 that the number of minimal unique substrings m is at most n.

structure of size $O(n)$ bits for the point SUS problem, returning *all* SUSs for a given query position. Notice that the data structure of Ganguly et al. [3] uses $4n + o(n)$ bits of space, but returns only one SUS for a point SUS query.

2   Preliminaries

Our model of computation is the word RAM with machine word size $\Omega(\log n)$.

2.1   Strings

Let Σ be an alphabet. An element of Σ^* is called a *string*. For $|\Sigma| = 2$, we call a string also a *bit array*. The length of a string T is denoted by $|T|$. The empty string ε is the string of length 0. Given a string T, the i-th character of T is denoted by $T[i]$, for an integer i with $1 \leq i \leq |T|$. For two integers i and j with $1 \leq i \leq j \leq |T|$, a substring of T starting at position i and ending at position j is denoted by $T[i..j]$. Namely, $T[i..j] = T[i]T[i+1]\cdots T[j]$. For two strings T and w, the number of occurrences of w in T is denoted by $\#T(w) := |\{i \mid T[i..i + |w| - 1] = w\}|$. For two intervals $[i, j]$ and $[x, y]$, let $cover([i, j], [x, y]) := [\min\{i, x\}, \max\{j, y\}]$ denote the shortest interval that contains the text positions i, j, x, and y. If the interval $[x, y]$ consists of a single point, i.e., $x = y$, $cover([i, j], [x, y])$ is denoted by $cover([i, j], x)$ when we want to emphasize on the fact that $x = y$.

In what follows, we fix a string T of length $n \geq 1$ whose characters are drawn from an integer alphabet Σ of size $\sigma = n^{O(1)}$.

2.2   MUSs and SUSs

Let u be a non-empty substring of T. u is called a *repeating substring* of T if $\#T(u) \geq 2$, and u is called a *unique substring* of T if $\#T(u) = 1$. Since every unique substring $u = T[i..j]$ of T occurs exactly once in T, we identify u with its corresponding interval $[i, j]$. We also say that the interval $[i, j]$ is unique iff the corresponding substring $T[i..j]$ is a unique substring of T.

A unique substring $u = T[i..j]$ of T is said to be a *minimal unique substring (MUS)* of T iff every proper substring of u is a repeating substring, i.e., $\#T(T[i'..j']) \geq 2$ for every integer i' and every integer j' with $[i', j'] \subset [i, j]$ and $j' - i' < j - i$. Let $\mathsf{MUS}_T := \{[i, j] \mid T[i..j]$ is a MUS of $T\}$ be the set of all intervals corresponding to the MUSs of T. From the definition of MUSs, the next lemma follows:

Lemma 1 ([13, Lemma 2]). *No element of* MUS_T *is nested in another element of* MUS_T, *i.e., two different MUSs* $[i, j], [k, l] \in \mathsf{MUS}_T$ *satisfy* $[i, j] \not\subset [k, l]$ *and* $[k, l] \not\subset [i, j]$. *Therefore,* $0 < |\mathsf{MUS}_T| \leq |T|$.

We use the following two sets containing interval and point SUSs, which were defined at the beginning of the introduction: Given an interval $[s, t] \subset [1, n]$,

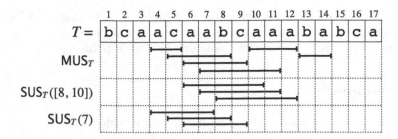

Fig. 1. The string $T = $ bcaacaabcaaababca, and its set $\text{MUS}_T = \{[4,5],[5,8],[6,9],$ $[7,11],[10,12],[13,14]\}$. MUS_T corresponds to the set $\{$ac, caab, aabc, abcaa, aaa, ba$\}$ of all MUSs of T. The substrings $T[6..10] = $ aabca, $T[7..11] = $ abcaa, and $T[8..12] = $ bcaaa are SUSs for the query interval $[8,10]$. Also, the substrings $T[4..7] = $ acaa, $T[5..8] = $ caab, and $T[6..9] = $ aabc are SUSs for the query position 7. The later defined leftmost/rightmost SUS and MUS (cf. Sect. 6) for $p = 7$ are $\text{lmSUS}_T^p = [4,7]$ $\text{lmSUS}_T^p = [4,7]$, $\text{lmMUS}_T^p = [4,5]$, $\text{rmSUS}_T^p = [6,9]$, and $\text{rmMUS}_T^p = [6,9]$.

$\text{SUS}_T([s,t])$ denotes the set of the interval SUSs of T for the interval $[s,t]$. Given a text position $p \in [1,n]$, $\text{SUS}_T(p)$ denotes the set of the point SUSs of T for the point p.

Given a query position $p \in [1,n]$ (resp. a query interval $[s,t] \subset [1,n]$), the *point* (resp. *interval*) *SUS problem* is to compute $\text{SUS}_T(p)$ (resp. $\text{SUS}_T([s,t])$). See Fig. 1 for an example depicting MUSs and SUSs.

3   Tools

In this section, we introduce the data structures needed for our approach solving both SUS problems.

3.1   Rank and Select

Given a string X of length n over the alphabet $[1,\sigma]$. For an integer i with $1 \le i \le n$ and a character $c \in [1,\sigma]$, the rank query $rank_X(c,i)$ returns the number of the character c in the prefix $X[1..i]$ of X. Also, the select query $select_X(c,i)$ returns the position of X containing the i-th occurrence of the character c (or returns the invalid symbol *nil* if such a position does not exist). For $\sigma = 2$ (i.e., X is a bit array), we can make use of the following lemma:

Lemma 2 ([1,8])**.** *We can endow a bit array X of length n with a data structure answering $rank_X$ and $select_X$ in constant time. This data structure takes $o(n)$ bits of space, and can be built on X in $O(n)$ time with $O(\log n)$ bits of additional working space.*

3.2 Predecessor and Successor

Let Y be an array of length k whose entries are positive integers in strictly increasing order. Further suppose that these integers are less than or equal to n. Given an integer d with $1 \leq d \leq n$, the *predecessor* and the *successor query* on Y with d are defined as $\mathsf{Pred}_Y(d) := \max\{i \mid Y[i] \leq d\}$ and $\mathsf{Succ}_Y(d) := \min\{i \mid Y[i] \geq d\}$, where we stipulate that $\min\{\} = \max\{\} = nil$.

Let BIT_Y be a bit array of length n marking all integers present in Y, i.e., $BIT_Y[i] = 1$ iff there is an integer j with $1 \leq j \leq k$ and $Y[j] = i$, for every i with $1 \leq i \leq n$. By endowing BIT_Y with a rank/select data structure, we yield an $n + o(n)$ bits data structure answering $\mathsf{Pred}_Y(d) = select_{BIT_Y}(1, rank_{BIT_Y}(1, d))$ and $\mathsf{Succ}_Y(d)^2$ in constant time for each d with $1 \leq d \leq n$.

3.3 RmQ and RMQ

Given an integer array Z of length n and an interval $[i, j] \subset [1, n]$, the range minimum query $\mathsf{RmQ}_Z(i, j)$ (resp. the range maximum query $\mathsf{RMQ}_Z(i, j)$) asks for the index p of a minimum element (resp. a maximum element) of the subarray $Z[i..j]$, i.e., $p \in \arg\min_{i \leq k \leq j} Z[k]$, or respectively $p \in \arg\max_{i \leq k \leq j} Z[k]$. We use the following well-known data structure to handle these kind of queries:

Lemma 3 ([2]). *Given an integer array Z of length n, there is an* RmQ *(resp.* RMQ*) data structure taking $2n + o(n)$ bits of space that can answer an* RmQ *(resp.* RMQ*) query on Z in constant time. This data structure can be constructed in $O(n)$ time with $o(n)$ bits of additional working space.*

3.4 Suffix Array, Inverse Suffix Array and LCP Array

We define the three integer arrays $\mathsf{SA}_T[1..n]$, $\mathsf{ISA}_T[1..n]$, and $\mathsf{LCP}_T[1..n + 1]$. The suffix array SA_T of T is the array with the property that $T[\mathsf{SA}_T[i]..n]$ is lexicographically smaller than $T[\mathsf{SA}_T[i+1]..n]$ for every i with $1 \leq i \leq n-1$ [10]. The inverse suffix array ISA_T of T is the inverse of SA_T, i.e., $\mathsf{SA}_T[\mathsf{ISA}_T[i]] = i$ for every i with $1 \leq i \leq n$. The LCP array LCP_T of T is the array with the property that $\mathsf{LCP}_T[1] = \mathsf{LCP}_T[n+1] = 0$ and $\mathsf{LCP}_T[i] = lcp(T[\mathsf{SA}_T[i]..n], T[\mathsf{SA}_T[i-1]..n])$ for every i with $2 \leq i \leq n$, where $lcp(P, Q)$ denotes the length of the longest common prefix of P and Q for two given strings P and Q.

4 Computing MUSs in Compact Space

For computing SUSs efficiently, it is advantageous to have a data structure available that can retrieve MUSs starting or ending at specific positions, as the following lemma gives a crucial connection between MUSs and SUSs:

Lemma 4 ([13, Lemma 2]). *Every point SUS contains exactly one MUS.*

---

[2] $\mathsf{Succ}_Y(d)$ can be computed similarly by considering the case whether $BIT_Y[d] = 1$.

Figure 2 gives an overview of our introduced data structure and shows the connections between this section and the following sections that focus on our two SUS problems. For our data structure retrieving MUSs, we propose a compact representation and an algorithm to compute this representation space-efficiently. Our data structure is based on the following two bit arrays MB_T and ME_T of length n with the properties that

- $\mathsf{MB}_T[i] = 1$ iff i is the beginning position of a MUS, and
- $\mathsf{ME}_T[i] = 1$ iff i is the ending position of a MUS.

For the rest of this paper, let m be the number of MUSs in T. We rank the MUSs by their starting positions in the text, such that the j-th MUS starts before the $(j+1)$-th MUS, for every integer j with $1 \le j \le m-1$.

Since MUSs are not nested (see Lemma 1), the number of 1's in MB_T and ME_T is exactly m. Hence, the starting position, the ending position, and the length of the j-th MUS can be computed with rank/select queries for every integer j with $1 \le j \le m$. How MB_T and ME_T can be computed is shown in the following lemma:

Lemma 5. *Let $\mathcal{D}_T$ be a data structure that can access $\mathsf{ISA}_T[i]$ and $\mathsf{LCP}_T[i]$ in $\pi_a(n)$ time for every position i with $1 \le i \le n$. Suppose that we can construct it in $\pi_c(n)$ time with $\pi_s(n)$ bits of working space including the space for $\mathcal{D}_T$. Then MB_T and ME_T can be computed in $O(\pi_c(n) + n \cdot \pi_a(n))$ total time while using $2n + \pi_s(n)$ bits of total working space including the space for MB_T and ME_T.*

Proof. Given a text position i with $1 \le i \le n$, $T[i..i + \ell_i - 1]$ with $\ell_i = \max\{\mathsf{LCP}_T[\mathsf{ISA}_T[i]], \mathsf{LCP}_T[\mathsf{ISA}_T[i]+1]\}$ is the longest repeating substring starting at i. If we extend this substring by the character to its right, it becomes unique. Thus, $T[i..i + \ell_i]$ is the shortest unique substring starting at i, except for the case that $i + \ell_i - 1 = n$ as we cannot extend it to the right (hence, there is no unique substring starting at i in this case). Additionally, the substring $T[i..i + \ell_i]$ is a MUS iff $T[i + 1..i + \ell_i]$ is not unique (we already checked that $T[i..i + \ell_i - 1]$ is not unique). $T[i + 1..i + \ell_i]$ is not unique iff $\ell_i \le \ell_{i+1}$ since $T[i + 1..i + 1 + \ell_{i+1}]$ is the smallest unique substring starting at $i + 1$. Since each ℓ_i can be computed in $O(\pi_a(n))$ time for every $1 \le i \le n$, the starting and ending positions of all MUSs (and hence, MB_T and ME_T) can be computed in $O(n \cdot \pi_a(n))$ time by a linear scan of the text. Therefore, the total computing time is $O(\pi_c(n) + n \cdot \pi_a(n))$ and the total working space is $2n + \pi_s(n)$ bits including the space for MB_T and ME_T. □

5   Compact Data Structure for the Interval SUS Problem

In this section, we propose a compact data structure for the interval SUS problem. It is based on the data structure of Mieno et al. [11], which we review in the following. We subsequently provide a compact representation of this data structure.

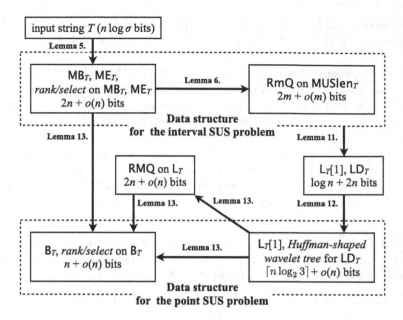

Fig. 2. Overview of the data structures proposed for solving the interval SUS and point SUS problem. Nodes are data structures. Edges of the same label (labeled by a certain lemma) describe an algorithm taking a set of input data structures to produce a data structure.

Data Structures. The data structure proposed by Mieno et al. [11] consists of three arrays, each of length m: X_T, Y_T, and MUSlen_T. The arrays X_T and Y_T store, respectively, the beginning positions and ending positions of all MUSs sorted by their beginning positions such that the interval $[X_T[i], Y_T[i]]$ is the i-th MUS, for every integer i with $1 \leq i \leq m$. Further, $\mathsf{MUSlen}_T[i] = Y_T[i] - X_T[i] + 1$ stores the length of i-th MUS. During a preprocessing phase, X_T and Y_T are endowed with a successor and a predecessor data structure, respectively. Further, MUSlen_T is endowed with an RmQ data structure.

Answering Queries. Given a query interval $[s, t]$, let $\ell = \mathsf{Pred}_{Y_T}(t)$ be the index in Y_T of the largest ending position of a MUS that is at most t, and $r = \mathsf{Succ}_{X_T}(s)$ be the index in X_T of the smallest starting position of a MUS that is at least s. Then, $\mathsf{SUS}_T([s, t]) \subset \{cover([s, t], [X_T[i], Y_T[i]]) \mid \ell \leq i \leq r\}$. That is because the shortest intervals in $\{cover([s, t], [X_T[i], Y_T[i]]) \mid \ell \leq i \leq r\}$ correspond to the shortest unique substrings (SUSs) among all substrings covering the interval $[s, t]$. Thus, one of the SUSs for $[s, t]$ can be detected by considering $cover([s, t], [X_T[\ell], Y_T[\ell]])$ (as a candidate for the leftmost SUS), $cover([s, t], [X_T[r], Y_T[r]])$ (as a candidate for the rightmost SUS), and $\mathsf{RmQ}_{\mathsf{MUSlen}_T}(\ell + 1, r - 1)$. To output all SUSs, it is sufficient to answer RmQ queries on subintervals of $\mathsf{MUSlen}_T[\ell+1..r-1]$ recursively. In detail, suppose that there is a MUS in $\mathsf{MUSlen}_T[\ell + 1..r - 1]$ that is a SUS for $[s, t]$. Further suppose

that this is the j-th MUS having length k. Then we query $\mathsf{MUSlen}_T[\ell+1..j-1]$ and $\mathsf{MUSlen}_T[j+1..r-1]$ for all other MUSs of minimal length k.

Compact Representation. Having the two bit arrays MB_T and ME_T of Sect. 4, we can simulate the three arrays X_T, Y_T, and MUSlen_T. By endowing these two bit arrays with rank/select data structures of Lemma 2, we can compute rank/select in constant time, which allows us to compute the value of $\mathsf{X}_T[p]$, $\mathsf{Y}_T[p]$, $\mathsf{MUSlen}_T[p]$, $\mathsf{Pred}_{\mathsf{Y}_T}(q)$ and $\mathsf{Succ}_{\mathsf{X}_T}(q)$ for every index p with $1 \le p \le m$ and every text position q with $1 \le q \le n$ in constant time while using only $2n + o(n)$ bits of total space. By endowing MUSlen_T with the RmQ data structure of Lemma 3, we can answer an RmQ query on MUSlen_T in constant time. This data structure takes $2m + o(m)$ bits of space. Altogether, with these data structures we yield the following theorem:

Theorem 1. *For the interval SUS problem, there exists a data structure of size $2n+2m+o(n)$ bits that can answer an interval SUS query in $O(occ)$ time, where occ is the number of SUSs of T for the respective query interval.*

Also, the data structure can be constructed space-efficiently:

Lemma 6. *Given MB_T and ME_T, the data structure proposed in Theorem 1 can be constructed in $O(n)$ time using $2m + o(n)$ bits of total working space, which includes the space for this data structure.*

Proof. The data structure proposed in Theorem 1 consists of the two bit arrays MB_T, ME_T, and an RmQ data structure on MUSlen_T, which is simulated by rank/select data structures on MB_T and ME_T. Since MB_T and ME_T are already given, it is left to endow MB_T and ME_T with rank/select data structures (using Lemma 2), and to compute the RmQ data structure on MUSlen_T (using Lemma 3). □

6   Compact Data Structure for the Point SUS Problem

Before solving the point SUS problem, we borrow some additional notations from Tsuruta et al. [13] to deal with point SUS queries. This is necessary since some of the MUSs never take part in finding a SUS such that there is no meaning to compute and store them. Since we want to provide an output-sensitive algorithm answering a query in optimal time, we only want to store MUSs that are candidates for being a SUS.

We say that the interval $[x, y] \in \mathsf{MUS}_T$ is a *meaningful* MUS if $T[x..y]$ is a substring of (or equal to) a point SUS, i.e., $cover([x, y], p) \in \mathsf{SUS}_T(p)$ for a position p. Also, we say that the interval $[x, y] \in \mathsf{MUS}_T$ is a *meaningless* MUS if $[x, y]$ is not a meaningful MUS. Let

$$\mathsf{MMUS}_T := \{[i, j] \in \mathsf{MUS}_T \mid \text{there exists a } p \text{ with } 1 \le p \le n$$
$$\text{such that } cover([i, j], p) \in \mathsf{SUS}_T(p)\}$$

denote the set of all meaningful MUSs of T.

Let lmSUS_T^p denote the interval in $\mathrm{SUS}_T(p)$ with the leftmost starting position, and let lmMUS_T^p denote the MUS contained in lmSUS_T^p. We say that lmSUS_T^p is the *leftmost SUS* for p, and lmMUS_T^p is the *leftmost MUS* for p. Similarly, we define the *rightmost SUS* rmSUS_T^p and the *rightmost MUS* rmMUS_T^p for p by symmetry. See Fig. 1 for an example for the leftmost/rightmost SUS and MUS.

Let L_T be an array of length n such that $\mathsf{L}_T[i]$ is the length of a SUS[3] of T containing i for each position i with $1 \le i \le n$. Let B_T be a bit array of length n such that $\mathsf{B}_T[i] = 1$ iff i is the beginning position of a meaningful MUS of T.

From the definition of L_T, we yield the following observation:

Observation 1. For every position p with $1 \le p \le n$ and every interval $[x, y] \in \mathrm{SUS}_T(p)$, $p - \mathsf{L}_T[p] + 1 \le x \le p \le y \le p + \mathsf{L}_T[p] - 1$.

Next, we define the following four functions related to L_T and B_T. For a position q with $1 \le q \le n$ let

- $pred1pos_{\mathsf{B}_T}(q) := \max\{i \mid i \le q \text{ and } \mathsf{B}_T[i] = 1\}$,
- $succ1pos_{\mathsf{B}_T}(q) := \min\{i \mid i \ge q \text{ and } \mathsf{B}_T[i] = 1\}$,
- $predneq_{\mathsf{L}_T}(q) := \max\{i \mid i < q \text{ and } \mathsf{L}_T[i] \ne \mathsf{L}_T[q]\}$, and
- $succneq_{\mathsf{L}_T}(q) := \min\{i \mid i > q \text{ and } \mathsf{L}_T[i] \ne \mathsf{L}_T[q]\}$.

For all four functions, we stipulate that $\min\{\} = \max\{\} = nil$. See Fig. 3 for an example of the arrays and functions defined above.

6.1 Finding SUSs with L and B

Our idea is to answer point SUS queries with L_T and B_T. For that, we first think about how to find the leftmost and rightmost SUS for a given query (Observation 1 gives us the range in which to search). Having this leftmost and the rightmost SUS, we can find all other SUSs with B_T marking the beginning positions of the meaningful MUSs that correspond to the SUSs we want to output. Before that, we need some properties of L_T that help us to prove the following lemmas in this section: Lemma 7 gives us a hint on the shape of L_T, while Lemma 8 shows us how to find SUSs based on two consecutive values of L_T with a connection to MUSs.

Lemma 7. $|\mathsf{L}_T[p] - \mathsf{L}_T[p+1]| \le 1$ *for every position p with $1 \le p \le n - 1$.*

Proof. Let $\ell = \mathsf{L}_T[p]$ and $\ell' = \mathsf{L}_T[p+1]$. From the definition of L_T, there exists a unique substring of length ℓ containing the position p. If $\ell < \ell'$, there is no unique substring of length ℓ containing $p + 1$. Thus, $T[p - \ell + 1..p]$ is unique, and consequently $T[p - \ell + 1..p + 1]$ is also unique. Hence, $\ell' = \ell + 1$. Similarly, in the case of $\ell > \ell'$, it can be proven that $\ell' = \ell - 1$. $\square$

---

[3] Although there can be multiple SUSs containing i, their lengths are all equal.

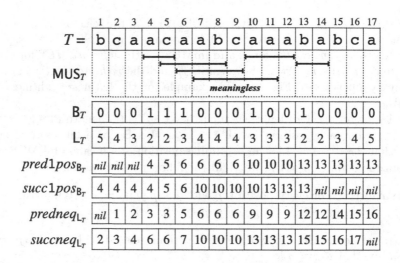

Fig. 3. $\mathsf{MUS}_T, \mathsf{B}_T, \mathsf{L}_T$, and the four functions defined in at the beginning of Sect. 6 for the string $T = \mathtt{bcaacaabcaaababca}$. $\mathsf{B}_T[7] = 0$ because the MUS $T[7..11] = \mathtt{abcaa}$ is meaningless.

Lemma 8. *Let p be a position with $1 \le p \le |T| - 1$, and let $\ell := \mathsf{L}_T[p]$. If $\mathsf{L}_T[p + 1] = \ell + 1$, then*

- $T[p - \ell + 1..p] \in \mathsf{SUS}_T(p)$,
- $T[p - \ell + 1..p + 1] \in \mathsf{SUS}_T(p + 1)$, *and*
- $p - \ell + 1$ *is the starting position of a MUS of T.*

If $\mathsf{L}[p + 1] = \ell - 1$ then

- $T[p..p + \ell - 1] \in \mathsf{SUS}_T(p)$,
- $T[p + 1..p + \ell - 1] \in \mathsf{SUS}_T(p + 1)$, *and*
- $p + \ell - 1$ *is the ending position of a MUS of T.*

Proof. First, we consider the case that $\mathsf{L}_T[p + 1] = \ell + 1$. From the proof of Lemma 7, $T[p - \ell + 1..p]$ and $T[p - \ell + 1..p + 1]$ are unique substrings in T. Thus, $T[p - \ell + 1..p] \in \mathsf{SUS}_T(p)$ and $T[p - \ell + 1..p + 1] \in \mathsf{SUS}_T(p + 1)$. Since every point SUS contains exactly one MUS (cf. Lemma 4), there exists a MUS $[b, e] \subset [p - \ell + 1, p]$. Assume that $b > p - \ell + 1$, then $T[b..p]$ is the shortest unique substring among all substrings containing the text position p. Its length is $p - b + 1 < \ell$. This contradicts that $T[p - \ell + 1..p] \in SUS_T(p)$, and therefore $b = p - \ell + 1$ must hold. The remaining case $\mathsf{L}_T[p + 1] = \ell - 1$ can be proven analogously by symmetry. □

In the following two lemmas (Lemmas 9 and 10), we focus on finding the leftmost SUS and the rightmost SUS for a given query point. That is because the leftmost SUS and the rightmost SUS give us an interval containing the starting positions of the remaining SUSs we want to report[4].

---

[4] The actual reporting of those SUSs is done in Lemma 14.

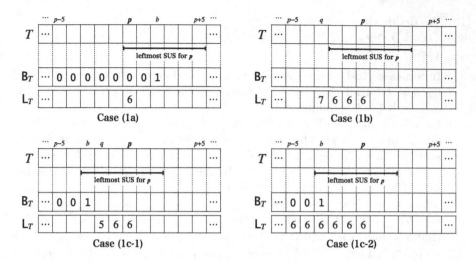

Fig. 4. Example of the proof of Lemma 9 with $L_T[p] = \ell = 6$. The example (as well as all later examples in this section) still works when replacing the number $\ell = 6$ with another number as long as the relative differences to the other entries in L_T and the search range in T is kept.

Lemma 9. *Let p be a position with $1 \le p \le n$, and let $\ell = L_T[p]$, $q = predneq_{L_T}(p)$, and $b = succ1pos_{B_T}(\max\{1, p-\ell+1\})$. Then, $b \le \min\{p+\ell-1, n\}$ and*

$$
lmSUS_T^p = \begin{cases} [p, p+\ell-1] & \text{if } b \ge p, & (1a) \\ [q+1, q+\ell] & \text{if } b < p \text{ and } q \ge p-\ell+1 \text{ and } L_T[q] > \ell, (1b) \\ [b, b+\ell-1] & \text{otherwise.} & (1c) \end{cases}
$$

Proof. If $\ell = 1$, it is clear that the interval $[p, p]$ of length 1 is a MUS of T, thus $b = p$ and $lmSUS_T^p = [p, p]$. For the rest of the proof, we focus on the case that $\ell \ge 2$. Since $L_T[p] = \ell$, there exists a unique substring of length ℓ containing the position p, and there exists at least one MUS that is a subinterval of $[p-\ell+1, p+\ell-1]$. Thus, $b \le \min\{p+\ell-1, n\}$. See Fig. 4 for an illustration of each of the above cases we consider in the following:

(1a) Assume that there exists a unique substring $T[p'..p'+\ell-1]$ containing the position p with $p' < p$. Since $b \ge p > p'$, $T[p'+1..p'+\ell-1]$ is also unique and contains position p. It contradicts $L_T[p] = \ell$; therefore, $lmSUS_T^p = [p, p+\ell-1]$.

(1b) From the definition of q and Lemma 7, $L_T[q] = \ell+1$ and $L_T[q+1] = \ell$. From Lemma 8, $[q+1, q+\ell]$ is unique. Also, $[q+1, q+\ell] \in SUS_T(p)$ because $p \in [q+1, q+\ell]$. Since $L_T[q] = \ell+1$, there is no unique substring that contains the position q and is shorter than $\ell+1$. Therefore, $lmSUS_T^p = [q+1, q+\ell]$.

(1c) We divide this case into two subcases:

(1c-1) $b < p$ and $q \geq p - \ell + 1$ and $\mathsf{L}_T[q] < \ell$, or

(1c-2) $b < p$ and $q < p - \ell + 1$.

In Subcase (1c-1), from the definition of q and Lemma 7, $\mathsf{L}_T[q] = \ell - 1$ and $\mathsf{L}_T[i] = \ell$ for all $i \in [q + 1, p]$. From Lemma 8, the interval $[q - \ell + 2, q]$ of length $\ell - 1$ is unique. Since $[p - \ell + 1, q] \subset [q - \ell + 2, q]$, $\mathsf{L}_T[i] \leq \ell - 1$ for all $i \in [p - \ell + 1, q]$. In Subcase (1c-2), it is clear that $\mathsf{L}_T[i] = \ell$ for all $i \in [p - \ell + 1, p]$. Therefore, $\mathsf{L}_T[i] \leq \ell$ for all $i \in [p - \ell + 1, p]$ in both subcases. Let e be the ending position of the meaningful MUS $[b, e]$ starting at the position b, and $\ell' = e - b + 1$ be the length of this MUS. We assume $\ell' > \ell$ for the sake of contradiction (and thus $[b, e]$ cannot be lmMUS_T^p whose length is at most ℓ). Since $b \geq p - \ell + 1$ and $\ell' > \ell$, $e > p$ must hold. Let $[b', e'] = \mathrm{lmMUS}_T^p$. Since (a) there is no interval $[x, y] \in \mathsf{SUS}_T(p)$ such that $x < \min\{b, p\}$, and (b) MUSs cannot be nested, we follow that $b' > b$ and $e' > e$. Thus, $\mathrm{lmSUS}_T^p = cover([b', e'], p) = [e' - \ell + 1, e']$ and $\mathsf{L}_T[i] \leq \ell$ for all $p \leq i \leq e'$. Since $[b, e] \subset [p - \ell + 1, e']$, $\mathsf{L}_T[i] \leq \ell$ for all $b \leq i \leq e$. This contradicts that the MUS $[b, e]$ of length $\ell' > \ell$ is a meaningful MUS. Therefore, $\ell' \leq \ell$ and $\mathrm{lmSUS}_T^p = cover([b, e], p) = [b, b + \ell - 1]$.   □

From Lemma 9 we yield the following corollary:

Corollary 1. *If we can compute* $\mathsf{L}_T[i]$, $predneq_{\mathsf{L}_T}(i)$ *and* $succ1pos_{\mathsf{B}_T}(i)$ *in constant time for each* i *with* $1 \leq i \leq n$, *we can compute* $lmSUS_T^p$ *in constant time for each position* p *with* $1 \leq p \leq n$.

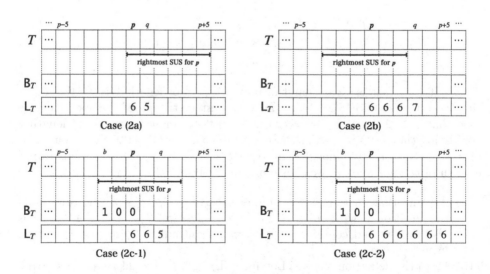

Fig. 5. Example of the proof of Lemma 10 with $\mathsf{L}_T[p] = \ell = 6$.

Lemma 10. *Let* p *be a position with* $1 \leq p \leq n$, *and let* $\ell = \mathsf{L}_T[p]$, $q = succneq_{\mathsf{L}_T}(p)$, *and* $b = pred1pos_{\mathsf{B}_T}(p)$. *Then,*

$$rmSUS_T^p = \begin{cases} [p, p+\ell-1] & \text{if } q = p+1 \text{ and } L_T[q] < \ell, & (2a) \\ [q-\ell, q-1] & \text{if } q \leq p+\ell-1 \text{ and } L_T[q] > \ell, & (2b) \\ [b, b+\ell-1] & \text{otherwise.} & (2c) \end{cases}$$

Proof. If $\ell = 1$, it is clear that the interval $[p, p]$ of length 1 is a MUS of T, thus $b = p$ and $rmSUS_T^p = [p, p]$. We consider the condition of $\ell \geq 2$. See Fig. 5 for an illustration of each of the above cases we consider in the following:

(2a) From Lemma 8, $[p, p+\ell-1]$ is a SUS for p, which is by definition the rightmost one.

(2b) In this case, $L_T[q] = \ell+1$ and $L_T[q-1] = \ell$. From Lemma 8, $[q-\ell, q-1]$ is unique. Since $p \in [q-\ell, q+1]$, $[q-\ell, q-1]$ is a SUS for p. Additionally, there is no unique interval $[x, y] \in SUS_T(p)$ such that $y \geq q$ because $L_T[q] = \ell+1$. Thus, $rmSUS_T^p = [q-\ell, q-1]$.

(2c) We divide this case into two subcases:
  (2c-1) $p+1 < q \leq p+\ell-1$ and $L_T[q] < \ell$, or
  (2c-2) $q > p+\ell-1$.

In Subcase (2c-1), $L_T[p+1] = \ell$ and $L_T[q] = \ell-1$ and $L_T[i] = \ell$ for all $p+2 \leq i \leq q-1$. From Lemma 8, $[q, q+\ell-2]$ (of length $\ell-1$) is unique. Since $[q, p+\ell-1] \subset [q, q+\ell-2]$, $L_T[i] \leq \ell-1$ for all $q \leq i \leq p+\ell-1$. In Subcase (2c-2), from the definition of q, $L_T[i] = \ell$ for all $p \leq i \leq p+\ell-1$. Therefore, $L_T[p+1] = \ell$ and $L_T[i] \leq \ell$ for all integers i with $p+2 \leq i \leq p+\ell-1$ in both subcases. For the sake of contradiction, assume that there is a MUS $[b', e']$ such that $b' > p$ and $cover([b', e'], p) = [p, e'] \in SUS_T(p)$. Since $L_T[p] = \ell$, $[p, e']$ is a unique substring of length ℓ. Hence, $cover([b', e'], p+1) = [p+1, e']$ is a unique substring of length $\ell-1$. It contradicts $L_T[p+1] = \ell$; therefore, the beginning position of the rightmost MUS for p is at most p. Next, we show that the MUS starting at b is the rightmost meaningful MUS for p. Let e be its ending position, and $\ell' = e - b + 1$ be its length. We assume that $\ell' > \ell$ for the sake of contradiction (and thus, $[b, e]$ is not $rmMUS_T^p$ whose length is at most ℓ). Since $L_T[p] = \ell$, $b \geq p - \ell + 1$ and $e > p$. Let $[b'', e''] = rmMUS_T^p$. Since MUSs cannot be nested, $b'' < b$. Since $e'' - b'' + 1 \leq \ell$, $L_T[i] \leq \ell$ for all i with $b'' \leq i \leq p+\ell-1$. We consider two cases to obtain a contradiction:

- If $e \leq p+\ell-1$ then it is clear that $L_T[i] \leq \ell$ for all i with $b \leq i \leq e$. This contradicts that the MUS $[b, e]$ of length ℓ' is a meaningful MUS.
- If $e > p+\ell-1$, it is clear that $|cover([b, e], p+\ell-1)| = |[b, e]| = \ell'$. Since $L_T[p+\ell-1] \leq \ell$, there exists a unique substring $[s, t]$ such that $s \leq p+\ell-1 \leq t$ and $t - s + 1 \leq \ell$. Hence, $L_T[i] \leq \ell$ for all i with $s \leq i \leq t$. Since $[b, e]$ is a MUS and $p \leq s$, $b < s < e < t$. Consequently, $L_T[i] \leq \ell$ for all i with $b \leq i \leq e$ and this contradicts that the MUS $[b, e]$ of length ℓ' is a meaningful MUS.

Therefore, $\ell' \leq \ell$ and $rmSUS_T^p = cover([b, e], p) = [b, b+\ell-1]$. $\square$

Corollary 2. *If we can compute* $L_T[i]$, $succneq_{L_T}(i)$ *and* $pred1pos_{B_T}(i)$ *in constant time for each* i *with* $1 \le i \le n$, *we can compute* $rmSUS_T^p$ *in constant time for each position* p *with* $1 \le p \le n$.

6.2   Compact Representations of L

We now propose a succinct representation of the array L_T consisting of the integer array LD_T of length n defined as $LD_T[1] = 0$ and $LD_T[i] = L_T[i] - L_T[i-1] \in \{-1, 0, 1\}$ for every i with $2 \le i \le n$.

Lemma 11. *The data structure of Theorem 1 can compute* $L_T[p]$ *in constant time with* $O(\log n)$ *bits of additional working space for each* p *with* $1 \le p \le n$.

Proof. Suppose that we have the data structure D of Theorem 1 and want to know $L_T[p]$. We query D with the interval $[p, p]$ to retrieve *one* SUS for the query interval $[p, p]$ in constant time. This can be achieved by stopping the retrieval after the first SUS $[i, j] \in \mathsf{SUS}_T([p, p])$ has been reported. Since all SUSs for $[p, p]$ have the same length, $L_T[p] = j - i + 1$. The additional working space is $O(\log n)$ bits. $\square$

Table 1. Working space used during the construction of the data structure proposed in Theorem 2. We can free up space of no longer needed data structures between several steps. See also Fig. 2 for the dependencies of the execution, and other possible ways to build the final data structure. However, these other ways need more maximum working space (at some step) than the way listed in this table.

No.	Process	Total working space in bits (excluding MB_T and ME_T)	
1	input MB_T, ME_T	-	
2	construct RmQ on $MUSlen_T$	$2m + o(n)$	Lemma 6
3	construct LD_T, $L_T[1]$	$2n + 2m + o(n)$	Lemma 11
4	free RmQ on $MUSlen_T$	$2n + o(n)$	
5	construct *Huffman-shaped Wavelet Tree* for LD_T	$2n + \lceil n \log_2 3 \rceil + o(n)$	Lemma 12
6	free LD_T	$\lceil n \log_2 3 \rceil + o(n)$	
7	construct RMQ on L_T	$2n + \lceil n \log_2 3 \rceil + o(n)$	Lemma 13
8	construct B_T	$3n + \lceil n \log_2 3 \rceil + o(n)$	Lemma 13

Lemma 11 allows us to compute LD_T in $O(n)$ time, which we represent as an integer array with bit width two, thus using $2n$ bits of space. In the following, we build a compressed rank/select data structure on LD_T. This data structure is a self-index such that we no longer need to keep LD_T in memory. With LD_T we can access L_T, as can be seen by the following lemma:

Lemma 12. *There exists a data structure of size $\lceil n \log_2 3 \rceil + o(n)$ bits that can access $L_T[i]$, and can compute $predneq_{L_T}(i)$ and $succneq_{L_T}(i)$ in constant time for each position i with $1 \leq i \leq n$. Given MB_T and ME_T, the data structure can be constructed in $O(n)$ time using $2n + \max\{\lceil n \log_2 3 \rceil, 2m\} + o(n)$ bits of total working space, which includes the space for this data structure.*

Proof. The following equations hold for every text position i with $1 \leq i \leq n$:

$$L_T[i] = L_T[1] + rank_{LD_T}(1, i) - rank_{LD_T}(-1, i),$$
$$predneq_{L_T}(i) = \max\{select_{LD_T}(c, rank_{LD_T}(c, i)) - 1 \mid c \in \{-1, 1\}\},$$
$$succneq_{L_T}(i) = \min\{select_{LD_T}(c, rank_{LD_T}(c, i) + 1) \mid c \in \{-1, 1\}\}.$$

We can compute the value of $L_T[1]$ and LD_T with Lemma 11. With a rank/select data structure on LD_T we can compute the above functions. Such a data structure is the Huffman-shaped wavelet tree [9]. This data structure can be constructed in linear time and takes $\lceil n \log_2 3 \rceil + o(n)$ bits of space, since the possible number of different values in LD_T is three. Therefore, it can also provide answers to rank/select queries in constant time.   □

Finally, we show how to compute B_T:

Lemma 13. *There exists a data structure of size $n + o(n)$ bits that can compute $succ1pos_{B_T}(i)$ and $pred1pos_{B_T}(i)$ in constant time for each position $1 \leq i \leq n$. Given MB_T and ME_T, this data structure can be constructed in $O(n)$ time using $3n + \lceil n \log_2 3 \rceil + o(n)$ bits of total working space including the space for this data structure.*

Proof. Our idea is to compute B_T since the following equations hold for every text position i with $1 \leq i \leq n$ (cf. BIT_Y in Sect. 3.2):

$$pred1pos_{B_T}(i) = \begin{cases} i & \text{if } B_T[i] = 1, \\ select_{B_T}(1, rank_{B_T}(1, i)) & \text{if } B_T[i] = 0. \end{cases}$$

$$succ1pos_{B_T}(i) = \begin{cases} i & \text{if } B_T[i] = 1, \\ select_{B_T}(1, rank_{B_T}(1, i) + 1) & \text{if } B_T[i] = 0. \end{cases}$$

In the following we show how to compute B_T from MB_T and ME_T in linear time with linear number of bits of working space. Let $b_i = select_{MB_T}(1, i)$ and $e_i = select_{ME_T}(1, i)$ be the starting position and the ending position of the i-th MUS respectively, for each $1 \leq i \leq m$. Given $x_i = RMQ_{L_T}(b_i, e_i)$, $L_T[x_i] \leq e_i - b_i + 1$ since $b_i \leq x_i \leq e_i$ and $[b_i, e_i]$ is unique. If $L_T[x_i] < e_i - b_i + 1$, there is no position p with $cover([b_i, e_i], p) \in SUS_T(p)$, i.e., $[b_i, e_i]$ is a meaningless MUS. Otherwise ($L_T[x_i] = e_i - b_i + 1$), $cover([b_i, e_i], x_i) = [b_i, e_i] \in SUS_T(x_i)$, i.e., $[b_i, e_i]$ is a meaningful MUS. Hence, it can be detected in constant time whether a MUS is meaningful by an RMQ query on L_T. We can compute the compact representation of L_T described in Lemma 12. The data structure takes $\lceil n \log_2 3 \rceil + o(n)$ bits and can be constructed with $2n + \max\{\lceil n \log_2 3 \rceil, 2m\} + o(n)$ bits of total working space. Subsequently, we endow it with the RMQ data

structure of Lemma 3 in $O(n)$ time using $2n + o(n)$ bits of space. Therefore, the computing time of B_T is $O(n)$ and the working space is, asides from the space for MB_T and ME_T, $3n + \lceil n \log_2 3 \rceil + o(n)$ bits, including the space for B_T. Finally, we can endow B_T with rank/select data structures, which allows us to compute each of the above two functions $pred1pos_{\mathsf{B}_T}$ and $succ1pos_{\mathsf{B}_T}$ in constant time.   □

Actually, having MB_T and ME_T available, we can simulate an access to $\mathsf{L}_T[i]$ in constant time with Lemma 11 by using the RmQ data structure on MUSlen_T. This allows us to compute the RMQ data structure on L_T directly without the need for computing B_T at first place, i.e., we can replace the working space of Lemma 13 with $2n + 2m + o(n)$ additional bits of working space. However, since our final data structure needs LD_T, computing B_T before LD_T would require more working space in the end than in the other way around, since we no longer need the RmQ data structure on MUSlen_T after having built the rank/select data structure of Lemma 12.

Before stating our final theorem, we need a property for meaningful MUSs:

Lemma 14. *On the one hand, $cover([s_i, e_i], p) \in \mathsf{SUS}_T(p)$ for every meaningful MUS $[s_i, e_i]$ starting with or after the leftmost MUS for p and starting before or with the rightmost MUS for p. On the other hand, each element (i.e., an interval) of $\mathsf{SUS}_T(p)$ starting with or after the leftmost MUS for p and starting before or with the rightmost MUS for p contains exactly one distinct MUS.*

Proof. The first part is shown by Tsuruta et al. [13, Lemma 3]. The second part is due to Lemma 4.   □

Theorem 2. *For the point SUS problem, there exists a data structure of size $n + \lceil n \log_2 3 \rceil + o(n)$ bits that can answer a point SUS query in $O(occ)$ time, where occ is the number of SUSs of T for the respective query point. Given MB_T and ME_T, the data structure can be constructed in $O(n)$ time using $3n + \lceil n \log_2 3 \rceil + o(n)$ bits of total working space, which includes the space for this data structure.*

Proof. Let p be a query position, and suppose that the number of SUSs for p is occ. Like the MUSs in Sect. 4, we rank the SUSs for p by their starting positions. Let $[s_j, e_j]$ be the j-th SUS for p with $1 \le j \le occ$ such that $[s_1, e_1]$ and $[s_{occ}, e_{occ}]$ are the leftmost SUS and the rightmost SUS for p, respectively. If $s_1 = p$ then $[s_1, e_1] = [s_{occ}, e_{occ}]$, and thus the output consists of this single interval. Otherwise ($s_1 \ne p$), we can compute s_i iteratively from s_{i-1} by $s_i = select_{\mathsf{B}_T}(1, rank_{\mathsf{B}_T}(1, s_{i-1}) + 1)$ in constant time for each i with $2 \le i \le occ - 1$, allowing us to answer the query in time linear to the number of SUSs. As occ is not known in advance, we stop the iteration whenever we computed an s_i that is larger than the starting position of the rightmost SUS for p. A detailed analysis of the claimed working space is given in Table 1.   □

Corollary 3. *The data structure of Theorem 2 can compute the number of SUSs for a query position in constant time.*

Proof. Let $[s_l, e_l]$ and $[s_r, e_r]$ be the leftmost and the rightmost SUS for a given query position, respectively. All MUSs starting between s_l and s_r (excluding s_l and s_r) are SUSs for this query position. Let occ' be their number. Therefore, the number we want to output is $occ = occ' + 2$. With Lemmas 9 and 10, we can find $[s_l, e_l]$ and $[s_r, e_r]$ in constant time. Further, we can compute occ' in constant time since $occ' = rank_{B_T}(1, s_r - 1) - rank_{B_T}(1, s_l)$. ☐

References

1. Clark, D.R.: Compact Pat Trees. Ph.D. thesis (1998), uMI Order No. GAXNQ-21335
2. Davoodi, P., Raman, R., Satti, S.R.: Succinct representations of binary trees for range minimum queries. In: Proceedings of the 18th Annual International Computing and Combinatorics Conference (COCOON 2012), pp. 396–407 (2012)
3. Ganguly, A., Hon, W.K., Shah, R., Thankachan, S.V.: Space-time trade-offs for finding shortest unique substrings and maximal unique matches. Theor. Comput. Sci. **700**, 75–88 (2017)
4. Haubold, B., Pierstorff, N., Möller, F., Wiehe, T.: Genome comparison without alignment using shortest unique substrings. BMC Bioinform. **6**(1), 123 (2005)
5. Hon, W.K., Thankachan, S.V., Xu, B.: An in-place framework for exact and approximate shortest unique substring queries. In: Proceedings of International Symposium on Algorithms and Computation (ISAAC), pp. 755–767 (2015)
6. Hu, X., Pei, J., Tao, Y.: Shortest unique queries on strings. In: Proceedings of String Processing and Information Retrieval (SPIRE), pp. 161–172 (2014)
7. İleri, A.M., Külekci, M.O., Xu, B.: A simple yet time-optimal and linear-space algorithm for shortest unique substring queries. Theor. Comput. Sci. **562**, 621–633 (2015)
8. Jacobson, G.: Space-efficient static trees and graphs. In: Proceedings of 30th Annual Symposium on Foundations of Computer Science (FOCS), pp. 549–554 (1989)
9. Mäkinen, V., Navarro, G.: Succinct suffix arrays based on run-length encoding. Nord. J. Comput. **12**(1), 40–66 (2005)
10. Manber, U., Myers, E.W.: Suffix arrays: a new method for on-line string searches. SIAM J. Comput. **22**(5), 935–948 (1993)
11. Mieno, T., Inenaga, S., Bannai, H., Takeda, M.: Shortest unique substring queries on run-length encoded strings. In: Proceedings of 41st International Symposium on Mathematical Foundations of Computer Science (MFCS), pp. 69:1–69:11 (2016)
12. Pei, J., Wu, W.C., Yeh, M.: On shortest unique substring queries. In: Proceedings of IEEE 29th International Conference on Data Engineering (ICDE), pp. 937–948 (2013)
13. Tsuruta, K., Inenaga, S., Bannai, H., Takeda, M.: Shortest unique substrings queries in optimal time. In: Proceedings of SOFSEM 2014: Theory and Practice of Computer Science, pp. 503–513 (2014)

Fast Cartesian Tree Matching

Siwoo Song[1], Cheol Ryu[1], Simone Faro[2], Thierry Lecroq[3],
and Kunsoo Park[1(✉)]

[1] Seoul National University, Seoul, Korea
{swsong,cryu,kpark}@theory.snu.ac.kr
[2] University of Catania, Catania, Italy
faro@dmi.unict.it
[3] Normandie University, Rouen, France
thierry.lecroq@univ-rouen.fr

Abstract. Cartesian tree matching is the problem of finding all substrings of a given text which have the same Cartesian trees as that of a given pattern. So far there is one linear-time solution for Cartesian tree matching, which is based on the KMP algorithm. We improve the running time of the previous solution by introducing new representations. We present the framework of a binary filtration method and an efficient verification technique for Cartesian tree matching. Any exact string matching algorithm can be used as a filtration for Cartesian tree matching on our framework. We also present a SIMD solution for Cartesian tree matching suitable for short patterns. By experiments we show that known string matching algorithms combined on our framework of binary filtration and efficient verification produce algorithms of good performances for Cartesian tree matching.

Keywords: Cartesian tree matching · Global-parent representation · Filtration algorithms

1 Introduction

String matching is one of fundamental problems in computer science. There are generalized matchings such as parameterized matching [3,5], swapped matching [1,4], overlap matching [2], jumbled matching [6], and so on. These problems are characterized by the way of defining a match, which depends on the application domains of the problems. In particular, order-preserving matching [17,18,20] and Cartesian tree matching [21] deal with the order relations between numbers.

The Cartesian tree [23] is a tree data structure that represents a string, focusing on the orders between elements of the string. Park et al. [21] introduced a metric of match called Cartesian tree matching. It is the problem of finding all substrings of a text T which have the same Cartesian trees as that of a pattern P. Cartesian tree matching can be applied to finding patterns in time series data such as share prices in stock markets, like order-preserving matching, but sometimes it may be more appropriate as indicated in [21].

© Springer Nature Switzerland AG 2019
N. R. Brisaboa and S. J. Puglisi (Eds.): SPIRE 2019, LNCS 11811, pp. 124–137, 2019.
https://doi.org/10.1007/978-3-030-32686-9_9

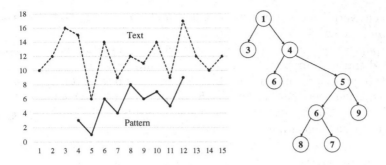

Fig. 1. Cartesian tree matching, and Cartesian tree corresponding to pattern.

Figure 1 shows an example of Cartesian tree matching. Suppose $T = (10, 12, 16, 15, 6, 14, 9, 12, 11, 14, 9, 17, 12, 10, 12)$ and $P = (3, 1, 6, 4, 8, 6, 7, 5, 9)$. The Cartesian tree of substring $u = (15, 6, 14, 9, 12, 11, 14, 9, 17)$ is the same as that of P. Note that if we use order-preserving matching instead of Cartesian tree matching as a metric, u does not match P.

String matching algorithms have been designed over the years. To speed up the search phase of string matching, algorithms based on automata and bit-parallelism were developed [12,14]. In recent years, the SIMD instruction set architecture gave rise to packed string matching, where one can compare packed data elements in parallel. In the last few years, many solutions for order-preserving matching have been proposed. Given a text of length n and a pattern of length m, Kubica et al. [20] and Kim et al. [18] gave $O(n + m \log m)$ time solutions based on the KMP algorithm. Cho et al. [11] presented an algorithm using the Boyer–Moore approach. Chhabra and Tarhio [10] presented a new practical solution based on filtration, and Chhabra et al. [9] gave a filtration algorithm using the Boyer-Moore-Horspool approach and SIMD instructions. Cantone et al. [7] proposed filtration methods using the q-neighborhood representation and SIMD instructions. These filtration methods [7,9,10] take sublinear time on average.

In this paper we introduce new representations, *prefix-parent representation* and *prefix-child representation*, which can be used to decide whether two strings have the same Cartesian trees or not. Using these representations, we improve the running time of the previous Cartesian tree matching algorithm in [21]. We also present a binary filtration method for Cartesian tree matching, and give an efficient verification technique for Cartesian tree matching based on the *global-parent representation*. On the framework of our binary filtration method and efficient verification technique, we can apply any known string matching algorithm [8,12,15] as a filtration for Cartesian tree matching. In addition, we present a SIMD solution for Cartesian tree matching based on the global-parent representation, which is suitable for short patterns. We conduct experiments comparing many algorithms for Cartesian tree matching, which show that known string matching algorithms combined on the framework of our binary filtration

and efficient verification for Cartesian tree matching produce algorithms of good performances for Cartesian tree matching.

This paper is organized as follows. In Sect. 2, we describe notations and the problem definition. In Sect. 3, we present an improved linear-time algorithm using new representations. In Sect. 4, we present the framework of binary filtration and efficient verification. In Sect. 5, we present a SIMD solution for short patterns. In Sect. 6, we give the experimental results of the previous algorithm and the proposed algorithms.

2 Preliminaries

2.1 Basic Notations

A string is defined as a finite sequence of elements in an alphabet Σ. In this paper, we will assume that Σ has a total order $<$. For a string S, $S[i]$ represents the ith element of S, and $S[i..j]$ represents a substring of S from the ith element to the jth element. If $i > j$ then $S[i..j]$ is an empty string.

We will say $S[i] \prec S[j]$, if and only if $S[i] < S[j]$, or $S[i]$ and $S[j]$ have the same value with $i < j$. Note that $S[i] = S[j]$ (as elements of the string) if and only if $i = j$. Unless stated otherwise, the minimum is defined by $\prec$.

2.2 Cartesian Tree Matching

A string S can be associated with its corresponding Cartesian tree $CT(S)$ [23] according to the following rules:

- If S is an empty string, then $CT(S)$ is an empty tree.
- If $S[1..n]$ is not empty and $S[i]$ is the minimum value among S, then $CT(S)$ is the tree with $S[i]$ as the root, $CT(S[1..i-1])$ as the left subtree, and $CT(S[i+1..n])$ as the right subtree.

Cartesian tree matching is to find all substrings of the text which have the same Cartesian trees as that of the pattern. Formally, Park et al. [21] define it as follows:

Definition 1. (Cartesian tree matching) Given two strings text $T[1..n]$ and pattern $P[1..m]$, find every $1 \leq i \leq n - m + 1$ such that $CT(T[i..i+m-1]) = CT(P[1..m])$.

Instead of building the Cartesian tree for every position in the text to solve Cartesian tree matching, Park et al. [21] use the following representation for a Cartesian tree.

Definition 2. (Parent-distance representation) Given a string $S[1..n]$, the parent-distance representation of S is a function $\mathcal{PD}_S$, which is defined as follows:

$$\mathcal{PD}_S(i) = \begin{cases} i - \max_{1 \leq j < i}\{j : S[j] \prec S[i]\} & \text{if such } j \text{ exists} \\ 0 & \text{otherwise.} \end{cases}$$

Since the parent-distance representation has a one-to-one mapping to the Cartesian tree [21], it can replace the Cartesian tree without any loss of information.

idx	1	2	3	4	5	6	7	8	9
S	3	1	6	4	8	6	7	5	9

$\mathcal{PP}_S$	1	2	2	2	4	4	6	4	8	$\mathcal{PP}_S$
$\mathcal{PC}_S$	1	1	3	3	5	5	7	6	9	$\mathcal{PC}_S$
$\mathcal{GP}_S$	2	2	4	2	6	8	6	4	8	$\mathcal{GP}_S$

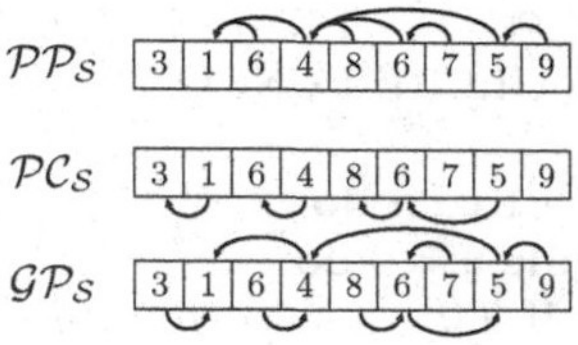

Fig. 2. $\mathcal{PP}_S, \mathcal{PC}_S, \mathcal{GP}_S$ for $S = (3, 1, 6, 4, 8, 6, 7, 5, 9)$.

3   Fast Linear Cartesian Tree Matching

The previous algorithm for Cartesian tree matching due to Park et al. [21] is based on the KMP algorithm [19]. They changed the pattern and the text to parent-distance representations and found matches using the KMP algorithm. To compute the parent-distance representations of substrings of the text using $O(m)$ space, however, they used a deque data structure. We improve the text search phase of the previous algorithm by removing the overhead of computing parent-distance representations including deque operations.

In the text search phase of the previous algorithm, the parent-distance of each element $T[i]$ in $T[i - q..i]$ is computed to check whether it matches $\mathcal{PD}_P(q + 1)$ when we know that $\mathcal{PD}_{P[1..q]}$ matches $\mathcal{PD}_{T[i-q..i-1]}$. We can do it directly without computing the parent-distances of text elements by using following representations: *prefix-parent representation* and *prefix-child representation*.

Definition 3. (prefix-parent representation) Given a string $S[1..n]$, the prefix-parent representation of S is a function $\mathcal{PP}_S$, which is defined as follows:

$$\mathcal{PP}_S(i) = \begin{cases} \max_{1 \le j < i}\{j : S[j] \prec S[i]\} & \text{if such } j \text{ exists} \\ i & \text{otherwise.} \end{cases}$$

Since $\mathcal{PP}_S(i) = i - \mathcal{PD}_S(i)$, the prefix-parent representation also has a one-to-one mapping to the Cartesian tree.

Definition 4. (prefix-child representation) Given a string $S[1..n]$, the prefix-child representation of S is a function $\mathcal{PC}_S$, which is defined as follows: $\mathcal{PC}_S(1) = 1$, and for $i \ge 2$,

$$\mathcal{PC}_S(i) = \begin{cases} j \text{ such that } S[j] \text{ is minimum for } 1 \le j < i & \text{if } \mathcal{PP}_S(i) = i \\ i & \text{if } \mathcal{PP}_S(i) = i - 1 \\ j \text{ such that } S[j] \text{ is minimum for } \mathcal{PP}_S(i) < j < i & \text{if } \mathcal{PP}_S(i) < i - 1. \end{cases}$$

In other words, $S[\mathcal{PC}_S(i)]$ is a child of $S[i]$, because $S[\mathcal{PC}_S(i)]$ is the root of $CT(S[\mathcal{PP}_S(i) + 1..i - 1])$ when $\mathcal{PP}_S(i) < i - 1$, and $S[\mathcal{PC}_S(i)]$ is the root of

---

Algorithm 1. Text search of Cartesian tree matching

```
1: procedure CARTESIAN-TREE-MATCH(T[1..n], P[1..m])
2:     (PP_P, PC_P) ← PREFIX-PARENT-CHILD-REP(P)
3:     π ← FAILURE-FUNC(P)
4:     q ← 0
5:     for i ← 1 to n do
6:         while q ≠ 0 do
7:             if T[i − q − 1 + PP_P(q + 1)] ⪯ T[i] ⪯ T[i − q − 1 + PC_P(q + 1)] then
8:                 break
9:             else
10:                 q ← π[q]
11:         q ← q + 1
12:         if q = m then
13:             print "Match occurred at i − m + 1"
14:             q ← π[q]
```

$CT(S[1..i − 1])$ when $PP_S(i) = i$. When $PP_S(i) = i − 1$, there is no child of $S[i]$ in $CT(S[1..i])$, and thus we set $PC_S(i)$ as i.

Figure 2 shows the prefix-parent representation (resp. the prefix-child representation) of string $S = (3, 1, 6, 4, 8, 6, 7, 5, 9)$ by arrows. The arrow starting from $S[i]$ indicates $PP_S(i)$ (resp. $PC_S(i)$). If $PP_S(i) = i$ (resp. $PC_S(i) = i$), we omit the arrow.

The advantage of using the prefix-child representation and the prefix-parent representation is that we can check whether each text element matches the corresponding pattern element in constant time without computing its parent-distance [21].

Theorem 1. Given two strings P and S, assume that $P[1..q]$ and $S[1..q]$ have the same prefix-parent representations. If $S[PP_P(q+1)] ⪯ S[q+1] ⪯ S[PC_P(q+1)]$, then $P[1..q+1]$ and $S[1..q+1]$ have the same prefix-parent representations, and vice versa.

Proof. ($\Longrightarrow$) If $q = 0$, $P[1]$ and $S[1]$ always have the same prefix-parent 1. Now let's assume $q \geq 1$. There are three cases, in each of which we show that $PP_P(q + 1) = PP_S(q + 1)$.

1. Case $PP_P(q + 1) = q + 1$: Since $P[PC_P(q + 1)]$ is the minimum element in $P[1..q]$ and $PP_P(i) = PP_S(i)$ for $1 \leq i \leq q$, $S[PC_P(q + 1)]$ is also the minimum element in $S[1..q]$. Therefore, if $S[q + 1] ⪯ S[PC_P(q + 1)]$ holds, then we have $PP_S(q + 1) = q + 1$.
2. Case $PP_P(q + 1) = q$: Since $S[q] ⪯ S[q + 1]$, we have $PP_S(q + 1) = q$.
3. Case $PP_P(q + 1) < q$: Since $P[PC_P(q + 1)]$ is the minimum element in $P[PP_P(q+1)+1..q]$ and $PP_P(i) = PP_S(i)$ for $1 \leq i \leq q$, $S[PC_P(q+1)]$ is also the minimum element in $S[PP_P(q+1) + 1..q]$. Therefore, if $S[PP_P(q+1)] ⪯ S[q + 1] ⪯ S[PC_P(q + 1)]$ holds, then $PP_S(q + 1) = PP_P(q + 1)$.

($\Longleftarrow$) It is trivial by definitions of PP and PC.   □

Algorithm 2. Computing prefix-parent and prefix-child representations

```
1:  procedure PREFIX-PARENT-CHILD-REP(P[1..m])
2:      ST ← an empty stack
3:      for i ← 1 to m do
4:          j_next ← i
5:          while ST is not empty do
6:              j ← ST.top
7:              if P[j] ≺ P[i] then
8:                  break
9:              ST.pop
10:             j_next ← j
11:         PC_P(i) ← j_next
12:         if ST is empty then
13:             PP_P(i) ← i
14:         else
15:             PP_P(i) ← j
16:         ST.push(i)
17:     return (PP_P, PC_P)
```

With the prefix-parent representation and the prefix-child representation of pattern P, we can simplify the text search. For each element $T[i]$, we can check $PP_P(q+1) = PP_{T[i-q..i]}(q+1)$ by comparing $T[i]$ with the elements in $T[i-q..i]$ whose indices correspond to $PP_P(q+1)$ and $PC_P(q+1)$ in P. Using this idea, we don't have to compute $PP_{T[i-q..i]}(q+1)$. Algorithm 1 describes the algorithm to do this. We compute the failure function π in the same way as [21] does.

Given a string $P[1..m]$, we can compute the prefix-child representation and the prefix-parent representation simultaneously in linear time using a stack. $PP_P(i) = j$ means that $P[j] \prec P[k]$ for $j < k < i$. The same is true for $PC_P(i)$. On the stack, therefore, we maintain only j's which satisfy $P[j] \prec P[k]$ for $j < k < i$ while scanning from $i = 1$ to m. Suppose that $j_1, j_2, \ldots, j_r$ are on the stack when we are computing $PP_P(i)$ and $PC_P(i)$. (We assume that $j_{r+1} = i$.) Then, $(P[j_1], P[j_2], \ldots, P[j_r])$ forms an increasing subsequence of P. When we consider a new index i, we pop the indices $j_r, j_{r-1}, \ldots, j_{t+1}$ repeatedly until we have $P[j_t] \prec P[i]$. If there exists such an index j_t, we set $PP_P(i) = j_t$ and $PC_P(i) = j_{t+1}$. (If $t = r$, then $PC_P(i) = j_{t+1} = i$.) Otherwise, $P[i]$ is the minimum element in $P[1..i]$, and thus $PP_P(i) = i$ and $PC_P(i) = j_1$. Finally, we push i onto the stack. Algorithm 2 describes the algorithm to compute PP_P and PC_P simultaneously.

4   Fast Cartesian Tree Matching with Filtration

In this section we present a practical solution based on filtration. Our solution for Cartesian tree matching consists of two phases: filtration and verification. First, the text is filtered with some exact string matching algorithm using a binary representation. In the second phase, the potential candidates are verified using a global-parent representation.

4.1   Filtration

In the filtration phase, a string S is translated into a *binary representation* β_S as follows.

Definition 5. (binary representation) Given a string $S[1..n]$, the binary representation of S is a binary string β_S of length $n - 1$, which is defined as follows:

$$\beta_S[i] = \begin{cases} 0 & \text{if } \mathcal{PP}_S(i+1) = i \\ 1 & \text{otherwise,} \end{cases}$$

for each $1 \leq i \leq n - 1$.

One can easily check whether $\mathcal{PP}_S(i + 1) = i$ is true or not by comparing $S[i]$ and $S[i+1]$: $\mathcal{PP}_S(i+1) = i$ if and only if $S[i] \prec S[i+1]$. The following theorem proves that the binary representation can be used to filter a text T to search for all Cartesian tree matching occurrences of a pattern P.

Theorem 2. Let P and T be two strings of lengths m and n, respectively, and let β_P and β_T be the binary representations associated with P and T, respectively. If $CT(P[1..m]) = CT(T[i..i + m - 1])$, then $\beta_P[j] = \beta_T[i + j - 1]$ for $1 \leq j \leq m - 1$.

Proof. The prefix-parent representation has a one-to-one mapping to the Cartesian tree. Therefore, if $CT(P[1..m]) = CT(T[i..i + m - 1])$, then $\mathcal{PP}_P(j + 1) = \mathcal{PP}_T(i + j)$ for $0 \leq j \leq m - 1$. If $\mathcal{PP}_P(j + 1) = \mathcal{PP}_T(i + j)$, then $\beta_P[j] = \beta_T[i + j - 1]$ for $1 \leq j \leq m - 1$.

Theorem 2 guarantees that any standard exact string matching algorithm can be used as a filtration procedure. As the exact string matching algorithm returns matches of β_P in β_T, these matches are only possible candidates of Cartesian tree matching which should be verified.

Cantone et al. [7] presented two filtration methods other than the binary representation to solve order-preserving matching. They used the property that T doesn't match P at position i if there are two positions j and k such that $P[j] \preceq P[k] \Leftrightarrow T[i+j-1] \preceq T[i+k-1]$ doesn't hold. Thus any comparison result between two positions can be used for filtration. In Cartesian tree matching, however, even if there exist such j and k, the corresponding Cartesian trees can be the same when $|j - k| > 1$. Therefore, we cannot use these filtration methods for Cartesian tree matching.

4.2   Verification

In the verification phase, we have to check whether the candidates found by the filtration phase are actual matches or not. This checking can be done using prefix-parent and prefix-child representations by Theorem 1, which takes 2 comparisons per element. In order to reduce the number of comparisons to 1, we introduce another representation as follows.

Definition 6. (Global-parent representation) Given a string $S[1..n]$, the global-parent representation of S is a function $\mathcal{GP}_S$, which is defined as follows:

$$\mathcal{GP}_S(i) = \begin{cases} j & \text{such that } \mathcal{PC}_S(j) = i \text{ for } j > i \\ \mathcal{PP}_S(i) & \text{if such } j \text{ doesn't exist.} \end{cases}$$

$\mathcal{GP}_S(i)$ is well-defined because there is at most one $j > i$ which satisfies $\mathcal{PC}_S(j) = i$. Figure 2 shows the global-parent representation by arrows. The arrow starting from $S[i]$ indicates the global parent of $S[i]$. If $\mathcal{GP}_S(i) = i$, we omit the arrow.

Theorem 3. Two strings $P[1..m]$ and $S[1..m]$ have the same Cartesian trees if and only if $S[\mathcal{GP}_P(i)] \preceq S[i]$ for all $1 \le i \le m$.

Proof. We will prove that $S[\mathcal{PP}_P(i)] \preceq S[i] \preceq S[\mathcal{PC}_P(i)]$ for all $1 \le i \le m$ if and only if $S[\mathcal{GP}_P(i)] \preceq S[i]$ for all $1 \le i \le m$.

($\Longrightarrow$) It is trivial by definition of $\mathcal{GP}$.

($\Longleftarrow$) Assume $S[\mathcal{GP}_P(i)] \preceq S[i]$ for all $1 \le i \le m$. For any $1 \le k \le m$, we first show $S[k] \preceq S[\mathcal{PC}_P(k)]$, and then we show $S[\mathcal{PP}_P(k)] \preceq S[k]$.

1. (Proof of $S[k] \preceq S[\mathcal{PC}_P(k)]$) There are two cases: $\mathcal{PC}_P(k) = k$ and $\mathcal{PC}_P(k) \ne k$. If $\mathcal{PC}_P(k) = k$, then $S[k] \preceq S[\mathcal{PC}_P(k)]$ holds trivially. Otherwise, since $\mathcal{GP}_P(\mathcal{PC}_P(k)) = k$, $S[k] = S[\mathcal{GP}_P(\mathcal{PC}_P(k))] \preceq S[\mathcal{PC}_P(k)]$. Therefore, $S[k] \preceq S[\mathcal{PC}_P(k)]$ holds.

2. (Proof of $S[\mathcal{PP}_P(k)] \preceq S[k]$) If $\mathcal{GP}_P(k) = \mathcal{PP}_P(k)$, then $S[\mathcal{PP}_P(k)] = S[\mathcal{GP}_P(k)] \preceq S[k]$. So we only have to consider the case that there is $k_1 > k$ which satisfies $\mathcal{PC}_P(k_1) = k$. Let $k = k_0 < k_1 < \cdots < k_r \le m$ be a sequence such that $\mathcal{PC}_P(k_{l+1}) = k_l$, and there is no $k_{r+1} > k_r$ which satisfies $\mathcal{PC}_P(k_{r+1}) = k_r$. Since $(k_0, k_1, \ldots, k_r)$ is a strictly increasing sequence, such k_r always exists. Note that $\mathcal{GP}_P(k_l) = k_{l+1}$ except for $\mathcal{GP}_P(k_r)$. On the sequence, there may or may not exist j such that $\mathcal{PP}_P(k_j) = k_j$.

Suppose that there exists some j such that $\mathcal{PP}_P(k_j) = k_j$. Since $k_{j-1} = \mathcal{PC}_P(k_j)$, $P[k_{j-1}]$ is the minimum element in $P[1..k_j - 1]$, and so $\mathcal{PP}_P(k_{j-1}) = k_{j-1}$. Proceeding inductively, $\mathcal{PP}_P(k_l) = k_l$ for all $l \le j$. Thus $S[\mathcal{PP}_P(k)] \preceq S[k]$ holds trivially.

Now we consider the case that $\mathcal{PP}_P(k_j) \ne k_j$ for all j. Then, we have $S[k_0] \succeq S[k_1] \succeq \cdots \succeq S[k_r] \succeq S[\mathcal{GP}_P(k_r)] = S[\mathcal{PP}_P(k_r)]$ by the assumption that $S[\mathcal{GP}_P(i)] \preceq S[i]$ for all i. We now show $\mathcal{PP}_P(k_r) = \mathcal{PP}_P(k)$ as follows. Since $\mathcal{PC}_P(k_r) = k_{r-1}$, $P[k_{r-1}]$ is the minimum element in $P[\mathcal{PP}_P(k_r) + 1..k_r - 1]$, and $P[k_{r-1}] \succeq P[\mathcal{PP}_P(k_r)]$. Hence, we have $\mathcal{PP}_P(k_{r-1}) = \mathcal{PP}_P(k_r)$. Inductively, we can show that $\mathcal{PP}_P(k_0) = \mathcal{PP}_P(k_1) = \cdots = \mathcal{PP}_P(k_r)$. Therefore, $S[\mathcal{PP}_P(k)] \preceq S[k]$ holds. $\square$

By Theorem 3, we only have to compare once for each element in the verification phase. For a potential candidate obtained from the filtration phase (say, it starts from $T[i]$), we compare $T[i + q - 1]$ and $T[i + \mathcal{GP}_P(q) - 1]$ from $q = 1$ to m. The candidate is discarded when there exists q such that $T[i + q - 1] \prec T[i + \mathcal{GP}_P(q) - 1]$.

Algorithm 3. Compare integers in parallel

1: **procedure** COMPAREUSINGSIMD($T[1..n], i$)
2:     _m128i $a \leftarrow$ _mm_loadu_si128((_m128i *)($T + i$))
3:     _m128i $b \leftarrow$ _mm_loadu_si128((_m128i *)($T + i + 1$))
4:     _m128i $r \leftarrow$ _mm_cmpgt_epi32(a, b)
5:     **return** _mm_movemask_ps(_mm_castsi128_ps(r))

We compute the global-parent representation using a stack, as in computing the prefix-parent and the prefix-child representations. The only difference is that first we set $\mathcal{GP}_P(i)$ as $\mathcal{PP}_P(i)$, and then if we find j such that $\mathcal{PC}_P(j) = i$ we update $\mathcal{GP}_P(i)$ to j.

4.3   Sublinear Time on Average

The proof of sublinearity is similar to the analysis of order-preserving matching with filtration [10]. Let's assume that the elements in the pattern P and the text T are independent of each other and the distribution is uniform. The verification phase takes time proportional to the pattern length times the number of potential candidates. When alphabet size is $|\Sigma|$, the probability that $\beta_P[i] = 0$ (i.e., probability that $P[i] \prec P[i+1]$) is $(|\Sigma|^2 + |\Sigma|)/(2|\Sigma|^2)$, since there are $|\Sigma|^2$ pairs and $|\Sigma|$ pairs among them have equal elements. Similarly, the probability that $\beta_P[i] = 1$ is $(|\Sigma|^2 - |\Sigma|)/(2|\Sigma|^2)$, and it is the same for $\beta_T[i]$. Therefore, the probability that $\beta_P[i] = \beta_T[i]$ is $((|\Sigma|^2 + |\Sigma|)/(2|\Sigma|^2))^2 + ((|\Sigma|^2 - |\Sigma|)/(2|\Sigma|^2))^2 = 1/2 + 1/(2|\Sigma|^2)$. As the pattern length increases, the number of potential candidates decreases exponentially, and the verification time approaches zero. Hence, the filtration time dominates. So if the filtration method takes a sublinear time in the average case, the total algorithm takes a sublinear time in the average case, too.

4.4   SIMD Instructions

When we use the Boyer-Moore-Horspool algorithm [15] and the Alpha skip search algorithm [8] as the filtration method, we pack four 32-bit numbers or sixteen 8-bit numbers into a register, as in order-preserving matching algorithms [7,9]. Each pair of two corresponding packed data elements can be compared in parallel using streaming SIMD extensions (SSE) [16]. In the case of 32-bit integers, for example, we compute $(T[i + 3] > T[i + 4])$, $(T[i + 2] > T[i + 3])$, $(T[i + 1] > T[i + 2])$, and $(T[i] > T[i + 1])$ in parallel as in Algorithm 3, where instruction _mm_loadu_si128((_m128i *)($T + i$)) loads four 32-bit integers from memory $T+i$ into a 128-bit register, instruction _mm_cmpgt_epi32(a, b) compares four pairs of packed 32-bit integers and returns the results of the comparisons into a 128-bit register, instruction _mm_castsi128_ps casts the integer type to the float type, and instruction _mm_movemask_ps selects only the most significant bits of the 4 floats. Comparing a pair of sixteen 8-bit numbers can be done similarly.

5   SIMD Solution for Short Patterns

In this section we present an algorithm that works when the alphabet consists of 1-byte characters and the pattern length m is at most 16. As shown in Sect. 4.2, we test $T[s+i-1] \succeq T[s+\mathcal{GP}_P(i)-1]$ for $1 \le i \le m$ to check for an occurrence at position s of the text T.

Let W be a word of 16 bytes containing the current window of the text, i.e., $W = T[s..s+15]$. For $1 \le i \le m$, we define W_i (word obtained from W by shifting $i - \mathcal{GP}_P(i)$ positions to the left or to the right, depending on the sign of $i - \mathcal{GP}_P(i)$) as follows:

$$W_i = \begin{cases} W \ll (\mathcal{GP}_P(i) - i) & \text{if } i < \mathcal{GP}_P(i) \\ W \gg (i - \mathcal{GP}_P(i)) & \text{if } i > \mathcal{GP}_P(i). \end{cases}$$

For fixed i, we can find the positions j which satisfy $W[j + i - 1] \succeq W[j + \mathcal{GP}_P(i) - 1]$ for $0 \le j \le 15$ in parallel by comparing W_i to W using SIMD instructions. The satisfying positions for all $1 \le i \le m$ are the occurrences of the pattern. The details of the algorithm are as follows. We test whether $W_i[j] \preceq W[j]$ for $0 \le j \le 15$ in parallel using the SIMD instruction $R_i = \_mm\_cmpgt\_epi8(W, W_i)$ for $i < \mathcal{GP}_P(i)$ or $R_i = \sim\_mm\_cmpgt\_epi8(W_i, W)$ for $i > \mathcal{GP}_P(i)$. (In order to get only significant bits when computing R_i, we use instruction $\_mm\_movemask\_epi8$.) Then we compute $q = \text{AND}_{i=1}^m (R_i \ll (i-1))$. Finally, we report a match at position $s + j$ of the text if $q[j] = 1$.

Example 1. Let's consider an example of the pattern $P = (3, 1, 6, 4, 8)$ and the window of the text $W = (10, 12, 16, 15, 6, 14, 9, 12, 11, 14, 9, 17, 12, 13, 12, 10)$. We observe that since $1 - \mathcal{GP}_P(1) = 3 - \mathcal{GP}_P(3)$, $R_1 = R_3$. Moreover we do not need to compute R_2, since $2 - \mathcal{GP}_P(2) = 0$. Hence we compute R_1, R_4, and R_5.

$$
\begin{array}{llllllllllllllllll}
W   & = 10, & 12, & 16, & 15, & 06, & 14, & 09, & 12, & 11, & 14, & 09, & 17, & 12, & 13, & 12, & 10 \\
W_1 & = 12, & 16, & 15, & 06, & 14, & 09, & 12, & 11, & 14, & 09, & 17, & 12, & 13, & 12, & 10 \\
R_1 & = 0, & 0, & 1, & 1, & 0, & 1, & 0, & 1, & 0, & 1, & 0, & 1, & 0, & 1, & 1, & - \\
W   & = 10, & 12, & 16, & 15, & 06, & 14, & 09, & 12, & 11, & 14, & 09, & 17, & 12, & 13, & 12, & 10 \\
W_4 & = & & & 10, & 12, & 16, & 15, & 06, & 14, & 09, & 12, & 11, & 14, & 09, & 17, & 12, & 13 \\
R_4 & = -, & -, & 1, & 1, & 0, & 0, & 1, & 0, & 1, & 1, & 0, & 1, & 1, & 0, & 1, & 1 \\
W   & = 10, & 12, & 16, & 15, & 06, & 14, & 09, & 12, & 11, & 14, & 09, & 17, & 12, & 13, & 12, & 10 \\
W_5 & = & & 10, & 12, & 16, & 15, & 06, & 14, & 09, & 12, & 11, & 14, & 09, & 17, & 12, & 13 \\
R_5 & = -, & 1, & 1, & 0, & 0, & 1, & 0, & 1, & 0, & 1, & 0, & 1, & 0, & 1, & 0, & 0 \\
\end{array}
$$

The final result q can be computed as follows:

$$
\begin{array}{lll}
R_1 & = 0, 0, 1, 1, 0, 1, 0, 1, 0, 1, 0, 1, 0, \ 1, 1, - \\
R_3 \ll 2 & = 1, 1, 0, 1, 0, 1, 0, 1, 0, 1, 0, 1, 1, \ -, 0, \ 0 \\
R_4 \ll 3 & = 1, 0, 0, 1, 0, 1, 1, 0, 1, 1, 0, 1, 1, \ 0, 0, \ 0 \\
R_5 \ll 4 & = 0, 1, 0, 1, 0, 1, 0, 1, 0, 1, 0, 0, 0, \ 0, 0, \ 0 \\
\hline
q & = 0, 0, 0, 1, 0, 1, 0, 0, 0, 1, 0, 0, 0, \ 0, 0, \ 0
\end{array}
$$

Table 1. Execution times in seconds for random patterns in texts (Random datasets: for 100 patterns, Seoul temperatures dataset: for 1000 patterns).

Dataset	m	KMPCT	IKMPCT	SBNDMCT			BMHCT				SKSCT				PMCT
				2	4	6	4	8	12	16	4	8	12	16	
Random int	5	10.52	6.84	4.99	4.42		4.17				**3.31**				
	9	10.71	6.83	2.71	2.31	1.95	1.95	**1.64**			1.91	2.26			
	17	10.69	6.83	1.39	1.34	0.95	1.31	0.80	0.86	1.60	1.13	**0.45**	0.61	3.91	
	33	10.69	6.83	0.72	0.70	0.65	1.07	0.51	0.51	1.01	0.76	0.32	**0.30**	0.48	
	65	10.71	6.83	0.72	0.71	0.66	0.98	0.44	0.43	0.71	0.61	0.27	**0.24**	0.28	
Seoul temp	5	5.08	3.07	2.67	2.91		2.52				**2.27**				
	9	5.11	3.14	1.56	1.45	1.55	1.55	**1.23**			1.27	1.77			
	17	5.51	3.12	0.89	0.81	0.71	1.10	0.62	0.63	0.84	0.88	**0.44**	0.49	2.55	
	33	5.56	3.12	0.49	0.48	0.45	0.84	0.40	0.34	0.41	0.68	0.32	**0.20**	0.25	
	65	5.52	3.11	0.48	0.48	0.46	0.77	0.26	0.19	0.28	0.57	0.25	0.13	**0.12**	
Random char	5	10.24	6.86	4.80	4.44		3.95				3.22				**0.50**
	7	10.32	6.86	3.53	2.89	4.47	2.39				2.40				**0.84**
	9	10.34	6.85	2.65	2.32	1.94	1.74	**1.24**			1.91	1.47			1.32
	13	10.32	6.85	1.75	1.68	1.10	1.23	0.70	0.68		1.34	**0.45**	1.15		3.76
	17	10.35	6.86	1.28	1.25	0.87	1.04	0.52	0.49	0.79	1.04	**0.27**	0.32	1.64	
	33	10.34	6.85	0.61	0.60	0.54	0.78	0.29	0.26	0.43	0.66	0.16	**0.09**	0.11	
	65	10.36	6.86	0.63	0.63	0.55	0.74	0.20	0.17	0.27	0.47	0.13	**0.04**	0.05	

Therefore, we can report 3 matches. After we have tested a window of the text, we shift the current window to the right by $17-m$ positions. This algorithm takes $O(mn/(17-m))$ SIMD instructions.

6   Experiments

In this section we conduct experiments comparing the following algorithms.

- KMPCT: algorithm of Park, Amir, Landau, and Park [21]
- IKMPCT: our improved linear-time algorithm based on prefix-parent and prefix-child representations (Sect. 3)
- PMCT: SIMD solution for short patterns (Sect. 5)
- SBNDMCTq: algorithm based on the SBNDMq filtration implemented by Faro and Lecroq [13] on the binary representations of the text and the pattern (Sect. 4.1) and verification using the global-parent representation (Sect. 4.2) [12] (The following algorithms have the same framework as SBNDMCTq; only SBNDMq is replaced by another filtration method.)
- BMHCTq: algorithm based on the q-gram Boyer-Moore-Horspool filtration using SIMD instructions [9,15,22]
- SKSCTq: algorithm based on the q-gram Alpha skip search filtration using SIMD instructions [7,8]

We tested for two random datasets and one real dataset, which is a time series of Seoul temperatures. The first random dataset consists of 10,000,000 random integers. The second random dataset consists of 10,000,000 random characters. The Seoul temperatures dataset consists of 658,795 integers referring to the

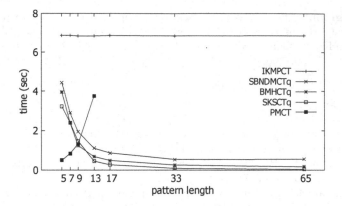

Fig. 3. Execution times for the random character dataset.

hourly temperatures in Seoul (multiplied by ten) in the years 1907-2019. In general, temperatures rise during the day and fall at night. Therefore, the Seoul temperatures dataset has more matches than random datasets. We picked 100 random patterns per pattern length from random datasets and 1000 random patterns per pattern length for the Seoul temperatures dataset.

The experimental environments and parameters are as follows. All algorithms were implemented in C++11 and compiled with GNU C++ compiler version 4.8.5, and O3 and msse4 options were used. The experiments were performed on a CentOS Linux 7 with 128 GB RAM and Intel Xeon CPU E5-2630 processor.

Table 1 shows the total execution times of Cartesian tree matching algorithms for random patterns (including the preprocessing). The best results are boldfaced. We choose the best results of the random character dataset from each algorithm regardless of q and present them in Fig. 3 (except KMPCT because of readability). Our linear-time algorithm IKMPCT improves upon algorithm KMPCT of [21] by about 35%. In the random character dataset, PMCT is the fastest algorithm for short patterns. However, as the pattern length grows, algorithms based on the filtration method are much faster in practice. It can be seen that SKSCT is the fastest algorithm in most cases. When the pattern length is equal to 9, BMHCT utilizing 8-grams is the fastest algorithm, irrespective of the datasets. As pattern length grows, SKSCT utilizing 12-grams becomes the fastest algorithm.

Regardless of the data type, the results are almost consistent. In details, however, there are several differences. First, filtration algorithms, especially SKSCT algorithms, are slower at the Seoul temperatures dataset relatively. It's because there are more matches in the Seoul temperatures dataset. Second, when q is large, BMHCT and SKSCT algorithms are faster in the random character dataset than in the random integer dataset. It's because the maximum number that we can compute in parallel is 16 in the character dataset while it is 4 in the integer dataset.

Acknowledgments. Song, Ryu and Park were supported by Collaborative Genome Program for Fostering New Post-Genome industry through the National Research Foundation of Korea(NRF) funded by the Ministry of Science ICT and Future Planning (No. NRF-2014M3C9A3063541).

References

1. Amir, A., Aumann, Y., Landau, G.M., Lewenstein, M., Lewenstein, N.: Pattern matching with swaps. J. Algorithms **37**(2), 247–266 (2000)
2. Amir, A., Cole, R., Hariharan, R., Lewenstein, M., Porat, E.: Overlap matching. Inf. Comput. **181**(1), 57–74 (2003)
3. Amir, A., Farach, M., Muthukrishnan, S.: Alphabet dependence in parameterized matching. Inf. Process. Lett. **49**(3), 111–115 (1994)
4. Amir, A., Lewenstein, M., Porat, E.: Approximate swapped matching. Inf. Process. Lett. **83**(1), 33–39 (2002)
5. Baker, B.S.: A theory of parameterized pattern matching: algorithms and applications. In: Proceedings of the Twenty-fifth Annual ACM Symposium on Theory of Computing, pp. 71–80. ACM (1993)
6. Burcsi, P., Cicalese, F., Fici, G., Liptak, Z.: Algorithms for jumbled pattern matching in strings. Int. J. Found. Comput. Sci. **23**(2), 357–374 (2012)
7. Cantone, D., Faro, S., Kulekci, M.O.: The order-preserving pattern matching problem in practice. Discrete Applied Mathematics (2018, in press)
8. Charras, C., Lecroq, T., Pehoushek, J.D.: A very fast string matching algorithm for small alphabets and long patterns. In: Combinatorial Pattern Matching, pp. 55–64 (1998)
9. Chhabra, T., Kulekci, M.O., Tarhio, J.: Alternative algorithms for order-preserving matching. In: Proceedings of the Prague Stringology Conference 2015, pp. 36–46 (2015)
10. Chhabra, T., Tarhio, J.: A filtration method for order-preserving matching. Inf. Process. Lett. **116**(2), 71–74 (2016)
11. Cho, S., Na, J.C., Park, K., Sim, J.S.: A fast algorithm for order-preserving pattern matching. Inf. Process. Lett. **115**(2), 397–402 (2015)
12. Durian, B., Holub, J., Peltola, H., Tarhio, J.: Improving practical exact string matching. Inf. Process. Lett. **110**(4), 148–152 (2010)
13. Faro, S., Lecroq, T., Borzi, S., Mauro, S.D., Maggio, A.: The string matching algorithms research tool. In: Proceedings of the Prague Stringology Conference 2016, pp. 99–113 (2016)
14. Fredriksson, K., Grabowski, S.: Practical and optimal string matching. In: String Processing and Information Retrieval, pp. 376–387 (2005)
15. Horspool, R.N.: Practical fast searching in strings. Softw. Pract. Experience **10**(6), 501–506 (1980)
16. Intel: Intel (R) 64 and IA-32 Architectures Optimization Reference Manual (2019)
17. Kim, J., Amir, A., Na, J.C., Park, K., Sim, J.S.: On representations of ternary order relations in numeric strings. Math. Comput. Sci. **11**(2), 127–136 (2017)
18. Kim, J., et al.: Order-preserving matching. Theoret. Comput. Sci. **525**, 68–79 (2014)
19. Knuth, D.E., Morris Jr., J.H., Pratt, V.R.: Fast pattern matching in strings. SIAM J. Comput. **6**(2), 323–350 (1977)

20. Kubica, M., Kulczynski, T., Radoszewski, J., Rytter, W., Walen, T.: A linear time algorithm for consecutive permutation pattern matching. Inf. Process. Lett. **113**(12), 430–433 (2013)

21. Park, S.G., Amir, A., Landau, G.M., Park, K.: Cartesian tree matching and indexing. In: Combinatorial Pattern Matching, pp. 16:1–16:14 (2019). https://arxiv.org/abs/1905.08974

22. Tarhio, J., Peltola, H.: String matching in the DNA alphabet. Software: Pract. Experience **27**(7), 851–861 (1997)

23. Vuillemin, J.: A unifying look at data structures. Commun. ACM **23**(4), 229–239 (1980)

Inducing the Lyndon Array

Felipe A. Louza[1]([📧])([ID]), Sabrina Mantaci[2]([ID]), Giovanni Manzini[3,4]([ID]),
Marinella Sciortino[2]([ID]), and Guilherme P. Telles[5]([ID])

[1] Faculty of Electrical Engineering,
Universidade Federal de Uberlândia, Uberlândia, Brazil
louza@ufu.br
[2] Dipartimento di Matematica e Informatica, University of Palermo, Palermo, Italy
{sabrina.mantaci,marinella.sciortino}@unipa.it
[3] University of Eastern Piedmont, Alessandria, Italy
[4] IIT CNR, Pisa, Italy
giovanni.manzini@uniupo.it
[5] Institute of Computing, University of Campinas, Campinas, Brazil
gpt@ic.unicamp.br

Abstract. In this paper we propose a variant of the induced suffix sorting algorithm by Nong (TOIS, 2013) that computes simultaneously the Lyndon array and the suffix array of a text in $O(n)$ time using $\sigma + O(1)$ words of working space, where n is the length of the text and σ is the alphabet size. Our result improves the previous best space requirement for linear time computation of the Lyndon array. In fact, all the known linear algorithms for Lyndon array computation use suffix sorting as a preprocessing step and use $O(n)$ words of working space in addition to the Lyndon array and suffix array. Experimental results with real and synthetic datasets show that our algorithm is not only space-efficient but also fast in practice.

Keywords: Lyndon array · Suffix array · Induced suffix sorting · Lightweight algorithms

1 Introduction

The suffix array is a central data structure for string processing. Induced suffix sorting is a remarkably powerful technique for the construction of the suffix array. Induced sorting was introduced by Itoh and Tanaka [10] and later refined by Ko and Aluru [11] and by Nong et al. [18,19]. In 2013, Nong [17] proposed a space efficient linear time algorithm based on induced sorting, called SACA-K, which uses only $\sigma + O(1)$ words of working space, where σ is the alphabet size and the working space is the space used in addition to the input and the output. Since a small working space is a very desirable feature, there have been many algorithms adapting induced suffix sorting to the computation of data structures related to the suffix array, such as the Burrows-Wheeler transform [21], the Φ-array [8], the LCP array [4,14], and the document array [13].

© Springer Nature Switzerland AG 2019
N. R. Brisaboa and S. J. Puglisi (Eds.): SPIRE 2019, LNCS 11811, pp. 138–151, 2019.
https://doi.org/10.1007/978-3-030-32686-9_10

The Lyndon array of a string is a powerful tool that generalizes the idea of Lyndon factorization. In the Lyndon array (LA) of string $T = T[1] \ldots T[n]$ over the alphabet Σ, each entry LA[i], with $1 \leq i \leq n$, stores the length of the longest Lyndon factor of T starting at that position i. Bannai *et al.* [2] used Lyndon arrays to prove the conjecture by Kolpakov and Kucherov [12] that the number of runs (maximal periodicities) in a string of length n is smaller than n. In [3] the authors have shown that the computation of the Lyndon array of T is strictly related to the construction of the Lyndon tree [9] of the string $\$T$ (where the symbol $\$$ is smaller than any symbol of the alphabet Σ).

In this paper we address the problem of designing a space economical linear time algorithm for the computation of the Lyndon array. As described in [5,15], there are several algorithms to compute the Lyndon array. It is noteworthy that the ones that run in linear time (cf. [1,3,5,6,15]) use the sorting of the suffixes (or a partial sorting of suffixes) of the input string as a preprocessing step. Among the linear time algorithms, the most space economical is the one in [5] which, in addition to the $n \log \sigma$ bits for the input string plus $2n$ words for the Lyndon array and suffix array, uses a stack whose size depends on the structure of the input. Such stack is relatively small for non pathological texts, but in the worst case its size can be up to n words. Therefore, the overall space in the worst case can be up to $n \log \sigma$ bits plus $3n$ words.

In this paper we propose a variant of the algorithm SACA-K that computes in linear time the Lyndon array as a by-product of suffix array construction. Our algorithm uses overall $n \log \sigma$ bits plus $2n + \sigma + O(1)$ words of space. This bound makes our algorithm the one with the best worst case space bound among the linear time algorithms. Note that the $\sigma + O(1)$ words of working space of our algorithm is optimal for strings from alphabets of constant size. Our experiments show that our algorithm is competitive in practice compared to the other linear time solutions to compute the Lyndon array.

2 Background

Let $T = T[1] \ldots T[n]$ be a string of length n over a fixed ordered alphabet Σ of size σ, where $T[i]$ denotes the i-th symbol of T. We denote $T[i,j]$ as the factor of T starting from the i-th symbol and ending at the j-th symbol. A suffix of T is a factor of the form $T[i,n]$ and is also denoted as T_i. In the following we assume that any integer array of length n with values in the range $[1, n]$ takes n words ($n \log n$ bits) of space.

Given $T = T[1] \ldots T[n]$, the *i-th rotation* of T begins with $T[i+1]$, corresponding to the string $T' = T[i+1] \ldots T[n]T[1] \ldots T[i]$. Note that, a string of length n has n possible rotations. A string T is a *repetition* if there exists a string S and an integer $k > 1$ such that $T = S^k$, otherwise it is called *primitive*. If a string is primitive, all of its rotations are different.

A primitive string T is called a *Lyndon word* if it is the lexicographical least among its rotations. For instance, the string $T = abanba$ is not a Lyndon word, while it is its rotation $aabanb$. A *Lyndon factor* of a string T is a factor of T that is a Lyndon word. For instance, anb is a Lyndon factor of $T = abanba$.

Definition 1. *Given a string $T = T[1] \ldots T[n]$, the Lyndon array (LA) of T is an array of integers in the range $[1, n]$ that, at each position $i = 1, \ldots, n$, stores the length of the longest Lyndon factor of T starting at i:*

$$\mathsf{LA}[i] = \max\{\ell \mid T[i, i + \ell - 1] \text{ is a Lyndon word}\}.$$

The suffix array (SA) [16] of a string $T = T[1] \ldots T[n]$ is an array of integers in the range $[1, n]$ that gives the lexicographic order of all suffixes of T, that is $T_{\mathsf{SA}[1]} < T_{\mathsf{SA}[2]} < \cdots < T_{\mathsf{SA}[n]}$. The inverse suffix array (ISA) stores the inverse permutation of SA, such that $\mathsf{ISA}[\mathsf{SA}[i]] = i$. The suffix array can be computed in $O(n)$ time using $\sigma + O(1)$ words of working space [17].

Usually when dealing with suffix arrays it is convenient to append to the string T a special end-marker symbol \$ (called sentinel) that does not occur elsewhere in T and \$ is smaller than any other symbol in Σ. Here we assume that $T[n] = \$$. Note that the values $\mathsf{LA}[i]$, for $1 \leq i \leq n - 1$ do not change when the symbol \$ is appended at the position n. Also, string $T = T[1] \ldots T[n - 1]\$$ is always primitive.

Given an array of integers A of size n, the next smaller value (NSV) array of A, denoted NSV_A, is an array of size n such that $\mathsf{NSV}_\mathsf{A}[i]$ contains the smallest position $j > i$ such that $\mathsf{A}[j] < \mathsf{A}[i]$, or $n + 1$ if such a position j does not exist. Formally:

$$\mathsf{NSV}_\mathsf{A}[i] = \min\big\{\{n + 1\} \cup \{i < j \leq n \mid \mathsf{A}[j] < \mathsf{A}[i]\}\big\}.$$

As an example, in Fig. 1 we consider the string $T = banaanananaanana\$$, and its Suffix Array (SA), Inverse Suffix Array (ISA), Next Smaller Value array of the ISA ($\mathsf{NSV}_{\mathsf{ISA}}$), and Lyndon Array (LA). We also show all the Lyndon factors starting at each position of T.

If the SA of T is known, the Lyndon array LA can be computed in linear time thanks to the following lemma that rephrases a result in [9]:

Lemma 1. *The factor $T[i, i + \ell - 1]$ is the longest Lyndon factor of T starting at i iff $T_i < T_{i+k}$, for $1 \leq k < \ell$, and $T_i > T_{i+\ell}$. Therefore, $\mathsf{LA}[i] = \ell$.*   □

Lemma 1 can be reformulated in terms of the inverse suffix array [5], such that $\mathsf{LA}[i] = \ell$ iff $\mathsf{ISA}[i] < \mathsf{ISA}[i + k]$, for $1 \leq k < \ell$, and $\mathsf{ISA}[i] > \mathsf{ISA}[i + \ell]$. In other words, $i + \ell = \mathsf{NSV}_{\mathsf{ISA}}[i]$. Since given ISA we can compute $\mathsf{NSV}_{\mathsf{ISA}}$ in linear time using an auxiliary stack [7,20] of size $O(n)$ words, we can then derive LA, in the same space of $\mathsf{NSV}_{\mathsf{ISA}}$, in linear time using the formula:

$$\mathsf{LA}[i] = \mathsf{NSV}_{\mathsf{ISA}}[i] - i, \text{ for } 1 \leq i \leq n. \tag{1}$$

Overall, this approach uses $n \log \sigma$ bits for T plus $2n$ words for LA and ISA, and the space for the auxiliary stack.

Alternatively, LA can be computed in linear time from the Cartesian tree [22] built for ISA [3]. Recently, Franek *et al.* [6] observed that LA can be computed in linear time during the suffix array construction algorithm by Baier [1] using overall $n \log \sigma$ bits plus $2n$ words for LA and SA plus $2n$ words for auxiliary

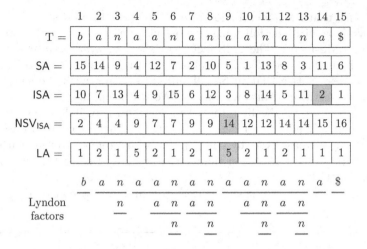

Fig. 1. SA, ISA, NSV$_{ISA}$, LA and all Lyndon factors for $T = banaananaanana\$$

integer arrays. Finally, Louza *et al.* [15] introduced an algorithm that computes LA in linear time during the Burrows-Wheeler inversion, using $n \log \sigma$ bits for T plus $2n$ words for LA and an auxiliary integer array, plus a stack with twice the size as the one used to compute NSV$_{ISA}$ (see Sect. 4).

Summing up, the most economical linear time solution for computing the Lyndon array is the one based on (1) that requires, in addition to T and LA, n words of working space plus an auxiliary stack. The stack size is small for non pathological inputs but can use n words in the worst case (see also Sect. 4). Therefore, considering only LA as output, the working space is $2n$ words in the worst case.

2.1 Induced Suffix Sorting

The algorithm SACA-K [17] uses a technique called induced suffix sorting to compute SA in linear time using only $\sigma + O(1)$ words of working space. In this technique each suffix T_i of $T[1, n]$ is classified according to its lexicographical rank relative to T_{i+1}.

Definition 2. *A suffix T_i is S-type if $T_i < T_{i+1}$, otherwise T_i is L-type. We define T_n as S-type. A suffix T_i is LMS-type (leftmost S-type) if T_i is S-type and T_{i-1} is L-type.*

The type of each suffix can be computed with a right-to-left scanning of T [18], or otherwise it can be computed on-the-fly in constant time during Nong's algorithm [17, Section 3]. By extension, the type of each symbol in T can be classified according to the type of the suffix starting with such symbol. In particular $T[i]$ is LMS-type if and only if T_i is LMS-type.

Definition 3. *An LMS-factor of T is a factor that begins with a LMS-type symbol and ends with the following LMS-type symbol.*

We remark that LMS-factors do not establish a factorization of T since each of them overlaps with the following one by one symbol. By convention, $T[n,n]$ is always an LMS-factor. The LMS-factors of $T = banaananaanana\$$ are shown in Fig. 2, where the type of each symbol is also reported. The LMS types are the grey entries. Notice that in SA all suffixes starting with the same symbol $c \in \Sigma$ can be partitioned into a c-bucket. We will keep an integer array $C[1, \sigma]$ where $C[c]$ gives either the first (head) or last (tail) available position of the c-bucket. Then, whenever we insert a value into the head (or tail) of a c-bucket, we increase (or decrease) $C[c]$ by one. An important remark is that within each c-bucket S-type suffixes are larger than L-type suffixes. Figure 2 shows a running example of algorithm SACA-K for $T = banaananaanana\$$.

Given all LMS-type suffixes of $T[1,n]$, the suffix array can be computed as follows:

Steps:

1. Sort all LMS-type suffixes recursively into SA^1, stored in $SA[1, n/2]$.
2. Scan SA^1 from right-to-left, and insert the LMS-suffixes into the tail of their corresponding c-buckets in SA.
3. Induce L-type suffixes by scanning SA left-to-right: for each suffix $SA[i]$, if $T_{SA[i]-1}$ is L-type, insert $SA[i] - 1$ into the head of its bucket.
4. Induce S-type suffixes by scanning SA right-to-left: for each suffix $SA[i]$, if $T_{SA[i]-1}$ is S-type, insert $SA[i] - 1$ into the tail of its bucket.

Step 1 considers the string T^1 obtained by concatenating the lexicographic names of all the consecutive LMS-factors (each different string is associated with a symbol that represents its lexicographic rank). Note that T^1 is defined over an alphabet of size $O(n)$ and that its length is at most $n/2$. The SACA-K algorithm is applied recursively to sort the suffixes of T^1 into SA^1, which is stored in the first half of SA. Nong et al. [18] showed that sorting the suffixes of T^1 is equivalent to sorting the LMS-type suffixes of T. We will omit details of this step, since our algorithm will not modify it.

Step 2 obtains the sorted order of all LMS-type suffixes from SA^1 scanning it from right-to-left and bucket sorting then into the tail of their corresponding c-buckets in SA. Step 3 induces the order of all L-type suffixes by scanning SA from left-to-right. Whenever suffix $T_{SA[i]-1}$ is L-type, $SA[i] - 1$ is inserted in its final (corrected) position in SA.

Finally, Step 4 induces the order of all S-type suffixes by scanning SA from right-to-left. Whenever suffix $T_{SA[i]-1}$ is S-type, $SA[i] - 1$ is inserted in its final (correct) position in SA.

Theoretical Costs. Overall, algorithm SACA-K runs in linear time using only an additional array of size $\sigma + O(1)$ words to store the bucket array [17].

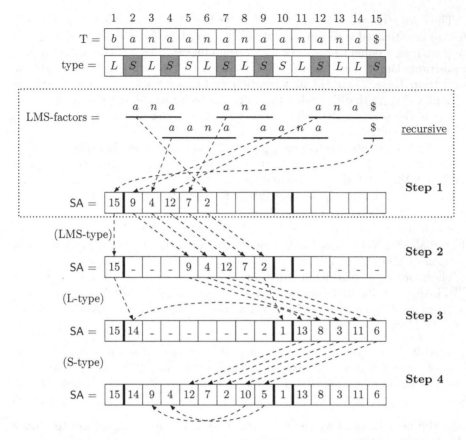

Fig. 2. Induced suffix sorting steps (SACA-K) for $T = banaananaanana\$$

3   Inducing the Lyndon Array

In this section we show how to compute the Lyndon array (LA) during Step 4 of algorithm SACA-K described in Sect. 2.1. Initially, we set all positions $LA[i] = 0$, for $1 \leq i \leq n$. In Step 4, when SA is scanned from right-to-left, each value $SA[i]$, corresponding to $T_{SA[i]}$, is read in its final (correct) position i in SA. In other words, we read the suffixes in decreasing order from $SA[n], SA[n-1], \ldots, SA[1]$. We now show how to compute, during iteration i, the value of $LA[SA[i]]$.

By Lemma 1, we know that the length of the longest Lyndon factor starting at position $SA[i]$ in T, that is $LA[SA[i]]$, is equal to ℓ, where $T_{SA[i]+\ell}$ is the next suffix (in text order) that is smaller than $T_{SA[i]}$. In this case, $T_{SA[i]+\ell}$ will be the first suffix in $T_{SA[i]+1}, T_{SA[i]+2} \ldots, T_n$ that has not yet been read in SA, which means that $T_{SA[i]+\ell} < T_{SA[i]}$. Therefore, during Step 4, whenever we read $SA[i]$, we compute $LA[SA[i]]$ by scanning $LA[SA[i]+1, n]$ to the right up to the first position $LA[SA[i] + \ell] = 0$, and we set $LA[SA[i]] = \ell$.

The correctness of this procedure follows from the fact that every position in $LA[1, n]$ is initialized with zero, and if $LA[SA[i]+1], LA[SA[i]+2], \ldots, LA[SA[i]+\ell-1]$ are no longer equal to zero, their corresponding suffixes has already been read in positions larger than i in $SA[i, n]$, and such suffixes are larger (lexicographically) than $T_{SA[i]}$. Then, the first position we find $LA[SA[i] + \ell] = 0$ corresponds to a suffix $T_{SA[i]+\ell}$ that is smaller than $T_{SA[i]}$, which was still not read in SA. Also, $T_{SA[i]+\ell}$ is the next smaller suffix (in text order) because we read $LA[SA[i] + 1, n]$ from left-to-right.

Figure 3 illustrates iterations $i = 15$, 9, and 3 of our algorithm for $T = banaananaanana\$$. For example, at iteration $i = 9$, the suffix T_5 is read at position $SA[9]$, and the corresponding value $LA[5]$ is computed by scanning $LA[6], LA[7], \ldots, LA[15]$ up to finding the first empty position, which occurs at $LA[7 = 5 + 2]$. Therefore, $LA[5] = 2$.

At each iteration $i = n, n - 1, \ldots, 1$, the value of $LA[SA[i]]$ is computed in additional $LA[SA[i]]$ steps, that is our algorithm adds $O(LA[i])$ time for each iteration of SACA-K.

Therefore, our algorithm runs in $O(n \cdot \text{avelyn})$ time, where $\text{avelyn} = \sum_{i=1}^{n} LA[i]/n$. Note that computing LA does not need extra memory on top of the space for $LA[1, n]$. Thus, the working space is the same as SACA-K, which is $\sigma + O(1)$ words.

Lemma 2. *The Lyndon array and the suffix array of a string $T[1, n]$ over an alphabet of size σ can be computed simultaneously in $O(n \cdot \text{avelyn})$ time using $\sigma + O(1)$ words of working space, where* avelyn *is equal to the average value in* $LA[1, n]$. $\qquad\square$

In the next sections we show how to modify the above algorithm to reduce both its running time and its working space.

3.1   Reducing the Running Time to $O(n)$

We now show how to modify the above algorithm to compute each LA entry in constant time. To this end, we store for each position $LA[i]$ the next smaller position ℓ such that $LA[\ell] = 0$. We define two additional pointer arrays $NEXT[1, n]$ and $PREV[1, n]$:

Definition 4. *For $i = 1, \ldots, n - 1$, $NEXT[i] = \min\{\ell | i < \ell \leq n \text{ and } LA[\ell] = 0\}$. In addition, we define $NEXT[n] = n + 1$.*

Definition 5. *For $i = 2, \ldots, n$, $PREV[i] = \ell$, such that $NEXT[\ell] = i$ and $LA[\ell] = 0$. In addition, we define $PREV[1] = 0$.*

The above definitions depend on LA and therefore NEXT and PREV are updated as we compute additional LA entries. Initially, we set $NEXT[i] = i + 1$ and $PREV[i] = i - 1$, for $1 \leq i \leq n$. Then, at each iteration $i = n, n - 1, \ldots, 1$, when we compute $LA[j]$ with $j = SA[i]$ setting:

$$LA[j] = NEXT[j] - j \qquad (2)$$

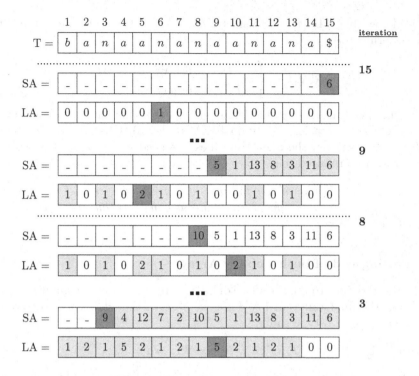

Fig. 3. Running example for $T = banaananaanana\$$.

we update the pointers arrays as follows:

$$\text{NEXT}[\text{PREV}[j]] = \text{NEXT}[j], \quad \text{if } \text{PREV}[j] > 0 \qquad (3)$$

$$\text{PREV}[\text{NEXT}[j]] = \text{PREV}[j], \quad \text{if } \text{NEXT}[j] < n + 1 \qquad (4)$$

The cost of computing each LA entry is now constant, since only two additional computations (Eqs. 3 and 4) are needed. Because of the use of the arrays PREV and NEXT the working space of our algorithm is now $2n + \sigma + O(1)$ words.

Theorem 1. *The Lyndon array and the suffix array of a string $T[1, n]$ over an alphabet of size σ can be computed simultaneously in $O(n)$ time using $2n + \sigma + O(1)$ words of working space.* ☐

3.2 Getting Rid of a Pointer Array

We now show how to reduce the working space of Sect. 3.1 by storing only one array, say $A[1, n]$, keeping NEXT/PREV information together. In a glace, we store NEXT initially into the space of $A[1, n]$, then we reuse $A[1, n]$ to store the (useful) entries of PREV.

Note that, whenever we write $\text{LA}[j] = \ell$, the value in $A[j]$, that is $\text{NEXT}[j]$ is no more used by the algorithm. Then, we can reuse $A[j]$ to store $\text{PREV}[j + 1]$.

Also, we know that if $\mathsf{LA}[j] = 0$ then $\mathsf{PREV}[j+1] = j$. Therefore, we can redefine PREV in terms of A:

$$\mathsf{PREV}[j] = \begin{cases} j - 1 & \text{if } \mathsf{LA}[j-1] = 0 \\ \mathsf{A}[j-1] & \text{otherwise.} \end{cases} \tag{5}$$

The running time of our algorithm remains the same since we have added only one extra verification to obtain $\mathsf{PREV}[j]$ (Eq. 5). Observe that whenever $\mathsf{NEXT}[j]$ is overwritten the algorithm does not need it anymore. The working space is therefore reduced to $n + \sigma + O(1)$ words.

Theorem 2. *The Lyndon array and the suffix array of a string $T[1,n]$ over an alphabet of size σ can be computed simultaneously in $O(n)$ time using $n+\sigma+O(1)$ words of working space.* $\square$

3.3   Getting Rid of both Pointer Arrays

Finally, we show how to use the space of $\mathsf{LA}[1,n]$ to store both the auxiliary array $\mathsf{A}[1,n]$ and the final values of LA. First we observe that it is easy to compute $\mathsf{LA}[i]$ when T_i is an L-type suffix.

Lemma 3. $\mathsf{LA}[j] = 1$ *iff T_j is an L-type suffix, or $i = n$.*

Proof. If T_j is an L-type suffix, then $T_j > T_{j+1}$ and $\mathsf{LA}[j] = 1$. By definition $\mathsf{LA}[n] = 1$. $\square$

Notice that at Step 4 during iteration $i = n, n-1, \ldots, 1$, whenever we read an S-type suffix T_j, with $j = \mathsf{SA}[i]$, its succeeding suffix (in text order) T_{j+1} has already been read in some position in the interval $\mathsf{SA}[i+1, n]$ (T_{j+1} have induced the order of T_j). Therefore, the LA-entries corresponding to S-type suffixes are always inserted on the left of a block (possibly of size one) of non-zero entries in $\mathsf{LA}[1,n]$.

Moreover, whenever we are computing $\mathsf{LA}[j]$ and we have $\mathsf{NEXT}[j] = j + k$ (stored in $\mathsf{A}[j]$), we know the following entries $\mathsf{LA}[j+1], \mathsf{LA}[j+2], \ldots, \mathsf{LA}[j+k-1]$ are no longer zero, and we have to update $\mathsf{A}[j+k-1]$, corresponding to $\mathsf{PREV}[j+k]$ (Eq. 5). In other words, we update PREV information only for rightmost entry of each block of non empty entries, which corresponds to a position of an L-type suffix because S-type are always inserted on the left of a block.

Then, at the end of the modified Step 4, if $\mathsf{A}[i] < i$ then T_i is an L-type suffix, and we know that $\mathsf{LA}[i] = 1$. On the other hand, the values with $\mathsf{A}[i] > i$ remain equal to $\mathsf{NEXT}[i]$ at the end of the algorithm. And we can use them to compute $\mathsf{LA}[i] = \mathsf{A}[i] - i$ (Eq. 2).

Thus, after the completion of Step 4, we sequentially scan $\mathsf{A}[1,n]$ overwriting its values with LA as follows:

$$\mathsf{LA}[j] = \begin{cases} 1 & \text{if } \mathsf{A}[j] < j \\ \mathsf{A}[j] - j & \text{otherwise.} \end{cases} \tag{6}$$

The running time of our algorithm is still linear, since we added only a linear scan over $A[1, n]$ at the end of Step 4. On the other hand, the working space is reduced to $\sigma + O(1)$ words, since we need to store only the bucket array $C[1, \sigma]$.

Theorem 3. *The Lyndon array and the suffix array of a string of length n over an alphabet of size σ can be computed simultaneously in $O(n)$ time using $\sigma + O(1)$ words of working space.* □

Note that the bounds on the working space given in the above theorems assume that the output consists of SA and LA. If one is interested in LA only, then the working space of the algorithm is $n + \sigma + O(1)$ words which is still smaller that the working space of the other linear time algorithms that we discussed in Sect. 2.

4 Experiments

We compared the performance of our algorithm, called SACA-K+LA, with algorithms to compute LA in linear time by Franek *et al.* [5,9] (NSV-LYNDON), Baier [1,6] (BAIER-LA), and Louza *et al.* [15] (BWT-LYNDON). We also compared a version of Baier's algorithm that computes LA and SA together (BAIER-LA+SA). We considered the three linear time alternatives of our algorithm described in Sects. 3.1, 3.2 and 3.3. We tested all three versions since one could be interested in the fastest algorithm regardless of the space usage. We used four bytes for each computer word so the total space usage of our algorithms was respectively $17n$, $13n$ and $9n$ bytes. We included also the performance of SACA-K [17] to evaluate the overhead added by the computation of LA in addition to the SA.

The experiments were conducted on a machine with an **Intel Xeon** Processor E5-2630 v3 20M Cache 2.40-GHz, 384 GB of internal memory and a 13 TB SATA storage, under a 64 bits **Debian GNU/Linux 8** (kernel 3.16.0-4) OS. We implemented our algorithms in ANSI C. The time was measured with clock() function of C standard libraries and the memory was measured using malloc_count library[1]. The source-code is publicly available at https://github.com/felipelouza/lyndon-array/.

We used string collections from the Pizza & Chili dataset[2]. In particular, the datasets einstein-de, kernel, fib41 and cere are highly repetitive texts[3], and the english.1G is the first 1GB of the original english dataset. We also created an artificial repetitive dataset, called bbba, consisting of a string T with 100×2^{20} copies of b followed by one occurrence of a, that is, $T = b^{n-2}a\$$. This dataset represents a worst-case input for the algorithms that use a stack (NSV-LYNDON and BWT-LYNDON).

---

[1] https://github.com/bingmann/malloc_count.
[2] http://pizzachili.dcc.uchile.cl/texts.html.
[3] http://pizzachili.dcc.uchile.cl/repcorpus.html.

Table 1. Running time (μs/input byte).

Dataset	σ	$n/2^{20}$	LA			LA and SA				SA
			NSV-Lyndon [9]	Baier-LA [1,6]	BWT-Lyndon [15]	Baier-LA+SA [1,6]	SACA-K+LA-17n	SACA-K+LA-13n	SACA-K+LA-9n	SACA-K [17]
pitches	133	53	**0.15**	0.20	0.20	0.26	0.26	0.22	**0.18**	0.13
sources	230	201	**0.26**	0.28	0.32	0.37	0.46	0.41	**0.34**	0.24
xml	97	282	**0.29**	0.31	0.35	0.42	0.52	0.47	**0.38**	0.27
dna	16	385	0.39	**0.28**	0.49	**0.43**	0.69	0.60	0.52	0.36
english.1GB	239	1,047	0.46	**0.39**	0.56	**0.57**	0.84	0.74	0.60	0.42
proteins	27	1,129	0.44	**0.40**	0.53	0.66	0.89	0.69	**0.58**	0.40
einstein-de	117	88	0.34	**0.28**	0.38	**0.39**	0.57	0.54	0.44	0.31
kernel	160	246	0.29	**0.29**	0.39	**0.38**	0.53	0.47	**0.38**	0.26
fib41	2	256	0.34	**0.07**	0.45	**0.18**	0.66	0.57	0.46	0.32
cere	5	440	0.27	**0.09**	0.33	**0.17**	0.43	0.41	0.35	0.25
bbba	2	100	0.04	**0.02**	0.05	**0.03**	0.05	0.04	**0.03**	0.03

Table 1 shows the running time of each algorithm in μs/input byte. The results show that our algorithm is competitive in practice. In particular, the version SACA-K+LA-9n was only about 1.35 times slower than the fastest algorithm (Baier-LA) for non-repetitive datasets, and 2.92 times slower for repetitive datasets. Also, the performance of SACA-K+LA-9n and Baier-LA+SA were very similar. Finally, the overhead of computing LA in addition to SA was small: SACA-K+LA-9n was 1.42 times slower than SACA-K, whereas Baier-LA+SA was 1.55 times slower than Baier-LA, on average. Note that SACA-K+LA-9n was consistently faster than SACA-K+LA-13n and SACA-K+LA-17n, so using more space does not yield any advantage.

Table 2 shows the peak space consumed by each algorithm given in bytes per input symbol. The smallest values were obtained by NSV-Lyndon, BWT-Lyndon and SACA-K+LA-9n. In details, the space used by NSV-Lyndon and BWT-Lyndon was $9n$ bytes plus the space used by the stack. The stack space was negligible (about 10KB) for almost all datasets, except for **bbba** where the stack used $4n$ bytes for NSV-Lyndon and $8n$ bytes for BWT-Lyndon (the number of stack entries is the same, but each stack entry consists of a pair of integers). On the other hand, our algorithm, SACA-K+LA-9n, used exactly $9n + 1024$ bytes for all datasets.

Table 2. Peak space (bytes/input size).

Dataset	σ	$n/2^{20}$	LA			LA and SA				SA
			NSV-Lyndon [9]	Baier-LA [1,6]	BWT-Lyndon [15]	Baier-LA+SA [1,6]	SACA-K+LA-17n	SACA-K+LA-13n	SACA-K+LA-9n	SACA-K [17]
pitches	133	53	**9**	17	**9**	17	17	13	**9**	5
sources	230	201	**9**	17	**9**	17	17	13	**9**	5
xml	97	282	**9**	17	**9**	17	17	13	**9**	5
dna	16	385	**9**	17	**9**	17	17	13	**9**	5
english.1GB	239	1,047	**9**	17	**9**	17	17	13	**9**	5
proteins	27	1,129	**9**	17	**9**	17	17	13	**9**	5
einstein-de	117	88	**9**	17	**9**	17	17	13	**9**	5
kernel	160	246	**9**	17	**9**	17	17	13	**9**	5
fib41	2	256	**9**	17	**9**	17	17	13	**9**	5
cere	5	440	**9**	17	**9**	17	17	13	**9**	5
bbba	2	100	**13**	17	17	17	17	13	**9**	5

5 Conclusions

We have introduced an algorithm for computing simultaneously the suffix array and Lyndon array (LA) of a text using induced suffix sorting. The most space-economical variant of our algorithm uses only $n + \sigma + O(1)$ words of working space making it the most space economical LA algorithm among the ones running in linear time; this includes both the algorithm computing the SA and LA and the ones computing only the LA. The experiments have shown our algorithm is only slightly slower than the available alternatives, and that computing the SA is usually the most expensive step of all linear time LA construction algorithms. A natural open problem is to devise a linear time algorithm to construct only the LA using $o(n)$ words of working space.

Acknowledgments. The authors thank Uwe Baier for kindly providing the source codes of algorithms Baier-LA and Baier-LA+SA, and Prof. Nalvo Almeida for granting access to the machine used for the experiments.

Funding. F.A.L. was supported by the grant #2017/09105-0 from the São Paulo Research Foundation (FAPESP). G.M. was partially supported by PRIN grant 2017WR7SHH, by INdAM-GNCS Project 2019 *Innovative methods for the solution*

of medical and biological big data and by the LSBC_19-21 Project from the University of Eastern Piedmont. S.M. and M.S. are partially supported by MIUR-SIR project CMACBioSeq *Combinatorial methods for analysis and compression of biological sequences* grant n. RBSI146R5L. G.P.T. acknowledges the support of Brazilian agencies Conselho Nacional de Desenvolvimento Científico e Tecnológico (CNPq) and Coordenação de Aperfeiçoamento de Pessoal de Nível Superior (CAPES).

References

1. Baier, U.: Linear-time suffix sorting – a new approach for suffix array construction. In: Proceedings of the Annual Symposium on Combinatorial Pattern Matching (CPM), pp. 23:1–23:12 (2016)
2. Bannai, H., Tomohiro, I., Inenaga, S., Nakashima, Y., Takeda, M., Tsuruta, K.: The "runs" theorem. SIAM J. Comput. **46**(5), 1501–1514 (2017)
3. Crochemore, M., Russo, L.M.: Cartesian and Lyndon trees. Theoret. Comput. Sci. (2018). https://doi.org/10.1016/j.tcs.2018.08.011
4. Fischer, J.: Inducing the LCP-array. In: Proceedings Workshop on Algorithms and Data Structures (WADS), pp. 374–385 (2011)
5. Franek, F., Islam, A.S.M.S., Rahman, M.S., Smyth, W.F.: Algorithms to compute the Lyndon array. In: Proceeding of the PSC, pp. 172–184 (2016)
6. Franek, F., Paracha, A., Smyth, W.F.: The linear equivalence of the suffix array and the partially sorted Lyndon array. In: Proceedings of the PSC, pp. 77–84 (2017)
7. Goto, K., Bannai, H.: Simpler and faster Lempel Ziv factorization. In: 2013 Data Compression Conference, DCC 2013, Snowbird, UT, USA, March 20–22, 2013, pp. 133–142 (2013)
8. Goto, K., Bannai, H.: Space efficient linear time Lempel-Ziv factorization for small alphabets. In: Proceedings of the IEEE Data Compression Conference (DCC), pp. 163–172 (2014)
9. Hohlweg, C., Reutenauer, C.: Lyndon words, permutations and trees. Theor. Comput. Sci. **307**(1), 173–178 (2003)
10. Itoh, H., Tanaka, H.: An efficient method for in memory construction of suffix arrays. In: Proceedings of the sixth Symposium on String Processing and Information Retrieval (SPIRE 1999), pp. 81–88. IEEE Computer Society Press (1999)
11. Ko, P., Aluru, S.: Space efficient linear time construction of suffix arrays. In: Baeza-Yates, R., Chávez, E., Crochemore, M. (eds.) CPM 2003. LNCS, vol. 2676, pp. 200–210. Springer, Heidelberg (2003). https://doi.org/10.1007/3-540-44888-8_15
12. Kolpakov, R.M., Kucherov, G.: Finding maximal repetitions in a word in linear time. In: Proceedings of the FOCS, pp. 596–604 (1999)
13. Louza, F.A., Gog, S., Telles, G.P.: Inducing enhanced suffix arrays for string collections. Theor. Comput. Sci. **678**, 22–39 (2017)
14. Louza, F.A., Gog, S., Telles, G.P.: Optimal suffix sorting and LCP array construction for constant alphabets. Inf. Process. Lett. **118**, 30–34 (2017)
15. Louza, F.A., Smyth, W.F., Manzini, G., Telles, G.P.: Lyndon array construction during Burrows-Wheeler inversion. J. Discrete Algorithms **50**, 2–9 (2018)
16. Manber, U., Myers, G.: Suffix arrays: a new method for on-line string searches. SIAM J. Comput. **22**(5), 935–948 (1993)
17. Nong, G.: Practical linear-time O(1)-workspace suffix sorting for constant alphabets. ACM Trans. Inf. Syst. **31**(3), 15 (2013)

18. Nong, G., Zhang, S., Chan, W.H.: Two efficient algorithms for linear time suffix array construction. IEEE Trans. Comput. **60**(10), 1471–1484 (2011)

19. Nong, G., Zhang, S., Chan, W.H.: Linear suffix array construction by almost pure induced-sorting. In: Proceedings of the IEEE Data Compression Conference (DCC), pp. 193–202 (2009)

20. Ohlebusch, E.: Bioinformatics Algorithms: Sequence Analysis, Genome Rearrangements, and Phylogenetic Reconstruction. Oldenbusch Verlag (2013)

21. Okanohara, D., Sadakane, K.: A linear-time Burrows-wheeler transform using induced sorting. In: Proceedings International Symposium on String Processing and Information Retrieval (SPIRE), pp. 90–101 (2009)

22. Vuillemin, J.: A unifying look at data structures. Commun. ACM **23**(4), 229–239 (1980)

Minimal Absent Words in Rooted
and Unrooted Trees

Gabriele Fici[1]([⊠]) and Paweł Gawrychowski[2]

[1] Dipartimento di Matematica e Informatica, Università di Palermo, Palermo, Italy
gabriele.fici@unipa.it
[2] Institute of Computer Science, University of Wrocław, Wrocław, Poland
gawry@cs.uni.wroc.pl

Abstract. We extend the theory of minimal absent words to (rooted and unrooted) trees, having edges labeled by letters from an alphabet Σ of cardinality σ. We show that the set $\mathrm{MAW}(T)$ of minimal absent words of a rooted (resp. unrooted) tree T with n nodes has cardinality $O(n\sigma)$ (resp. $O(n^2\sigma)$), and we show that these bounds are realized. Then, we exhibit algorithms to compute all minimal absent words in a rooted (resp. unrooted) tree in output-sensitive time $O(n + |\mathrm{MAW}(T)|)$ (resp. $O(n^2 + |\mathrm{MAW}(T)|)$ assuming an integer alphabet of size polynomial in n.

1 Introduction

Minimal absent words (a.k.a. minimal forbidden words or minimal forbidden factors) are a useful combinatorial tool for investigating words (strings). A word u is absent from a word w if u does not occur (as a factor) in w, and it is minimal if all its proper factors occur in w. This definition naturally extends to languages of words closed under taking factors.

The theory of minimal absent words has been developed in a series of papers [3,5,14,25,27] (the reader is pointed to [18] for a survey on these results). Minimal absent words have then found applications in several areas, e.g., data compression [15–17,28], on-line pattern matching [13], sequence comparison [10,11], sequence assembly [20,26], bioinformatics [9,19,31], musical data extraction [12].

Bounds on the number of minimal absent words have been extensively investigated. The upper bound on the number of minimal absent words of a word of length n over an alphabet of size σ is $O(n\sigma)$ [14,27], and this is tight for integer alphabets [10]; in fact, for large alphabets, such as when $\sigma \geq \sqrt{n}$, this bound is also tight even for minimal absent words having the same length [1].

Several algorithms are known to compute the set of minimal absent words of a word. State-of-the-art algorithms compute all minimal absent words of a word of length n over an alphabet of size σ in time $O(n\sigma)$ [2,14] or in output-sensitive $O(n + |\mathrm{MAW}(w)|)$ time [11,22] for integer alphabets. Space-efficient data structures based on the Burrows-Wheeler transform can also be applied for this computation [6,7].

© Springer Nature Switzerland AG 2019
N. R. Brisaboa and S. J. Puglisi (Eds.): SPIRE 2019, LNCS 11811, pp. 152–161, 2019.
https://doi.org/10.1007/978-3-030-32686-9_11

For a finite set of words P over an alphabet of size σ, the minimal absent words of the factorial closure of P can be computed in $O(|P|^2\sigma)$ [3], where $|P|$ is the sum of the lengths of the words of P. Generalizations of minimal absent words have been considered for circular words [10,21] and multi-dimensional shifts [4].

In this paper, we extend the theory of minimal absent words to trees. We consider trees with edges labeled by letters from an integer alphabet Σ of cardinality σ polynomial in n. In the case of a rooted tree T, every node v is associated with a word $\mathsf{str}(v)$, defined as the sequence of edge labels from v to the root. A rooted tree T can therefore be seen as a set of words $L_T = \{\mathsf{str}(v) \mid v \text{ in } T\}$, that we call the *language* of T. If T has n nodes, then L_T contains at most n distinct words, each of which has length at most n. We call a rooted tree T *proper* when the edges from a node to its children are labeled by pairwise distinct letters. Throughout the paper we will assume that all rooted trees are proper, which can be ensured without losing the generality thank to the following lemma.

Lemma 1. *Given a rooted T on n nodes we can construct in $O(n)$ time a proper rooted tree T' with the same set of corresponding words.*

Proof. The depth of a node of T is its distance from the root. We start with sorting, for every $d = 1, 2, ..$, the set of nodes $S(d)$ at depth d according to the labels of the edges leading to their parents. This can be done in $O(n)$ total time with counting sort. Then, we construct T' by processing $S(0), S(1), S(2), \ldots$ Assuming that we have already identified, for every node $u \in S(d)$, its corresponding node $f(u)$ of T', we need to construct and identify the nodes $f(u')$ for every $u' \in S(d+1)$. We process all nodes $u' \in S(d+1)$ in groups corresponding to the same letter a on the edge leading to their parent (because of the initial sorting we already have these groups available). Denoting by u the parent of u' in T, we check if $f(u)$ has been already accessed while processing the group of a, and if so we set $f(u')$ to be the already created node of T'. Otherwise, we create a new edge outgoing from $f(u)$ to a new node v in T' and labeled with a, and set $f(u')$ to be v. To check if $f(u)$ has been already accessed while processing the current group (and retrieve the corresponding $f(u')$ if this is the case) we simply allocate an array A of size n indexed by nodes of T' identified by number from $\{1, 2, \ldots, n\}$. For every entry of A we additionally store a timestamp denoting the most recent group for which the corresponding entry has been modified, and increase the timestamp after having processed the current group. □

One could also define the set of words corresponding to a rooted tree T by considering a set of words from the root to every node v (in the literature this is sometimes called a *forward trie*, as opposed to a *backward trie*, cf. [24]). In our context, this distinction is meaningless, as the obtained languages are the same up to reversing all the words.

We say that a word aub, with $a, b \in \Sigma$, is a minimal absent word of a rooted tree T if aub is not a factor of any word $\mathsf{str}(v)$ in L_T but there exist words $\mathsf{str}(v_1)$ and $\mathsf{str}(v_2)$ in L_T (not necessarily distinct) such that au is a factor of $\mathsf{str}(v_1)$ and ub is a factor of $\mathsf{str}(v_2)$. That is, the set $\mathrm{MAW}(T)$ of minimal absent words of T

is the set of minimal absent words of the factorial closure of the language L_T. Since any word of length n can be transformed into a unary rooted tree with $n + 1$ nodes, some of the properties of minimal absent words for usual words can be transferred to rooted trees. Indeed, rooted trees are a strict generalization of words.

For unrooted trees, the definition of minimal absent words is analogous: We identify an unrooted tree T with the language of words $L(T)$ corresponding to all (concatenations of labels of) simple paths that can be read in T from any of its nodes. The language $L(T)$ contains $O(n^2)$ words, each of which has length at most n. We therefore define the set MAW(T) of minimal absent words of T as the set of minimal absent words of the language $L(T)$, which in this case is already closed under taking factors by definition.

Our Results. We prove that for any rooted tree with n nodes there are $O(n\sigma)$ minimal absent words, and we show that this bound is tight. For unrooted trees, we prove that the previous bound becomes $O(n^2\sigma)$, and we give an explicit construction that achieves this bound. We also consider the case of minimal absent words of fixed length and generalize a previously-known construction.

Furthermore, we present an algorithm that computes all the minimal absent words in a rooted tree T with n nodes in output-sensitive time $O(n+|\text{MAW}(T)|)$. This also yields an algorithm that computes all the minimal absent words in an unrooted tree T with n nodes in time $O(n^2 + |\text{MAW}(T)|)$. Note that while it is plausible that an efficient algorithm could have been designed, as in the case of words, from a DAWG [22], the size of the DAWG of a backward/forward tree is superlinear [24], so it is not immediately clear if such an approach would lead to an optimal algorithm. Excluding the space necessary to store all the results, our algorithms need $O(n)$ and $O(n^2)$ space, respectively.

Our algorithms are designed in the word-RAM model with $\Omega(\log n)$-bit words.

2   Bounds on the Number of Minimal Absent Words

Let T be a rooted tree with n nodes and edges labeled by letters from an integer alphabet Σ of cardinality σ polynomial in n. Let the language of T be $L_T = \{\text{str}(v) \mid v \text{ in } T\}$, where $\text{str}(v)$ is the sequence of edge labels from node v to the root.

For convenience, we add a new root to T and an edge labeled by a new letter \$ not belonging to Σ from the new root to the old root. This corresponds to appending \$ at the end of each word of L_T. We then arrange all the words of L_T in a trie. Each node u of this trie corresponds to a word obtained by concatenating the edges from the root of the trie to node u, so in this paper we will implicitly identify a node of the trie with the corresponding word in the set of prefixes of L_T. Following a standard approach, if we compact this trie by collapsing maximal chains of edges with every inner node having exactly one child and edges labeled by words, we obtain the suffix tree ST of T. The nodes in ST (the branching nodes) are called explicit nodes, while the nodes of the trie

that have been collapsed (the non-branching nodes) are called implicit. Because $ does not belong to the original alphabet, the leaves of ST are in one-to-one correspondence with the nodes of T.

A word aub, with $a, b \in \Sigma$, is a minimal absent word for T if it is a minimal absent word for the factorial closure of L_T, that is, if both au and ub but not aub are factors of some words in L_T. The set of minimal absent words of T is denoted by MAW(T).

If $aub \in$ MAW(T), then au occurs as a factor in some word of L_T but never followed by letter b, hence there exists a letter $b' \in \Sigma \cup \{\$\}$ such that ub and ub' can be read in ST spelled from the root (possibly terminating in an implicit node). This implies that u corresponds to an explicit node in ST, and b is the first letter on its outgoing edge. Consequently, ub can be identified with an edge of ST, so the number of minimal absent words of T is upper-bounded by the product of σ and the number of edges of ST.

Theorem 1. *The number of minimal absent words of a rooted tree with n nodes whose edges are labeled by letters from an alphabet of size σ is $O(n\sigma)$.*

Therefore, the same upper bound that holds for words also holds for rooted trees. As a consequence, we have that all known upper bounds for words, and constructions that realize them, are still valid for rooted trees.

In particular, one question that has been studied is whether the upper bound $O(n\sigma)$ is still tight when one considers minimal absent words of a fixed length. Almirantis et al. [1, Lemma 2] showed that the upper bound $O(n\sigma)$ for a fixed length of minimal absent words is tight if $\sqrt{n} < \sigma \leq n$. Actually, they showed that it is possible to construct words of any length n, with $\sigma \leq n \leq \sigma(\sigma - 1)$, having $\Omega(n\sigma)$ minimal absent words of length 3. We now give a construction that generalizes this result.

Let $\Sigma = \{1, 2, \ldots, \sigma\}$. For every n, let $k > 1$ be such that $\sigma^k \leq n < \sigma^{k+1}$. Let $\Sigma^k = \{s_1, s_2, \ldots, s_{\sigma^k}\}$. For every $1 \leq i \leq \sigma^k$ we define the word

$$w_i = \$1s_i\$s_i1\$2s_i\$s_i2\$ \cdots \$\sigma s_i\$s_i\sigma\$,$$

where $\$$ is a new symbol not belonging to Σ. The length of each word w_i is $2\sigma(k + 2) + 1$, which is smaller than n up to excluding small cases [1].

Let $\ell = \lfloor n/|w_i| \rfloor$ and set $w = w_1 w_2 \cdots w_\ell$, so that $|w| > n/2$. We have that as_ib is a minimal absent word of w for every $a, b \in \Sigma$ and $1 \leq i \leq \ell$. So, w has length $\Theta(k\sigma\ell)$ and there are $\Theta(\sigma^2\ell)$ minimal absent words of w of length $k + 2$.

Thus, we have proved the following theorem.

Theorem 2. *A word of length n over an alphabet of size σ can have $\Omega(n\sigma/\log_\sigma n)$ minimal absent words all of the same length.*

Observe that for $\sqrt{n} < \sigma \leq n$, $\log_\sigma n = \Theta(1)$, therefore Theorem 2 strictly generalizes Almirantis et al.'s result.

---

[1] The reader may verify that for $k = 2$, $|w_i| \leq \sigma^k$ as soon as $\sigma \geq 9$; for $k > 2$, $|w_i| \leq \sigma^k$ as soon as $\sigma + k \geq 7$.

Let now T be an unrooted tree. Then the number of distinct simple paths in T is $O(n^2\sigma)$, and this is thus an upper bound on the number of minimal absent words of T.

Theorem 3. *The number of minimal absent words of an unrooted tree with n nodes whose edges are labeled by letters from an alphabet of size σ is $O(n^2\sigma)$.*

We now provide an example of an unrooted tree realizing this bound. Let $\Sigma = \{0, \bar{0}, 1, \ldots, s\}$. Our unrooted tree T is built as follows:

- We first build a sequence of $2N + 1$ nodes such that every other node is connected to s terminal nodes with edges labeled by $1, 2, \ldots, s$ and is connected to the next node of the sequence with an edge labeled by 0 and to the previous node of the sequence with an edge labeled by $\bar{0}$;
- Then, we attach to each of the last nodes of the previous sequence s simple paths composed of $2N$ nodes with edges labeled by alternating 0 and $\bar{0}$.

See Fig. 1 for an illustration.

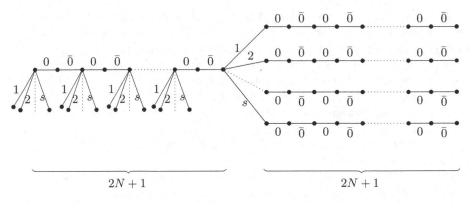

Fig. 1. An unrooted tree realizing the upper bound on the number of minimal absent words.

In total, T has $(s+1)(2N+1)+sN$ nodes. We therefore set $n = (s+1)(2N+1) + sN$, so that $n = \Theta(sN)$.

It is readily verified that for every a, b, c in $\Sigma \setminus \{0, \bar{0}\}$ and for every $0 < j, k \leq N$, there is a minimal absent word of the form $a(0\bar{0})^j b(0\bar{0})^k c$ (the prefix $a(0\bar{0})^j b(0\bar{0})^k$ can be found reading from the left part to the right part of the figure, while the suffix $(0\bar{0})^j b(0\bar{0})^k c$ can be found reading from the right part to the left part, the letter b being one of the labels of the edges joining the left and the right part). Hence, the number of minimal absent words of T is $\Omega(s^3 N^2) = \Omega(n^2\sigma)$.

Remark 1. The previous construction can be simplified by merging adjacent edges labeled by 0 and $\bar{0}$ into one edge labeled by 0, if one does not require the condition that edges adjacent to a node must have distinct labels.

3   Algorithms for Computing Minimal Absent Words

We now present an algorithm that computes the set MAW(T) of all minimal absent words of a rooted tree T with n nodes in output-sensitive time $O(n + |MAW(T)|)$.

We construct the suffix tree ST of T in time $O(n)$ [30]. Recall that the leaves of ST are in one-to-one correspondence with the nodes of T and we can assume that every node u of T stores a pointer to the leaf of ST corresponding to $str(u)$.

Definition 1. *For every (implicit or explicit) node u of ST, we define the set $A(u)$ as the set of all letters $a \in \Sigma$ such that au can be spelled from the root of ST, i.e., there exists a node v of T such that $str(v) = auz$ for some (possibly empty) word z.*

As already noted before, if aub is a minimal absent word of T, then au occurs as a factor in some word of L_T followed by a letter $b' \in \Sigma \cup \{\$\}$ different from b, hence u is an explicit node of ST.

Lemma 2. *Let u be an explicit node of the suffix tree ST of the tree T. Let $u_1, u_2, \ldots, u_k$ be the children of u in the non-compacted trie from which we obtained ST, and let $b_1, b_2, \ldots, b_k$ be the labels of the corresponding edges. Then, for every $1 \le i \le k$ and every letter*

$$a_j \in (A(u_1) \cup \ldots \cup A(u_k)) \setminus A(u_i),$$

the word $a_j u b_i$ is a minimal absent word of T.

Conversely, every minimal absent word of T is of the form $a_j u b_i$ described above.

Proof. Since a_j does not belong to $A(u_i)$, then by definition the word $a_j u b_i$ does not belong to L_T, but there exists $\ell \ne i$ such that $a_j \in A(u_\ell)$, that is, $a_j u b_\ell$ is a factor of a word in L_T. Hence, $a_j u$ is a factor of a word in L_T. Since $u b_i$ is also a factor of a word in L_T by construction, we have that $a_j u b_i$ is a minimal absent word of T.

Conversely, if $a_j u b_i$ is a minimal absent word of T, then u occurs as a factor in some word of L_T followed by different letters in $\Sigma \cup \{\$\}$, hence it corresponds to an explicit node in ST, so all minimal absent words of T are found in this way.   □

Definition 2. *For every leaf u of ST we define the set $B(u)$ as the set of all letters $a \in \Sigma$ such that $au = str(v)$ for some node v in T.*

Lemma 3. *For every (implicit or explicit) node u of ST, we have $A(u) = \bigcup\{B(u') \mid u'$ is a leaf in the subtree of ST rooted at $u\}$.*

Proof. Let u' be a leaf in the subtree of ST rooted at u. Thus, the word u is a prefix of the word u', i.e., $u' = uz$ for some word z. By definition, $B(u')$ is the set of all letters $a \in \Sigma$ such that $au' = str(v')$ for some node v' in T. That is,

the set of all letters $a \in \Sigma$ such that $au' = auz$ is a word in L_T. On the other hand, by definition, $A(u)$ is the set of all letters $a \in \Sigma$ such that $\text{str}(v) = auz$ for some node v of T and word z. That is, the set of all letters $a \in \Sigma$ such that auz is a word in L_T for some word z.   □

We now show how to compute, in time proportional to the output size, the set MAW(T).

We start with creating, for every letter $a \in \Sigma$, a list $L(a)$ of all leaves u such that $a \in B(u)$ sorted in preorder. The lists can be obtained in linear time by traversing all the non-root nodes $v \in T$, following the edge labeled by a from v to its parent v', and finally following the pointer from v' to the leaf v'' of ST corresponding to $\text{str}(v')$ and adding v'' to $L(a)$. Finally, because the preorder numbers are from $[n]$ the lists can be sorted in linear time with counting sort.

Now we iterate over all letters $a \in \Sigma$. Due to Lemma 2, the goal is to extract all explicit nodes $u \in ST$ such that, for some child u_i of u such that the b_i is the first letter on the edge from u to u_i, aub_i is a minimal absent word. By Lemma 3, this is equivalent to u having a descendant $u' \in L(a)$ (where possibly $u = u'$) and u_i not having any such descendant. This suggests that we should work with the subtree of ST, denoted $ST(a)$, induced by all leaves $v \in L(a)$. Formally, $u \in ST(a)$ when $u' \in L(a)$ for some leaf u' in the subtree of u. Even though ST does not contain nodes with just one child, this is no longer the case for $ST(a)$. Thus, we actually work with its compact version, denoted $ST(a)$. Every node of $ST(a)$ stores a pointer to its corresponding node of ST. Assuming that ST has been preprocessed for constant-time Lowest Common Ancestor queries (which can be done in linear time and space [8,29]), we can construct $ST(a)$ efficiently due to the following lemma.

Lemma 4. *Given $L(a)$, we can construct $ST(a)$ in $O(|L(a)|)$ time.*

Proof. The procedure follows the general idea used in the well-known linear time procedure for creating a Cartesian tree [23]. We process the nodes $u \in L(a)$ in preorder and maintain a compact version of the subtree of ST induced by all the already-processed nodes. Additionally, we maintain a stack storing the edges on its rightmost path. Processing $u \in L(a)$ requires popping a number of edges from the stack, possibly splitting the topmost edge into two (with one immediately popped as well), and pushing a new edge ending at u. Checking if an edge should be popped, and also determining if (and how) an edge should be split, can be implemented with LCA queries on ST, assuming that we maintain pointers to the corresponding nodes of ST.   □

Having constructed $ST(a)$, we need to consider two cases corresponding to u being an explicit or an implicit node of $ST(a)$. In the former case, we need to extract the edges outgoing from u in ST such that there is no edge outgoing from the corresponding node in $ST(a)$ starting with the same letter b, and output aub as a minimal absent word. Assuming that the outgoing edges are sorted by their first letters, this can be easily done in time proportional to the degree of u plus the number of extracted letters. In the latter case, let the implicit node belong

to an edge connecting u to v in $ST(a)$, and let u' and v' be their corresponding nodes in ST with u' being an ancestor of v'. We iterate through all explicit nodes between u' and v' in ST and for each such node we extract all of its outgoing edges. For each such edge we check if v' belongs to the subtree rooted at its endpoint other than u, and if not, extract its first letter b to output aub as a minimal absent word.

The overall time for every letter $a \in \Sigma$ can be bounded by the sum of the size of $ST(a)$ and the number of generated minimal absent words. Because $\sum_{a \in \Sigma} |L(a)| = O(n)$ and the size of $ST(a)$ can be bounded by $O(|L(a)|)$, the total time complexity is $O(n + |\mathsf{MAW}(\mathsf{T})|)$.

The previous algorithm can be used to design an algorithm that outputs all the minimal absent words of an unrooted tree T with n nodes in time $O(n^2 + |\mathsf{MAW}(\mathsf{T})|)$ as follows. For every node u of T, we create a rooted tree T_u by fixing u as the root. Then we merge all trees T_u into a single tree T of size $O(n^2)$ by identifying their roots. Finally, we apply Lemma 1 to make T proper and apply our algorithm for rooted trees in $O(n^2)$ total time.

Acknowledgments. This research was carried out during a visit of the first author to the Institute of Computer Science of the University of Wrocław, supported by grant CORI-2018-D-D11-010133 of the University of Palermo. The first author is also supported by MIUR project PRIN 2017K7XPAN "Algorithms, Data Structures and Combinatorics for Machine Learning".

We thank anonymous reviewers for helpful comments.

References

1. Almirantis, Y., et al.: On avoided words, absent words, and their application to biological sequence analysis. Algorithms Mol. Biol. **12**(1), 51–512 (2017)
2. Barton, C., Héliou, A., Mouchard, L., Pissis, S.P.: Linear-time computation of minimal absent words using suffix array. BMC Bioinform. **15**, 388 (2014)
3. Béal, M., Crochemore, M., Mignosi, F., Restivo, A., Sciortino, M.: Computing forbidden words of regular languages. Fundam. Inform. **56**(1–2), 121–135 (2003)
4. Béal, M., Fiorenzi, F., Mignosi, F.: Minimal forbidden patterns of multidimensional shifts. IJAC **15**(1), 73–93 (2005)
5. Béal, M.-P., Mignosi, F., Restivo, A.: Minimal forbidden words and symbolic dynamics. In: Puech, C., Reischuk, R. (eds.) STACS 1996. LNCS, vol. 1046, pp. 555–566. Springer, Heidelberg (1996). https://doi.org/10.1007/3-540-60922-9_45
6. Belazzougui, D., Cunial, F.: A framework for space-efficient string kernels. Algorithmica **79**(3), 857–883 (2017)
7. Belazzougui, D., Cunial, F., Kärkkäinen, J., Mäkinen, V.: Versatile succinct representations of the bidirectional burrows-wheeler transform. In: Bodlaender, H.L., Italiano, G.F. (eds.) ESA 2013. LNCS, vol. 8125, pp. 133–144. Springer, Heidelberg (2013). https://doi.org/10.1007/978-3-642-40450-4_12
8. Bender, M.A., Farach-Colton, M.: The LCA problem revisited. In: Gonnet, G.H., Viola, A. (eds.) LATIN 2000. LNCS, vol. 1776, pp. 88–94. Springer, Heidelberg (2000). https://doi.org/10.1007/10719839_9
9. Chairungsee, S., Crochemore, M.: Using minimal absent words to build phylogeny. Theoret. Comput. Sci. **450**, 109–116 (2012)

10. Charalampopoulos, P., Crochemore, M., Fici, G., Mercaş, R., Pissis, S.P.: Alignment-free sequence comparison using absent words. Inf. Comput. **262**(1), 57–68 (2018)
11. Charalampopoulos, P., Crochemore, M., Pissis, S.P.: On extended special factors of a word. In: Gagie, T., Moffat, A., Navarro, G., Cuadros-Vargas, E. (eds.) SPIRE 2018. LNCS, vol. 11147, pp. 131–138. Springer, Cham (2018). https://doi.org/10.1007/978-3-030-00479-8_11
12. Crawford, T., Badkobeh, G., Lewis, D.: Searching page-images of early music scanned with OMR: a scalable solution using minimal absent words. In: ISMIR, pp. 233–239 (2018)
13. Crochemore, M., Héliou, A., Kucherov, G., Mouchard, L., Pissis, S.P., Ramusat, Y.: Minimal absent words in a sliding window and applications to on-line pattern matching. In: Klasing, R., Zeitoun, M. (eds.) FCT 2017. LNCS, vol. 10472, pp. 164–176. Springer, Heidelberg (2017). https://doi.org/10.1007/978-3-662-55751-8_14
14. Crochemore, M., Mignosi, F., Restivo, A.: Automata and forbidden words. Inf. Process. Lett. **67**(3), 111–117 (1998)
15. Crochemore, M., Mignosi, F., Restivo, A., Salemi, S.: Data compression using antidictionaries. Proc. IEEE **88**(11), 1756–1768 (2000)
16. Crochemore, M., Navarro, G.: Improved antidictionary based compression. In: 22nd International Conference of the Chilean Computer Science Society (SCCC 2002), 6–8 November 2002, Copiapo, Chile, pp. 7–13 (2002)
17. Fiala, M., Holub, J.: DCA using suffix arrays. In: 2008 Data Compression Conference (DCC 2008), 25–27 March 2008, Snowbird, UT, USA, p. 516. IEEE Computer Society (2008)
18. Fici, G.: Minimal Forbidden Words and Applications. Ph.D. thesis, Université de Marne-la-Vallée (2006)
19. Fici, G., Langiu, A., Lo Bosco, G., Rizzo, R.: Bacteria classification using minimal absent words. AIMS Med. Sci. **5**(1), 23–32 (2018)
20. Fici, G., Mignosi, F., Restivo, A., Sciortino, M.: Word assembly through minimal forbidden words. Theoret. Comput. Sci. **359**(1), 214–230 (2006)
21. Fici, G., Restivo, A., Rizzo, L.: Minimal forbidden factors of circular words. Theoret. Comput. Sci., to appear
22. Fujishige, Y., Tsujimaru, Y., Inenaga, S., Bannai, H., Takeda, M.: Computing DAWGs and minimal absent words in linear time for integer alphabets. In: Faliszewski, P., Muscholl, A., Niedermeier, R. (eds.) 41st International Symposium on Mathematical Foundations of Computer Science, MFCS 2016, August 22–26, 2016 - Kraków, Poland. LIPIcs, vol. 58, pp. 38:1–38:14. Schloss Dagstuhl - Leibniz-Zentrum fuer Informatik (2016)
23. Gabow, H.N., Bentley, J.L., Tarjan, R.E.: Scaling and related techniques for geometry problems. In: Proceedings of the Sixteenth Annual ACM Symposium on Theory of Computing, STOC 1984, pp. 135–143. ACM, New York (1984)
24. Inenaga, S.: Suffix Trees, DAWGs and CDAWGs for Forward and Backward Tries. CoRR, abs/1904.04513 (2019)
25. Mignosi, F., Restivo, A., Sciortino, M.: Forbidden factors in finite and infinite words. In: Karhumäki, J., Maurer, H., Păun, G., Rozenberg, G. (eds.) Jewels are Forever, pp. 339–350. Springer, Heidelberg (1999). https://doi.org/10.1007/978-3-642-60207-8_30
26. Mignosi, F., Restivo, A., Sciortino, M.: Forbidden factors and fragment assembly. ITA **35**(6), 565–577 (2001)

27. Mignosi, F., Restivo, A., Sciortino, M.: Words and forbidden factors. Theor. Comput. Sci. **273**(1–2), 99–117 (2002)
28. Ota, T., Morita, H.: On the adaptive antidictionary code using minimal forbidden words with constant lengths. In: Proceedings of the International Symposium on Information Theory and its Applications, ISITA 2010, 17–20 October 2010, Taichung, Taiwan, pp. 72–77. IEEE (2010)
29. Schieber, B., Vishkin, U.: On finding lowest common ancestors: simplification and parallelization. SIAM J. Comput. **17**(6), 1253–1262 (1988)
30. Shibuya, T.: Constructing the suffix tree of a tree with a large alphabet. IEICE Trans. Fundam. Electron. Commun. Comput. Sci. **E86–A**(5), 1061–1066 (2003)
31. Silva, R.M., Pratas, D., Castro, L., Pinho, A.J., Ferreira, P.J.S.G.: Three minimal sequences found in Ebola virus genomes and absent from human DNA. Bioinformatics **31**(15), 2421–2425 (2015)

On Longest Common Property Preserved Substring Queries

Kazuki Kai[1], Yuto Nakashima[1], Shunsuke Inenaga[1], Hideo Bannai[1], Masayuki Takeda[1], and Tomasz Kociumaka[2,3]

[1] Department of Informatics, Kyushu University, Fukuoka, Japan
kai.kazuki.640@s.kyushu-u.ac.jp,
{yuto.nakashima,inenaga,bannai,takeda}@inf.kyushu-u.ac.jp
[2] Department of Computer Science, Bar-Ilan University, Ramat Gan, Israel
[3] Institute of Informatics, University of Warsaw, Warsaw, Poland
kociumaka@mimuw.edu.pl

Abstract. We revisit the problem of longest common property preserving substring queries introduced by Ayad et al. (SPIRE 2018, arXiv 2018). We consider a generalized and unified on-line setting, where we are given a set X of k strings of total length n that can be pre-processed so that, given a query string y and a positive integer $k' \leq k$, we can determine the longest substring of y that satisfies some specific property and is common to at least k' strings in X. Ayad et al. considered the longest square-free substring in an on-line setting and the longest periodic and palindromic substring in an off-line setting. In this paper, we give efficient solutions in the on-line setting for finding the longest common square, periodic, palindromic, and Lyndon substrings. More precisely, we show that X can be pre-processed in $O(n)$ time resulting in a data structure of $O(n)$ size that answers queries in $O(|y| \log \sigma)$ time and $O(1)$ working space, where σ is the size of the alphabet, and the common substring must be a square, a periodic substring, a palindrome, or a Lyndon word.

Keywords: Squares · Periodic substrings · Palindromes · Lyndon words

1 Introduction

The longest common substring of two strings x and y is a longest string that is a substring of both x and y. It is well known that the problem can be solved in linear time, using the generalized suffix tree of x and y [12,19].

Ayad et al. [1,2] proposed a class of problems called *longest common property preserved substring*, where the aim is to find the longest substring that has some

This work was supported by JSPS KAKENHI Grant Numbers JP18K18002 (YN), JP17H01697 (SI), JP16H02783 (HB), and JP18H04098 (MT). Tomasz Kociumaka was supported by ISF grants no. 824/17 and 1278/16 and by an ERC grant MPM under the EU's Horizon 2020 Research and Innovation Programme (grant no. 683064).

N. R. Brisaboa and S. J. Puglisi (Eds.): SPIRE 2019, LNCS 11811, pp. 162–174, 2019.
https://doi.org/10.1007/978-3-030-32686-9_12

property and is common to a subset of the input strings. They considered several problems in two different settings.

In the first *on-line* setting, one is given a string x that can be pre-processed, and the problem is to answer, for any given query string y, the longest square-free substring that is common to both x and y. Their solution takes $O(|x|)$ time for preprocessing and $O(|y| \log \sigma)$ time for queries, where σ is the alphabet size.

In the second *off-line* setting, one is given a set of k strings of total length n and a positive integer $k' \leq k$, and the problem is to find the longest periodic substring, as well as the longest palindromic substring, that is common to at least k' of the strings. Their solution works in $O(n)$ time and space. However, it does not (at least directly) give a solution for the on-line setting.

In this paper, we consider a generalized and unified on-line setting, where we are given a set X of k strings with total length n that can be pre-processed, and the problem is to answer, for any given query string y and positive integer $k' \leq k$, the longest property preserved substring that is common to y and at least k' of the strings. We give solutions to the following properties in this setting, all working in $O(n)$ time and space preprocessing, and $O(|y| \log \sigma)$ time and $O(1)$ working space for answering queries: squares, periodic substrings, palindromes, and Lyndon words. Furthermore, we note that solutions for the off-line setting can be obtained by using our solutions for the on-line setting. We also note that our algorithms can be modified to remove the $\log \sigma$ factor in the off-line setting.

As related work, the off-line version of property preserved subsequences have been considered for some properties. The longest common square subsequence between two strings can be computed in $O(n^6)$ time [14]. The longest common palindromic subsequence between two strings of length n can be computed in $O(n^4)$ time [3].

2 Preliminaries

2.1 Strings

Let Σ be a set of symbols, or alphabet, and Σ^* the set of strings over Σ. We assume a constant or linearly-sortable integer alphabet[1] and use σ to denote the size of the alphabet, i.e. $|\Sigma| = \sigma$. For any string $x \in \Sigma^*$, let $|x|$ denote its length. For any integer $1 \leq i \leq |x|$, $x[i]$ is the ith symbol of x, and for any integers $1 \leq i \leq j \leq |x|$, $x[i..j] = x[i] \cdots x[j]$. For convenience, $x[i..j]$ is the empty string when $i > j$. If a string w satisfies $w = xyz$, then, strings x, y, z are respectively called a prefix, substring, and suffix of w. A prefix (resp. substring, suffix) is called a *proper* prefix (resp. substring, suffix) if it is shorter than the string.

Let x^R denote the reverse of x, i.e., $x^R = x[|x|] \cdots x[1]$. A string x is said to be a *palindrome* if $x = x^R$. A string y is a *square* if $y = xx$ for some string x, called the *root* of y. A string y is *primitive* if there does not exist any x such that $y = x^k$ for some integer $k \geq 2$. A square is called a *primitively rooted* square, if

---

[1] Note that a string of length n on a general ordered alphabet can be transformed into a string on an integer alphabet in $O(n \log \sigma)$ time.

its root is primitive. An integer p, $1 \leq p \leq |x|$, is called a *period* of a string x, if $x[i] = x[i + p]$ for all $1 \leq i \leq |x| - p$. A string x is *periodic*, if the smallest period p of x is at most $|x|/2$. A *run* in a string is a maximal periodic substring, i.e., a periodic substring $x[i..j]$ with smallest period p is a run, if the period of any string $x[i'..j']$ with $i' \leq i \leq j \leq j'$ and either $i' \neq i$ or $j' \neq j$, is not p. The following is the well known (weak) periodicity lemma concerning periods:

Lemma 1 ((Weak) Periodicity Lemma [8]). *If p and q are two periods of a string w, and $p + q \leq |w|$, then, $\gcd(p, q)$ is also a period of w.*

Let $\prec$ denote a total order on Σ, as well as the lexicographic order on Σ^* it induces. That is, for any two strings x, y, $x \prec y$ if and only if either x is a prefix of y, or there exist strings $w, x', y' \in \Sigma^*$ such that $x = wx'$ and $y = wy'$, and $x'[1] \prec y'[1]$. A string is a *Lyndon word* [17] if it is lexicographically smaller than any of its non-empty proper suffixes.

2.2   Suffix Trees

The suffix tree [19] $ST(x)$ of a string x is a compacted trie of the set of non-empty suffixes of $x\$$, where $\$$ denotes a unique symbol that does not occur in x. More precisely, it is (1) a rooted tree where edges are labeled by non-empty strings, (2) the concatenation of the edge-labels on root to leaf paths correspond to all and only suffixes of $x\$$, 3) any non-leaf node has at least two children, and the first letter of the label on the edges to its children are distinct.

A node in $ST(x)$ is called an *explicit* node, while a position on the edges corresponding to proper prefixes of the edge label are called *implicit* nodes. For a (possibly implicit) node v in $ST(x)$, let $str(v)$ denote the string obtained by concatenating the edge labels on the path from the root to v. The *locus* of a string p in $ST(x)$ is a (possibly implicit) node v in $ST(x)$ such that $str(v) = p$. Each explicit node v of the suffix tree can be augmented with a *suffix link*, that points to the node u, such that $str(v) = str(v)[1]str(u)$. It is easy to see that because v is an explicit node, u is also always an explicit node.

It is well known that the suffix tree (and suffix links) can be represented in $O(|x|)$ space and constructed in $O(|x| \log \sigma)$ time [18] for general ordered alphabets, or in $O(|x|)$ time for constant [19] or linearly-sortable integer alphabets [7]. The suffix tree can also be defined for a set of strings $X = \{x_1, \ldots, x_k\}$, and again can be constructed in $O(n \log \sigma)$ time for general ordered alphabets or in $O(n)$ time for constant or linearly-sortable integer alphabets, where n is the total length of the strings. This is done by considering and building the suffix tree for the string $x_1\$ \cdots x_k\$$ and pruning edges below any $\$$. It is also easy to process $ST(X)$ to compute for each explicit node v, the length $|str(v)|$, as well as an occurrence (s, b) of $str(v)$ in X, where $x_s[b..b + |str(v)| - 1] = str(v)$. Also, these values can be computed in constant time for any implicit node, given the values for the closest descendant explicit node.

We will later use the following lemma to efficiently find the loci of a given set of substrings, and to make these loci explicit (by subdividing the edges of the suffix tree containing loci that were originally implicit).

Lemma 2 ([16, **Corollary 7.3**]). *Given a collection of substrings $s_1, \ldots, s_k$ of a string w of length n, each represented by an occurrence in w, in $O(n + k)$ total time we can compute the locus of each substring s_i in the suffix tree of w. Moreover, these loci can be turned into explicit nodes in $O(n + k)$ extra time.*

For a string x of length n, a *longest common extension query*, given positions $1 \leq i, j \leq n$ asks for the longest common prefix between $x[i..n]$ and $x[j..n]$. It is known that the string can be pre-processed in $O(n)$ time so that the longest common extension query can be answered in $O(1)$ time for any i, j (e.g. [9]).

2.3   Matching Statistics

For two strings x and y, the *matching statistics* of y with respect to x is an array $MS_{y,x}[1..|y|]$, where

$$MS_{y,x}[i] = \max\{l \geq 0 : \exists_{j \in \{1,\ldots,|x|\}}\, x[j..j + l - 1] = y[i..i + l - 1]\}$$

for any $1 \leq i \leq |y|$. That is, for each position i of y, $MS_{y,x}[i]$ is the length of the longest prefix of $y[i..|y|]$ that occurs in x. The concept of matching statistics can be generalized to a set of strings. For a set $X = \{x_1, \ldots, x_k\}$ of k strings and a string y, the k'-matching statistics of y with respect to X is an array $MS_{y,X}^{k'}[1..|y|]$ where

$$MS_{y,X}^{k'}[i] = \max\{l \geq 0 : |\{x \in X : \exists_{j \in \{1,\ldots,|x|\}}\, x[j..j+l-1] = y[i..i+l-1]\}| \geq k'\}.$$

That is, for each position i of y, $MS_{y,X}^{k'}[i]$ is the length of the longest prefix of $y[i..|y|]$ that occurs in at least k' of the strings in X.

2.4   Longest Common Property Preserved Substring Queries

Let a function $P : \Sigma^* \rightarrow \{\texttt{true}, \texttt{false}\}$ be called a *property function*. In this paper, we will consider the following property functions P_{sqf}, $P_{\mathrm{sqr}}, P_{\mathrm{per}}, P_{\mathrm{pal}}, P_{\mathrm{Lyn}}$, which return $\texttt{true}$ if and only if a string is respectively a square-free, square, periodic string, palindrome, or a Lyndon word.

The following is the on-line version of the problem considered in [1], where a solution was given for the longest common square free substring, i.e., $P = P_{\mathrm{sqf}}$.

Problem 3 (Longest common property preserved substring query). Let P be a property function. Consider a string x which can be pre-processed. For a query string y, compute a longest string z that is a substring of both x and y, and also satisfies $P(z) = \texttt{true}$.

The following is the generalized version of the on-line setting that we consider in this paper.

Problem 4 (Generalized longest common property preserved substring query). Let P be a property function. Consider a set of strings $X = \{x_1, \ldots, x_k\}$ that can be pre-processed. For a query string y and positive integer $k' \leq k$, compute a longest substring z of y that is a substring of at least k' strings in X, and also satisfies $P(z) = \texttt{true}$.

Below is a summary of our results. Here, the *working space* of the query is the amount of memory that is required in excess to the data structure constructed in the pre-processing. All memory other than the working space can be considered as read-only.

Theorem 5. *For any property function* $P \in \{P_\mathrm{sqf}, P_\mathrm{sqr}, P_\mathrm{per}, P_\mathrm{pal}, P_\mathrm{Lyn}\}$, *Problem 4 can be answered in* $O(|y| \log \sigma)$ *time and* $O(1)$ *working space by constructing an* $O(n)$ *space data structure in* $O(n)$ *time, where* $n = \sum_{i=1}^{k} |x_i|$ *is the total length of the strings in* X.

We further note that our algorithms do not require random access to y during the query and thus work in the same time/space bounds even if each symbol of y is given in a streaming fashion.

The proof of the Theorem for each property function is given in the next section.

3   Algorithms

In this section we present our algorithms, starting from the common outline.
    The preprocessing consists of the following steps:

1. Construct the generalized suffix tree $ST(X)$ of X.
2. For each explicit node v of $ST(X)$, compute the number of strings in X that contain $str(v)$ as a substring.
3. Process $ST(X)$ and construct a data structure so that, given any position on $ST(X)$, we can efficiently find a *candidate* for the solution.

    Then, queries are answered as follows:

4. For each position i in y, compute $MS_{y,X}^{k'}[i]$, i.e., the k'-matching statistics of y with respect to X, as the locus v_i of $y[i..e_i]$ in $ST(X)$.
5. For each such locus v_i, compute a candidate using the data structure computed in Step 3 of the pre-processing.
6. Output the longest string computed in the previous step.

As mentioned in Sect. 2.2, Step 1 can be performed in $O(n)$ time and space. The task of Step 2 is known as the color set size problem, and it can also be executed in $O(n)$ time [13].
    Using $ST(X)$, the outcome of Step 4, i.e., the locus v_i of the substring $y[i..e_i]$ where $e_i - i + 1 = MS_{y,X}^{k'}[i]$, can be computed for all $1 \leq i \leq |y|$ in $O(|y| \log \sigma)$ total time and $O(1)$ working space, with a minor modification to the algorithm for computing the matching statistics with respect to a single string [12, Theorem 7.8.1]. The algorithm for a single string first searches the longest prefix of $y[1..|y|]$ in the suffix tree to compute the locus corresponding to $MS_{y,x}[1]$. Let this prefix be $y[1..e_1]$. Given a locus v_i of $y[i..e_i]$ for some $1 \leq i < |y|$, the suffix link of the closest ancestor of v_i is used in order to first efficiently find the locus of $y[i + 1..e_i]$. Then, the suffix tree is further traversed to obtain the locus of $y[i + 1..e_{i+1}]$. The time bound for the traversal

follows from a well known amortized analysis which considers the depth of the traversed nodes, similar to that in the online construction of suffix trees [18]. For computing $MS^{k'}_{y,X}$, we can simply imagine that subtrees below edges leading to a node v of $ST(X)$ are pruned if $str(v)$ is contained in less than k' strings of X. This can be simulated by aborting the traversal of $y[i..|y|]$ when we encounter such an edge, detected using the information obtained in Step 2. It is easy to see that the algorithm still works in the same time bound since the suffix link of any remaining node still points to a node that is not pruned (i.e., if a string is contained in at least k' strings, then its suffix will also be contained in the same strings). Thus, we can visit each locus corresponding to $MS^{k'}_{y,X}$ in $O(|y| \log \sigma)$ total time and $O(1)$ working space.

The crux of our algorithm is therefore in the details of Step 3 and Step 5: designing what the candidates are and how to compute them given v_i. Notice that the solution is the longest string z which satisfies $P(z) = \texttt{true}$ and is a substring of $y[i..e_i]$ for some $i = 1, \ldots, n$.

3.1 Square-Free Substrings

Ayad et al. [1,2] gave a solution to the on-line longest common square-free substring query problem for a single string (Problem 3), in $O(|x|)$ time and space preprocessing and $O(|y| \log \sigma)$ time and $O(1)$ space query. We note that their algorithm easily extends to the generalized version (Problem 4). The only difference lies in that $MS^{k'}_{y,X}$ is computed instead of $MS_{y,x}$, which can be done in $O(|y| \log \sigma)$ time and $O(1)$ space, as described above. Details can be found in [1,2].

3.2 Squares

As mentioned in the introduction, Ayad et al. [1,2] also considered longest common periodic substrings, but in the off-line setting. However, their algorithm is not readily extendible to the on-line setting. It relies on the fact that the ending position of a longest common periodic substring must coincide with an ending position of some run in the set of strings, and solved the problem by computing all loci corresponding to runs in X. To utilize this observation in the on-line setting, the loci of all runs in the query string y must be identified in $ST(X)$, which seems difficult to do in time not depending on X.

Here, we first show how to efficiently solve the problem in the on-line setting for squares, and then we extend that solution to obtain an efficient solution for periodic substrings. Below is an important property of squares which we exploit.

Lemma 6 ([10, Theorem 1]). *A given position can be the right-most occurrence of at most two distinct squares.*

It follows from the above lemma that the number of distinct squares in a given string is linear in its length [10]. Also, it gives us the following Corollary.

Corollary 7. *On a given edge of a suffix tree, there can only be at most two implicit nodes which correspond to a square.*

Proof. The right-most occurrence of a square is the maximum position corresponding to leaves in the subtree rooted at the square. Since implicit nodes that correspond to squares on the same edge share the right-most occurrence, a third implicit node would contradict Lemma 6.   □

For squares, we first compute the locus in $ST(X)$ of all distinct squares that are substrings of strings in X, and we make them explicit nodes in $ST(X)$. Note that these additional nodes will not have suffix links, but since there are only a constant number of them on each edge of the original suffix tree, it will only add a constant factor to the amortized analysis when computing Step 4. The loci of all squares can be computed in $O(n)$ total time using the algorithm of [4]. We further add to each explicit node in $ST(X)$ (including the ones we introduced above) a pointer to the nearest ancestor that is the locus of a square (see Fig. 1 for an example of the pointers). Notice that a node that is the locus of a square is explicit and points to itself. This can be also done in linear time by a simple depth-first traversal on $ST(X)$.

The candidate, which we will compute in Step 5 for each locus v_i, is the longest square that is a prefix of $y[i..e_i]$. It can be determined for each locus v_i by using the pointers. When v_i is an explicit node, the pointer of v_i is the answer. When v_i is an implicit node, the pointer of the parent of v_i is the answer. The longest such square for all loci is the answer to the query. This is because the longest common square must be a prefix of the string corresponding to the k'-matching statistics of some position. Thus, we have a solution as claimed in Theorem 5 for $P = P_{\mathrm{sqr}}$.

3.3   Periodic Substrings

Next, we extend the solution for squares to periodic substrings as follows.

We first explain the data structure, which is again an augmented $ST(X)$. For each primitively rooted square substring w, we make the locus of w in $ST(X)$ an explicit node. (The non-primitively rooted squares are redundant since they will lead to the same periodic substrings.) Furthermore, we also make explicit the deepest locus v in $ST(X)$ obtained by periodically extending a primitively rooted square w, i.e., w is a prefix of $str(v)$ and the smallest period of $str(v)$ is $\frac{1}{2}|w|$. We add to each explicit node, a pointer to the nearest explicit ancestor (including itself) that is a locus of some square or its extension. If an explicit node is an extension of a square, it will also hold information to identify which square it is an extension of (in our case, the smallest period of the square suffices). Note that the pointer of an explicit node that lies between a square and its extension will point to itself. Figure 2 shows an example of $ST(X)$ for a single string $X = \{\texttt{aababababbbabababab\$}\}$, where loci corresponding to squares and their rightmost-maximal extensions are depicted.

We first show how the above augmentation of $ST(X)$ can be executed in $O(n)$ time. We first make explicit all loci of squares as in Sect. 3.2, which can be done

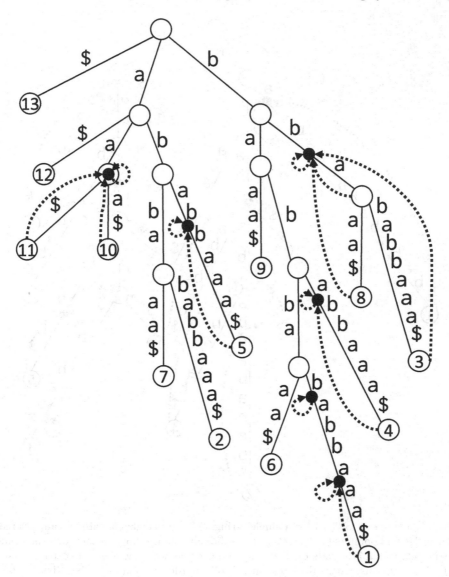

Fig. 1. Example of $ST(X)$ for a single string $X = \{\texttt{babbababbaaa\$}\}$ augmented with nodes and pointers for $P = P_{\mathrm{sqr}}$. The solid dark circles show the loci corresponding to squares which are made explicit. The dotted arrows show the pointers which point to the nearest ancestor which is a square. Pointers which point to the root (i.e., there is no non-empty prefix that is a square) are omitted.

in $O(n)$ time. Then, we start from the locus of each primitively rooted square and extend the square periodically towards the leaves of $ST(X)$ as deep as possible. For each explicit node we encounter during this extension, the pointer will point to itself, and the node will also store the period of the underlying square.

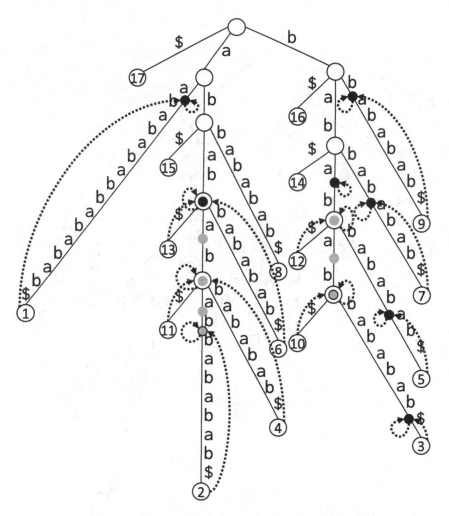

Fig. 2. Example of $ST(X)$ for a single string $X = \{\texttt{aababababbbababab\$}\}$ augmented with nodes and pointers for $P = P_{\mathrm{per}}$. The solid dark circles show the loci corresponding to squares, which are made explicit, and the grey circles show the loci corresponding to their extensions, where the ones with a solid border are made explicit. The dotted arrows show the pointers which point to the nearest ancestor (longest prefix) which is periodic. The pointers which point to the root are omitted. The total number of solid dark circles, as well as grey circles with a solid border is $O(n)$. The total number of implicit grey circles without the solid border is not necessarily $O(n)$, but since they occur consecutively from a solid node, they can be represented in $O(n)$ space.

The total cost of this extension can be bounded as follows. Due to the periodicity lemma (Lemma 1), any locus of a primitively rooted square or its extension cannot be an extension of a shorter square; if it was, a (proper) divisor of the period of the longer square would also be a period of the square, and would

contradict that the longer square is primitively rooted. Thus, we can naturally discard all non-primitive squares by doing the extensions in increasing order of the length of squares (which can be obtained in linear time by radix sort). During the extension, if the square is non-primitive, then the pointers of nodes will already have been determined by a square of a shorter period, and this situation can be detected. From the above arguments, any edge of $ST(X)$ is traversed at most once, and thus the extension can be done by traversing a total of $O(n)$ nodes and edges. Because we know the period of the square we wish to extend, and for any locus v in $ST(X)$, an occurrence b in some $x_s \in X$, we can compute the extension in $O(1)$ time per edge of $ST(X)$ by using longest common extension queries. Therefore, the total time for building the augmented $ST(X)$ can be done in $O(n)$ time.

Queries can be answered in the same way as for squares, except for a small modification. For any locus v_i, if v_i is an explicit node, then the pointer of v_i gives the answer. When v_i is an implicit node, we check v_i's parent and its immediate descendant. If these nodes are both extensions originating from the same square (i.e., their labels have the same period), then, the answer is v_i itself, since it is also an extension of the square. Otherwise, the pointer of the parent node provides the answer.

Thus, we have a solution as claimed in Theorem 5 for $P = P_{\mathrm{per}}$.

3.4 Palindromes

It is well known that the number of non-empty distinct palindromes in a string of length n is at most n, since only the longest palindrome starting at some position can be the right-most occurrence of that palindrome in the string.

Lemma 8 ([6, **Proposition 1**]). *A given position can be the right-most occurrence of at most one distinct palindrome.*

All distinct palindromes in a string can be computed in linear time [11]. The locus of each palindrome in $ST(X)$ can be computed in $O(n)$ time by Lemma 2. The rest is the same as for squares; we make all loci corresponding to a palindrome an explicit node, and do a linear time depth-first traversal on $ST(X)$ to make pointers to the nearest ancestor that is a palindrome. As in the case of squares, we can bound the number of palindromes which will be on an edge of the original suffix tree.

Corollary 9. *On a given edge of a suffix tree, there can only be at most one implicit node that corresponds to a palindrome.*

Proof. Analogous to the proof of Corollary 7.   □

The rest of the algorithm and analysis is the same, thus we obtain a solution for $P = P_{\mathrm{pal}}$ of Theorem 5.

3.5 Lyndon Words

For Lyndon words, we use the following result by Kociumaka [15]. A *minimal suffix query* (MSQ) on a string T, given indices ℓ, r such that $1 \leq \ell \leq r \leq |T|$, determines the lexicographically smallest suffix of $T[\ell..r]$.

Lemma 10 ([15, **Theorem 17**]). *For any string T of length n, there exists a data structure of size $O(n)$ which can be constructed in $O(n)$ time, which can answer minimal suffix queries on T in constant time.*

Using Lemma 10, we can find the longest Lyndon word ending at a given position.

Lemma 11. *For any string w, the lexicographically smallest suffix is the longest Lyndon word that is a suffix of w.*

Proof. From the definition of Lyndon words, it is clear that the minimal suffix must be a Lyndon word, and that a longer suffix cannot be a Lyndon word. □

For Step 3, we process all strings in X so that MSQ can be answered in constant time.

For Step 5, the situation is a bit different from squares, periodic strings, and palindromes, in that we can have multiple candidates for each v_i rather than just one. For convenience, let $e_0 = 0$. For each $0 \leq i < n$, suppose we have obtained the locus v_i of $y[i..e_i]$, and the next locus v_{i+1} of $y[i+1..e_{i+1}]$. Notice that $e_i \leq e_{i+1}$ and for all positions e' such that $e_i < e' \leq e_{i+1}$, $y[i+1..e']$ is the longest substring of y that ends at e' and is a substring of $y[j..e_j]$ for some $j = 1, \ldots, n$. As mentioned in Sect. 2.2, we can obtain an occurrence (s, b) of $y[i+1..e_{i+1}]$ such that $y[i+1..e_{i+1}] = x_s[b..b + MS_{y,X}^{k'}[i] - 1]$. Then, we use MSQ on substrings $x_s[b..r]$ such that $b+e_i-i \leq r \leq b+e_{i+1}-(i+1)$, which is equivalent to using MSQ on substrings $y[i+1..e']$ for all $e_i < e' \leq e_{i+1}$. The longest suffix Lyndon word obtained in all the queries is therefore the longest Lyndon word that is a substring of $y[j..e_j]$ for some $j = 1, \ldots, n$. Since we perform $e_{i+1} - e_i$ MSQs for each position $i+1$ of y, the total number of MSQs is $|y|$, which takes $O(|y|)$ time. Thus, we have a solution as claimed in Theorem 5 for $P = P_{\mathrm{Lyn}}$.

3.6 Solutions in the Off-Line Setting

We note that a solution for the on-line setting gives a solution for the off-line setting, since, for any $X = \{x_1, \ldots, x_k\}$, we can consider the string $y = x_1 \# \cdots \# x_k$, where $\#$ is again a symbol that doesn't appear elsewhere. Since $|y| = O(n)$, the preprocessing time is $O(n)$, and the query time is $O(n \log \sigma)$.

Furthermore, we can remove the $\log \sigma$ factor by processing $ST(X)$ for *level ancestor* queries. Level ancestor queries, given a node v of tree T and integer d, answer the ancestor of v at (node) depth d. It is known that level ancestor queries can be answered in constant time, after linear time preprocessing of T (e.g. [5]). The $\log \sigma$ factor came from determining which child of a branching node we needed to follow when traversing $ST(X)$ with some suffix of y. Since, in this

case, we can identify the leaf in $ST(X)$ that corresponds to the current suffix of y (i.e. some suffix of x_i in X) that is being traversed, we can use level ancestor query to determine, in constant time, the child of the branching node that is an ancestor of that leaf, thus getting rid of the $\log \sigma$ factor.

4 Conclusion

We considered the generalized on-line variant of the longest common property preserved substring problem proposed by Ayad et al. [1,16], and (1) unified the two problem settings, and (2) proposed algorithms for several properties, namely, squares, periodic substrings, palindromes, and Lyndon words. For all these properties, we can answer queries in $O(|y| \log \sigma)$ time and $O(1)$ working space, with $O(n)$ time and space preprocessing.

References

1. Ayad, L.A.K., et al.: Longest property-preserved common factor. In: Gagie, T., Moffat, A., Navarro, G., Cuadros-Vargas, E. (eds.) SPIRE 2018. LNCS, vol. 11147, pp. 42–49. Springer, Cham (2018). https://doi.org/10.1007/978-3-030-00479-8_4
2. Ayad, L.A.K., et al.: Longest property-preserved common factor (2018). http://arxiv.org/abs/1810.02099
3. Bae, S.W., Lee, I.: On finding a longest common palindromic subsequence. Theor. Comput. Sci. **710**, 29–34 (2018). https://doi.org/10.1016/j.tcs.2017.02.018
4. Bannai, H., Inenaga, S., Köppl, D.: Computing all distinct squares in linear time for integer alphabets. In: Kärkkäinen, J., Radoszewski, J., Rytter, W. (eds.) 28th Annual Symposium on Combinatorial Pattern Matching, CPM 2017. LIPIcs, vol. 78, pp. 22:1–22:18. Schloss Dagstuhl-Leibniz-Zentrum für Informatik (2017). https://doi.org/10.4230/LIPIcs.CPM.2017.22
5. Bender, M.A., Farach-Colton, M.: The level ancestor problem simplified. Theor. Comput. Sci. **321**(1), 5–12 (2004). https://doi.org/10.1016/j.tcs.2003.05.002
6. Droubay, X., Justin, J., Pirillo, G.: Episturmian words and some constructions of de Luca and Rauzy. Theor. Comput. Sci. **255**(1–2), 539–553 (2001). https://doi.org/10.1016/S0304-3975(99)00320-5
7. Farach-Colton, M., Ferragina, P., Muthukrishnan, S.: On the sorting-complexity of suffix tree construction. J. ACM **47**(6), 987–1011 (2000). https://doi.org/10.1145/355541.355547
8. Fine, N.J., Wilf, H.S.: Uniqueness theorems for periodic functions. Proc. Am. Math. Soc. **16**(1), 109–114 (1965). https://doi.org/10.1090/s0002-9939-1965-0174934-9
9. Fischer, J., Heun, V.: Space-efficient preprocessing schemes for range minimum queries on static arrays. SIAM J. Comput. **40**(2), 465–492 (2011). https://doi.org/10.1137/090779759
10. Fraenkel, A.S., Simpson, J.: How many squares can a string contain? J. Comb. Theory Ser. A **82**(1), 112–120 (1998). https://doi.org/10.1006/jcta.1997.2843
11. Groult, R., Prieur, É., Richomme, G.: Counting distinct palindromes in a word in linear time. Inf. Process. Lett. **110**(20), 908–912 (2010). https://doi.org/10.1016/j.ipl.2010.07.018

12. Gusfield, D.: Algorithms on Strings, Trees, and Sequences: Computer Science and Computational Biology. Cambridge University Press, Cambridge (1997). https://doi.org/10.1017/cbo9780511574931
13. Chi, L., Hui, K.: Color Set Size problem with applications to string matching. In: Apostolico, A., Crochemore, M., Galil, Z., Manber, U. (eds.) CPM 1992. LNCS, vol. 644, pp. 230–243. Springer, Heidelberg (1992). https://doi.org/10.1007/3-540-56024-6_19
14. Inoue, T., Inenaga, S., Hyyrö, H., Bannai, H., Takeda, M.: Computing longest common square subsequences. In: Navarro, G., Sankoff, D., Zhu, B. (eds.) 29th Annual Symposium on Combinatorial Pattern Matching, CPM 2018. LIPIcs, vol. 105, pp. 15:1–15:13. Schloss Dagstuhl-Leibniz-Zentrum für Informatik (2018). https://doi.org/10.4230/LIPIcs.CPM.2018.15
15. Kociumaka, T.: Minimal suffix and rotation of a substring in optimal time. In: Grossi, R., Lewenstein, M. (eds.) 27th Annual Symposium on Combinatorial Pattern Matching, CPM 2016. LIPIcs, vol. 54, pp. 28:1–28:12. Schloss Dagstuhl-Leibniz-Zentrum für Informatik (2016). https://doi.org/10.4230/LIPIcs.CPM.2016.28
16. Kociumaka, T., Kubica, M., Radoszewski, J., Rytter, W., Waleń, T.: A linear time algorithm for seeds computation (2019). http://arxiv.org/abs/1107.2422v2
17. Lyndon, R.C.: On Burnside's problem. Trans. Am. Math. Soc. **77**(2), 202–215 (1954). https://doi.org/10.1090/s0002-9947-1954-0064049-x
18. Ukkonen, E.: On-line construction of suffix trees. Algorithmica **14**(3), 249–260 (1995). https://doi.org/10.1007/BF01206331
19. Weiner, P.: Linear pattern matching algorithms. In: 14th Annual Symposium on Switching and Automata Theory, SWAT 1973, pp. 1–11. IEEE Computer Society (1973). https://doi.org/10.1109/SWAT.1973.13

Online Algorithms on Antipowers
and Antiperiods

Mai Alzamel[2,3], Alessio Conte[1], Daniele Greco[1], Veronica Guerrini[1],
Costas Iliopoulos[2], Nadia Pisanti[1], Nicola Prezza[1], Giulia Punzi[1],
and Giovanna Rosone[1(✉)]

[1] Department of Computer Science, University of Pisa, Pisa, Italy
{conte,veronica.guerrini,pisanti,nicola.prezza}@di.unipi.it,
grc.daniele.cs@gmail.com, giuliagpunzi@gmail.com,
giovanna.rosone@unipi.it
[2] Department of Informatics, King's College London, London, UK
{mai.alzamel,costas.iliopoulos}@kcl.ac.uk
[3] Department of Computer Science, King Saud University,
Riyadh, Kingdom of Saudi Arabia

Abstract. The definition of *antipower* introduced by Fici et al. (ICALP
2016) captures the notion of being the opposite of a *power*: a sequence
of k pairwise distinct blocks of the same length. Recently, Alamro et al.
(CPM 2019) defined a string to have an *antiperiod* if it is a prefix of an
antipower, and gave complexity bounds for the offline computation of
the minimum antiperiod and all the antiperiods of a word. In this paper,
we address the same problems in the *online* setting. Our solutions rely
on new arrays that compactly and incrementally store antiperiods and
antipowers as the word grows, obtaining in the process this information
for all the word's prefixes. We show how to compute those arrays online
in $O(n \log n)$ space, $O(n \log n)$ time, and $o(n^\epsilon)$ delay per character, for
any constant $\epsilon > 0$. Running times are worst-case and hold with high
probability. We also discuss more space-efficient solutions returning the
correct result with high probability, and small data structures to support
random access to those arrays.

Keywords: Antiperiod · Antipower · Power · Periodicity ·
Repetition · Regularities · Online algorithms

1 Introduction

String properties that highlight regularities such as periodicity, powers, repetitions, palindromes, as well as properties that—dually—highlight diversity (e.g. being square-free, non-periodic, etc.), have been extensively investigated in literature. They have been studied both in terms of combinatorial properties [25,26] and of algorithmic methods to detect or certify them in a single string [9,10], or finding maximal common factors that share such properties [2,3,5,8,18,19]. In addition to having a combinatorial interest on their own, such string properties

© Springer Nature Switzerland AG 2019
N. R. Brisaboa and S. J. Puglisi (Eds.): SPIRE 2019, LNCS 11811, pp. 175–188, 2019.
https://doi.org/10.1007/978-3-030-32686-9_13

are also very relevant for several applications [9, 27, 28]. For instance, they are at the core of many problems arising in biological sequence analysis [22, 24].

Arguably, one of the most natural notions of regularity is that of an exact repetition (power), i.e. a substring consisting of two or more consecutive identical factors. The number of these factors, or blocks, is called the *order* of the repetition. The study of powers began in the early 1900s with the work of Thue [29], who studied a class of strings that do not contain any substrings that are powers. Recently, a new notion of string diversity in terms of powers has been introduced: an *antipower* of order k (k-antipower) is a string that can be decomposed into k pairwise-distinct blocks of identical length [14, 15]. An antipower is the opposite of a power; i.e., a concatenation of blocks that have the same length but are all different.[1] Likewise, the concept of *antiperiod* is symmetrical to that of period: an integer p is a *period* for a word w if and only if w is a prefix of a power with blocks of length p, whereas ℓ is an antiperiod for w if and only if it is a prefix of an antipower with blocks of length ℓ [1].

With respect to *infinite* words, in [14] the authors prove that regardless of the alphabet size, every infinite word must contain powers of any order or antipowers of any order; i.e., the existence of powers or antipowers is an unavoidable regularity (cf. also [15]). Inspired by this seminal work, Defant [11] studied the sequence of lengths of the shortest prefixes of the Thue-Morse word that are k-antipowers, and proved that this grows linearly in k. The latter result is further extended in [7] to a generalization of k-antipowers defined in [15].

For *finite* words, the first algorithmic approach concerning antipowers appeared in [4]. In this work, the authors tackle the problem of finding all the factors of a string w that are k-antipowers. Specifically, they prove that the number of such factors over an alphabet of any size is $\Omega(n^2/k)$, and provide an algorithm that finds them all in $O(n^2/k)$ time and linear space. The latter results are improved in [21], where the authors give an algorithm that counts and reports the number C of substrings of a word w of length n that are k-antipowers, in $O(nk \log k)$ and $O(nk \log k + C)$ time, respectively. Moreover, they are also able to test whether a factor $w[i, j]$ is a k-antipower (i.e, answering an *antipower query* (i, j, k)) in $O(r)$ time, by constructing a data structure of size $O(n^2/r)$ in $O(n^2/r)$ time, for any $r \in \{1, \ldots, n\}$. As far as antiperiods are concerned, Alamro et al. [1] are the first to give algorithmic results. Specifically, they compute all antiperiods of a string of length n in $O(n \log n)$ time, by employing a split-find data structure with initialization time $O(n)$ and linear space, which quickly answers monotone weighted level ancestor queries over the suffix tree. Furthermore, applying recursion to the same solution, they show how to compute just the smallest antiperiod t in $O(n \log^* t)$ time.

In this paper, we extend the problems considered in [1] to the online setting. We show how to efficiently update all the antiperiods (in particular, the minimum one) of a word upon single character extensions. To achieve this, we introduce the notion of *purely antiperiodic* array—i.e. the array containing, for each word's

---

[1] We remark that a word may be a power/antipower for different orders, even though in some cases [1] the focus is on the smallest such order.

prefix $w[1, i]$, the smallest block length ℓ such that the prefix is an antipower of order i/ℓ—and the more relaxed notion of *antiperiodic* array, which allows the last block to have length less than ℓ. In addition, we provide the more powerful notion of *complete antipower* array, containing for each possible block length ℓ, the greatest i such that the word's prefix $w[1, i]$ is an antipower of order i/ℓ. The complete antipower array stores, implicitly, all antiperiods of all the word's prefixes, as well as the value i/ℓ itself, that is the maximum order for which a prefix of w is an antipower of period i. We show that these arrays can be computed online in $O(n \log n)$ space, $O(n \log n)$ time, and $o(n^\epsilon)$ delay per character, for any constant $\epsilon > 0$. We also show that if we allow a small (inverse-polynomial) probability of failure, then we can compute online the antiperiodic array in $O(n)$ space, $O(n \log t)$ time and $O(n \log t)$ delay (all running times are worst-case and hold with high probability). Finally, we describe small data structures supporting fast random access to those three arrays, without the need to explicitly store all of them.

2   Preliminaries and Problem Definitions

Let $\Sigma = \{c_1, c_2, \ldots, c_\sigma\}$ be a finite ordered alphabet of size σ with $c_1 < c_2 < \ldots < c_\sigma$. Given a word (or string) $w = w[1]w[2] \cdots w[n] \in \Sigma^*$ we denote by $|w|$ its length n. We use ϵ to denote the empty word. A *factor* (or substring) of w is written as $w[i, j] = w[i] \cdots w[j]$ with $1 \leq i \leq j \leq n$. A factor of type $w[1, j]$ is called a *prefix* of w, while a factor of type $w[i, n]$ is called a *suffix* of w.

An integer $p \geq 1$ is a *period* of a word $w[1]w[2] \cdots w[n]$ where $w[i] \in \Sigma$ if $w[i] = w[i + p]$ for $i = i, \ldots, n - p$. The smallest period of w is called *the period* of w.

A *power of order k* is a string that is the concatenation of k identical equal-length blocks of letters. More formally, given a finite word w, w^k denotes the word obtained by concatenating k copies of w. For a power of order k and length n, the integer n/k is a period.

Definition 1 ([14]). *An antipower of order k, or simply a k-antipower, is defined as a concatenation of k consecutive pairwise distinct blocks of the same length.*

The length of the pairwise distinct blocks is called *antiperiod*. The notion of antiperiod to words that are not antipowers has been extended in [1]: w has an antiperiod ℓ if it is a prefix of some k-antipower w whose antiperiod is ℓ. Holding this intuition, we formalize a definition of *antiperiodic* words as follows:

Definition 2. *A word $w = u_1 \ldots u_k$ is called ℓ-antiperiodic if (i) $u_i \neq u_j$, for all $i \neq j$; (ii) $|u_i| = \ell$, for all $i < k$; (iii) $|u_k| \leq \ell$. The number $\ell = \lfloor \frac{n}{k} \rfloor$ is an antiperiod for w.*

Note that a k-antipower of length n is $\frac{n}{k}$-antiperiodic. Therefore, we also call it *purely $\frac{n}{k}$-antiperiodic.*

Example 1. The word abbbaa is a 3-antipower, therefore it is also purely 2-antiperiodic: the distinct-factors partition ab|bb|aa testifies this.
The word ababaab is not 2-antiperiodic, but it is 3-antiperiodic: aba|baa|b.

We now formally define the three arrays that will be the output of our online algorithms. The first two arrays respectively store, for each word's prefix, its minimum antiperiod and its minimum pure antiperiod:

ANTIPERIODIC ARRAY (APD)
INPUT: A word w of length n
OUTPUT: An array of length n where APD$[i]$ is the smallest ℓ such that $w[1, i]$ is ℓ-antiperiodic.

PURELY ANTIPERIODIC ARRAY (pAPD)
INPUT: A word w of length n
OUTPUT: An array of length n where pAPD$[i]$ is the smallest ℓ such that $w[1, i]$ is a $\frac{i}{\ell}$-antipower.

We further consider the following question: given i and ℓ is the prefix $w[1, i]$ (purely) ℓ-antiperiodic? We can answer this question by building our third array, the complete array cAP of w:

COMPLETE ANTIPOWER ARRAY (cAP)
INPUT: A word w of length n
OUTPUT: An array of length n where cAP$[\ell]$ is the maximum index i such that the prefix $w[1, i]$ is purely ℓ-antiperiodic.

Example 2. Let $w = $ abaabaab. The antiperiodic array of w is given by APD $= [1, 1, 2, 2, 2, 2, 2, 4]$, its purely antiperiodic array pAPD $= [1, 1, 3, 2, 5, 2, 7, 4]$, and its complete antipower array is cAP $= [2, 6, 3, 8, 5, 6, 7, 8]$.

In the next sections, we show how to build these arrays online, and how to access them using little space (i.e. less space than the three explicit arrays).

2.1   Basic Properties of Antiperiodic Words

First, we provide some properties of (purely) ℓ-antiperiodic words.

Lemma 1. *If ℓ is an antiperiod for some word w of length $> \ell$, then ℓ is also antiperiod for $w' = w[1, |w| - 1]$.*

Proof. First, let us assume ℓ does not divide $|w|$. By hypothesis, all blocks $u_s = w[(s - 1)\ell + 1, s\ell]$ for $s = 1, ..., \lfloor \frac{|w|}{\ell} \rfloor = k - 1$ and $u_k = w[\lfloor \frac{|w|}{\ell} \rfloor \ell + 1, |w|]$ are pairwise distinct. Let $u'_k = u_k[1, |u_k| - 1]$ (possibly the empty word); then the blocks $u_1, ..., u_{k-1}$ are still distinct, and they are all also distinct from u'_k since they are of different lengths. Therefore, since $w' = w[1, |w| - 1] = u_1 \cdots u_{k-1} u'_k$, ℓ is an antiperiod for w'. If ℓ divides $|w|$, the proof still holds with last block being u_{k-1} instead of u_k.   □

Proposition 1. *The following properties hold:*

(i) *If a word w is ℓ-antiperiodic, then for all $i \geq \ell$ the prefix $w[1, i]$ is ℓ-antiperiodic.*

(ii) *If a word w is purely ℓ-antiperiodic, then for all $i \geq \ell$ such that ℓ divides i, the prefix $w[1, i]$ is purely ℓ-antiperiodic.*

(iii) *If a prefix $w[1, j]$ is purely ℓ-antiperiodic, then $w[1, i]$ is ℓ-antiperiodic for all $\ell \leq i < j + \ell$.*

Proof. Properties (i) and (ii) follow by recursively applying Lemma 1. In order to prove Property (iii), we assume that $w[1, j]$ is purely ℓ-antiperiodic. By part (i) we have that the thesis holds for $i \in [\ell, j]$. On the other hand, if $w[1, j]$ is purely ℓ-antiperiodic, then the blocks $u_s = w[(s-1)\ell+1, s\ell]$ for $s = 1, \ldots, \frac{j}{\ell}$ are pairwise distinct. Therefore they are also trivially distinct from $u_h = w[j + 1, h]$ for all $h < j + \ell$, since they are of different lengths. This ensures the ℓ-antiperiodicity of $w[1, i]$ for $j < i < j + \ell$. ∎

2.2 Algorithmic and Data Structures Toolkits

We make use of Karp-Rabin fingerprinting [20] in the more modern variant, where the fingerprint is a polynomial modulo a prime number evaluated at a random point. To make the paper self-contained, we recapitulate these notions.

Definition 3 (Karp-Rabin fingerprinting [20]). *The Karp-Rabin hash function (over integer alphabet) is defined as $h_{q,x}(w) = \sum_{i=1}^{|w|} w[i] \cdot x^{|w|-i} \mod q$, where q is a prime and x is a random integer in $[1, q]$.*

The value $h_{q,x}(w)$ will be called *hash value* or *fingerprint* of w interchangeably. In the following, *with high probability* (or inverse-polynomial probability) means with probability at least $1 - n^{-c}$ for an arbitrarily large constant c fixed at construction time, where n is the input's size. We will often say that *running times* of our algorithms *hold with high probability*. This means that the algorithm terminates in the claimed running time (or has the claimed delay) with probability $1 - n^{-c}$ for an arbitrarily large constant c fixed at construction time. On the other hand, with probability n^{-c} the algorithm (or single operations) could take polynomial time to terminate.

A crucial property of Karp-Rabin fingerprinting is that, with high probability, no collision occurs among the factors of a given word. To see this, consider two words $s \neq t$ of length n. The polynomial $h_{q,x}(s) - h_{q,x}(t)$ has at most n roots modulo (prime) q, so the two words collide (i.e. $h_{q,x}(s) - h_{q,x}(t) = 0$) with probability n/q. For big enough q we take care of all possible $O(n^2)$ collisions between the factors of a word of length n:

Lemma 2. *For a sufficiently large prime q such that $\log q \in \Theta(\log n)$, the hash function $h_{q,x}$ is collision-free over all factors of any fixed word $w[1, n]$ with high-probability.*

If n is not known in advance (e.g. in the online setting), in the above lemma we can take instead $\log q \in \Theta(\omega)$, where ω is the machine word size. We thus work in the word RAM model, where the word size always satisfies $\omega = \Omega(\log n)$ for any input size n and standard arithmetic operations between $\Theta(\omega)$-bits numbers take constant time.

In our application we will need to check collisions between factors of the current word in an online fashion, with small delay per added character. The following standard extension of Karp-Rabin fingerprinting (see also [6]) will allow us to reach this goal.

Definition 4 (Extended Karp-Rabin fingerprinting). *The extended Karp-Rabin hash function is defined as*

$$\kappa_{q,x}(w) = \Big\langle h_{q,x}(w[1, 2^{\lfloor \log_2 |w| \rfloor}]), h_{q,x}(w[|w| - 2^{\lfloor \log_2 |w| \rfloor} + 1, |w|]), |w| \Big\rangle$$

We say that $\kappa_{q,x}$ is *collision-free* on w if $\kappa_{q,x}$ does not generate collisions among factors of w. Note that $\kappa_{q,x}$ is collision-free on w if and only if $h_{q,x}$ is collision-free among the factors of w whose length is a power of two.

A *dictionary* D is a data structure implementing a set of objects (e.g. integers). Each object x is associated with some satellite information y (e.g. another integer), which is retrieved using the notation $y \leftarrow D[x]$. A dynamic dictionary supports the insertion of pairs (object, satellite information) $\langle x, y \rangle$ as $D[x] \leftarrow y$. In our algorithms we will use the dictionary design of Dietzfelbinger et al. [12]:

Lemma 3 (Dynamic Dictionaries [12]). *There exists a linear-space dynamic dictionary data structure supporting insertion and retrieval operations in constant worst-case time w.h.p.*

The probabilities involved in the performance of our algorithms will depend solely on the random choices they make (e.g. choosing the hash function), not on the input word w (which may be arbitrary). However, we do require that the word w is fixed before the algorithm starts, i.e. before the random choices are made. This is a standard assumption with randomized dynamic data structures; violating this assumption is equivalent to using fully-deterministic structures, for which strong lower-bounds apply (see, e.g., the case of dictionaries considered by Dietzfelbinger et al. [13], who also make the same assumption).

3   Online Algorithms

In this section, we first discuss how to compute the antiperiodic array APD (by using two different approaches), and then extend the solution to pAPD and to cAP arrays. The final goal of this section will be proving the following theorems:

Theorem 1. *The antiperiodic array (APD) of a word w of length n can be computed with an online solution working in $O(n)$ space, $O(n \log t)$ time, and $O(n \log t)$ delay per character, where t is the smallest antiperiod of w. Running times are worst-case and hold w.h.p. The returned solution is correct w.h.p.*

Theorem 2. *The antiperiodic array (*APD*), purely antiperiodic array (*pAPD*) and complete antipower array (*cAP*) of a word w of length n can be computed with an online solution working in $O(n \log n)$ space, $O(n \log n)$ time, and $o(n^\epsilon)$ delay per character for any constant $\epsilon > 0$. Running times are worst-case and hold w.h.p.*

The proof of Theorem 1 is given in Sect. 3.1 and the proof of Theorem 2 follows immediately from Lemmas 6, 7, and 8.

Whenever we say we use a dictionary or an array, it is assumed we use the data structure of Lemma 3. This way, we do not need to assume that the final size n of the text is known. Note that this dictionary offers stronger guarantees than classic hash tables, where query times are constant only in expectation. Similarly, a simple array can be used to represent a dynamic text with right-append operations; however, also in this case resizing operations (e.g. doubling techniques) require $O(n)$ delay in the worst case. In order to achieve a small delay in our online algorithms, we cannot afford paying such overheads and thus opt for the structure of Lemma 3.

The following dynamic string data structure stands at the core of our results. It extends standard techniques (see e.g. [6]) to the online setting to efficiently compute the extended Karp-Rabin fingerprint of any factor and check collisions among them.

Lemma 4. *There is a $O(n \log n)$-space data structure on a word $w[1, n]$ that computes $\kappa_{q,x}(w[i, j])$ in constant time for any $1 \le i \le j \le n$, where $\kappa_{q,x}$ is collision-free. The structure can be updated in $O(\log n)$ time by appending a new character to the end of w, possibly changing function $\kappa_{q,x}$ if a collision is detected. Space and update time can be reduced to $O(n)$ and $O(1)$, respectively, at the cost of using a hash function that is collision-free with high probability only. In all cases, running times are worst-case and hold w.h.p.*

Proof. Similarly to Bille et al. [6] (where they consider more space-efficient variants), we store the hash value of every prefix of w. Then, the fingerprint value of any factor can be computed by subtracting (modulo q) the hashes of two prefixes (the hash of the shortest prefix need also to be multiplied by a suitable power of x, so we compute also all powers of x modulo q). To extend the structure for the word w by one character c, it is sufficient to compute the fingerprint of wc. This can be done with one multiplication and one addition (mod q) to combine the fingerprints of w and c. This is already sufficient to obtain the linear-space version of the structure that is correct with high probability and that supports queries and updates in constant time.

To make the structure collision-free, we check collisions of $h_{q,x}$ among factors whose length is a power of 2. This suffices to ensure that $\kappa_{q,x}$ is collision-free among factors of any length. For each $e = 1, \ldots, \log n$ we keep a dictionary C_e storing all fingerprints of factors of length 2^e. To each such fingerprint X, we associate a position i such that $h_{q,x}(w[i, i + 2^e - 1]) = X$. Assume that the function is collision-free among factors of w whose length is a power of 2. To extend the property to $w' = wc$, for a character c, first we extend the

fingerprinting structure as seen above. Then, for $e = 1, \ldots, \log n$ (in this order) we do as follows. Assume that the suffix of length 2^{e-1} of w' does not generate collisions, which at the beginning ($e = 1$) can be checked in constant time by accessing dictionary C_1 and the word. We look for $X = h_{q,x}(w'[|w'|-2^e+1, |w'|])$ in C_e. If it does not occur, then the suffix of w' of length 2^e does not generate collisions. We insert $\langle X, |w'| - 2^e + 1 \rangle$ in the dictionary and we proceed with $e+1$. Otherwise, let $i < |w'| - 2^e + 1$ be the position found in the dictionary such that $h_{q,x}(w'[i, i+2^e-1]) = X$. Since by assumption $h_{q,x}$ is collision-free among factors of length 2^{e-1}, we can check if $w[i, i+2^e-1] = w'[|w'| - 2^e + 1, |w'|]$ in constant time by comparing the collision-free fingerprints of their two halves. If we detect a collision, then we re-build the whole structure with a new random x. This happens with probability at most n^{-c} for an arbitrarily large constant c, so updates take $O(\log n)$ time w.h.p.   □

The *optimal dynamic strings* of Gawrychowski et al. [17] could also be used to replace the structure of Lemma 4. However, they support factor comparison in $O(\log n)$ time[2], whereas Lemma 4 achieves $O(1)$ time (at the price of being less space-efficient and supporting less powerful queries). This is crucial to obtain the claimed time bounds in our algorithms.

3.1   Online Antiperiodic Array: Linear Space Construction

Our first solution to build the antiperiodic array relies on its monotonicity.

Proposition 2. *The antiperiodic array* APD *is non-decreasing; that is, for all* $i < |w|$, APD$[i] \leq$ APD$[i + 1]$.

Proof. We know that APD$[i + 1]$ is an antiperiod for $w[1, i + 1]$. By Lemma 1, APD$[i+1]$ is also an antiperiod for $w[1, i]$. Since APD$[i]$ is the minimum antiperiod for this word, APD$[i] \leq$ APD$[i + 1]$ holds.   □

We now describe a simple online solution for computing array APD that achieves linear space and $O(n \log t)$ running time, where t is the smallest antiperiod of the word, proving Theorem 1.

Proof (Theorem 1). Assume we have processed $w[1, i]$ and computed APD$[1, i]$, with APD$[i] = \ell$. We keep the dynamic linear-space data structure of Lemma 4 on w (i.e. the version that is correct w.h.p. only). We also keep a dictionary R containing the fingerprints of all blocks of length ℓ up to position i. If ℓ does not divide $i+1$, then we write APD$[i+1] = \ell$ and proceed to $i+2$. Otherwise, $w[i, i+1]$ ends a block of length ℓ. We insert the fingerprint of the last block of length ℓ in the dictionary R. If the fingerprint is not already in the dictionary, we can write APD$[i + 1] = \ell$ and proceed to $i + 2$. Otherwise, ℓ is not an antiperiod of $i + 1$. We empty the dictionary R and look for a new antiperiod $\ell' = \ell + 1, \ell + 2, \ldots$.

---

[2] Using their interface, factor comparisons can be achieved by extracting (splitting) the factors in logarithmic time, then comparing them in constant time (or, alternatively, by navigating their grammar in logarithmic time without performing splits).

Proposition 2 guarantees that only these antiperiods need to be tried, since APD is non-decreasing. To check antiperiod ℓ', we insert the fingerprints of all blocks of length ℓ' in R in $O((i+1)/\ell')$ time. If a duplicate is found, we empty R and proceed to $\ell'+1$. We stop at the smallest integer ℓ' that does not generate duplicates in R (and we keep the current elements in R).

Let t be the smallest antiperiod of the whole word $w[1, n]$. Overall, we spend time $O(\sum_{\ell=1}^{t} n/\ell) = O(n \log t)$. Since we keep only one dictionary at a time, the space is $O(n)$. In the worst case, it could be that we are processing a position $i \in \Theta(n)$ and that we need to check most of the antiperiods ℓ' smaller than t, so also the delay is $\Theta(n \log t)$ in the worst case. Moreover, since we do not check collisions between Karp-Rabin fingerprints, the solution is correct with high probability only.   □

3.2 Online Antiperiodic Array: Reducing the Delay

In this section we improve the delay of the solution described in the previous section with an algorithm returning always the correct solution (at the cost of increasing space usage and running time).

Our solution will require to compute the divisors of all numbers $i = 1, \ldots, n$. With the next lemma we show how to achieve this goal in an online fashion and constant time per element. Running time is w.h.p. because we assume that n is not known in advance and we use the dictionary of Lemma 3 to implement a dynamic array; otherwise, if n is known one can use a simple array and remove randomization.

Lemma 5. *We can list all $d(i)$ divisors of the integers $i = 1, \ldots, n$ in $O(d(i))$ time per integer and $O(n \log n)$ total space. The running time is worst-case and holds w.h.p.*

Proof. We show how to build a dictionary D such that $D[i]$ is the multi-set of the $O(\log i)$ prime divisors of i: if d is the largest integer such that p^d divides i and p is prime, $D[i]$ contains d copies of p. $D[i]$ can be implemented as a simple array (to re-size it, we can simply double its current size: since at most $O(\log i)$ primes can divide i, $D[i]$ can be re-sized with delay $O(\log i)$, which is acceptable for our purposes). At the beginning, we have $D[1] = \emptyset$. Given $D[i]$, all divisors of i can be enumerated in constant time per element by multiplying the integers of any subset of $D[i]$ (with backtracking, to avoid repeating the same multiplications). We also return the trivial divisors 1 and i (or just one of them if $i = 1$).

Suppose we have computed $D[1, i]$. To proceed to the next position $i+1$, for each $p \in D[i]$ we insert p in $D[j]$, where $j = ((i/p)+1) \cdot p$ is the next multiple of p. If $D[i+1]$ is empty, then we insert $i+1$ in $D[i+1]$. It is easy to see that now $D[i+1]$ contains all primes that divide $i+1$: let p be a prime dividing $i+1$. We have two cases. (i) $i+1 = q \cdot p$, with $q > 1$. Then, by definition of our procedure when we processed $D[(q-1)\cdot p]$ we also inserted p in $D[(q-1+1)\cdot p] = D[i+1]$. (ii) $i+1 = p$. Then, $D[i+1]$ is empty when we visit it and our procedure inserts p in $D[i+1]$. Now, if a prime p divides $i+1$ we can find the largest power p^d dividing

$i + 1$ by simply computing the remainder between $i + 1$ and p^e, for $e = 1, \ldots, d$. We insert $d - 1$ further copies of p in $D[i + 1]$. Note that we compute each $D[i]$ in $O(\log i) \subseteq O(d(i))$ time and delay, and that at any time D contains at most $\Theta(n)$ multi-sets of size $O(\log n)$ each (when processing position i, the furthest cell of D that we touch is $D[2 \cdot i]$; this happens precisely when i is prime).   □

We can now prove the main result of this section regarding the APD:

Lemma 6. *Theorem 2 holds with respect to the antiperiodic array.*

Proof. Assume we have processed $w[1, i]$ and computed APD$[1, i]$. We keep the $O(n \log n)$-space data structure of Lemma 4 on w (i.e. the version that always returns the correct result and supports $O(\log n)$-time updates).

We build i dictionaries $H_1, \ldots, H_i$, where H_ℓ stores all the fingerprints $\kappa_{q,x}(w[\ell \cdot (j - 1) + 1, \ell \cdot j])$ for $j = 1, \ldots, \lfloor i/\ell \rfloor$. At each time step, note that the dictionaries store overall $O(n \log n)$ elements.

We also keep a log-time successor data structure $\mathcal{B}$ (e.g. a red-black tree) storing all the values $\{\ell_1, \ldots, \ell_q\} \subseteq [1, i]$ such that ℓ_j is an antiperiod of $w[1, i]$.

Now, suppose we extend $w[1, i]$ with a character, obtaining $w[1, i + 1]$, and assume that the last computed value in the antiperiodic array is APD$[i] = \ell$. To update our structures, we perform the following operations:

1. First, we insert $i + 1$ in $\mathcal{B}$, since $i + 1$ is a valid antiperiod of $w[1, i + 1]$.
2. Then, we update the dictionary H_j for all j that divide $i + 1$, i.e. such that a block of length j ends at the end of $w[1, i + 1]$. Each such dictionary H_j is updated by inserting $X = \kappa_{q,x}(w[i - j + 2, i + 1])$, i.e. the hash of the last block of length j. If X is already in the dictionary, then j cannot be anymore an antiperiod, and we remove it from $\mathcal{B}$. In Lemma 5 we show how to list all $d(i + 1)$ divisors of $i + 1$ in constant time per item. Any integer x has at most $e^{O(\log x / \log \log x)} = o(x^\epsilon)$ divisors, for any constant $\epsilon > 0$, so $o(n^\epsilon)$ will become our delay. Overall, these operations amortize to $O(n \log n)$ time (since $\sum_{i=1}^{n} d(i) = O(n \log n)$, where $d(i)$ is the number of divisors of i).
3. If APD$[i] = \ell$ does not divide $i + 1$, then we set APD$[i + 1] = \ell$. Otherwise, ℓ divides $i + 1$. Then, ℓ is the minimum antiperiod of $w[1, i + 1]$ iff $\ell \in \mathcal{B}$. If it is, we set APD$[i + 1] = \ell$. Otherwise, we find the successor ℓ' of ℓ in $\mathcal{B}$ and we set APD$[i + 1] = \ell'$. All these operations take $O(\log n)$ time.   □

3.3   Purely Antiperiodic Array

In this section we extend the solution for computing APD of Sect. 3.2 to the computation of the *purely antiperiodic* array pAPD.

Lemma 7. *Theorem 2 holds with respect to the purely antiperiodic array.*

Proof. We build the dictionaries $H_1, \ldots, H_i$ as in the proof of Theorem 6. We mark each dictionary H_ℓ with a flag (one bit) recording whether ℓ is no longer an antiperiod for the current position i. When processing position $i + 1$, we scan all divisors of $i + 1$ (using Lemma 5) and (i) opportunely update the flags

whenever for some divisor ℓ' of $i+1$, the current prefix $w[1, i+1]$ is no longer a $\frac{i+1}{\ell'}$-antipower (but the prefix $w[1, (i+1) - \ell']$ was), and (ii) find the minimum divisor ℓ for which $w[1, i+1]$ is a $\frac{i+1}{\ell}$-antipower (i.e. the minimum ℓ such that H_ℓ is not marked). This ℓ is the value we write in $p\text{APD}[i+1]$. Our time bounds follow since we spend constant time per divisor of $i+1$. $\square$

3.4   Complete Antipower Array

In this section we show how to build the complete antipower array $c\text{AP}$. Recall that $c\text{AP}[\ell]$ stores the maximum index j such that the prefix $w[1, j]$ is purely ℓ-antiperiodic. Thus, for any $\ell > \lfloor n/2 \rfloor$, $c\text{AP}[\ell]$ is trivially equal to its index ℓ.

With an easy adaptation of Theorem 7 we obtain:

Lemma 8. *Theorem 2 holds with respect to the complete antipower array.*

Proof. We proceed as in the proof of Theorem 7. We keep i dictionaries H_ℓ, $\ell = 1, \ldots, i$ storing the fingerprints of blocks of length ℓ up to position i. We also mark each dictionary H_ℓ with a flag (one bit) recording (when true) whether ℓ is no longer an antiperiod for the current position i. Assume we have computed $c\text{AP}[1, i]$. When processing position $i+1$, we generate all divisors of $i+1$ with Lemma 5. Whenever for some divisor ℓ' of $i+1$, the current prefix $w[1, i+1]$ is no longer a $\frac{i+1}{\ell'}$-antipower (but the prefix $w[1, (i+1) - \ell']$ was), we set the flag associated with dictionary $H_{\ell'}$. Then, for all divisors ℓ'' of $i+1$ such that the flag of $H_{\ell''}$ is false we write $c\text{AP}[\ell''] \leftarrow i+1$. Our time and space bounds follow since the number of divisors is always upper-bounded by $o(n^\epsilon)$ and the sum of the number of divisors of all $i = 1, \ldots, n$ is $O(n \log n)$. $\square$

4   Relationships Between cAP, APD and pAPD

We observe that the array $c\text{AP}$ stores more information than arrays APD and $p\text{APD}$, which only record the *minimum* ℓ for each word's prefix. Moreover, the following lemma shows the relation between the array $c\text{AP}$ and the property of being ℓ-antiperiodic.

Lemma 9. *A prefix $w[1, i]$ is ℓ-antiperiodic if and only if $\ell \le i < c\text{AP}[\ell] + \ell$.*

Proof. If a prefix $w[1, i]$ is ℓ-antiperiodic, it must hold that the length i of the prefix is at least ℓ. Moreover, by definition of $c\text{AP}$, since $c\text{AP}[\ell] = x$ is the rightmost index of w such that $w[1, x]$ is purely ℓ-antiperiodic, it also must hold $i < x + \ell$. Conversely, suppose we have an index i such that $\ell \le i < x + \ell$, where $x = c\text{AP}[\ell]$. Then the fact that $w[1, i]$ is ℓ-antiperiodic follows by the definition of $c\text{AP}$ and part (*iii*) of Proposition 1. $\square$

Here we show how $c\text{AP}$ can be used to directly obtain or access efficiently the other two structures.

Accessing APD from cAP. We show that the array $c\text{AP}$ allows fast random access to APD. Indeed, the entry $\text{APD}[i]$ is the minimum length ℓ such that the

prefix $w[1, i]$ is ℓ-antiperiodic. By Lemma 9, the value APD$[i]$ is the smallest ℓ such that $\ell \leq i < c$AP$[\ell] + \ell$. To find such value we need to find the leftmost entry in cAP$[1, i]$ such that cAP$[\ell] + \ell$ is greater than i.

Given cAP, we build a constant-time range maximum data structure (RMQ) on the array A $= [c$AP$[1] + 1, c$AP$[2] + 2, \ldots, c$AP$[n] + n]$, which requires only $2n + o(n)$ bits and can be built in linear time [16]. The structure returns, for every range $[i, j]$, the index containing the maximum element in A. Then, accessing APD$[i]$ corresponds to finding the leftmost entry in A that exceeds i. This can be solved in $O(\log n)$ time with binary search using the RMQ on A.

Obtaining APD from cAP. We show that cAP may be used to build APD in $O(n)$ time. Since APD$[i] = \min\{\ell : w[1, i]$ is $\ell - $antiperiodic$\}$, by Lemma 9 it holds APD$[i] = \min\{\ell : \ell \leq i < cAP[\ell] + \ell\}$. We can thus obtain APD by iterating over cAP$[\ell]$ and each time setting values of APD that have not already been set. More formally, we start from $\ell = 1$ and set an auxiliary index $j = 1$. The index j prevents us from writing twice on the same cell. Given ℓ, we mark APD$[i] = \ell$ for all indices i in the range $[j, c$AP$[\ell] + \ell - 1]$. Then, we update the index j by setting $j = c$AP$[\ell] + \ell$ if $j < c$AP$[\ell] + \ell$, and we repeat the procedure for $\ell + 1$. Note that if cAP$[\ell] + \ell - 1$ is smaller than j for some ℓ, we do nothing apart from increasing ℓ. We stop when APD and cAP are of the same length.

When we set APD$[i] = \ell$, ℓ is the smallest value such that $i < c$AP$[\ell] + \ell$. The cost is $O(n)$ as each cell of APD is only written once, and each cell of cAP only read once, both arrays having length n.

Accessing pAPD from cAP. We further note that cAP allows random access to pAPD by performing just $o(n^\epsilon)$ accesses to cAP (for any $\epsilon > 0$). Indeed, pAPD$[i] = \min\{\ell : w[1, i]$ is purely $\ell - $antiperiodic$\}$. Note that $i \mod \ell$ must be 0. This means we can find pAPD$[i]$ by iterating over each divisor x of i and finding the lowest one such that cAP$[x] \geq i$. As any integer number $i \leq n$ has $o(n^\epsilon)$ divisors, the process takes $o(n^\epsilon)$ accesses to cAP.

To obtain $o(n^\epsilon)$ running time, one needs to list the divisors of the index $i \leq n$ in constant time per element. One solution could be dividing i by all numbers $j \leq \sqrt{i}$. This solution successfully finds all divisors in $O(\sqrt{n})$ time. This can be improved to $o(n^\epsilon)$ by explicitly factoring i: integer factorization algorithms like Schnorr-Seysen-Lenstra's [23] find all factors of $i \leq n$ in $o(n^\epsilon)$ expected time.

Obtaining pAPD from cAP. Using the above solution to access each cell of pAPD, running time amortizes to $O(n \log n)$, i.e. the sum of the number of divisors of all numbers $i \leq n$. In this case, the divisors can be found using Lemma 5.

5   Final Remarks

In this paper, we showed how to efficiently compute in online fashion the antiperiodic array, the purely antiperiodic array, and the complete antipower array.

Moreover, using the complete antipower array, we can answer in constant time the question posed in Sect. 2: given i and ℓ is the prefix $w[1, i]$ (purely)

ℓ-antiperiodic? Indeed, $w[1, i]$ is purely ℓ-antiperiodic if $i \bmod \ell = 0$ and $i \leq$ $c_{\text{AP}}[\ell]$, and ℓ-antiperiodic if $\ell \leq i < c_{\text{AP}}[\ell] + \ell$.

Note that the definition of ℓ-antiperiodic word $w = u_1 \ldots u_k$ poses no constraint on the last block u_k when its length is strictly less than ℓ. We may think of extending/restricting the ℓ-antiperiodic notion by imposing conditions on this incomplete block, such as u_k not being a prefix (or suffix, factor...) of any u_i.

Acknowledgements. The authors would like to thank Roberto Grossi for useful conversations on the topic. GR, NPi are partially, and DG, VG, NPr are supported by the project MIUR-SIR CMACBioSeq ("Combinatorial methods for analysis and compression of biological sequences") grant n. RBSI146R5L.

References

1. Alamro, H., Badkobeh, G., Belazzougui, D., Iliopoulos, C.S., Puglisi, S.J.: Computing the Antiperiod(s) of a string. In: 30th Annual Symposium on Combinatorial Pattern Matching (CPM). LIPIcs (2019, to appear)
2. Alzamel, M., et al.: Quasi-linear-time algorithm for longest common circular factor. In: 30th Annual Symposium on Combinatorial Pattern Matching (CPM). LIPIcs (2019, to appear)
3. Ayad, L.A.K., et al.: Longest property-preserved common factor. In: Gagie, T., Moffat, A., Navarro, G., Cuadros-Vargas, E. (eds.) SPIRE 2018. LNCS, vol. 11147, pp. 42–49. Springer, Cham (2018). https://doi.org/10.1007/978-3-030-00479-8_4
4. Badkobeh, G., Fici, G., Puglisi, S.J.: Algorithms for anti-powers in strings. Inf. Process. Lett. **137**, 57–60 (2018)
5. Bae, S.W., Lee, I.: On finding a longest common palindromic subsequence. Theor. Comput. Sci. **710**, 29–34 (2018)
6. Bille, P., Gørtz, I.L., Knudsen, M.B.T., Lewenstein, M., Vildhøj, H.W.: Longest common extensions in sublinear space. In: Cicalese, F., Porat, E., Vaccaro, U. (eds.) CPM 2015. LNCS, vol. 9133, pp. 65–76. Springer, Cham (2015). https://doi.org/10.1007/978-3-319-19929-0_6
7. Burcroff, A.: (k, λ)-anti-powers and other patterns in words. Electron. J. Comb. **25**, P4.41 (2018)
8. Chowdhury, S., Hasanl, M., Iqbal, S., Rahman, M.: Computing a longest common palindromic subsequence. Fundam. Inform. **129**(4), 329–340 (2014)
9. Crochemore, M., Ilie, L., Rytter, W.: Repetitions in strings: algorithms and combinatorics. Theor. Comput. Sci. **410**(50), 5227–5235 (2009). Mathematical Foundations of Computer Science (MFCS 2007)
10. Crochemore, M., Rytter, W.: Jewels of Stringology. World Scientific, Singapore (2002)
11. Defant, C.: Anti-power prefixes of the Thue-Morse word. Electron. J. Comb. **24**, P1.32 (2017)
12. Dietzfelbinger, M., Meyer auf der Heide, F.: A new universal class of hash functions and dynamic hashing in real time. In: Paterson, M.S. (ed.) ICALP 1990. LNCS, vol. 443, pp. 6–19. Springer, Heidelberg (1990). https://doi.org/10.1007/BFb0032018
13. Dietzfelbinger, M., Karlin, A., Mehlhorn, K., Meyer auF der Heide, F., Rohnert, H., Tarjan, R.E.: Dynamic perfect hashing: upper and lower bounds. SIAM J. Comput. **23**(4), 738–761 (1994)

14. Fici, G., Restivo, A., Silva, M., Zamboni, L.Q.: Anti-powers in infinite words. In: ICALP 2016. Leibniz International Proceedings in Informatics (LIPIcs), vol. 55, pp. 124:1–124:9. Schloss Dagstuhl-Leibniz-Zentrum fuer Informatik, Dagstuhl, Germany (2016)

15. Fici, G., Restivo, A., Silva, M., Zamboni, L.Q.: Anti-powers in infinite words. J. Comb. Theory Ser. A **157**, 109–119 (2018)

16. Fischer, J., Heun, V.: Space-efficient preprocessing schemes for range minimum queries on static arrays. SIAM J. Comput. **40**(2), 465–492 (2011)

17. Gawrychowski, P., Karczmarz, A., Kociumaka, T., Łacki, J., Sankowski, P.: Optimal dynamic strings. In: Proceedings of the Twenty-Ninth Annual ACM-SIAM Symposium on Discrete Algorithms, SODA 2018, pp. 1509–1528. Society for Industrial and Applied Mathematics (2018)

18. Inenaga, S., Hyyrö, H.: A hardness result and new algorithm for the longest common palindromic subsequence problem. Inf. Process. Lett. **129**, 11–15 (2018)

19. Inoue, T., Inenaga, S., Hyyrö, H., Bannai, H., Takeda, M.: Computing longest common square subsequences. In: 29th Symposium on Combinatorial Pattern Matching (CPM). LIPIcs, vol. 105, pp. 15:1–15:13 (2018)

20. Karp, R.M., Rabin, M.O.: Efficient randomized pattern-matching algorithms. IBM J. Res. Dev. **31**(2), 249–260 (1987)

21. Kociumaka, T., Radoszewski, J., Rytter, W., Straszyński, J., Waleń, T., Zuba, W.: Efficient representation and counting of antipower factors in words. In: Martín-Vide, C., Okhotin, A., Shapira, D. (eds.) LATA 2019. LNCS, vol. 11417, pp. 421–433. Springer, Cham (2019). https://doi.org/10.1007/978-3-030-13435-8_31

22. Kolpakov, R., Bana, G., Kucherov, G.: mreps: efficient and flexible detection of tandem repeats in DNA. Nucl. Acids Res. **31**(13), 3672–3678 (2003). https://doi.org/10.1093/nar/gkg617

23. Lenstra, H.W., Pomerance, C.: A rigorous time bound for factoring integers. J. Am. Math. Soc. **5**(3), 483–516 (1992)

24. Li, L., Jin, R., Kok, P.L., Wan, H.: Pseudo-periodic partitions of biological sequences. Bioinformatics **20**(3), 295–306 (2004)

25. Lothaire, M.: Combinatorics on Words. Cambridge University Press, Cambridge (1997)

26. Lothaire, M.: Algebraic Combinatorics on Words. Cambridge University Press, Cambridge (2002)

27. Lothaire, M.: Applied Combinatorics on Words. Encyclopedia of Mathematics and its Applications, Cambridge University Press (2005). https://doi.org/10.1017/CBO9781107341005

28. Lothaire, M.: Review of applied combinatorics on words. SIGACT News **39**(3), 28–30 (2008)

29. Thue, A.: Uber unendliche zeichenreihen. Norske Vid Selsk. Skr. I Mat-Nat Kl. (Christiana) **7**, 1–22 (1906)

Polynomial-Delay Enumeration
of Maximal Common Subsequences

Alessio Conte[1], Roberto Grossi[1], Giulia Punzi[1(✉)], and Takeaki Uno[2]

[1] Università di Pisa, Pisa, Italy
{conte,grossi}@di.unipi.it, giuliagpunzi@gmail.com
[2] National Institute of Informatics, Tokyo, Japan
uno@nii.ac.jp

Abstract. A *Maximal Common Subsequence* (MCS) between two strings X and Y is an inclusion-maximal subsequence of both X and Y. MCSs are a natural generalization of the classical concept of Longest Common Subsequence (LCS), which can be seen as a longest MCS. We study the problem of efficiently listing all the *distinct* MCSs between two strings. As discussed in the paper, this problem is algorithmically challenging as the same MCS cannot be listed multiple times: for example, dynamic programming [Fraser et al., CPM 1998] incurs in an exponential waste of time, and a recent algorithm for finding an MCS [Sakai, CPM 2018] does not seem to immediately extend to listing. We follow an alternative and novel graph-based approach, proposing the first output-sensitive algorithm for this problem: it takes polynomial time in n per MCS found, where $n = \max\{|X|, |Y|\}$, with polynomial preprocessing time and space.

1 Introduction

The widely known Longest Common Subsequence (LCS) is a special case of the general notion of (inclusion-)Maximal Common Subsequence (MCS) between two strings X and Y. Defined formally below, the MCS is a subsequence S of both X and Y such that inserting any character at any position of S no longer yields a common subsequence. We believe that the enumeration of the distinct MCSs is an intriguing problem from the point of view of string algorithms, for which we offer a novel graph-theoretic approach in this paper.

Problem Definition. Let Σ be an alphabet of size σ. A string S over Σ is a concatenation of any number of its characters. A string S is a *subsequence* of a string X, denoted $S \subset X$, if there exist $0 \leq i_0 < \ldots < i_{|S|-1} < |X|$ such that $X[i_k] = S[k]$ for all $k \in [0, |S| - 1]$.

Definition 1. *Given two strings X, Y, a string S is a* Maximal Common Subsequence *of X and Y, denoted $S \in MCS(X, Y)$, if*

1. $S \subset X$ and $S \subset Y$; that is, S is a common subsequence;

© Springer Nature Switzerland AG 2019
N. R. Brisaboa and S. J. Puglisi (Eds.): SPIRE 2019, LNCS 11811, pp. 189–202, 2019.
https://doi.org/10.1007/978-3-030-32686-9_14

2. *there is no other string W satisfying the above condition 1 such that $S \subset W$, namely, S is inclusion-maximal as a common subsequence.*

Example 1. Consider $X =$ TGACGA and $Y =$ ATCGTA, where $MCS(X,Y) = \{$TCGA, ACGA$\}$. A greedy left-to-right common sequence is not necessarily a MCS: iteratively finding the nearest common character in X and Y, from left to right, gives $W =$ TGA, which is not in $MCS(X,Y)$ as TGA $\subset$ TCGA.

The focus of this paper is on the enumeration of $MCS(X,Y)$ between two strings X and Y, stated formally below.

*Problem 1 (**MCS enumeration**).* Given two strings X, Y of length $O(n)$ over an alphabet Σ of size σ, list all maximal common subsequences $S \in MCS(X,Y)$.

In enumeration algorithms, the aim is to list all objects of a given set. The time complexity of these type of algorithms depends on the cardinality of the set, which is often exponential in the size of the input. This motivates the need to define a different complexity class, based on the time required to output one solution.

Definition 2. *An enumeration algorithm is* polynomial delay *if it generates the solutions one after the other, in such a way that the delay between the output of any two consecutive solutions is bounded by a polynomial in the input size.*

Our aim will be to provide a polynomial delay MCS enumeration algorithm, more specifically we will prove the following result.

Theorem 1. *There is a polynomial-delay enumeration algorithm for Problem 1, with polynomial preprocessing time and space.*

In the full version of this paper, we show how to get a small $O(n\sigma(\sigma + \log n))$ polynomial delay, with $O(n^2(\sigma + \log n))$ preprocessing time and $O(n^2)$ space.

Motivation and Relation to Previous Work. Maximal common subsequences were first introduced in the mid 90s by Fraser et al. [5]. Here, the concept of MCS was a stepping stone for one of the main problems addressed by the authors: the computation of the length of the Shortest Maximal Common Subsequence (SMCS) (i.e. the shortest string length in $MCS(X,Y)$), introduced in the context of LCS approximation. For this, a dynamic programming algorithm was given to find the length of a SMCS of two strings in cubic time.

While LCSs have thoroughly been studied [3,6,10,12], little is known for MCSs. In general, LCSs only provide with information about the longest possible alignment between two strings, while MCSs offer a different range of information, possibly revealing slightly smaller but alternative alignments. A recent paper by Sakai [11] presents an algorithm that deterministically finds one MCS between two strings of length $O(n)$ in $O(n \log \log n)$ time, in contrast with the computation of the length of the LCS, for which a quadratic conditional lower bound based on SETH has been proved [1]. This same algorithm can also be

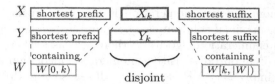

Fig. 1. Visual representation of Sakai's characterization

used to extend a given sequence to a maximal one in the same time. Furthermore, an $O(n)$-time algorithm to check whether a given subsequence is maximal is described in the paper. To this end, Sakai gives a neat characterization of MCSs, which will be useful later, as stated in Lemma 1 and illustrated in Fig. 1.

Lemma 1 (MCS Characterization [11]). *Given a common subsequence W of X and Y, we define X_k (resp. Y_k) as the remaining substring obtained from X (resp. Y) by deleting both the shortest prefix containing $W[0, k)$, and the shortest suffix containing $W[k, |W|)$. Substrings X_k, Y_k are called the k-th interspaces. With this notion, W is maximal if and only if $X_k \cap Y_k = \emptyset$ for all $k \in [0, |W|]$.*

The aforementioned two results seem to be of little help in our case, as neither of the two can be directly employed to obtain a polynomial-delay enumeration algorithm to solve Problem 1, which poses a quite natural question.

Consider the dynamic programming approach in [5]: even if the dynamic programming table can be modified to list the lengths of all MCS in polynomial time, this result cannot be easily generalized to Problem 1. Indeed we show below that any incremental approach, including dynamic programming, leads to an *exponential*-delay enumeration algorithm.

Example 2. Consider $X = \texttt{TAATAATAAT}$, $Z = \texttt{TATATATATAT}$ and $Y = Z\,Z$. Since $X \subset Y$, the only string in $MCS(X, Y)$ is the whole X. But if we were to proceed incrementally over Y, at half way we would compute $MCS(X, Z)$, which can be shown to have size $O(\exp(|X|))$. This means that it would require an exponential time in the size of the input to provide just a single solution as output.

As for the approach in [11], the algorithm cannot be easily generalized to solve Problem 1, since the specific choices it makes are crucial to ensure maximality of the output, and the direct iterated application of Lemma 1 does not lead to an efficient algorithm for Problem 1, as shown next.

Example 3. For a given common subsequence W to start with, first find all values of $k \leq |W|$ such that $X_k \cap Y_k \neq \emptyset$. Then, for these values, compute all distinct characters $c \in X_k \cap Y_k$, and for each of these recur on the extended sequences $W' = W[0, ..., k-1]\, c\, W[k, ..., |W|-1]$. For instance, given the strings $X = \texttt{ACACA}$ and $Y = \texttt{ACACACA}$ with starting sequences $W = \texttt{A}$ and $W = \texttt{C}$, this algorithm would recur on almost every subsequence of X, just to end up outputting the single $MCS(X, Y) = \{X\}$ an exponential (in the size of X) number of times.

Getting polynomial-delay enumeration is therefore an intriguing question. The fact that one maximal solution can be found in polynomial time does not directly imply an enumeration algorithm: there are cases where this is possible, but the existence of an output-sensitive enumeration algorithm is an open problem [8], or would imply $P = NP$ [9]. As we will see, solving Problem 1 can lead to further pitfalls that we circumvent in this paper.

2   MCS as a Graph Problem

As a starting point we reduce Problem 1 to a graph problem in order to give a theoretic characterization and get some insight on how to combine MCS. Afterwards, this characterization will be reformulated in an operative way, leading to an algorithm for MCS enumeration.

2.1   Graph $G(X, Y)$

Definition 3. *Given two strings X, Y, their* string bipartite graph $G(X, Y)$ *has vertex set $V = [0, |X| - 1] \cup [0, |Y| - 1]$ representing the positions inside X and Y, and edge set $E = \{(i, j) \mid X[i] = Y[j]\}$ where each edge, called* pairwise occurrence, *connects positions with the same character in different strings.*

Definition 4. *A mapping of $G(X, Y)$ is a subset $\mathcal{P}$ of its edges such that for any two edges $(i, j), (h, k) \in \mathcal{P}$ we have $i < h$ iff $j < k$. That is, a mapping is a non-crossing matching of the string graph.*

Each mapping of the string graph spells a common subsequence. Vice versa, each common subsequence has at least one corresponding mapping. Thus one might incorrectly think that MCS correspond to inclusion-maximal mappings; as a counterexample consider $X = \text{AGG}$ and $Y = \text{AGAG}$, with $MCS(X, Y) = \{\text{AGG}\}$. $G(X, Y)$ has an inclusion-maximal mapping corresponding to $\text{AG} \notin MCS(X, Y)$.

For a string S, let $next_S(i)$ be the smallest $j > i$ with $S[j] = S[i]$ (if any), and $next_S(i) = |S| - 1$ otherwise; we use the shorthand $\mathcal{I}_S(i) = S[i+1, \ldots, next_S(i)]$.

Definition 5. *A mapping of $G(X, Y)$ is called* rightmost *if for each edge (i, j) of the mapping, corresponding to character $c \in \Sigma$, the next edge (i', j') of the mapping is such that $next_X(i) \geq i'$ and $next_Y(j) \geq j'$. That is, there are no occurrences of c in $X[i+1, \ldots, i'-1]$ and $Y[j+1, \ldots, j'-1]$, the portions between edges (i, j) and (i', j'). We can symmetrically define a* leftmost *mapping.*

In order to design an efficient and correct enumeration algorithm that uses also Definition 5, we first need to study how $MCS(X, Y)$ and $MCS(X', Y')$ relate to $MCS(X X', Y Y')$ for any two pairs of strings X, Y and X', Y'.

Remark 1. A simple concatenation of the pairwise MCS fails: consider for example $X = \text{AGA}$, $X' = \text{TGA}$, $Y = \text{TAG}$ and $Y' = \text{GAT}$, with $MCS(X X', Y Y') = \{\text{AGGA}, \text{AGAT}, \text{TGA}\}$. We have $MCS(X, Y) = \{\text{AG}\}$ and $MCS(X', Y') = \{\text{GA}, \text{T}\}$. Combining the latter two sets we find the sequence AGGA, which is in fact maximal, but also AGT, which is not maximal as $\text{AGT} \subset \text{AGAT}$.

The correct condition for combining MCS is a bit more sophisticated, as stated in Theorem 2. Here, for a position i of a string S, we denote by $S_{<i}$ the prefix of S up to position $i - 1$, and by $S_{>i}$ the suffix of S from position $i + 1$.

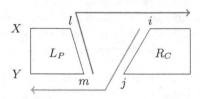

Fig. 2. For their concatenation to be a MCS, P has to be maximal in the red part and C in the orange one (Color figure online)

Theorem 2. *(MCS Combination) Let P and C be common subsequences of X, Y. Let (l, m) be the last edge of the* leftmost *mapping L_P of P, and (i, j) be the first edge of the* rightmost *mapping R_C of C (see Fig. 2). Then*

$$PC \in MCS(X, Y) \text{ iff } P \in MCS(X_{<i}, Y_{<j}) \text{ and } C \in MCS(X_{>l}, Y_{>m}).$$

Proof. To ensure the equivalence, it is sufficient to show that Sakai's interspaces for string PC over X, Y are the same as the ones for either P over $X_{<i}, Y_{<j}$, or for C over $X_{>l}, Y_{>m}$. Let $|P| = s$, $|C| = r$, and consider an index k.

Case $k < s$: the shortest suffixes of X, Y containing $p_{k+1}, ..., p_s, c_1, ..., c_r$ are unchanged from the shortest suffixes of $X_{<i}, Y_{<j}$ containing $p_{k+1}, ..., p_s$, since C is already in rightmost form starting exactly at (i, j). The shortest prefixes containing $p_1, ..., p_k$ are simply the first k edges of the leftmost mapping of P, both in X, Y and $X_{<i}, Y_{<j}$. Therefore, the interspaces for the whole strings are unchanged from the interspaces for P over $X_{<i}, Y_{<j}$.

Case $k > s$: this case is symmetrical to the previous one: the interspaces for the whole strings are unchanged from the interspaces for C over $X_{>l}, Y_{>m}$.

Case $k = s$: The last interspaces for P and the first for C coincide, and they are $X[l + 1, ..., i - 1]$ and $Y[m + 1, ..., j - 1]$. Since P is in leftmost form ending at (l, m) and C is in rightmost form beginning at (i, j), these two strings also coincide with the k-th interspaces for PC. □

Theorem 2 gives a precise characterization on how to combine maximal subsequences, but it cannot be blindly employed to design an enumeration algorithm for a number of reasons.

Let a string P be called a *valid prefix* if there exists $W \in MCS(X, Y)$ such that P is a prefix of W. Suppose that the leftmost mapping for P ends with edge (l, m), and that we want to expand P by appending characters to it so that it remains valid. These characters correspond to the edges (i, j) related to (l, m) as stated by Theorem 2, for some maximal sequence C. The rest of the paper describes how to perform this task without explicitly knowing C.

Remark 2. It does not work to consider every edge (i, j) satisfying the first condition in Theorem 2, that is, $P \in MCS(X_{<i}, Y_{<j})$. As a counterexample,

consider $X = \text{AGAGAT}$ and $Y = \text{TAGGA}$. Note that $P = \text{AG}$ is a valid prefix, since $\text{AGGA} \in MCS(X,Y)$, and its leftmost mapping ends at edge $(l,m) = (1,2)$, labeled with G. Edge $(i,j) = (2,4)$ corresponding to character A is such that $P \in MCS(X_{<i}, Y_{<j})$; but $PA = \text{AGA}$ is not a valid prefix (and $\text{AGA} \subset \text{AGGA}$). Along the same lines, Sakai's algorithm cannot help here. It generates a MCS that contains P as a subsequence, but *not* necessarily as a prefix. Therefore, it cannot be easily employed to identify the edges (i,j).

We need a more in-depth study of the properties of graph $G(X,Y)$ to characterize the relationship between (l,m) and (i,j). First, we give the notion of *unshiftable edges*, and show that edge (i,j) needs to be unshiftable. Second, as being unshifable is only a necessary condition, we discuss how to single out the (i,j)'s suitable for our given (l,m).

2.2   Unshiftable Edges

Definition 6. *An edge (i,j) of the bipartite graph $G(X,Y)$ is called* unshiftable *if it belongs to at least one maximal rightmost mapping of $G(X,Y)$. The set of unshiftable edges is denoted $\mathcal{U}$. An edge is called* shiftable *if it is not unshiftable.*

Example 4. Consider $X = \text{ACCGTTA}$ and $Y = \text{TAAGGACTG}$. The unshiftable edges for these two strings are the following ones.

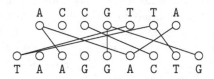

Unshiftable edges[1] can be characterized in a more immediate way, stated in Proposition 1.

Proposition 1. *An edge (i,j) is unshiftable if and only if either (i) it corresponds to the rightmost pairwise occurrence of $X[i] = Y[j]$ in the strings, or (ii) there is at least one unshiftable edge in the subgraph $G(\mathcal{I}_X(i), \mathcal{I}_Y(j))$.*

Proof. (Only if) It is sufficient to show that all edges of a maximal rightmost mapping satisfy one of the two conditions. Let $\mathcal{R} = r_1, ..., r_N$ be a rightmost maximal mapping of $G(X,Y)$. By definition of rightmost mapping, r_N corresponds to the last pairwise occurrence of some character, thus it satisfies the base case. Consider now $p < N$, and let r_p correspond to some character c. By definition of unshiftability, $r_k \in \mathcal{U}$ for all k, specifically $r_{p+1} \in \mathcal{U}$. Since the mapping is rightmost maximal, there can be no occurrences of c between r_p and r_{p+1}; therefore the unshiftable edge r_{p+1} belongs to the subgraph $G(\mathcal{I}_X(i), \mathcal{I}_Y(j))$.

---

[1] A symmetric definition of *left*-unshiftable edges can be given by considering maximal leftmost mappings. The k-dominant edges for LCS [2,4,7] are a subset of left-unshiftable edges.

(*If*) Let (i, j) satisfy one of the two conditions. If it satisfies the base case, then (i, j) is in rightmost form, and we can extend it to the left to a rightmost maximal mapping. On the other hand, let (i, j) satisfy the second condition. Then, there is an edge $(h, k) \in \mathcal{U}$ that belongs to the subgraph $G(\mathcal{I}_X(i), \mathcal{I}_Y(j))$. Consider the rightmost maximal mapping $\mathcal{R}$ that contains (h, k); if it also contains (i, j) we are done. Otherwise, let $\mathcal{R}' \subset \mathcal{R}$ be the restriction that only contains (h, k) and subsequent edges. Consider the rightmost mapping $(i, j) \cup \mathcal{R}'$; we can extend it to the left until it is rightmost maximal. In any case, we have obtained a rightmost maximal mapping containing (i, j), which is then unshiftable.   □

Remark 3. Although every MCS has a corresponding rightmost maximal mapping, and the edges in the latter are unshiftable by Definition 6, it is incorrect to conclude that the opposite holds too. Not all rightmost maximal mappings give MCS: consider for example $X = $ AAGAAG and $Y = $ AAGA. In $G(X, Y)$ we have a maximal rightmost mapping for AAG, but AAG $\subset$ AAGA $\in MCS(X, Y)$.

2.3   Candidate Extensions

We finalize the characterization of the relationship between edges (l, m) and (i, j) of Theorem 2, where (l, m) is the last edge of the leftmost mapping in $G(X, Y)$ of a valid prefix P. We would like to single out a priori the corresponding possible (i, j)'s, without explicitly knowing their Cs. This in turn will lead to the incremental discovery of such Cs one character c at a time.

Specifically, we look for edges (i, j) corresponding to the characters $c \in \Sigma$ such that Pc is still a valid prefix.

Definition 7. *Given an edge (l, m), its cross $\chi_{(l,m)} = \langle e, f \rangle$ (see Fig. 3) is given by (at most) two unshiftable edges $e = (e_1, e_2), f = (f_1, f_2)$ such that*

$$e_1 = \min\{h_1 > l \mid \exists h_2 > m : (h_1, h_2) \in \mathcal{U}\} \text{ and } e_2 = \min\{h_2 > m : (e_1, h_2) \in \mathcal{U}\},$$

$$f_2 = \min\{h_2 > m \mid \exists h_1 > l : (h_1, h_2) \in \mathcal{U}\} \text{ and } f_1 = \min\{h_1 > l : (h_1, f_2) \in \mathcal{U}\}.$$

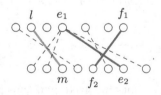

Fig. 3. Graphical representation of the cross $\langle e, f \rangle$ for edge (l, m), drawn in purple: $e = (e_1, e_2), f = (f_1, f_2)$ are the first unshiftable edges soon after (l, m). (Color figure online)

Definition 8. *Given an edge* (l, m), *let* $\chi_{(l,m)} = \langle e, f \rangle$ *be its cross. We define the set of its* mikado edges *as the unshiftable edges of* $G(X[e_1, ..., f_1], Y[f_2, ..., e_2])$,

$$Mk_{(l,m)} = \{(i,j) \in \mathcal{U} \mid e_1 \leq i \leq f_1 \text{ and } f_2 \leq j \leq e_2\},$$

and the subset of candidate extensions *for* (l, m) *as*

$$Ext_{(l,m)} = \{(i,j) \in Mk_{(l,m)} \mid \nexists (h,k) \in Mk_{(l,m)} \setminus (i,j) \text{ such that } h \leq i \text{ and } k \leq j\}.$$

It follows immediately from the definition that no two edges in $Ext_{(l,m)}$ have a common endpoint, and thus $|Ext_{(l,m)}| \leq n$.

Definitions 7 and 8 find their application in identifying a valid prefix extension, as shown in Fig. 4 and discussed next.

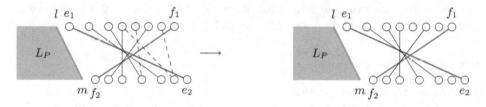

Fig. 4. Extraction of $Ext_{(l,m)}$ from the set $Mk_{(l,m)}$, pictured on the left. The edges belonging to $Mk_{(l,m)} \setminus Ext_{(l,m)}$ are dashed.

2.4   Valid Prefix Extensions

Let P be a valid prefix with leftmost mapping L_P ending at edge (l, m). We use shorthands for $Mk_P = Mk_{(l,m)}$ and $Ext_P = Ext_{(l,m)}$. The candidates in $Ext_P \subseteq Mk_P$ are the unshiftable edges soon after L_P such that no other unshiftable edge lies completely delimited between L_P and any of them, as illustrated in Fig. 4.

We thus are ready to give our algorithmic characterization of valid extensions of prefixes to relate edges (l, m) and (i, j) from Theorem 2.

Theorem 3. *Let* P *be a valid prefix of some* $M \in MCS(X, Y)$, *with leftmost mapping* L_P *ending at edge* (l, m). *Then* Pc *is a valid prefix if and only if the following two conditions hold.*

(1) There exists $(i, j) \in Ext_P$ *corresponding to character* c, *and*
(2) $P \in MCS(X_{<i}, Y_{<j})$.

The proof of Theorem 3 is quite involved, and thus postponed to Sect. 4. This result is crucial for our polynomial-delay binary partition algorithm, as the latter recursively enumerates $MCS(X, Y)$ by building increasingly long valid prefixes and avoiding unfruitful recursive calls.

3   Polynomial-Delay MCS Enumeration Algorithm

The characterization given in Theorem 3 immediately gives prefix-expanding enumeration Algorithm 1, which progressively augments prefixes with characters that keep them valid, until whole MCSs are recursively generated. It is worth noting that Theorem 3 guarantees that each recursive call yields at least one MCS; moreover, all the MCSs are listed once.

Algorithm 1 employs a binary partition scheme. First, it builds the necessary data structures and finds the set of unshiftable edges in a polynomial preprocessing phase, using FINDUNSHIFTABLES as described in detail in Sect. 3.1. Then, it begins a recursive computation BINARYPARTITION where, at each step, it considers the enumeration of the MCSs that start with some valid prefix P. The partition is made through over characters $c \in \Sigma$ such that Pc is valid.

For convenience, we add a dummy character $\# \notin \Sigma$ at the beginning of both strings; i.e. at positions $(-1, -1)$. The recursive computation then starts with $P = \#$, and leftmost mapping $L_P = \{(-1, -1)\}$. At each step, the procedure

---

Algorithm 1. EnumerateMCS

1: **procedure** ENUMERATEMCS(X, Y, Σ)
2:     $\mathcal{U} = $ FINDUNSHIFTABLES$((|X|, |Y|))$
3:     BINARYPARTITION($\#, \{(-1, -1)\}$)
4: **end procedure**

5: **procedure** FINDUNSHIFTABLES$((i, j))$
6:     **for** $c \in \Sigma$ **do**
7:         $l_X \leftarrow$ the right-most occurrence of c in $X_{<i}$
8:         $l_Y \leftarrow$ the right-most occurrence of c in $Y_{<j}$
9:         **if** $l_X \neq -1$ and $l_Y \neq -1$ and $(l_X, l_Y) \notin \mathcal{U}$ **then**
10:             yield (l_X, l_Y) // which is added to $\mathcal{U}$
11:             FINDUNSHIFTABLES$((l_X, l_Y))$
12:         **end if**
13:     **end for**
14: **end procedure**

15: **procedure** BINARYPARTITION(P, L_P)
16:     compute the set of extensions Ext_P using $\mathcal{U}$
17:     **if** $Ext_P = \emptyset$ **then Output** P
18:     **else**
19:         **for** $(i, j) \in Ext_P$ corresponding to some $c \in \Sigma$ **do**
20:             **if** $P \in MCS(X_{<i}, Y_{<j})$ **then**
21:                 let (l, m) be the last edge of L_P
22:                 find leftmost mapping edge (l_c, m_c) for c in $G(X_{>l}, Y_{>m})$
23:                 BINARYPARTITION($Pc, L_P \cup (l_c, m_c)$)
24:             **end if**
25:         **end for**
26:     **end if**
27: **end procedure**

---

finds the valid extensions Ext_P for the given prefix P using the unshiftable edges from $\mathcal{U}$. If Ext_P is empty, then P is an MCS, and is returned. Otherwise, for each character $c \in \Sigma$ corresponding to an edge in Ext_P (i.e. condition *(1)* of Theorem 3), it checks whether Pc satisfies condition *(2)* of Theorem 3. If it does, given the last edge (l, m) of L_P, the algorithm finds the leftmost mapping (l_c, m_c) for character c in $G(X_{>l}, Y_{>m})$, as to update $L_{Pc} = L_P \cup (l_c, m_c)$. Then, it partitions the MCSs to enumerate into the ones that have Pc as a prefix, and recursively proceeds on Pc, and $L_P \cup (l_c, m_c)$. The correctness of Algorithm 1 immediately follows from Theorem 3.

3.1 Finding Unshiftable Edges

We compute unshiftable edges by going backwards in the strings X and Y, and exploiting the observation below, which follows immediately from Proposition 1.

Fact 1. *Let $(i, j) \in \mathcal{U}$ and $c \in \Sigma$. If i', j' are the rightmost occurrences of c respectively in $X_{<i}$ and $Y_{<j}$, then edge (i', j') is also unshiftable.*

Symmetrically as we did in the previous section, let us add a special character $\$ \notin \Sigma$ at the end of both strings, as to obtain an unshiftable edge at the last positions $(|X|, |Y|)$. Starting from this edge, we have a natural recursive visiting procedure that finds unshiftable edges based on the Fact 1. For each character $c \in \Sigma$, candidate unshiftable edges are found by taking the rightmost occurrences of c before the current edge in both strings. Then, we recur in these new edges, unless already visited. This originates our FINDUNSHIFTABLES procedure, whose pseudocode is shown in Algorithm 1.

All unshiftable edges are found in this fashion. The last pairwise occurrences of every character are visited from edge $(|X|, |Y|)$. If an unshiftable edge (i, j) is not the last pairwise occurrence, then by Proposition 1 there is at least one unshiftable edge in $G(\mathcal{I}_X(i), \mathcal{I}_Y(j))$. Edge (i, j) will then surely be visited from the leftmost of these edges, and therefore it will be marked as unshiftable.

3.2 Polynomial Delay

We now show that Algorithm 1 has polynomial delay by analyzing the two main components of ENUMERATEMCS.

The following remark is crucial for our complexity analysis:

Remark 4. Unshiftable edges can be dense in $G(X, Y)$. For example, consider $X = \mathtt{A}^n(\mathtt{CA})^n$ and $Y = \mathtt{A}^n\mathtt{C}^n$. In this situation, every $\mathtt{A}$ of Y has out-degree of unshiftable edges equal to the number of $\mathtt{C}$s in X, that is $O(n)$. The total number of unshiftable edges is therefore $|\mathcal{U}| = O(n \cdot n) = O(n^2)$.

In the rest of the section, we assume that the next and previous occurrences of a given character with respect to some position of the strings can be performed in logarithmic time, with a linear-space search tree.

Preprocessing Phase of FindUnshiftables. This algorithm examines every unshiftable edge exactly once, adding it to a set if not already found (see Line 9). For each of these edges it finds the previous pairwise occurrences of every character and checks whether they are already in the unshiftable set. This takes $O(|\mathcal{U}| \cdot \sigma \log n) = O(n^2 \log n)$ time.

Recursive BinaryPartition. The height of the recursive binary partition tree is at most the length of the longest MCS, which is at most $\min\{|X|, |Y|\} = O(n)$.

The first operation at each step consists in computing the set Ext_P: by scanning the unshiftable edges we can trivially find the cross in $O(|\mathcal{U}|) = O(n^2)$ time, and by another scan the mikado and Ext_P.

When it is nonempty, we check for maximality of P for each of its elements (recalling $|Ext_P| = O(n)$), which takes $O(n)$ time by employing Sakai's maximality test [11]. If the test is positive, we only need to perform a leftward re-map of the new edge, which can be done in logarithmic time, totalizing $O(|\mathcal{U}| + |Ext_P|(n + \log n)) = O(n^2)$ time.

Overall, the delay of BINARYPARTITION is the cost of a root-to-leaf path in the recursion tree, which has depth $\leq n$. This leads to a polynomial-delay algorithm with delay $O(n^3)$ and a polynomial-time preprocessing cost of $O(n^2 \log n)$. The space required is $O(n^2)$, as we need to store the set Ext_P for all recursive calls in a root-to-leaf path plus the set of unshiftable edges $\mathcal{U}$.

In the full version of the paper we provide a more refined algorithm which achieves $O(n \log n)$ delay, with the same preprocessing time and space. An ideal method would yield each distinct MCS in time proportional to its length: as the latter can be $\Theta(n)$, this would take linear time in n. In our refined algorithm we only spend a further logarithmic time factor per solution, so it is quite close to the ideal method. As for space and preprocessing time, the quadratic factor is unavoidable when employing the possibly quadratic unshiftable edges in $\mathcal{U}$.

4   Proof of Theorem 3

In this section we finalize the proof of Theorem 3, at the heart of our results. We introduce the concept of *certificate edges*, and use it to show sufficiency and necessity of the two conditions *(1)* and *(2)* in Theorem 3.

4.1   Certificate Edges

Certificate edges are defined as follows, and illustrated in Fig. 5.

Definition 9. *An edge* $(i', j') \in \mathcal{U}$ *is a* certificate *for another edge* (i, j) *if* $(i', j') \in G(\mathcal{I}_X(i), \mathcal{I}_Y(j))$ *and no* $(x, y) \in \mathcal{U} \setminus \{(i', j')\}$ *has* $x \in (i, i']$, $y \in (j, j']$. *In this case we say that* (i', j') certifies (i, j). *We denote with* $\mathcal{C}_{(i,j)}$ *the set of certificates of edge* (i, j). *An edge* $(i, j) \in \mathcal{U}$ *is called a* root *iff* $\mathcal{C}_{(i,j)} = \emptyset$.

Definition 10. *A* certificate mapping *is a mapping in which the rightmost edge is a root, and each edge except the leftmost is a certificate for the one to its left.*

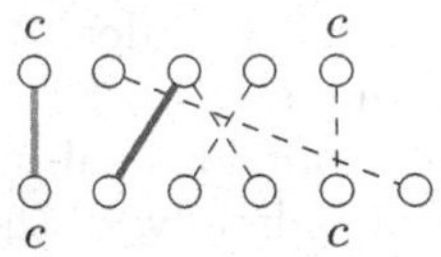

Fig. 5. The only certificate for the green edge is drawn in blue. (Color figure online)

Lemma 2. *Let $M = \{r_1, ..., r_N\}$ be a maximal certificate mapping in $G(X', Y')$ of a common subsequence $S = S_1 \cdots S_N$ between X' and Y', where $r_1 = (i_1, j_1)$.*

1. M is a rightmost maximal mapping of unshiftable edges in $G(X', Y')$, and
2. if $G(X'_{<i_1}, Y'_{<j_1}) \cap \mathcal{U} = \emptyset$, then $M \in MCS(X', Y')$.   *(proof in full version)*

We now define the FIND$_R$ procedure, used to generate certificate mappings. This procedure implicitly finds the C from Theorem 2. Given an unshiftable edge, FIND$_R$ chooses one of its certificates and recurs until it gets to a root edge.

$$\text{FIND}_R(i, j) = \{(i, j)\} \cup \left(\cup_{(h,k) \in \mathcal{C}_{(i,j)}} \text{FIND}_R(h, k) \right)$$

Proposition 2. *Let (l, m) be any edge of the graph, and $(i, j) \in Ext_{(l,m)}$ in the set of extensions of (l, m). Then FIND$_R(i, j)$ returns a certificate mapping with first edge (i, j), such that the corresponding subsequence is $M \in MCS(X_{>l}, Y_{>m})$.*

Proof. The procedure FIND$_R(i, j)$ generates a certificate mapping starting with edge (i, j) by definition. Since $(i, j) \in Ext_{(l,m)}$, there cannot be any unshiftable edges in the subgraph $G(X[l+1, ..., i], Y[m+1, ..., j])$. By setting $X' = X_{>l}$ and $Y' = Y_{>m}$ in Lemma 2, $M \in MCS(X_{>l}, Y_{>m})$ and is rightmost.   □

4.2   Necessary and Sufficient Conditions

Necessity. First of all, we will prove that conditions *(1)* and *(2)* of Theorem 3 are necessary. Let Pc be a valid prefix of some $W \in MCS(X, Y)$.

First, we show that condition *(1)* holds, namely, there exists $(i, j) \in Ext_P$ corresponding to character c. We use contradiction below, supposing that none of the edges in Ext_P correspond to c.

By Sakai's characterization of maximality, for all indices $k \leq |W|$ we have $X_k \cap Y_k = \emptyset$. Let $\hat{k} = |P|$, and thus $W = P W_{>\hat{k}}$ and $W_{>\hat{k}}$ starts with c because Pc is a valid prefix of W. By definition, $X_{\hat{k}} \cap Y_{\hat{k}} = \emptyset$, where $X_{\hat{k}}$ and $Y_{\hat{k}}$ are given by the parts of the strings between the leftmost mapping L_P of P and the rightmost mapping of $W_{>\hat{k}}$. The first edge of the latter mapping is $(i, j) \in \mathcal{U}$ corresponding to character c as $W_{>\hat{k}}$ starts with c. By contradiction, suppose $(i, j) \notin Ext_P$. We now have two cases.

Case $(i, j) \notin Mk_P$: this implies that $i > f_1$ or $j > e_2$, where f_1 and e_1 are those given in Definition 8. Therefore $X_{\hat{k}} \cap Y_{\hat{k}} \neq \emptyset$ as there would be at least the character corresponding respectively to f or e. This is a contradiction.

Case $(i, j) \in Mk_P \setminus Ext_P$: this implies that $\exists (h, k) \in Mk_P \setminus (i, j)$ such that $h \leq i$ and $k \leq j$. Then $X_{\hat{k}} \cap Y_{\hat{k}} \neq \emptyset$ as we would have edge (h, k) in $G(X_{\hat{k}}, Y_{\hat{k}})$, giving a contradiction.

Second, we prove the necessity of condition (2), namely, $P \in MCS(X_{<i}, Y_{<j})$.

To this end, we need a brief remark on the restriction of maximals: let $W \in MCS(X, Y)$ and $\{(x_1, y_1), ..., (x_{|W|}, y_{|W|})\}$ any mapping spelling W in the two strings. Given any $k \leq |W|$, we have $W_{<k} \in MCS(X_{<x_k}, Y_{<y_k})$.

Let $P c$ be a valid prefix of some $W \in MCS(X, Y)$, and $\hat{k} = |P|$. In the first part of the proof we have shown that the first edge of the rightmost mapping of $W_{>\hat{k}}$ is some $(i, j) \in Ext_P$ corresponding to c. Therefore, let us consider the mapping for W consisting of P in leftmost form, and $W_{>\hat{k}}$ in rightmost form. Applying the above remark for $k = \hat{k} + 1$ we get $W_{<\hat{k}+1} = P \in MCS(X_{<i}, Y_{<j})$.

Sufficiency. Suppose that conditions (1) and (2) of Theorem 3 hold. By Proposition 2, $\text{FIND}_R(i, j) = C \in MCS(X_{>l}, Y_{>m})$. Since $P \in MCS(X_{<i}, Y_{<j})$ by hypothesis, we have $P C \in MCS(X, Y)$ by Theorem 2. The latter string starts with $P c$, which is therefore a good prefix.

5   Conclusions and Acknowledgements

In this paper we have studied the Maximal Common Subsequences (MCSs), and investigated their combinatorial nature by familiarizing with some of their properties. Circumventing various pitfalls, we ultimately provided an efficient, binary partition-based, polynomial-delay algorithm for listing all MCSs on an equivalent bipartite graph problem.

The work done in this paper was partially funded by NII, Japan and JST CREST, Grant Number JPMJCR1401. The work was partially done at author TU's laboratory when author AC was a postdoc at NII, author RG was on leave from the University of Pisa to visit NII, and author GP was visiting NII.

References

1. Abboud, A., Backurs, A., Williams, V.V.: Tight hardness results for LCS and other sequence similarity measures. In: 2015 IEEE 56th Annual Symposium on Foundations of Computer Science, pp. 59–78. October 2015
2. Apostolico, A.: Improving the worst-case performance of the Hunt-Szymanski strategy for the longest common subsequence of two strings. Inf. Process. Lett. **23**(2), 63–69 (1986)
3. Bergroth, L., Hakonen, H., Raita, T.: A survey of longest common subsequence algorithms. In: Proceedings Seventh International Symposium on String Processing and Information Retrieval. SPIRE 2000, pp. 39–48. September 2000

4. Eppstein, D., Galil, Z., Giancarlo, R., Italiano, G.F.: Sparse dynamic programming I: linear cost functions. J. ACM **39**(3), 519–545 (1992)
5. Fraser, C.B., Irving, R.W., Middendorf, M.: Maximal common subsequences and minimal common supersequences. Inf. Comput. **124**(2), 145–153 (1996)
6. Hirschberg, D.S.: Algorithms for the longest common subsequence problem. J. ACM **24**(4), 664–675 (1977)
7. Hunt, J.W., Szymanski, T.G.: A fast algorithm for computing longest common subsequences. Commun. ACM **20**(5), 350–353 (1977)
8. Kanté, M.M., Limouzy, V., Mary, A., Nourine, L.: On the enumeration of minimal dominating sets and related notions. SIAM J. Discrete Math. **28**(4), 1916–1929 (2014)
9. Lawler, E.L., Lenstra, J.K., Rinnooy Kan, A.H.G.: Generating all maximal independent sets: NP-hardness and polynomial-time algorithms. SIAM Journal on Computing **9**(3), 558–565 (1980)
10. Masek, W.J., Paterson, M.S.: A faster algorithm computing string edit distances. J. Comput. Syst. Sci. **20**(1), 18–31 (1980)
11. Sakai, Y.: Maximal common subsequence algorithms. In: Navarro, G., Sankoff, D., Zhu, B. (eds.) Annual Symposium on Combinatorial Pattern Matching (CPM 2018), vol. 105, Leibniz International Proceedings in Informatics (LIPIcs), pp. 1:1–1:10. Dagstuhl, Germany, Schloss Dagstuhl-Leibniz-Zentrum fuer Informatik (2018)
12. Wagner, R.A., Fischer, M.J.: The string-to-string correction problem. J. ACM **21**(1), 168–173 (1974)

Searching Runs in Streams

Oleg Merkurev and Arseny M. Shur[✉]

Ural Federal University, Ekaterinburg, Russia
o.merkuryev@gmail.com, arseny.shur@urfu.ru

Abstract. Runs, or maximal periodic substrings, capture the whole picture of periodic fragments in a string. Computing all runs of a string in the usual RAM model is a well-studied problem. We approach this problem in the streaming model, where input symbols arrive one by one and the available space is sublinear. We show that no streaming algorithm can identify all squares, and hence all runs, in a string. Our main contribution is a Monte Carlo algorithm which approximates the set of runs in the length-n input stream S. Given the error parameter $\varepsilon = \varepsilon(n) \in (0, \frac{1}{2}]$, the algorithm reports a set of substrings such that for each run of exponent $\beta \geq 2 + \varepsilon$ in S a single its substring is reported, with the same period and the exponent at least $\beta - \varepsilon$; for runs of smaller exponent, zero or one substring with the same period and exponent at least 2 is reported. The algorithm uses $O(\frac{\log^2 n}{\varepsilon})$ words of memory and spends $O(\log n)$ operations, including dictionary operations, per symbol.

1 Introduction

Recall that a string is periodic if its exponent, which is the ratio between its length and its minimum period, is at least 2. Maximal periodic substrings, or runs, are quite well studied in both algorithmic and combinatorial setting. Kolpakov and Kucherov [15] showed that a length-n string has $O(n)$ runs and conjectured that this number is less than n. After several intermediate results, this conjecture was proved by Bannai et al. [3]; for further studies on the number of runs see, e.g., [9]. The first run-searching algorithm was proposed in [18]. In [15], a linear-time algorithm for computing all runs in a string over a polynomial integer alphabet was presented, based on the Lempel-Ziv factorization (a different approach was later proposed in [3]). However, over a general ordered alphabet of size σ this factorization requires $\Theta(n \log \sigma)$ time [16], while all runs can be computed faster as was shown in a series of papers [11,17] culminating in $\mathcal{O}(n \cdot \alpha(n))$-time algorithm by Crochemore et al. [6], where α is the inverse Ackermann function.

Not much is known about computing runs with restricted memory. The only paper we are aware of is [10], where two algorithms, computing all runs and using a small amount of memory in addition to an input string, were given. An $\mathcal{O}(n \log n)$ algorithm uses $\mathcal{O}(1)$ memory and works over a general alphabet, while a $\mathcal{O}(n)$ algorithm uses $o(n)$ memory over a constant alphabet. In this paper we go further and attack the streaming version of the problem. We recall that in the

© Springer Nature Switzerland AG 2019
N. R. Brisaboa and S. J. Puglisi (Eds.): SPIRE 2019, LNCS 11811, pp. 203–220, 2019.
https://doi.org/10.1007/978-3-030-32686-9_15

streaming model the input symbols arrive in the online fashion and cannot be accessed after reading; algorithms should somehow "sketch" the processed part of the input using sublinear space. For most problems, streaming algorithms are randomized and approximate. Related results about regular structures in streams include Monte Carlo approximated algorithms for longest palindrome [12], longest palindrome with d mismatches [13], longest repeated substring and longest repeated reversed substring [19].

We show that no streaming algorithm can, w.h.p., detect (or count) all runs in a stream. So we define the approximate version approxRuns of the problem of computing all runs as follows: *given an input string S and a parameter $\varepsilon = \varepsilon(n) \in (0, \frac{1}{2}]$, report a set of substrings of S such that (i) for each run of exponent $\beta \geq 2 + \varepsilon$ in S a single its substring is reported, with the same period and the exponent at least $\beta - \varepsilon$, and (ii) for runs of smaller exponent, zero or one substring is reported, with the same period and the exponent at least 2.* Our main aim is to prove the following result.

Theorem 1. *There is a streaming algorithm that solves* approxRuns *performing* $\mathcal{O}(\log n)$ *operations[1] per read and using* $\mathcal{O}(\frac{\log^2 n}{\varepsilon})$ *words of memory.*

In Sect. 2 we recall notation, some basic results and prove that no streaming algorithm can count runs exactly. In Sect. 3 we introduce the main tools and constructions; the algorithm announced in Theorem 1 is presented in Sect. 4.

2   Preliminaries

Let S denote a string of length n over an alphabet $\Sigma = \{1, \ldots, \sigma\}$, where σ is polynomial in n. We write $S[i]$ for the ith symbol of S and $S[i..j]$ for its *substring* (or *factor*) $S[i]S[i+1]\cdots S[j]$; thus, $S[1..n] = S$, $S[i..i-1]$ is an empty string. A *prefix* (resp. *suffix*) of S is a substring of the form $S[1..j]$ (resp., $S[j..n]$). If $w = S[i..j]$, we say that w *occurs* (or *has an occurrence*) in S at position i. A string is *primitive* if it is not a concatenation of two or more copies of a shorter string. A *period* of S is a positive integer p such that $S[1..n-p] = S[p+1..n]$; the *exponent* of S is the ratio $\exp(S) = |S|/p$, where p is the minimum period of S. We call a period p of S *primitive* if $S[1..p]$ is primitive, and *short* if $p \leq |S|/2$. The next lemma is the classical Fine–Wilf theorem [8] written in a suitable form (the second statement is immediate from the first one).

Lemma 1. *(1) A string with primitive periods p and q has length less than*
$p + q - \gcd(p, q)$.
(2) All short periods of a string are multiples of its minimum period.

A *repetition of period p* is a string having a minimum short period p. Thus, S is a repetition iff $\exp(S) \geq 2$; repetitions of exponent 2 are called *squares*. A repetition $S[i..j]$ of period p is called a *run* (in S) if both $S[i-1..j]$ and

---

[1] We extend the set of elementary operations with dictionary operations (insert, delete, lookup). The optimal choice of dictionary depends on ε and is discussed in Remark 3.

$S[i..j+1]$, whenever exist, have no period p. By Lemma 1, it is equivalent to say that $S[i-1..j]$ and $S[i..j+1]$ are not repetitions. For a fixed S, we denote a repetition $S[l..r]$ of period p by the triple (l, p, r). The definitions immediately imply

Observation 1. *If (l, p, r) and (h, p, i) are two repetitions and $l < h \leq h + p - 1 \leq r < i$, then (l, p, i) is a repetition.*

We work in the *streaming model* of computation: the input string $S[1..n]$ (the *stream*) is read from left to right, one symbol at a time, and cannot be stored, because the available space is sublinear in n. The output string can be of size $\Omega(n)$ but cannot be accessed. The space is counted as the number of $\mathcal{O}(\log n)$-bit machine words (in this paper, log stands for the binary logarithm).

A *Monte Carlo algorithm* gives a correct answer with high probability (at least $1 - \frac{1}{n}$ on a length n input) and has deterministic working time and space. For the problems of finding longest palindromes and longest repeats in streams [12, 19], Monte Carlo approximated algorithms were used. For computing runs in a stream, we use the same approach for the same reason: other approaches do not work. The next theorem demonstrates that no streaming algorithm can correctly identify, w.h.p., all squares, and hence all runs, in a string. Let $\mathsf{midSquare}(\Sigma, n)$ be the problem of finding the longest "middle" square, which is a square of the form $S[\frac{n}{2}-i+1..\frac{n}{2}+i]$, in the input stream of even length n over the alphabet Σ.

Theorem 2. *There is a constant γ such that every algorithm solving the problem $\mathsf{midSquare}(\Sigma, n)$ exactly with probability at least $1 - \frac{1}{n}$ uses at least $\gamma n \log \sigma$ bits of memory.*

Proof. We use the same scheme as in [12, Lemma 2]. First we prove that if a Monte Carlo streaming algorithm solves $\mathsf{midSquare}$ exactly using less than $\lfloor \frac{n}{2} \log \sigma \rfloor$ bits of memory, then its error probability is at least $\frac{1}{n\sigma}$. According to Yao's minimax principle [20], it is sufficient to construct a probability distribution $\mathcal{Q}$ over Σ^n such that for any deterministic algorithm D using less than $\lfloor \frac{n}{2} \log \sigma \rfloor$ bits of memory, the expected probability of error on a string chosen according to $\mathcal{Q}$ is at least $\frac{1}{n\sigma}$.

Let $a \in \Sigma$, $n' = n/2$. For arbitrary $x \in \Sigma^{n'}$, $c \in \Sigma$, and $k \in \{1, \ldots, n'\}$ denote $w(x, k, c) = x[1..n']cx[n'-k+2..n']a^{n'-k}$. Let $\mathcal{Q}$ be the uniform distribution over all strings $w(x, k, c)$.

Choose an arbitrary maximal set of disjoint pairs (x, x') of strings from $\Sigma^{n'}$ such that D is in the same state after reading either x or x'. Let $x = vcs$, $x' = v'c's$, where $v, v', s \in \Sigma^*$, $c, c' \in \Sigma$, and $c \neq c'$. Then D returns the same answer on $w(x, |s|+1, c)$ and $w(x', |s|+1, c)$, because the right halves of these two strings coincide. However, the correct answers are different. Indeed, $w(x, |s|+1, c)$ has a middle square of period $|s|$, while $w(x', |s|+1, c)$ has no such square; and if one of two strings has a longer middle square, then the other one has not, because a letter in the right half cannot match simultaneously c from x and c' from x'. Therefore, D errs on at least one of the analysed inputs; similarly, it errs on either $w(x, |s|+1, c')$ or $w(x', |s|+1, c')$.

Since the memory of D has no more than $\sigma^{n'}/2$ states, and at most one string per state is unpaired, the number of pairs is at least $\sigma^{n'}/4$. So the number of errors is at least $\sigma^{n'}/2$ for the total of $\sigma^{n'} \cdot n' \cdot \sigma$ strings, which implies that the probability of error is $\geq \frac{1}{n\sigma}$.

Now assume that some Monte Carlo streaming algorithm A solves midSquare exactly with error probability $\varepsilon \leq \frac{1}{n}$ using $s(n)$ bits of memory. Then we can run its k instances simultaneously and return the most frequent answer. The new algorithm A_k uses $\mathcal{O}(k \cdot s(n))$ bits of memory and its error probability ε_k satisfies the inequality $\varepsilon_k \leq \sum_{2i<k} \binom{k}{i}(1-\varepsilon)^i \varepsilon^{k-i} \leq 2^k \cdot \varepsilon^{k/2} = (4\varepsilon)^{k/2}$. Recall that $\sigma = \mathcal{O}(n^p)$ for some constant p. Let $k = 2p+3$ and take any positive $\gamma \leq \frac{1}{2k}$. If $s(n) < \gamma n \log \sigma$, then algorithm A_k uses less than $\lfloor \frac{n}{2} \log \sigma \rfloor$ bits of memory and has, for n big enough, the error probability less than $\frac{1}{n\sigma}$; as shown above, this is impossible, so the theorem holds for the chosen value of γ. $\square$

3   Tools

3.1   Fingerprints, Frames, Checkpoints

Our algorithm for approxRuns makes use of Karp–Rabin fingerprints [14], which is a hash function common for many streaming string algorithms. Let p be a fixed prime from the range $[n^4, n^5]$, and r be a fixed integer randomly chosen from $\{1, \ldots, p-1\}$. For a string S, its hash is defined as $\phi(S) = (\sum_{i=1}^{n} S[i] \cdot r^i) \bmod p$. The probability of hash collision for two strings of length m is at most m/p. Our algorithm compares hashes of strings having equal lengths of the form 2^j. The probability that a pair of such strings collide is less than n^3/p and thus less than the allowed error probability for Monte Carlo algorithms. Hence all further considerations assume that no collisions happen. For a string A, its *frame* is the tuple $(|A|, \phi(A), r^{|A|} \bmod p, r^{-|A|} \bmod p)$. The crucial property of frames is the following.

Lemma 2 ([4]). *If the frames of any two of the strings A, B, AB are known, the frame of the third string can be computed in $\mathcal{O}(1)$ time.*

All definitions below refer to the input stream S. For any i, the ith iteration of a streaming algorithm processing S begins with reading $S[i]$ and ends just before reading $S[i+1]$. We write $I(i)$ for the frame of $S[1..i-1]$. Lemma 2 implies that one can compute $I(i+1)$ in $\mathcal{O}(1)$ time from $I(i)$ and $S[i]$.

All information currently stored by the algorithm is associated with *check-points*, which form a subset of all positions. Each position k becomes a checkpoint at kth iteration and "lives" during $\mathsf{ttl}(k)$ iterations, where the time-to-live function is defined by $\mathsf{ttl}(k) = 2^{t_\varepsilon + 2 + \beta(k)}$ with $t_\varepsilon = \lceil \log \frac{2}{\varepsilon} \rceil$ and $\beta(k)$ being the maximum power of 2 dividing k. If $k + \mathsf{ttl}(k) = i$, then at the start of ith iteration k "dies" (loses the status of checkpoint) and all associated information is deleted.

Lemma 3. *(1) The number of checkpoints at i'th iteration is $\mathcal{O}(\frac{\log i}{\varepsilon})$.*
(2) At most one checkpoint dies at each iteration.

Proof. (1) Partition the range $[1..i]$ into segments (see an example in Fig. 1): two rightmost segments are of length $2^{t_\varepsilon+2}$ each, and each of the other segments is twice longer than its right neighbor (the leftmost segment can be incomplete). A position k from the jth segment, counting from the right, is a checkpoint iff $\beta(k) \geq j - 1$. Thus we have $\mathcal{O}(\log i)$ segments with $\mathcal{O}(\frac{1}{\varepsilon})$ checkpoints in each, whence the result.

(2) Note that $\beta(k) = \beta(k + \mathsf{ttl}(k))$ for any k. Hence if $k + \mathsf{ttl}(k) = i$, one has $\beta(k) = \beta(i)$ and then $k = i - 2^{t_\varepsilon+2+\beta(i)}$; if k is positive, then it is a unique checkpoint to die at the ith iteration; otherwise, no such checkpoints exist. $\square$

Fig. 1. The checkpoints (black) after the iteration $i = 105$ ($t_\varepsilon = 2$). For example, $\mathsf{ttl}(52) = 2^{2+2+2} = 64$, so the position 52 is a checkpoint until the iteration 116.

3.2 Visible Repetitions and Watch List

Our algorithm works with substrings of length 2^j, $j = 0, \ldots, \lfloor \log n \rfloor$. Such a substring occurring at a checkpoint is called a *j-block*. Now we introduce the key notion. Let (h, p, i) be a repetition, $j = \lfloor \log p \rfloor$, $f = i - p - 2^j + 1$. The repetition (h, p, i) is *visible* if $\mathsf{ttl}(h), \mathsf{ttl}(f) \geq 2^{j+2}$, and $f - h \leq 2^j$ (see Fig. 2). The intuition is that such a repetition is covered by pairs of occurrences of two overlapping (or touching) j-blocks; we will show that this repetition can be detected at the ith iteration using the fact that both h and f are checkpoints at that moment.

Lemma 4. *Every repetition (l, p, r) with exponent $\geq 2 + \varepsilon$ contains a visible repetition of period p as a substring.*

Proof. Let $j = \lfloor \log p \rfloor$, $z = 2^{\max(0, j - t_\varepsilon)}$. Then k is divisible by z iff $\mathsf{ttl}(k) \geq 2^{j+2}$. Let $i = r - (r - 2^j - p + 1) \bmod z$, $h = i - 2p + 1 - (i - 2p + 1) \bmod z$, and consider the substring $S[h..i]$. Its length is $\geq 2p$, and $r - z + 1 \leq i \leq r$. Further,

$$h \geq (r - z + 1) - 2p + 1 - (z - 1) = r - 2p - 2z + 3$$
$$\geq r - 2p - 2^{j+1-t_\varepsilon} + 3 \geq r - 2p - \lceil \varepsilon p \rceil + 1 = l.$$

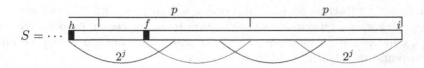

Fig. 2. Visible repetition covered by overlapping occurrences of two j-blocks.

So, $S[h..i]$ is a substring of $S[l..r]$, has period p and length $\geq 2p$; since p is the minimum period of $S[l..r]$, it is primitive. Hence (h, p, i) is a repetition by Lemma 1(2). Next note that z divides both $f = i - p - 2^j + 1$ and h, so $\mathsf{ttl}(f), \mathsf{ttl}(h) \geq 2^{j+2}$. Finally, $f - h = p - 2^j + (i - 2p + 1) \bmod z$ is the smallest multiple of z which exceeds $p - 2^j < 2^j$, so $f - h \leq 2^j$. Therefore, (h, p, i) is visible by definition. $\qquad\square$

Lemma 5. *Suppose that (h, p, i) is a visible repetition and $(h{+}z, p, i{+}z)$ is a repetition, where $j = \lfloor \log p \rfloor$, $z = 2^{\max(0, j - t_\varepsilon)}$. Then (i) $(h{+}z, p, i{+}z)$ is visible, (ii) $(h, p, i{+}z)$ is a repetition, and (iii) both (i) and (ii) hold if we replace $(h{+}z, p, i{+}z)$ by $(h{-}z, p, i{-}z)$.*

Proof. Both h and $f = i - p - 2^j + 1$ are divisible by z by definition of a visible repetition; so the same is true for $h + z$, $f + z$, implying $\mathsf{ttl}(h{+}z), \mathsf{ttl}(f{+}z) \geq 2^{j+2}$. Since $(f + z) - (h + z) = f - h \leq 2^j$, we have (i) by definition. Since $z < p$, one has $i - (h{+}z) \geq p$; now (ii) follows from Observation 1. Almost the same argument works for (iii). $\qquad\square$

By *watch list* we mean a data structure W containing a list of repetitions as described below. Initially W is empty. At ith iteration, it is updated as follows. For all visible repetitions (h, p, i) in the stream, if W contains a repetition (l, p, r) with the same period, this repetition is updated to (l, p, i); otherwise, (h, p, i) is added to W. After that, all repetitions (l, p, r) such that $r + z = i$, where z is defined as in Lemma 5, are deleted.

Lemma 6. *Any streaming algorithm that*

- *finds all visible repetitions of the form (h, p, i) during ith iteration;*
- *maintains the watch list;*
- *outputs repetitions after their deletion from the watch list, solves the problem* approxRuns.

Proof. Let (l, p, r) be a run of exponent $\alpha \geq 2 + \varepsilon$ in S, and let (h, p, i) be the leftmost visible repetition of period p inside $S[l..r]$. Define z as above; then $h - z < l$: otherwise, the repetition $(h{-}z, p, i{-}z)$ would be visible by Lemma 5, contradicting the choice of (h, p, i). Let k be such that $i + kz \leq r < i + (k{+}1)z$. Then $(h + z, p, i + z), \ldots, (h + kz, p, i + kz)$ are visible repetitions by Lemma 5; moreover, no other visible repetitions of period p are substrings of $S[l..r]$, because it is necessary to add at least z to get the positions with required ttl. Consider the ith iteration. The algorithm finds (h, p, i) and looks into W. If W contains a repetition of period p, this repetition was added or updated at most z iterations ago, so S contains a visible repetition (h', p, i'), where $i - z \leq i' < i$. This repetition is not a substring of $S[l..r]$ but overlaps it by at least $2p - z > p$ symbols. Hence (h', p, r) is a repetition by Observation 1; this repetition properly contains a run, contradicting definitions. Thus, the watch list contains no repetition of period p and the algorithm adds (h, p, i) to it.

Now note that the repetition of period p in W will be updated at the iterations $i + z, \ldots, i + kz$, and deleted at the iteration $i + (k{+}1)z$. So the algorithm will

output the repetition $(h, p, i+kz)$, which is shorter than (l, p, r) by at most $2z - 2 \leq \varepsilon p$ symbols and thus has the exponent at least $\alpha - \varepsilon$.

A run of exponent less than $2 + \varepsilon$ may contain no visible repetition inside; the remaining argument for such runs is the same as above. $\square$

Lemma 6 reduces Theorem 1 to the construction of the algorithm which detects visible repetitions immediately after their appearance, maintains the watch list, and satisfies the required time/space limitations.

Remark 1. An algorithm satisfying the conditions of Lemma 6 finds all runs of periods $p \leq 4\lceil \log 1/\varepsilon \rceil$ exactly (if a square of such period p is a suffix of $S[1..i]$, all its positions are checkpoints, so the start and the end of periodicity can be detected exactly).

3.3 Groups

We are interested in recent occurrences of blocks, and define two types of such occurrences. At ith iteration, an occurrence of a j-block T at position h of $S[1..i]$ is *fresh* if $h > i - 2^{j+1} + 1$ and *stale* if $i - 3 \cdot 2^j + 1 < h \leq i - 2^{j+1} + 1$. Equivalently, an occurrence of T is fresh iff it was read less than $|T|$ iterations ago and stale iff it was fresh $|T|$ iterations ago. *Regular* occurrences are those that are not fresh (stale occurrences are regular). Next observation, obvious from Fig. 2, clarifies one use of fresh occurrences:

Observation 2. *If (h, p, i) is a visible repetition, $j = \lfloor \log p \rfloor$, then at the ith iteration (i) the suffix T of length 2^j of $S[1..i]$ is a j-block occurring at the checkpoint $f = i - p - 2^j + 1$ and (ii) the substring U of length 2^j at position $h+p$ is a fresh occurrence of the j-block occurring at the checkpoint h.*

Lemma 7. *(1) Every two fresh occurrences of T overlap.*
(2) If T has at least three fresh occurrences, all positions of fresh occurrences form an arithmetic sequence with the difference equal to the minimum period of T.
(3) The analogs of 1 and 2 hold for stale occurrences as well.

Proof. Since the range $[i - 2^{j+1} + 2..i]$ contains less than $2|T|$ positions, (1) is obvious. Further note that two of any three fresh occurrences of T overlap by at least $|T|/2$, so $|T|$ has a short period; hence (2) follows from Lemma 1, and the difference is the minimum period of T. Finally, (1) and (2) for ith iteration imply (3) for $(i+|T|)$th iteration. $\square$

We say that each j-block has a (possibly empty) *series* of fresh (resp., of stale) occurrences. By Lemma 7, storing such a series requires $\mathcal{O}(1)$ space.

3.4   Data Structures

Let us describe the data structures we use. For each j-block T we maintain a basic structure B_T called "group" and consisting of

- frame of T (we store only hash, other elements are common to all j-blocks);
- doubly-connected lists $flist$ and $rlist$ of all checkpoints, in increasing order, that are positions, respectively, of fresh and regular occurrences of T;
- the series $fseries$ and $sseries$ of fresh and stale occurrences of T.

In each node in $flist$ and $rlist$ we store, apart from the checkpoint and links to next and previous nodes, two auxiliary numbers: $period$ and $extension$. These numbers for the checkpoint k in the group B_T are computed at the $(k+|T|)$th iteration after adding a fresh occurrence of T to B_T. We set $period$ to the period of $B_T.fseries$; if the series is one-element, we set $(period, extension) = (0,0)$. If $period = p > 0$, then $S[1..k+|T|]$ has a repetition of period p as a suffix. If the watch list contains such a repetition, we set $extension$ to its position; otherwise, $extension$ is set to the position of the first fresh occurrence of T.

Remark 2. A group can be viewed as a constant-size structure (frame, two series, links to beginnings and ends of lists) plus a set of constant-size nodes (position, links to next and previous elements of the list, period, extension). We can store all groups in an array of constant-size cells endowed with a stack of empty cells. This allows creating a new group/node and deleting an existing group/node in $\mathcal{O}(1)$ time. The size of the array is proportional to the number of groups plus the number of occurrences of j-blocks; both numbers are $O(\frac{\log^2 n}{\varepsilon})$.

We also need to discuss how to store and update $fseries$ and $sseries$. For each series, we store a tuple $(first, period, last, frame1, frame2)$, consisting of the positions of the first and the last occurrences, the distance between consecutive occurrences ("period"), and the frames for positions of the first and the second occurrence. Empty and one-element series are stored as (0) and $(first, 0, frame1)$ respectively. The series of a j-block T should be updated at ith iteration whenever either a new fresh occurrence of T appears, or a fresh occurrence becomes stale, or a stale occurrence expires.

Lemma 8. *Each update of $fseries$ and $sseries$ requires $\mathcal{O}(1)$ time.*

Proof. First consider deletion. The occurrence to be deleted is first in the series. For one-element series the deletion is trivial. For two-element series it suffices to set $first$ to $(first + period)$ and $frame1$ to $frame2$. If the series is longer, we also need to compute the new value of $frame2$ ($period$ and $last$ stay the same). By Lemma 2, we compute, in $\mathcal{O}(1)$ time, the frame of the string $A = S[first..first+period-1] = S[first+period..first+2\cdot period-1]$ from $frame1$ and $frame2$; then we again use Lemma 2 to compute the new $frame2$, which is $I(first+2\cdot period)$, from $frame2$ and the frame of A.

Now consider insertion. The new occurrence is the last one, so if the series contained ≥ 2 elements, we just assign a new value to $last$. For shorter series, we

also need to know the frame for the position of this occurrence. If we insert to $sseries$, this is the $frame1$ just deleted from $fseries$. If we insert to $fseries$, the new occurrence of T is the suffix of $S[1..i]$, so it can be computed by Lemma 2 from the frames of $S[1..i]$ and T. Thus, a constant number of operations is performed. $\square$

For navigation we use six dictionaries, described in the following table. The values in the first five dictionaries are stored as links.

Id	Key	Value
H_1	j, hash F	group of the j-block with hash F
H_2	j, checkpoint k	group of the j-block occurring at k
H_3	j, position k	group of the j-block having the first fresh occurrence at k
H_4	j, position k	group of the j-block having the first stale occurrence at k
HH	j, checkpoint k	node for k in the group of the j-block occurring at k
HC	checkpoint k	frame $I(k)$

We also store the watch list W as a doubly-connected list ordered by periods of repetitions.

4   Algorithm

Each iteration starts with reading a new symbol $S[i]$ from the stream and computing the frame $I(i+1)$; the new frame is stored in the variable I (the old value of I is put to HC). The subsequent operations are grouped in four stages:

- *Checkpoint deletion.* Here we delete all data about the checkpoint that dies at this iteration. (It is unique by Lemma 3(2)).
- *Groups update.* Here we add j-blocks of the form $S[h..i]$ to the existing groups or create new groups for them. In addition, we move the checkpoints of expired fresh occurrences from $flist$ to $rlist$ and delete expired stale occurrences from $sseries$.
- *Repetition detection.* Here we detect visible repetitions of the form (h, p, i). As a side effect, we finalize the update of existing groups, adding fresh occurrences to $fseries$ and moving expired fresh occurrences to $sseries$.
- *Watch list update.* Here we add detected visible repetitions to the watch list, and then delete and output the repetitions which provably cannot be extended to the position i.

4.1   Checkpoint Deletion

We process all j-blocks at the checkpoint position, deleting the checkpoints from their groups and from dictionaries; groups without checkpoints are also deleted.

Lemma 9. *Algorithm 1 works in $\mathcal{O}(\log i)$ operations.*

Proof. The loop in line 4 runs $O(\log i)$ times, $\mathsf{ttl}(i)$ is computed in $\mathcal{O}(\log i)$ time, a group can be deleted in $O(1)$ time (see Remark 2).   □

4.2   Groups Update

The stage is a sequence of three loops. The first loop processes the j-blocks of the form $S[h..i]$, where h is a checkpoint. For each block we compute its hash and extract its group B from the table H_1; if B does not exist, it is created. The occurrence at h is fresh, so a node for h is added to $B.flist$. The period and extension are computed for this node according to definitions.

The second loop moves checkpoints from fresh lists to regular lists. The checkpoint to move depends on j only, so the corresponding group can be extracted from the table H_2.

Algorithm 1. ith iteration, Checkpoint deletion

1: compute $\mathsf{ttl}(i)$; $h \leftarrow i - \mathsf{ttl}(i)$
2: **if** $h \geq 1$ **then**
3:     delete $HC(h)$
4:     **for** $\{j \leftarrow 0; h + 2^j - 1 < i\}$ **do**
5:         $B \leftarrow H_2(j, h)$; delete $H_2(j, h)$
6:         $node \leftarrow HH(j, h)$; delete $HH(j, h)$
7:         delete $node$ from $B.flist$ or $B.rlist$
8:         **if** $B.flist$ and $B.rlist$ are empty **then**
9:             delete B from H_1, H_3, H_4 ▷ all keys are stored in B
10:             delete group B

Algorithm 2. ith iteration, Groups update

1: **for** $\{j \leftarrow 0; h \leftarrow i - 2^j + 1$ && $h \geq 1$ && $h + \mathsf{ttl}(h) > i\}$ **do**
2:     $I_h \leftarrow HC(h)$; $F \leftarrow \phi(S[h..i])$; $B \leftarrow H_1(j, F)$ ▷ $\phi(S[h..i])$ is computed from I_h, I
3:     **if** $B = null$ **then**
4:         $B \leftarrow$ new group; $H_1(j, F) \leftarrow B$; $B.frame \leftarrow F$
5:     $N \leftarrow$ new node; add N to $B.flist$; $N.checkpoint \leftarrow h$
6:     $HH(j, h) \leftarrow N$; add h to $B.fseries$; $N.period \leftarrow B.fseries.period$
7:     $N.extension \leftarrow \{t$, if $(t, N.period, i') \in W$ for some i'; $B.fseries.first$ otherwise$\}$
8: **for** $\{j \leftarrow 0; h \leftarrow i - 2^{j+1} + 1$ && $h \geq 1$ && $h + \mathsf{ttl}(h) > i\}$ **do**
9:     $B \leftarrow H_2(j, h)$
10:     move $B.flist.first$ to $B.rlist$
11: **for** $\{j \leftarrow 0; h \leftarrow i - 3 \cdot 2^j + 1$ && $h \geq 1\}$ **do**
12:     $B \leftarrow H_4(j, h)$
13:     **if** $B \neq null$ **then**
14:         $H_4(j, h + B.sseries.period) \leftarrow B$; delete $H_4(j, h)$
15:         delete $B.sseries.first$ ▷ see Lemma 8

The third loop deletes first elements of stale series if these elements are no longer stale occurrences. The position of such an element is determined by j, so the corresponding group can be extracted from the table H_4.

Lemma 10. *Algorithm 2 works in $\mathcal{O}(\log i)$ operations.*

Proof. Each loop runs $\mathcal{O}(\log i)$ times and contains constant number of calls to dictionaries. A group is created in $\mathcal{O}(1)$ time (Remark 2). Computing $\mathsf{ttl}(h)$ in lines 1,8 requires constant time if the value $\beta(i+1)$ is computed once (in logarithmic time) before the loops. In line 7, we consult the watch list W to compute *extension*. Since the periods of extensions increase as j increases, in total we need a single pass over W. By Lemma 12 below, $|W| = \mathcal{O}(\log i)$. Finally, a series in line 15 is modified in $\mathcal{O}(1)$ time (Lemma 8). $\qquad\square$

4.3 Repetition Detection

At this stage we solve first main task: find visible repetitions that are suffixes of $S[1..i]$ (see Lemma 6). For each j, we try to detect repetitions with periods in the range $[2^j..2^{j+1}-1]$. For such a repetition to exist, the suffix $T = S[i-2^j+1..i]$ of the stream should occur at some checkpoint f (see Fig. 2). In particular, there is a group for T and we have to find it not knowing the hash $\phi(T)$. Note that there is no way to find $\phi(T)$ if T has no occurrence at a checkpoint.

Let T occur at checkpoint k. Then $T' = S[k..k+2^{j-1}-1] = S[i-2^j+1..i-2^{j-1}]$ is the prefix of T of length 2^{j-1}. Since T' occurs at k, it has a group, say B', and a fresh occurrence at $i-2^j+1$. This occurrence is first in $B'.fseries$, because it becomes stale at this iteration. Then we get $I(i-2^j+1) = B'.fseries.frame1$ and compute $F = \phi(T)$ by Lemma 2; see Algorithm 3,

---

Algorithm 3. ith iteration, Repetition detection I (searching groups)

```
 1: for {j ← 0; k = i − 2^j + 1 && k ≥ 1} do
 2:    if j = 0 then
 3:       F ← φ(S[i])
 4:    else
 5:       B' ← H₃(j − 1, k)
 6:       if B' = null then
 7:          continue
 8:       F ← φ(S[k..i])                              ▷ from I(i) and B'.fseries.first
 9:       add k to B'.sseries; H₄(j, B'.sseries.first) ← B'
10:       H₃(j, k+B'.fseries.period) ← B'; delete H₃(j, k); delete B'.fseries.first
11:    B ← H₁(j, F)
12:    if B = null then
13:       continue
14:    add k to B.fseries; H₃(j, B.fseries.first) ← B
15:    search a visible repetition using j, B                            ▷ Algorithm 4
```

lines 5–8. Note that in lines 9–10 we move the mentioned occurrence of T' from fresh to stale, and in line 14 we add a fresh occurrence of T.

Now we know the group B of T. To detect visible repetitions, we find all candidates to the checkpoint f (see Fig. 2; note that T has a stale occurrence at f), compute p and h for each f, and check whether (h, p, i) is indeed a repetition. Below we show how to perform all computation in $\mathcal{O}(1)$ time. The two steps are

(i) Choose at most two candidates for f, guaranteeing that no other occurrence of T can lead to a repetition of the form (h, p, i) with $p \in [2^j .. 2^{j+1}-1]$;

(ii) For a given f, compute p, h, and check whether (h, p, i) is a repetition.

We start with (ii). From i, j, f we get the period $p = i - f - 2^j + 1$, $z = 2^{\max\{0, j - t_\varepsilon\}}$ and the checkpoint $h = \left\lfloor \frac{i-2p+1}{z} \right\rfloor \cdot z$. In the group $B_h = H_2(j, h)$ we check, in constant number of arithmetic operations, whether a fresh occurrence exists at position $h+p$. If no, there is no repetition of period p. If yes, we know that $S[h..i]$ has short period p. It remains to check that p is primitive. If it is not, we can write $S[i-2p+1..i] = U^{2t}$, where U is primitive, $|U| = q$, $t > 1$, $p = qt$. Then U is a suffix of T and T is a suffix of U^t. Hence T has occurrences ending at $i, i-q, \ldots, i-tq$ and no occurrences "in between", because U, as a primitive word, occurs only twice in each substring U^2. Since $q \leq \frac{p}{2} < 2^j$, the occurrence of T ending at $i-q$ is fresh by definition, so $B.fseries.period = q$. The occurrence of T at f, which ends at $i-tq$, is stale, so the last stale occurrence of T is at distance q from its first fresh occurrence: $B.fseries.first - B.sseries.last = q$. On the other hand, if $q = B.fseries.period$ divides p and equals $B.fseries.first - B.sseries.last$, then the string $S[i-p+1..i]$ has period q which divides its length, and thus p is not primitive. So we check these two conditions and report the repetition (h, p, i) iff at least one of them fails. The total number of operations used is clearly $\mathcal{O}(1)$.

Now we approach (i). Let f_1 be the top checkpoint in $B.rlist$. If the occurrence of T at f_1 is not stale, then T has no stale occurrences and thus there are no candidates. So we assume that this occurrence is stale and set f_1 as a candidate. If $|B.sseries| \leq 2$, only two top nodes in $B.rlist$ can be candidates. So below we assume $|B.sseries| \geq 3$ and make use of periods: now T is periodic with minimum period $q = B.sseries.period$. Let $p = i - f_1 - 2^j + 1$. Then the possible periods of repetitions, corresponding to stale occurrences of T, are exactly the numbers $p + mq$, where $m \geq 0$, $p + mq < 2^{j+1}$ (we should consider the worst case, where all stale occurrences are at checkpoints).

If $S[1..i]$ ends with a repetition of period p, the strings $S[i-2p+1..i-p]$ and $S[i-p+1..i]$ are equal and thus share the longest suffix of period q. Let V (resp., V') be the longest suffix of $S[1..i]$ (resp., $S[1..i-p]$) of period q. Note that the suffix of $S[1..i]$ of length $p+q$ has period p due to two occurrences of T at this distance; if it also has period q, which is primitive, then by Lemma 1 q divides p. Then neither of the periods $p + mq$ is primitive, so no visible repetition in the analyzed range of periods exists. Thus below we assume $|V| < p+q$.

On the other hand, T is a repetition of period q and a suffix of $S[i-p+1..i]$, so V is a repetition too; let $V = (x, q, i)$. Let l be such that the watch list W contains a repetition of the form (l, q, r); if W contains no such repetition, let

$l = i - 2^j + 1$ be the position of the suffix T of $S[1..i]$. From Lemma 4 we conclude that $x \in [l - \lfloor q\varepsilon \rfloor + 1..l]$. Similarly, $V' = (x', q, i - p)$ is a repetition with suffix T. The analog of the value l used above is stored as $l' = f_1.extension$, and we again use Lemma 4 to conclude $x' \in [l' - \lfloor q\varepsilon \rfloor + 1..l']$.

If p is a period of a repetition, then either $|V|, |V'| \geq p$ or $V = V'$. One can check in $\mathcal{O}(1)$ arithmetic operations whether any of the two conditions may hold. Depending on the answer yes/no, f_1 is/is not a candidate for f. Further, consider some other checkpoint f_2 which corresponds to the period $p+mq$ for some $m \geq 1$, and let V'' be the longest suffix of $S[1..i-p-mq+1]$ having period q. Since $|V| < p + mq$, $V'' = V$ is a necessary condition for having a repetition of period $p+mq$. Note that V'' is a prefix of V' and thus its position is x'. The known ranges for both x and x' have length $\lfloor q\varepsilon \rfloor$, and $2 \cdot \lfloor q\varepsilon \rfloor - 1 \leq q$. Hence there exists at most one period of the form $p + mq$ such that $|V''| = |V|$ is possible for some values of x and x'; all other stale occurrences of T do not correspond to repetitions. We compute the value of m by arithmetic operations; the position $f_2 = f_1 - mq$ is the only possible second candidate for f. We check (e.g., in H_2) whether f_2 is a checkpoint; if yes, we make f_2 a candidate. Thus we are done with (i). The above argument justifies the correctness of Algorithm 4 below; an example is given in Fig. 3. Altogether, we have the following lemma.

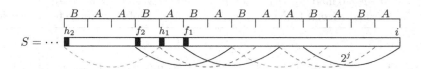

Fig. 3. Example of repetition detection. A, B are strings of length 2^{j-2}. The suffix $T = BABA$ has two stale occurrences at checkpoints f_1 and f_2 (candidate periods $p_1 = 5 \cdot 2^{j-2}$, $p_2 = 7 \cdot 2^{j-2}$, respectively). Repetitions (h_1, p_1, i) and (h_2, p_2, i) are detected. The stale series has short period $q = 2^{j-1}$. One has $V = V'' = uABABA$, $V' = uABABABA$, where $u = lcs(A, B)$; a candidate checkpoint f_1 satisfies $|V|, |V'| \geq p_1$.

Lemma 11. *All visible repetitions which are suffixes of $S[1..i]$ can be computed in $\mathcal{O}(\log i)$ operations.*

Proof. As j increases, the periods q (line 10 of Algorithm 4) do not decrease. Thus all requests to the watch list W (line 11 of Algorithm 4) can be done in $\mathcal{O}(|W|)$ time using a pointer. By Lemma 12, $|W| = \mathcal{O}(\log i)$. Other than that, for each j Algorithms 3 and 4 perform a constant number of operations. □

4.4 Updating the Watch List

After the previous stage, we have the list New of visible repetitions detected at this iteration. We now need to merge New with the watch list W and then delete from W the "old" repetitions. Our aim is to prove the following.

Algorithm 4. ith iteration, Repetition detection II (search with j, B)

```
1:  f₁ ← B.rlist.top                    ▷ position of rightmost regular occurrence of T
2:  if f₁ ≤ i − 3 · 2ʲ then
3:      continue        ▷ no stale occurrences at checkpoints; proceed to next j in Alg. 3
4:  else
5:      C ← {f₁}              ▷ always include f₁ to the list C of candidates
6:      if |B.sseries| ≤ 2 then
7:          if f₁.previous > i − 3 · 2ʲ then  ▷ stale occurrence at previous checkpoint
8:              add f₁.previous to C
9:      else                                          ▷ define borders of q-periodicity
10:         p ← i − f₁ − 2ʲ + 1; q ← B.sseries.period; l′ ← f₁.extension
11:         l ← {t, if (t, q, i′) ∈ W for some i′; B.fseries.first otherwise}
12:         if ⌈(l−l′+1−⌊qε⌋−p)/q⌉ = ⌊(l−e−1+⌊qε⌋−p)/q⌋ then
13:             f₂ ← f₁ − ⌊(l−l′−1+⌊qε⌋−p)/q⌋ · q  ▷ occurrence possibly satisfying V″ = V
14:             if H₂(j, f₂) exists then                        ▷ f₂ is a checkpoint
15:                 add f₂ to C
16: for {f ∈ C} do
17:     p ← i − f − 2ʲ + 1; z ← 2^max{0,j−tε}; h ← ⌊(i−2p)/z⌋ · z; Bₕ ← H₂(j, h)
18:     if h + p ∈ Bₕ.fseries then                            ▷ period p detected
19:         q ← B.fseries.period                    ▷ check primitiveness of p
20:         if {p ≠ 0 mod q || B.fseries.first − B.sseries.last ≠ q} then
21:             add (h, p, i) to New          ▷ update list of newly detected repetitions
```

Lemma 12. *At ith iteration, the watch list (i) has length $\mathcal{O}(\log i)$ and (ii) can be updated in $\mathcal{O}(\log i)$ operations.*

Proof of (i) requires deleting some repetitions from W earlier than it is prescribed by definition, if such repetitions cannot be updated later. The algorithm for (ii) is Algorithm 5 below. We need two lemmas.

Lemma 13 (Three Square Lemma, [5]). *If three squares with primitive periods $p_1 < p_2 < p_3$ end at the same position of a string, then $p_3 \geq p_1 + p_2$.*

Lemma 14. *Given repetitions $(h, p, i) \in New$ and $(l, q, r) \in W$, where $r < i$ and $\frac{2}{3}p < q < \frac{3}{2}p$, it is possible to determine, in $\mathcal{O}(1)$ operations, whether (l, q, i) is a repetition.*

Proof. Let $j = \lfloor \log p \rfloor$, $T = S[i − 2^j + 1..i]$. Since (h, p, i) is added at the current iteration, T is the j-block at checkpoint $f = i − p − 2^j + 1$ (solid arcs in Fig. 4).

First consider the case $q > p$. Since $q < \frac{3}{2}p \leq p + 2^j$, the mutual location of the repetitions (h, p, i) and (l, q, r) is as in Fig. 4a. The repetition (l, q, r) extends to (l, q, i) iff the suffix T occurs at position $i − q − 2^j + 1$ (dashed arc in Fig. 4a). Since $q − p < 2^j$, this occurrence overlaps the occurrence at f. Hence, in the group of T, $f.extension$ is nontrivial. Namely, $f.period$ divides $q − p$ and $f.extension \leq i − q − 2^j + 1$. So we extract the node of f as $HH(j, f)$ and check $f.period$ and $f.extension$; in total, $\mathcal{O}(1)$ operations are performed.

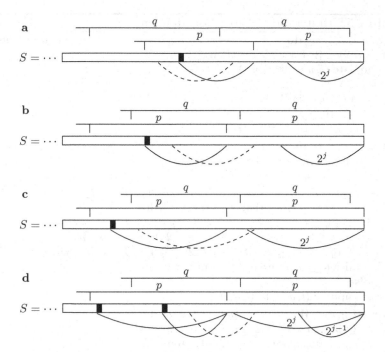

Fig. 4. Mutual location of a new repetition of period p and a live repetition of period q, where q is close to p. Black rectangles denote live checkpoints.

Now let $q < p$. If $r - q \leq i - p$, then (l, q, i) is not a repetition: otherwise, the suffix of length $p + q$ of $S[1..i]$ would have primitive periods p and q, contradicting Lemma 1. Then the repetitions (h, p, i), (l, q, r) and the occurrences of T are mutually located as in Fig. 4b–d. If $2^j < q$ (Fig. 4b), then (l, q, i) is a repetition iff T has a stale occurrence at position $i - q - 2^j + 1$ (dashed arc). If $q \leq 2^j \leq q + i - r$ (Fig. 4c), the existence of the repetition (l, q, i) is equivalent to the fresh occurrence of T at $i - q - 2^j + 1$ (dashed arc). Finally, if $2^j > q + i - r$ (Fig. 4d), we no longer can rely on the occurrences of T. Indeed, it may happen that $l > i - q - 2^j + 1$; if we then see no occurrence of T at position $i - q - 2^j + 1$, which is outside the repetition with period q, we can derive no useful conclusions. However, in this case $2^j > q > \frac{2}{3}p$, and thus $f' = i - p - 2^{j-1} + 1$ is a checkpoint; indeed, either $\beta(f') = j - 1$ and $\mathrm{ttl}(f') \geq 2^{j+1} > p + 2^{j-1}$ or $\beta(f') = \beta(f)$ and these two positions have the same ttl. Let $T' = S[i - 2^{j-1} + 1..i]$; this is the $(j-1)$-block occurring at the checkpoint f'. Now (l, q, i) is a repetition iff T' has a stale occurrence at position $i - q - 2^{j-1} + 1$ (dashed arc in Fig. 4d). Hence, in all three cases it is enough to check whether a given block has a (fresh or stale) occurrence at a given position, which requires $\mathcal{O}(1)$ operations. The lemma is proved. □

Algorithm 5. ith iteration, Watch List Maintenance

1: **for** $(h, p, i) \in New$ **do** $\qquad\qquad\qquad\qquad\qquad\qquad$ ▷ adding new repetitions
2: $\quad$ **if** $(l, p, r) \in W$ for some l, r **then**
3: $\qquad$ update (l, p, r) to (l, p, i)
4: $\quad$ **else**
5: $\qquad$ add (h, p, i) to W

6: **for** $(l, p, r) \in W$ **do**
7: $\quad j \leftarrow \lfloor \log p \rfloor;\ z \leftarrow 2^{\max\{0, j - t_\varepsilon\}}$
8: $\quad$ **if** $r + z = i$ **then** $\qquad\qquad\qquad\qquad\qquad$ ▷ repetition missed an update
9: $\qquad$ delete (l, p, r) from W; output (l, p, r)

10: **for** $(h, p, i) \in W$ **do** $\qquad\qquad\qquad\qquad$ ▷ deletions/updates by Lemma 14
11: $\quad j \leftarrow \lfloor \log p \rfloor;\ f \leftarrow i - p - 2^j + 1;\ B \leftarrow H_2(j, f);\ B' \leftarrow H_2(j-1, i-p-2^{j-1}+1)$
12: $\quad$ **for** $\{(l, q, r) \in W$ such that $2p/3 < q < p\}$ **do**
13: $\qquad$ **if** $(r - q \leq l - p)\ ||\ (q > 2^j\ \&\&\ i - q - 2^j + 1 \notin B.sseries)$
14: $\qquad\quad ||\ (q \leq 2^j \leq q + i - r\ \&\&\ i - q - 2^j + 1 \notin B.fseries)$
15: $\qquad\quad ||\ (q + i - r \leq 2^j\ \&\&\ i - q - 2^{j-1} + 1 \notin B'.sseries)$ **then**
16: $\qquad\qquad$ delete (l, q, r) from W; output (l, q, r)
17: $\qquad$ **else**
18: $\qquad\qquad$ update (l, q, r) to (l, q, i)

19: $\quad$ **for** $\{(l, q, r) \in W$ such that $p < q < 3p/2\}$ **do**
20: $\qquad$ **if** $(f.period = 0)\ ||\ (q - p \neq 0 \bmod f.period)$
21: $\qquad\quad ||\ (f.extension > i - q - 2^j + 1)$ **then**
22: $\qquad\qquad$ delete (l, q, r) from W; output (l, q, r)
23: $\qquad$ **else**
24: $\qquad\qquad$ update (l, q, r) to (l, q, i)

Proof (of Lemma 12). We prove by induction on i the following fact:

($*$) *for any positive integer r, after i'th iteration the watch list contains at most two repetitions with periods in the range $[r..\frac{3}{2}r]$.*

The base case is obvious. For the step case consider a range $[r..\frac{3}{2}r]$ containing the period of at least one new repetition (otherwise, it is nothing to prove). After applying the third outer loop of Algorithm 5, all remaining repetitions with periods in $[r..\frac{3}{2}r]$ end at position i, and there are at most two of them because of Lemma 13. Thus ($*$) is proved. Clearly, ($*$) implies statement (i).

$\qquad$ Each of three outer loops of Algorithm 5 can be performed in the time proportional to the total length of involved list(s) with the use of two pointers. The fact ($*$) ensures that the pointer for q in the third outer loop never goes back by more than two elements. Since $|New| = \mathcal{O}(\log i)$ by Lemma 13, (i) implies (ii). $\qquad\qquad\qquad\qquad\qquad\qquad\qquad\qquad\qquad\qquad\qquad\qquad\qquad\qquad$ □

Proof (of Theorem 1). The algorithm described in Sect. 4 satisfies Lemma 6 and thus solves approxRuns. The space bound stems from Lemma 3(1): the number of keys in each dictionary, the total number of groups, and the total length of lists of checkpoints are bounded by the number of checkpoints times $\log n$. The size of the watch list is negligible. The time bound follows from Lemmas 8–12. For the insertion/deletion of groups and nodes see Remark 2. $\qquad\qquad\qquad\qquad$ □

In conclusion we say a few words about the choice of dictionaries.

Remark 3. If ε is small (inverse polynomial), it makes sense to use dynamic perfect hash tables [2, 7] as dictionaries. Both cited versions guarantee that with probability $1 - m^{-c}$, where m is the dictionary size and c is an arbitrary constant, all dictionary operations will take $\mathcal{O}(1)$ time. Thus the total probability of a failed run of an algorithm can still be kept below $1/n$ with $\mathcal{O}(\log n)$ elementary operations between reads. However, this is not the case for big (such as constant or inverse polylog) values of ε. So in this case we suggest to use deterministic dictionaries by Anderson and Thorup [1] which give us $\mathcal{O}(\sqrt{\frac{\log\log n}{\log\log\log n}} \cdot \log n)$ elementary operations between reads.

References

1. Andersson, A., Thorup, M.: Dynamic ordered sets with exponential search trees. J. ACM **54**(3), 13 (2007)
2. Arbitman, Y., Naor, M., Segev, G.: De-amortized cuckoo hashing: provable worst-case performance and experimental results. In: Albers, S., Marchetti-Spaccamela, A., Matias, Y., Nikoletseas, S., Thomas, W. (eds.) ICALP 2009. LNCS, vol. 5555, pp. 107–118. Springer, Heidelberg (2009). https://doi.org/10.1007/978-3-642-02927-1_11
3. Bannai, H., I, T., Inenaga, S., Nakashima, Y., Takeda, M., Tsuruta, K.: The "runs" theorem. SIAM J. Comput. **46**(5), 1501–1514 (2017)
4. Breslauer, D., Galil, Z.: Real-time streaming string-matching. In: Giancarlo, R., Manzini, G. (eds.) CPM 2011. LNCS, vol. 6661, pp. 162–172. Springer, Heidelberg (2011). https://doi.org/10.1007/978-3-642-21458-5_15
5. Crochemore, M., Rytter, W.: Squares, cubes, and time-space efficient string searching. Algorithmica **13**, 405–425 (1995)
6. Crochemore, M., et al.: Near-optimal computation of runs over general alphabet via non-crossing LCE queries. In: Inenaga, S., Sadakane, K., Sakai, T. (eds.) SPIRE 2016. LNCS, vol. 9954, pp. 22–34. Springer, Cham (2016). https://doi.org/10.1007/978-3-319-46049-9_3
7. Dietzfelbinger, M., Meyer auf der Heide, F.: Dynamic hashing in real time. In: Buchmann, J., Ganzinger, H., Paul, W.J. (eds.) Informatik. TEUBNER-TEXTE zur Informatik, vol. 1, pp. 95–119. Springer, Wiesbaden (1992). https://doi.org/10.1007/978-3-322-95233-2_7
8. Fine, N.J., Wilf, H.S.: Uniqueness theorems for periodic functions. Proc. Amer. Math. Soc. **16**, 109–114 (1965)
9. Fischer, J., Holub, Š., I, T., Lewenstein, M.: Beyond the runs theorem. In: Iliopoulos, C., Puglisi, S., Yilmaz, E. (eds.) SPIRE 2015. LNCS, vol. 9309, pp. 277–286. Springer, Cham (2015). https://doi.org/10.1007/978-3-319-23826-5_27
10. Gasieniec, L., Kolpakov, R.M., Potapov, I.: Space efficient search for maximal repetitions. Theor. Comput. Sci. **339**(1), 35–48 (2005)
11. Gawrychowski, P., Kociumaka, T., Rytter, W., Walen, T.: Faster longest common extension queries in strings over general alphabets. In: 27th Annual Symposium on Combinatorial Pattern Matching, CPM 2016. LIPIcs, vol. 54, pp. 5:1–5:13. Schloss Dagstuhl - Leibniz-Zentrum fuer Informatik (2016)

12. Gawrychowski, P., Merkurev, O., Shur, A.M., Uznański, P.: Tight tradeoffs for real-time approximation of longest palindromes in streams. In: 27th Annual Symposium on Combinatorial Pattern Matching, CPM 2016. LIPIcs, vol. 54, pp. 18:1–18:13 (2016)

13. Grigorescu, E., Azer, E.S., Zhou, S.: Streaming for aibohphobes: Longest palindrome with mismatches. In: 37th IARCS Annual Conference on Foundations of Software Technology and Theoretical Computer Science, FSTTCS 2017. LIPIcs, vol. 93, pp. 31:1–31:13 (2017)

14. Karp, R.M., Rabin, M.O.: Efficient randomized pattern-matching algorithms. IBM J. Res. Dev. **31**(2), 249–260 (1987)

15. Kolpakov, R.M., Kucherov, G.: Finding maximal repetitions in a word in linear time. In: 40th Annual Symposium on Foundations of Computer Science, FOCS 1999, pp. 596–604. IEEE Computer Society (1999)

16. Kosolobov, D.: Lempel-Ziv factorization may be harder than computing all runs. In: 32nd International Symposium on Theoretical Aspects of Computer Science, STACS 2015, Garching, Germany, 4–7 March 2015. LIPIcs, vol. 30, pp. 582–593. Schloss Dagstuhl - Leibniz-Zentrum fuer Informatik (2015)

17. Kosolobov, D.: Computing runs on a general alphabet. Inf. Process. Lett. **116**(3), 241–244 (2016)

18. Main, M.G.: Detecting leftmost maximal periodicities. Discrete Appl. Math. **25**(1–2), 145–153 (1989)

19. Merkurev, O., Shur, A.M.: Searching long repeats in streams. In: 30th Annual Symposium on Combinatorial Pattern Matching CPM 2019. LIPIcs, vol. 128, pp. 31:1–31:14 (2019)

20. Yao, A.: Probabilistic computations: Toward a unified measure of complexity. In: Proceedings of the 18th IEEE Symposium on Foundations of Computer Science (FOCS), pp. 222–227 (1977)

Weighted Shortest Common Supersequence Problem Revisited

Panagiotis Charalampopoulos[1] ⓘ, Tomasz Kociumaka[2,3] ⓘ, Solon P. Pissis[4] ⓘ,
Jakub Radoszewski[3(✉)] ⓘ, Wojciech Rytter[3] ⓘ, Juliusz Straszyński[3] ⓘ,
Tomasz Waleń[3] ⓘ, and Wiktor Zuba[3] ⓘ

[1] Department of Informatics, King's College London, London, UK
panagiotis.charalampopoulos@kcl.ac.uk
[2] Department of Computer Science, Bar-Ilan University, Ramat Gan, Israel
[3] Institute of Informatics, University of Warsaw, Warsaw, Poland
{kociumaka,jrad,rytter,jks,walen,w.zuba}@mimuw.edu.pl
[4] CWI, Amsterdam, The Netherlands
solon.pissis@cwi.nl

Abstract. A weighted string, also known as a position weight matrix, is a sequence of probability distributions over some alphabet. We revisit the Weighted Shortest Common Supersequence (WSCS) problem, introduced by Amir et al. [SPIRE 2011], that is, the SCS problem on weighted strings. In the WSCS problem, we are given two weighted strings W_1 and W_2 and a threshold $\frac{1}{z}$ on probability, and we are asked to compute the shortest (standard) string S such that both W_1 and W_2 match subsequences of S (not necessarily the same) with probability at least $\frac{1}{z}$. Amir et al. showed that this problem is NP-complete if the probabilities, including the threshold $\frac{1}{z}$, are represented by their logarithms (encoded in binary).

We present an algorithm that solves the WSCS problem for two weighted strings of length n over a constant-sized alphabet in $\mathcal{O}(n^2\sqrt{z} \log z)$ time. Notably, our upper bound matches known conditional lower bounds stating that the WSCS problem cannot be solved in $\mathcal{O}(n^{2-\varepsilon})$ time or in $\mathcal{O}^*(z^{0.5-\varepsilon})$ with time, where the $\mathcal{O}^*$ notation suppresses factors polynomial with respect to the instance size (with numeric values encoded in binary), unless there is a breakthrough improving upon long-standing upper bounds for fundamental NP-hard problems (CNF-SAT and SUBSET SUM, respectively).

We also discover a fundamental difference between the WSCS problem and the Weighted Longest Common Subsequence (WLCS) problem, introduced by Amir et al. [JDA 2010]. We show that the WLCS problem cannot be solved in $\mathcal{O}(n^{f(z)})$ time, for any function $f(z)$, unless P = NP.

Tomasz Kociumaka was supported by ISF grants no. 824/17 and 1278/16 and by an ERC grant MPM under the EU's Horizon 2020 Research and Innovation Programme (grant no. 683064).

Jakub Radoszewski and Juliusz Straszyński were supported by the "Algorithms for text processing with errors and uncertainties" project carried out within the HOMING program of the Foundation for Polish Science co-financed by the European Union under the European Regional Development Fund.

N. R. Brisaboa and S. J. Puglisi (Eds.): SPIRE 2019, LNCS 11811, pp. 221–238, 2019.
https://doi.org/10.1007/978-3-030-32686-9_16

1   Introduction

Consider two strings X and Y. A common supersequence of X and Y is a string S such that X and Y are both subsequences of S. A shortest common supersequence (SCS) of X and Y is a common supersequence of X and Y of minimum length. The SHORTEST COMMON SUPERSEQUENCE problem (the SCS problem, in short) is to compute an SCS of X and Y. The SCS problem is a classic problem in theoretical computer science [18,23,25]. It is solvable in quadratic time using a standard dynamic-programming approach [13], which also allows computing a shortest common supersequence of any constant number of strings (rather than just two) in polynomial time. In case of an arbitrary number of input strings, the problem becomes NP-hard [23] even when the strings are binary [25].

A weighted string of length n over some alphabet Σ is a type of uncertain sequence. The uncertainty at any position of the sequence is modeled using a subset of the alphabet (instead of a single letter), with every element of this subset being associated with an occurrence probability; the probabilities are often represented in an $n \times |\Sigma|$ matrix. These kinds of data are common in various applications where: (i) imprecise data measurements are recorded; (ii) flexible sequence modeling, such as binding profiles of molecular sequences, is required; (iii) observations are private and thus sequences of observations may have artificial uncertainty introduced deliberately [2]. For instance, in computational biology they are known as position weight matrices or position probability matrices [26].

In this paper, we study the WEIGHTED SHORTEST COMMON SUPERSE-QUENCE problem (the WSCS problem, in short) introduced by Amir et al. [5], which is a generalization of the SCS problem for weighted strings. In the WSCS problem, we are given two weighted strings W_1 and W_2 and a probability threshold $\frac{1}{z}$, and the task is to compute the shortest (standard) string such that both W_1 and W_2 match subsequences of S (not necessarily the same) with probability at least $\frac{1}{z}$. In this work, we show the first efficient algorithm for the WSCS problem.

A related problem is the WEIGHTED LONGEST COMMON SUBSEQUENCE problem (the WLCS problem, in short). It was introduced by Amir et al. [4] and further studied in [14] and, very recently, in [20]. In the WLCS problem, we are also given two weighted strings W_1 and W_2 and a threshold $\frac{1}{z}$ on probability, but the task is to compute the longest (standard) string S such that S matches a subsequence of W_1 with probability at least $\frac{1}{z}$ and S matches a subsequence of W_2 with probability at least $\frac{1}{z}$. For standard strings S_1 and S_2, the length of their shortest common supersequence $|\mathrm{SCS}(S_1, S_2)|$ and the length of their longest common subsequence $|\mathrm{LCS}(S_1, S_2)|$ satisfy the following folklore relation:

$$|\mathrm{LCS}(S_1, S_2)| + |\mathrm{SCS}(S_1, S_2)| = |S_1| + |S_2|. \tag{1}$$

However, an analogous relation does not connect the WLCS and WSCS problems, even though both problems are NP-complete because of similar reductions,

which remain valid even in the case that both weighted strings have the same length [4,5]. In this work, we discover an important difference between the two problems.

Kociumaka et al. [21] introduced a problem called WEIGHTED CONSENSUS, which is a special case of the WSCS problem asking whether the WSCS of two weighted strings of length n is of length n, and they showed that the WEIGHTED CONSENSUS problem is NP-complete yet admits an algorithm running in pseudo-polynomial time $\mathcal{O}(n + \sqrt{z} \log z)$ for constant-sized alphabets[1]. Furthermore, it was shown in [21] that the WEIGHTED CONSENSUS problem cannot be solved in $\mathcal{O}^*(z^{0.5-\varepsilon})$ time for any $\varepsilon > 0$ unless there is an $\mathcal{O}^*(2^{(0.5-\varepsilon)n})$-time algorithm for the SUBSET SUM problem. Let us recall that the SUBSET SUM problem, for a set of n integers, asks whether there is a subset summing up to a given integer. Moreover, the $\mathcal{O}^*(2^{n/2})$ running time for the SUBSET SUM problem, achieved by a classic meet-in-the-middle approach of Horowitz and Sahni [15], has not been improved yet despite much effort; see e.g. [6].

Abboud et al. [1] showed that the LONGEST COMMON SUBSEQUENCE problem over constant-sized alphabets cannot be solved in $\mathcal{O}(n^{2-\varepsilon})$ time for $\varepsilon > 0$ unless the Strong Exponential Time Hypothesis [16,17,22] fails. By (1), the same conditional lower bound applies to the SCS problem, and since standard strings are a special case of weighted strings (having one letter occurring with probability equal to 1 at each position), it also applies to the WSCS problem.

The following theorem summarizes the above conditional lower bounds on the WSCS problem.

Theorem 1 (Conditional hardness of the WSCS problem; see [1,21]). *Even in the case of constant-sized alphabets, the* WEIGHTED SHORTEST COMMON SUPERSEQUENCE *problem is NP-complete, and for any $\varepsilon > 0$ it cannot be solved:*

1. *in $\mathcal{O}(n^{2-\varepsilon})$ time unless the Strong Exponential Time Hypothesis fails;*
2. *in $\mathcal{O}^*(z^{0.5-\varepsilon})$ time unless there is an $\mathcal{O}^*(2^{(0.5-\varepsilon)n})$-time algorithm for the* SUBSET SUM *problem.*

Our Results. We give an algorithm for the WSCS problem with pseudo-polynomial running time that depends polynomially on n and z. Note that such algorithms have already been proposed for several problems on weighted strings: pattern matching [9,12,21,24], indexing [3,7,8,11], and finding regularities [10]. In contrast, we show that no such algorithm is likely to exist for the WLCS problem.

Specifically, we develop an $\mathcal{O}(n^2 \sqrt{z} \log z)$-time algorithm for the WSCS problem in the case of a constant-sized alphabet[2]. This upper bound matches the conditional lower bounds of Theorem 1. We then show that unless $P = NP$, the WLCS problem cannot be solved in $\mathcal{O}(n^{f(z)})$ time for any function $f(\cdot)$.

---

[1] Note that in general $z \notin \mathcal{O}^*(1)$ unless z is encoded in unary.

[2] We consider the case of $|\Sigma| = \mathcal{O}(1)$ just for simplicity. For a general alphabet, our algorithm can be modified to work in $\mathcal{O}(n^2|\Sigma|\sqrt{z} \log z)$ time.

Model of Computations. We assume the word RAM model with word size $w = \Omega(\log n + \log z)$. We consider the log-probability representation of weighted sequences, that is, we assume that the non-zero probabilities in the weighted sequences and the threshold probability $\frac{1}{z}$ are all of the form $c^{\frac{p}{2dw}}$, where c and d are constants and p is an integer that fits in $\mathcal{O}(1)$ machine words.

2 Preliminaries

A *weighted string* $W = W[1] \cdots W[n]$ of length $|W| = n$ over alphabet Σ is a sequence of sets of the form

$$W[i] = \{(c,\ \pi_i^{(W)}(c))\ :\ c \in \Sigma\}.$$

Here, $\pi_i^{(W)}(c)$ is the occurrence probability of the letter c at the position $i \in [1 .. n]$.[3] These values are non-negative and sum up to 1 for a given index i.

By $W[i .. j]$ we denote the weighted *substring* $W[i] \cdots W[j]$; it is called a prefix if $i = 1$ and a suffix if $j = |W|$.

The *probability of matching* of a string S with a weighted string W, with $|S| = |W| = n$, is

$$\mathcal{P}(S, W) = \prod_{i=1}^{n} \pi_i^{(W)}(S[i]) = \prod_{i=1}^{n} \mathcal{P}(S[i] = W[i]).$$

We say that a (standard) string S *matches a weighted string W with probability at least* $\frac{1}{z}$, denoted by $S \approx_z W$, if $\mathcal{P}(S, W) \geq \frac{1}{z}$. We also denote

$$\mathsf{Matched}_z(W) = \{S \in \Sigma^n : \mathcal{P}(S, W) \geq \tfrac{1}{z}\}.$$

For a string S we write $W \subseteq_z S$ if $S' \approx_z W$ for some subsequence S' of S. Similarly we write $S \subseteq_z W$ if $S \approx_z W'$ for some subsequence W' of W.

Our main problem can be stated as follows.

Weighted Shortest Common Supersequence ($\mathrm{WSCS}(W_1, W_2, z)$)

Input: Weighted strings W_1 and W_2 of length up to n and a threshold $\frac{1}{z}$.

Output: A shortest standard string S such that $W_1 \subseteq_z S$ and $W_2 \subseteq_z S$.

Example 2. If the alphabet is $\Sigma = \{\mathsf{a}, \mathsf{b}\}$, then we write the weighted string as $W = [p_1, p_2, \ldots, p_n]$, where $p_i = \pi_i^{(W)}(\mathsf{a})$; in other words, p_i is the probability that the ith letter $W[i]$ is a. For

$$W_1 = [1,\ 0.2,\ 0.5],\ W_2 = [0.2,\ 0.5,\ 1],\ \text{and}\ z = \tfrac{5}{2},$$

we have $\mathrm{WSCS}(W_1,\ W_2,\ z) = \mathsf{baba}$ since $W_1 \subseteq_z \underline{\mathsf{ba}}\mathsf{ba}$, $W_2 \subseteq_z \mathsf{ba}\underline{\mathsf{ba}}$ (the witness subsequences are underlined), and baba is a shortest string with this property.

---

[3] For any two integers $\ell \leq r$, we use $[\ell .. r]$ to denote the integer range $\{\ell, \ldots, r\}$.

We first show a simple solution to WSCS based on the following facts.

Observation 3 (Amir et al. [3]). *Every weighted string W matches at most z standard strings with probability at least $\frac{1}{z}$, i.e., $|\mathsf{Matched}_z(W)| \leq z$.*

Lemma 4. *The set $\mathsf{Matched}_z(W)$ can be computed in $\mathcal{O}(nz)$ time if $|\Sigma| = \mathcal{O}(1)$.*

Proof. If $S \in \mathsf{Matched}_z(W)$, then $S[1..i] \in \mathsf{Matched}_z(W[1..i])$ for every index i. Hence, the algorithm computes the sets $\mathsf{Matched}_z$ for subsequent prefixes of W. Each string $S \in \mathsf{Matched}_z(W[1..i])$ is represented as a triple (c, p, S'), where $c = S[i]$ is the last letter of S, $p = \mathcal{P}(S, W[1..i])$, and $S' = S[1..i-1]$ points to an element of $\mathsf{Matched}_z(W[1..i-1])$. Such a triple is represented in $\mathcal{O}(1)$ space.

Assume that $\mathsf{Matched}_z(W[1..i-1])$ has already been computed. Then, for every $S' = (c', p', S'') \in \mathsf{Matched}_z(W[1..i-1])$ and every $c \in \Sigma$, if $p := p' \cdot \pi_i^{(W)}(c) \geq \frac{1}{z}$, then the algorithm adds (c, p, S') to $\mathsf{Matched}_z(W[1..i])$.

By Observation 3, $|\mathsf{Matched}_z(W[1..i-1])| \leq z$ and $|\mathsf{Matched}_z(W[1..i])| \leq z$. Hence, the $\mathcal{O}(nz)$ time complexity follows.   □

Proposition 5. *The WSCS problem can be solved in $\mathcal{O}(n^2 z^2)$ time if $|\Sigma| = \mathcal{O}(1)$.*

Proof. The algorithm builds $\mathsf{Matched}_z(W_1)$ and $\mathsf{Matched}_z(W_2)$ using Lemma 4. These sets have size at most z by Observation 3. The result is the shortest string in

$$\{\mathrm{SCS}(S_1, S_2) \,:\, S_1 \in \mathsf{Matched}_z(W_1),\, S_2 \in \mathsf{Matched}_z(W_2)\}.$$

Recall that the SCS of two strings can be computed in $\mathcal{O}(n^2)$ time using a standard dynamic programming algorithm [13].   □

We substantially improve upon this upper bound in Sects. 3 and 4.

2.1 Meet-in-the-Middle Technique

In the decision version of the KNAPSACK problem, we are given n items with weights w_i and values v_i, and we seek for a subset of items with total weight up to W and total value at least V. In the classic meet-in-the-middle solution to the KNAPSACK problem by Horowitz and Sahni [15], the items are divided into two sets S_1 and S_2 of sizes roughly $\frac{1}{2}n$. Initially, the total value and the total weight is computed for every subset of elements of each set S_i. This results in two sets A, B, each with $\mathcal{O}(2^{n/2})$ pairs of numbers. The algorithm needs to pick a pair from each set such that the first components of the pairs sum up to at most W and the second components sum up to at least V. This problem can be solved in linear time w.r.t. the set sizes provided that the pairs in both sets A and B are sorted by the first component.

Let us introduce a modified version this problem.

MERGE(A, B, w)

Input: Two sets A and B of points in 2 dimensions and a threshold w.

Output: Do there exist $(x_1, y_1) \in A$, $(x_2, y_2) \in B$ such that $x_1 x_2, y_1 y_2 \geq w$?

A linear-time solution to this problem is the same as for the problem in the meet-in-the-middle solution for KNAPSACK. However, for completeness we prove the following lemma (see also [21, Lemma 5.6]):

Lemma 6 (Horowitz and Sahni [15]). *The* MERGE *problem can be solved in linear time assuming that the points in A and B are sorted by the first component.*

Proof. A pair (x, y) is *irrelevant* if there is another pair (x', y') in the same set such that $x' \geq x$ and $y' \geq y$. Observe that removing an irrelevant point from A or B leads to an equivalent instance of the MERGE problem.

Since the points in A and B are sorted by the first component, a single scan through these pairs suffices to remove all irrelevant elements. Next, for each $(x, y) \in A$, the algorithm computes $(x', y') \in B$ such that $x' \geq w/x$ and additionally x' is smallest possible. As the irrelevant elements have been removed from B, this point also maximizes y' among all pairs satisfying $x' \geq w/x$. If the elements (x, y) are processed by non-decreasing values x, the values x' do not increase, and thus the points (x', y') can be computed in $\mathcal{O}(|A| + |B|)$ time in total.   □

3   Dynamic Programming Algorithm for WSCS

Our algorithm is based on dynamic programming. We start with a less efficient procedure and then improve it in the next section. Henceforth, we only consider computing the length of the WSCS; an actual common supersequence of this length can be recovered from the dynamic programming using a standard approach (storing the parent of each state).

For a weighted string W, we introduce a data structure that stores, for every index i, the set $\{\mathcal{P}(S, W[1..i]) : S \in \mathsf{Matched}_z(W[1..i])\}$ represented as an array of size at most z (by Observation 3) with entries in the increasing order. This data structure is further denoted as $Freq_i(W, z)$. Moreover, for each element $p \in Freq_{i+1}(W, z)$ and each letter $c \in \Sigma$, a pointer to $p' = p/\pi_{i+1}^{(W)}(c)$ in $Freq_i(W, z)$ is stored provided that $p' \in Freq_i(W, z)$. A proof of the next lemma is essentially the same as of Lemma 4.

Lemma 7. *For a weighted string W of length n, the arrays $Freq_i(W, z)$, with $i \in [1..n]$, can be constructed in $\mathcal{O}(nz)$ total time if $|\Sigma| = \mathcal{O}(1)$.*

Proof. Assume that $Freq_i(W, z)$ is computed. For every $c \in \Sigma$, we create a list

$$L_c = \{p \cdot \pi_{i+1}^{(W)}(c) : p \in Freq_i(W, z), \, p \cdot \pi_{i+1}^{(W)}(c) \geq \tfrac{1}{z}\}.$$

The lists are sorted since $Freq_i(W, z)$ was sorted. Then $Freq_{i+1}(W, z)$ can be computed by merging all the lists L_c (removing duplicates). This can be done in $\mathcal{O}(z)$ time since $\sigma = \mathcal{O}(1)$. The desired pointers can be computed within the same time complexity. $\qquad\square$

Let us extend the WSCS problem in the following way:

WSCS$'(W_1, W_2, \ell, p, q)$:

Input: Weighted strings W_1, W_2, an integer ℓ, and probabilities p, q.

Output: Is there a string S of length ℓ with subsequences S_1 and S_2 such that $\mathcal{P}(S_1, W_1) = p$ and $\mathcal{P}(S_2, W_2) = q$?

In the following, a *state* in the dynamic programming denotes a quadruple (i, j, ℓ, p), where $i \in [0 .. |W_1|]$, $j \in [0 .. |W_2|]$, $\ell \in [0 .. |W_1| + |W_2|]$, and $p \in Freq_i(W_1, z)$.

Observation 8. *There are $\mathcal{O}(n^3 z)$ states.*

In the dynamic programming, for all states (i, j, ℓ, p), we compute

$$\mathbf{DP}[i, j, \ell, p] = \max\{q : \text{WSCS}'(W_1[1 .. i], W_2[1 .. j], \ell, p, q) = \textbf{true}\}. \qquad (2)$$

Let us denote $\pi_i^k(c) = \pi_i^{(W_k)}(c)$. Initially, the array $\mathbf{DP}$ is filled with zeroes, except that the values $\mathbf{DP}[0, 0, \ell, 1]$ for $\ell \in [0 .. |W_1| + |W_2|]$ are set to 1. In order to cover corner cases, we assume that $\pi_0^1(c) = \pi_0^2(c) = 1$ for any $c \in \Sigma$ and that $\mathbf{DP}[i, j, \ell, p] = 0$ if (i, j, ℓ, p) is not a state. The procedure Compute implementing the dynamic-programming algorithm is shown as Algorithm 1.

Algorithm 1. Compute(W_1, W_2, z)

---

for $\ell = 0$ **to** $|W_1| + |W_2|$ **do**
　　$\mathbf{DP}[0, 0, \ell, 1] := 1$;
foreach *state (i, j, ℓ, p) in lexicographic order* **do**
　　foreach $c \in \Sigma$ **do**
　　　　$x := \pi_i^1(c)$; $y := \pi_j^2(c)$;
　　　　$\mathbf{DP}[i, j, \ell, p] := \max\{$
　　　　　　$\mathbf{DP}[i, j, \ell, p]$,
　　　　　　$\mathbf{DP}[i - 1, j, \ell - 1, \frac{p}{x}]$,
　　　　　　$y \cdot \mathbf{DP}[i, j - 1, \ell - 1, p]$,
　　　　　　$y \cdot \mathbf{DP}[i - 1, j - 1, \ell - 1, \frac{p}{x}]$
　　　　$\}$;
　　return $\min\{\ell : \mathbf{DP}[|W_1|, |W_2|, \ell, p] \geq \frac{1}{z}$ *for some* $p \in Freq_{|W_1|}(W_1, z)\}$;

The correctness of the algorithm is implied by the following lemma:

Lemma 9 (Correctness of Algorithm 1). *The array* **DP** *satisfies* (2). *In particular,* Compute(W_1, W_2, z) = WSCS(W_1, W_2, z).

Proof. The proof that **DP** satisfies (2) goes by induction on $i + j$. The base case of $i + j = 0$ holds trivially. It is simple to verify the cases that $i = 0$ or $j = 0$. Let us henceforth assume that $i > 0$ and $j > 0$.

We first show that

$$\mathbf{DP}[i, j, \ell, p] \leq \max\{q : \text{WSCS}'(W_1[1 \mathinner{.\,.} i], W_2[1 \mathinner{.\,.} j], \ell, p, q) = \mathbf{true}\}.$$

The value $q = \mathbf{DP}[i, j, \ell, p]$ was derived from $\mathbf{DP}[i - 1, j, \ell - 1, p/x] = q$, or $\mathbf{DP}[i, j - 1, \ell - 1, p] = q/y$, or $\mathbf{DP}[i - 1, j - 1, \ell - 1, p/x] = q/y$, where $x = \pi_i^1(c)$ and $y = \pi_j^2(c)$ for some $c \in \Sigma$. In the first case, by the inductive hypothesis, there exists a string T that is a solution to WSCS'($W_1[1 \mathinner{.\,.} i-1], W_2[1 \mathinner{.\,.} j], \ell-1, p/x, q$). That is, T has subsequences T_1 and T_2 such that

$$\mathcal{P}(T_1, W_1[1 \mathinner{.\,.} i - 1]) = p/x \quad \text{and} \quad \mathcal{P}(T_2, W_2[1 \mathinner{.\,.} j]) = q.$$

Then, for $S = Tc$, $S_1 = T_1c$, and $S_2 = T_2$, we indeed have

$$\mathcal{P}(S_1, W_1[1 \mathinner{.\,.} i]) = p \quad \text{and} \quad \mathcal{P}(S_2, W_2[1 \mathinner{.\,.} j]) = q.$$

The two remaining cases are analogous.

Let us now show that

$$\mathbf{DP}[i, j, \ell, p] \geq \max\{q : \text{WSCS}'(W_1[1 \mathinner{.\,.} i], W_2[1 \mathinner{.\,.} j], \ell, p, q) = \mathbf{true}\}.$$

Assume a that string S is a solution to WSCS'($W_1[1 \mathinner{.\,.} i], W_2[1 \mathinner{.\,.} j], \ell, p, q$). Let S_1 and S_2 be the subsequences of S such that $\mathcal{P}(S_1, W_1) = p$ and $\mathcal{P}(S_2, W_2) = q$.

Let us first consider the case that $S_1[i] = S[\ell] \neq S_2[j]$. Then $T_1 = S_1[1 \mathinner{.\,.} i-1]$ and $T_2 = S_2$ are subsequences of $T = S[1 \mathinner{.\,.} \ell - 1]$. We then have

$$p' := \mathcal{P}(T_1, W_1[1 \mathinner{.\,.} i - 1]) = p/\pi_i^1(S_1[i]).$$

By the inductive hypothesis, $\mathbf{DP}[i - 1, j, \ell - 1, p'] \geq q$. Hence, $\mathbf{DP}[i, j, \ell, p] \geq q$ because $\mathbf{DP}[i - 1, j, \ell - 1, p']$ is present as the second argument of the maximum in the dynamic programming algorithm for $c = S[\ell]$.

The cases that $S_1[i] \neq S[\ell] = S_2[j]$ and that $S_1[i] = S[\ell] = S_2[j]$ rely on the values $\mathbf{DP}[i, j - 1, \ell - 1, p] \geq q/y$ and $\mathbf{DP}[i - 1, j - 1, \ell - 1, p/x] \geq q/y$, respectively.

Finally, the case that $S_1[i] \neq S[\ell] \neq S_2[j]$ is reduced to one of the previous cases by changing $S[\ell]$ to $S_1[i]$ so that S is still a supersequence of S_1 and S_2 and a solution to WSCS'($W_1[1 \mathinner{.\,.} i], W_2[1 \mathinner{.\,.} j], \ell, p, q$). $\qquad\square$

Proposition 10. *The* WSCS *problem can be solved in* $\mathcal{O}(n^3 z)$ *time if* $|\Sigma| = \mathcal{O}(1)$.

Proof. The correctness follows from Lemma 9. As noted in Observation 8, the dynamic programming has $\mathcal{O}(n^3 z)$ states. The number of transitions from a single state is constant provided that $|\Sigma| = \mathcal{O}(1)$.

Before running the dynamic programming algorithm of Proposition 10, we construct the data structures $Freq_i(W_1, z)$ for all $i \in [1..n]$ using Lemma 7. The last dimension in the $\mathbf{DP}[i, j, \ell, p]$ array can then be stored as a position in $Freq_i(W_1, z)$. The pointers in the arrays $Freq_i$ are used to follow transitions. □

4 Improvements

4.1 First Improvement: Bounds on ℓ

Our approach here is to reduce the number of states (i, j, ℓ, p) in Algorithm 1 from $\mathcal{O}(n^3 z)$ to $\mathcal{O}(n^2 z \log z)$. This is done by limiting the number of values of ℓ considered for each pair of indices i, j from $\mathcal{O}(n)$ to $\mathcal{O}(\log z)$.

For a weighted string W, we define $\mathcal{H}(W)$ as a standard string generated by taking the most probable letter at each position, breaking ties arbitrarily. The string $\mathcal{H}(W)$ is also called the *heavy* string of W. By $d_H(S, T)$ we denote the Hamming distance of strings S and T. Let us recall an observation from [21].

Observation 11 ([21, **Observation 4.3**]). *If $S \approx_z W$ for a string S and a weighted string W, then $d_H(S, \mathcal{H}(W)) \leq \log_2 z$.*

The lemma below follows from Observation 11.

Lemma 12. *If strings S_1 and S_2 satisfy $S_1 \approx_z W_1$ and $S_2 \approx_z W_2$, then*

$$|\mathrm{SCS}(S_1, S_2) - \mathrm{SCS}(\mathcal{H}(W_1), \mathcal{H}(W_2))| \leq 2 \log_2 z.$$

Proof. By Observation 11,

$$d_H(S_1, \mathcal{H}(W_1)) \leq \log_2 z \quad \text{and} \quad d_H(S_2, \mathcal{H}(W_2)) \leq \log_2 z.$$

Due to the relation (1) between LCS and SCS, it suffices to show the following.

Claim. Let S_1, H_1, S_2, H_2 be strings such that $|S_1| = |H_1|$ and $|S_2| = |H_2|$. If $d_H(S_1, H_1) \leq d$ and $d_H(S_2, H_2) \leq d$, then $|\mathrm{LCS}(S_1, S_2) - \mathrm{LCS}(H_1, H_2)| \leq 2d$.

Proof. Notice that if S_1', S_2' are strings resulting from S_1, S_2 by removing up to d letters from each of them, then $\mathrm{LCS}(S_1', S_2') \geq \mathrm{LCS}(S_1, S_2) - 2d$.

We now create strings S_k' for $k = 1, 2$, by removing from S_k letters at positions i such that $S_k[i] \neq H_k[i]$. Then, according to the observation above, we have

$$\mathrm{LCS}(S_1', S_2') \geq \mathrm{LCS}(S_1, S_2) - 2d.$$

Any common subsequence of S_1' and S_2' is also a common subsequence of H_1 and H_2 since S_1' and S_2' are subsequences of H_1 and H_2, respectively. Consequently,

$$\mathrm{LCS}(H_1, H_2) \geq \mathrm{LCS}(S_1, S_2) - 2d.$$

In a symmetric way, we can show that $\mathrm{LCS}(S_1, S_2) \geq \mathrm{LCS}(H_1, H_2) - 2d$. This completes the proof of the claim. □

We apply the claim for $H_1 = \mathcal{H}(W_1)$, $H_2 = \mathcal{H}(W_2)$, and $d = \log_2 z$.   □

Let us make the following simple observation.

Observation 13. *If $S = \mathrm{WSCS}(W_1, W_2, z)$, then $S = \mathrm{SCS}(S_1, S_2)$ for some strings S_1 and S_2 such that $W_1 \subseteq_z S_1$ and $W_2 \subseteq_z S_2$.*

Using Lemma 12, we refine the previous algorithm as shown in Algorithm 2.

Algorithm 2. Improved1(W_1, W_2, z)

In the beginning, we apply the classic $\mathcal{O}(n^2)$-time dynamic-programming solution to the standard SCS problem on $H_1 = \mathcal{H}(W_1)$ and $H_2 = \mathcal{H}(W_2)$. It computes a 2D array T such that

$$T[i,j] = \mathrm{SCS}(H_1[1 \ldots i], H_2[1 \ldots j]).$$

Let us denote an interval

$$L[i,j] = [T[i,j] - \lfloor 2 \log_2 z \rfloor \ldots T[i,j] + \lfloor 2 \log_2 z \rfloor].$$

We run the dynamic programming algorithm Compute restricted to states (i, j, ℓ, p) with $\ell \in L[i,j]$.

Let $\mathbf{DP}'$ denote the resulting array, restricted to states satisfying $\ell \in L[i,j]$. We return $\min \{\ell : \mathbf{DP}'[|W_1|, |W_2|, \ell, p] \geq \frac{1}{z}$ for some $p \in Freq_{|W_1|}(W_1, z)\}$.

Lemma 14 (Correctness of Algorithm 2). *For every state (i, j, ℓ, p), an inequality $\mathbf{DP}'[i,j,\ell,p] \leq \mathbf{DP}[i,j,\ell,p]$ holds. Moreover, if $S = \mathrm{SCS}(S_1, S_2)$, $|S| = \ell$, $\mathcal{P}(S_1, W_1[1 \ldots i]) = p \geq \frac{1}{z}$ and $\mathcal{P}(S_2, W_2[1 \ldots j]) = q \geq \frac{1}{z}$, then $\mathbf{DP}'[i,j,\ell,p] \geq q$. Consequently, Improved1$(W_1, W_2, z) = \mathrm{WSCS}(W_1, W_2, z)$.*

Proof. A simple induction on $i+j$ shows that the array $\mathbf{DP}'$ is lower bounded by $\mathbf{DP}$. This is because Algorithm 2 is restricted to a subset of states considered by Algorithm 1, and because $\mathbf{DP}'[i,j,\ell,p]$ is assumed to be 0 while $\mathbf{DP}[i,j,\ell,p] \geq 0$ for states (i,j,ℓ,p) ignored in Algorithm 2.

We prove the second part of the statement also by induction on $i + j$. The base cases satisfying $i = 0$ or $j = 0$ can be verified easily, so let us henceforth assume that $i > 0$ and $j > 0$.

First, consider the case that $S_1[i] = S[\ell] \neq S_2[j]$. Let $T = S[1 \ldots \ell - 1]$ and $T_1 = S_1[1 \ldots i - 1]$. We then have

$$p' := \mathcal{P}(T_1, W_1[1 \ldots i - 1]) = p/\pi_i^1(S_1[i]).$$

Claim. If $S_1[i] = S[\ell] \neq S_2[j]$, then $T = \mathrm{SCS}(T_1, S_2)$.

Proof. Let us first show that T is a common supersequence of T_1 and S_2. Indeed, if T_1 was not a subsequence of T, then $T_1 S_1[i] = S_1$ would not be a subsequence of $T S_1[i] = S$, and if S_2 was not a subsequence of T, then it would not be a subsequence of $T S_1[i] = S$ since $S_1[i] \neq S_2[j]$.

Finally, if T_1 and S_2 had a common supersequence T' shorter than T, then $T'S_1[i]$ would be a common supersequence of S_1 and S_2 shorter than S.   □

By the claim and the inductive hypothesis, $\mathbf{DP}'[i-1, j, \ell-1, p'] \geq q$. Hence, $\mathbf{DP}'[i, j, \ell, p] \geq q$ due to the presence of the second argument of the maximum in the dynamic programming algorithm for $c = S[\ell]$. Note that (i, j, ℓ, p) is a state in Algorithm 2 since $\ell \in L[i, j]$ follows from Lemma 12.

The cases that $S_1[i] \neq S[\ell] = S_2[j]$ and that $S_1[i] = S[\ell] = S_2[j]$ use the values $\mathbf{DP}'[i, j-1, \ell-1, p] \geq q/y$ and $\mathbf{DP}'[i-1, j-1, \ell-1, p/x] \geq q/y$, respectively. Finally, the case that $S_1[i] \neq S[\ell] \neq S_2[j]$ is impossible as $S = \mathrm{SCS}(S_1, S_2)$.   □

Example 15. Let $W_1 = [1, 0]$, $W_2 = [0]$ (using the notation from Example 2), and $z \geq 1$. The only strings that match W_1 and W_2 are $S_1 = \mathsf{ab}$ and $S_2 = \mathsf{b}$, respectively. We have $\mathbf{DP}[2, 1, 3, 1] = 1$ which corresponds, in particular, to a solution $S = \mathsf{abb}$ which is not an SCS of S_1 and S_2. However, $\mathbf{DP}[2, 1, 2, 1] = \mathbf{DP}'[2, 1, 2, 1] = 1$ which corresponds to $S = \mathsf{ab} = \mathrm{SCS}(S_1, S_2)$.

Proposition 16. *The* WSCS *problem can be solved in* $\mathcal{O}(n^2 z \log z)$ *time if* $|\Sigma| = \mathcal{O}(1)$.

Proof. The correctness of the algorithm follows from Lemma 14. The number of states is now $\mathcal{O}(n^2 z \log z)$ and thus so is the number of considered transitions. □

4.2   Second Improvement: Meet in the Middle

The second improvement is to apply a meet-in-the-middle approach, which is possible due to following observation resembling Observation 6.6 in [21].

Observation 17. *If* $S \approx_z W$ *for a string* S *and weighted string* W *of length* n, *then there exists a position* $i \in [1..n]$ *such that*

$$S[1..i-1] \approx_{\sqrt{z}} W[1..i-1] \quad and \quad S[i+1..n] \approx_{\sqrt{z}} W[i+1..n].$$

Proof. Select i as the maximum index with $S[1..i-1] \approx_{\sqrt{z}} W[1..i-1]$.   □

We first use dynamic programming to compute two arrays, $\overrightarrow{\mathbf{DP}}$ and $\overleftarrow{\mathbf{DP}}$. The array $\overrightarrow{\mathbf{DP}}$ contains a subset of states from $\mathbf{DP}'$; namely the ones that satisfy $p \geq \frac{1}{\sqrt{z}}$. The array $\overleftarrow{\mathbf{DP}}$ is an analogous array defined for suffixes of W_1 and W_2. Formally, we compute $\overrightarrow{\mathbf{DP}}$ for the reversals of W_1 and W_2, denoted as $\overrightarrow{\mathbf{DP}}^R$, and set $\overleftarrow{\mathbf{DP}}[i, j, \ell, p] = \overrightarrow{\mathbf{DP}}^R[|W_1|+1-i, |W_2|+1-j, \ell, p]$. Proposition 16 yields

Observation 18. *Arrays* $\overrightarrow{\mathbf{DP}}$ *and* $\overleftarrow{\mathbf{DP}}$ *can be computed in* $\mathcal{O}(n^2 \sqrt{z} \log z)$ *time.*

Henceforth, we consider only a simpler case in which there exists a solution S to $\mathrm{WSCS}(W_1, W_2, z)$ with a decomposition $S = S_L \cdot S_R$ such that

$$W_1[1 \mathinner{.\,.} i] \subseteq_{\sqrt{z}} S_L \quad \text{and} \quad W_1[i+1 \mathinner{.\,.} |W_1|] \subseteq_{\sqrt{z}} S_R \tag{3}$$

holds for some $i \in [0 \mathinner{.\,.} |W_1|]$.

In the pseudocode, we use the array $L[i,j]$ from the first improvement, denoted here as $\overrightarrow{L}[i,j]$, and a symmetric array $\overleftarrow{L}$ from right to left, i.e.:

$$\overleftarrow{T}[i,j] = \mathrm{SCS}(\mathcal{H}(W_1)[i \mathinner{.\,.} |W_1|], \mathcal{H}(W_2)[j \mathinner{.\,.} |W_2|]),$$
$$\overleftarrow{L}[i,j] = [\overleftarrow{T}[i,j] - \lfloor 2\log_2 z \rfloor \mathinner{.\,.} \overleftarrow{T}[i,j] + \lfloor 2\log_2 z \rfloor].$$

Algorithm 3 is applied for every $i \in [0 \mathinner{.\,.} |W_1|]$ and $j \in [0 \mathinner{.\,.} |W_2|]$.

Algorithm 3. Improved2(W_1, W_2, z, i, j)

$res := \infty$;
foreach $\ell_L \in \overrightarrow{L}[i,j],\ \ell_R \in \overleftarrow{L}[i+1, j+1]$ **do**
  $A := \{(p,q) : \overrightarrow{\mathbf{DP}}[i,j,\ell_L,p] = q\}$;
  $B := \{(p,q) : \overleftarrow{\mathbf{DP}}[i+1, j+1, \ell_R, p] = q\}$;
  if MERGE(A, B, z) **then**
    $res := \min(res, \ell_L + \ell_R)$;
return res;

Lemma 19 (Correctness of Algorithm 3). *Assuming that there is a solution S to $\mathrm{WSCS}(W_1, W_2, z)$ that satisfies (3), we have*

$$\mathrm{WSCS}(W_1, W_2, z) = \min_{i,j}(\mathsf{Improved2}(W_1, W_2, z, i, j)).$$

Proof. Assume that $\mathrm{WSCS}(W_1, W_2, z)$ has a solution $S = S_L \cdot S_R$ that satisfies (3) for some $i \in [0 \mathinner{.\,.} |W_1|]$ and denote $\ell_L = |S_L|$, $\ell_R = |S_R|$. Let S'_L and S'_R be subsequences of S_L and S_R such that

$$p_L := \mathcal{P}(S'_L, W_1[1 \mathinner{.\,.} i]) \geq \tfrac{1}{\sqrt{z}} \quad \text{and} \quad p_R := \mathcal{P}(S'_R, W_1[i+1 \mathinner{.\,.} |W_1|]) \geq \tfrac{1}{\sqrt{z}}.$$

Let S''_L and S''_R be subsequences of S_L and S_R such that

$$\mathcal{P}(S''_L, W_2[1 \mathinner{.\,.} j]) = q_L \quad \text{and} \quad \mathcal{P}(S''_R, W_2[j+1 \mathinner{.\,.} |W_2|]) = q_R$$

for some j and $q_L q_R \geq \tfrac{1}{z}$.

By Lemma 14, $\overrightarrow{\mathbf{DP}}[i, j, \ell_L, p_L] \geq q_L$ and $\overleftarrow{\mathbf{DP}}[i+1, j+1, \ell_R, p_R] \geq q_R$. Hence, the set A will contain a pair (p_L, q'_L) such that $q'_L \geq q_L$ and the set B will contain a pair (p_R, q'_R) such that $q'_R \geq q_R$. Consequently, MERGE(A, B, z) will return a positive answer.

Similarly, if $\text{MERGE}(A, B, z)$ returns a positive answer for given i, j, ℓ_L and ℓ_R, then

$$\overrightarrow{\mathbf{DP}}[i, j, \ell_L, p_L] \geq q_L \quad \text{and} \quad \overleftarrow{\mathbf{DP}}[i+1, j+1, \ell_R, p_R] \geq q_R$$

for some $p_L p_R, q_L q_R \geq \frac{1}{z}$. By Lemma 14, this implies that

$$\text{WSCS}'(W_1[1 .. i], W_2[1 .. j], \ell_L, p_L, q_L)$$

and

$$\text{WSCS}'(W_1[i+1 .. |W_1|], W_2[j+1 .. |W_2|], \ell_R, p_R, q_R)$$

have a positive answer, so

$$\text{WSCS}'(W_1, W_2, \ell_L + \ell_R, p_L p_R, q_L q_R)$$

has a positive answer too. Due to $p_L p_R, q_L q_R \geq \frac{1}{z}$, this completes the proof. $\quad\square$

Proposition 20. *The WSCS problem can be solved in $\mathcal{O}(n^2 \sqrt{z} \log^2 z)$ time if $|\Sigma| = \mathcal{O}(1)$.*

Proof. We use the algorithm Improved2, whose correctness follows from Lemma 19 in case (3) is satisfied. The general case of Observation 17 requires only a minor technical change to the algorithm. Namely, the computation of $\overrightarrow{\mathbf{DP}}$ then additionally includes all states (i, j, ℓ, p) such that $\ell \in \overrightarrow{L}[i, j]$, $p \geq \frac{1}{z}$, and $p = \pi_i^1(c)p'$ for some $c \in \Sigma$ and $p' \in Freq_{i-1}(W_1, \sqrt{z})$. Due to $|\Sigma| = \mathcal{O}(1)$, the number of such states is still $\mathcal{O}(n^2 \sqrt{z} \log z)$.

For every i and j, the algorithm solves $\mathcal{O}(\log^2 z)$ instances of MERGE, each of size $\mathcal{O}(\sqrt{z})$. This results in the total running time of $\mathcal{O}(n^2 \sqrt{z} \log^2 z)$. $\quad\square$

4.3   Third Improvement: Removing One log z Factor

The final improvement is obtained by a structural transformation after which we only need to consider $\mathcal{O}(\log z)$ pairs (ℓ_L, ℓ_R).

For this to be possible, we compute prefix maxima on the ℓ-dimension of the $\overrightarrow{\mathbf{DP}}$ and $\overleftarrow{\mathbf{DP}}$ arrays in order to guarantee monotonicity. That is, if $\text{MERGE}(A, B, z)$ returns true for ℓ_L and ℓ_R, then we make sure that it would also return true if any of these two lengths increased (within the corresponding intervals).

This lets us compute, for every $\ell_L \in \overrightarrow{L}[i, j]$ the smallest $\ell_R \in \overleftarrow{L}[i, j]$ such that $\text{MERGE}(A, B, z)$ returns true using $\mathcal{O}(\log z)$ iterations because the sought ℓ_R may only decrease as ℓ_L increases. The pseudocode is given in Algorithm 4.

Algorithm 4. Improved3(W_1, W_2, z, i, j)

---

foreach *state* (i, j, ℓ, p) *of* $\overrightarrow{\mathbf{DP}}$ *in lexicographic order* **do**
$\quad \overrightarrow{\mathbf{DP}}[i, j, \ell, p] := \max(\overrightarrow{\mathbf{DP}}[i, j, \ell, p], \overrightarrow{\mathbf{DP}}[i, j, \ell - 1, p]);$

foreach *state* (i, j, ℓ, p) *of* $\overleftarrow{\mathbf{DP}}$ *in lexicographic order* **do**
$\quad \overleftarrow{\mathbf{DP}}[i, j, \ell, p] := \max(\overleftarrow{\mathbf{DP}}[i, j, \ell, p], \overleftarrow{\mathbf{DP}}[i, j, \ell - 1, p]);$

$[a \mathinner{\ldotp\ldotp} b] := \overrightarrow{L}[i, j]; \; [a' \mathinner{\ldotp\ldotp} b'] := \overleftarrow{L}[i + 1, j + 1];$

$\ell_L := a; \; \ell_R := b' + 1; \; res := \infty;$

while $\ell_L \leq b$ **and** $\ell_R \geq a'$ **do**
$\quad A := \{(p, q) : \overrightarrow{\mathbf{DP}}[i, j, \ell_L, p] = q\};$
$\quad B := \{(p, q) : \overleftarrow{\mathbf{DP}}[i + 1, j + 1, \ell_R - 1, p] = q\};$
$\quad$ **if** MERGE(A, B, z) **then**   ▷ ℓ_R is too large for the current ℓ_L
$\qquad \ell_R := \ell_R - 1;$
$\quad$ **else**   ▷ ℓ_R reached the target value for the current ℓ_L
$\qquad$ **if** $\ell_R \leq b'$ **then** $res := \min(res, \ell_L + \ell_R);$
$\qquad \ell_L := \ell_L + 1;$
return *res*;

Theorem 21. *The* WSCS *problem can be solved in* $\mathcal{O}(n^2 \sqrt{z} \log z)$ *time if* $|\Sigma| = \mathcal{O}(1)$.

Proof. Let us fix indices i and j. Let us denote $Freq_i(W, z)$ by $\overrightarrow{Freq}_i(W, z)$ and introduce a symmetric array

$$\overleftarrow{Freq}_i(W, z) = \{\mathcal{P}(S, W[i \mathinner{\ldotp\ldotp} |W|]) : S \in \mathsf{Matched}_z(W[i \mathinner{\ldotp\ldotp} |W|])\}.$$

In the first loop of prefix maxima computation, we consider all $\ell \in \overrightarrow{L}[i, j]$ and $p \in \overrightarrow{Freq}_i(W_1, \sqrt{z})$, and in the second loop, all $\ell \in \overleftarrow{L}[i, j]$ and $p \in \overleftarrow{Freq}_i(W_1, \sqrt{z})$. Hence, prefix maxima take $\mathcal{O}(\sqrt{z} \log z)$ time to compute.

Each step of the while-loop in Improved3 increases ℓ_L or decreases ℓ_R. Hence, the algorithm produces only $\mathcal{O}(\log z)$ instances of MERGE, each of size $\mathcal{O}(\sqrt{z})$. The time complexity follows.   □

5   Lower Bound for WLCS

Let us first define the WLCS problem as it was stated in [4,14].

WEIGHTED LONGEST COMMON SUBSEQUENCE (WLCS($W_1, W_2, z)$)
Input: Weighted strings W_1 and W_2 of length up to n and a threshold $\frac{1}{z}$.
Output: A longest standard string S such that $S \subseteq_z W_1$ and $S \subseteq_z W_2$.

We consider the following well-known NP-complete problem [19]:

SUBSET SUM

Input: A set S of positive integers and a positive integer t.

Output: Is there a subset of S whose elements sum up to t?

Theorem 22. *The* WLCS *problem cannot be solved in* $\mathcal{O}(n^{f(z)})$ *time if* $\mathrm{P} \neq \mathrm{NP}$.

Proof. We show the hardness result by reducing the NP-complete SUBSET SUM problem to the WLCS problem with a constant value of z.

For a set $S = \{s_1, s_2, \ldots, s_n\}$ of n positive integers, a positive integer t, and an additional parameter $p \in [2 \mathinner{.\,.} n]$, we construct two weighted strings W_1 and W_2 over the alphabet $\Sigma = \{\mathsf{a}, \mathsf{b}\}$, each of length n^2.

Let $q_i = \frac{s_i}{t}$. At positions $i \cdot n$, for all $i = [1 \mathinner{.\,.} n]$, the weighted string W_1 contains letter a with probability 2^{-q_i} and b otherwise, while W_2 contains a with probability $2^{\frac{1}{p-1}(q_i-1)}$ and b otherwise. All the other positions contain letter b with probability 1. We set $z = 2$.

We assume that S contains only elements smaller than t (we can ignore the larger ones and if there is an element equal to t, then there is no need for a reduction). All the weights of a are then in the interval $(\frac{1}{2}, 1)$ since $-q_i \in (-1, 0)$ and $\frac{1}{p-1}(q_i - 1) \in (-1, 0)$. Thus, since $z = 2$, letter b originating from a position $i \cdot n$ can never occur in a subsequence of W_1 or in a subsequence of W_2. Hence, every common subsequence of W_1 and W_2 is a subsequence of $(\mathsf{b}^{n-1}\mathsf{a})^n$.

For $I \subseteq [1 \mathinner{.\,.} n]$, we have

$$\prod_{i \in I} \pi_{i \cdot n}^{(W_1)}(\mathsf{a}) = \prod_{i \in I} 2^{-s_i/t} \geq 2^{-1} = \tfrac{1}{z} \iff \sum_{i \in I} s_i \leq t$$

and

$$\prod_{i \in I} \pi_{i \cdot n}^{(W_2)}(\mathsf{a}) = \prod_{i \in I} 2^{\frac{1}{p-1}(s_i/t-1)} \geq 2^{-1} = \tfrac{1}{z} \iff$$

$$\frac{1}{t(p-1)} \left(\sum_{i \in I} s_i \right) - \frac{|I|}{p-1} \geq -1 \iff \sum_{i \in I} s_i \geq t(1 - p + |I|).$$

If I is a solution to the instance of the SUBSET SUM problem, then for $p = |I|$ there is a weighted common subsequence of length $n(n-1) + p$ obtained by choosing all the letters b and the letters a that correspond to the elements of I.

Conversely, suppose that the constructed WLCS instance with a parameter $p \in [2 \mathinner{.\,.} n]$ has a solution of length at least $n(n-1)+p$. Notice that a at position $i \cdot n$ in W_1 may be matched against a at position $i' \cdot n$ in W_2 only if $i = i'$. (Otherwise, the length of the subsequence would be at most $(n - |i - i'|)n \leq (n-1)n < n(n-1) + p$.) Consequently, the solution yields a subset $I \subseteq [1 \mathinner{.\,.} n]$ of at least p indices i such that a at position $i \cdot n$ in W_1 is matched against a at position $i \cdot n$ in W_2. By the relations above, we have (a) $|I| \geq p$, (b) $\sum_{i \in I} s_i \leq t$,

and (c) $\sum_{i \in I} s_i \geq t(1 - p + |I|)$. Combining these three inequalities, we obtain $\sum_{i \in I} s_i = t$ and conclude that the SUBSET SUM instance has a solution.

Hence, the SUBSET SUM instance has a solution if and only if there exists $p \in [2 .. n]$ such that the constructed WLCS instance with p has a solution of length at least $n(n-1) + p$. This concludes that an $\mathcal{O}(n^{f(z)})$-time algorithm for the WLCS problem implies the existence of an $\mathcal{O}(n^{2f(2)+1}) = \mathcal{O}(n^{\mathcal{O}(1)})$-time algorithm for the SUBSET SUM problem. The latter would yield $P = NP$.   □

Example 23. For $S = \{3, 7, 11, 15, 21\}$ and $t = 25 = 3 + 7 + 15$, both weighted strings W_1 and W_2 are of the form:

$$\mathsf{b^4 * b^4 * b^4 * b^4 * b^4 *},$$

where each $*$ is equal to either a or b with different probabilities.

The probabilities of choosing a's for W_1 are equal respectively to

$$\left(2^{-\frac{3}{25}}, 2^{-\frac{7}{25}}, 2^{-\frac{11}{25}}, 2^{-\frac{15}{25}}, 2^{-\frac{21}{25}}\right),$$

while for W_2 they depend on the value of p, and are equal respectively to

$$\left(2^{-\frac{22}{25(p-1)}}, 2^{-\frac{18}{25(p-1)}}, 2^{-\frac{14}{25(p-1)}}, 2^{-\frac{10}{25(p-1)}}, 2^{-\frac{4}{25(p-1)}}\right).$$

For $p = 3$, we have: $\mathrm{WLCS}(W_1, W_2, 2) = \mathsf{b^4\,a\,b^4\,a\,b^4\,b^4\,a\,b^4}$, which corresponds to taking the first, the second, and the fourth a. The length of this string is equal to $23 = n(n-1) + p$, and its probability of matching is $\frac{1}{2} = 2^{-\frac{22}{50}} \cdot 2^{-\frac{18}{50}} \cdot 2^{-\frac{10}{50}}$. Thus, the subset $\{3, 7, 15\}$ of S consisting of its first, second, and fourth element is a solution to the SUBSET SUM problem.

References

1. Abboud, A., Backurs, A., Williams, V.V.: Tight hardness results for LCS and other sequence similarity measures. In: Guruswami, V. (ed.) 56th IEEE Annual Symposium on Foundations of Computer Science, pp. 59–78. IEEE Computer Society (2015). https://doi.org/10.1109/FOCS.2015.14
2. Aggarwal, C.C., Yu, P.S.: A survey of uncertain data algorithms and applications. IEEE Trans. Knowl. Data Eng. **21**(5), 609–623 (2009). https://doi.org/10.1109/TKDE.2008.190
3. Amir, A., Chencinski, E., Iliopoulos, C.S., Kopelowitz, T., Zhang, H.: Property matching and weighted matching. Theor. Comput. Sci. **395**(2–3), 298–310 (2008). https://doi.org/10.1016/j.tcs.2008.01.006
4. Amir, A., Gotthilf, Z., Shalom, B.R.: Weighted LCS. J. Discrete Algorithms 8(3), 273–281 (2010). https://doi.org/10.1016/j.jda.2010.02.001
5. Amir, A., Gotthilf, Z., Shalom, B.R.: Weighted shortest common supersequence. In: Grossi, R., Sebastiani, F., Silvestri, F. (eds.) SPIRE 2011. LNCS, vol. 7024, pp. 44–54. Springer, Heidelberg (2011). https://doi.org/10.1007/978-3-642-24583-1_6
6. Bansal, N., Garg, S., Nederlof, J., Vyas, N.: Faster space-efficient algorithms for subset sum, k-sum, and related problems. SIAM J. Comput. **47**(5), 1755–1777 (2018). https://doi.org/10.1137/17M1158203

7. Barton, C., Kociumaka, T., Liu, C., Pissis, S.P., Radoszewski, J.: Indexing weighted sequences: neat and efficient. Inf. Comput. (2019). https://doi.org/10.1016/j.ic.2019.104462
8. Barton, C., Kociumaka, T., Pissis, S.P., Radoszewski, J.: Efficient index for weighted sequences. In: Grossi, R., Lewenstein, M. (eds.) 27th Annual Symposium on Combinatorial Pattern Matching, CPM 2016. LIPIcs, vol. 54, pp. 4:1–4:13. Schloss Dagstuhl-Leibniz-Zentrum für Informatik (2016). https://doi.org/10.4230/LIPIcs.CPM.2016.4
9. Barton, C., Liu, C., Pissis, S.P.: Linear-time computation of prefix table for weighted strings & applications. Theor. Comput. Sci. **656**, 160–172 (2016). https://doi.org/10.1016/j.tcs.2016.04.029
10. Barton, C., Pissis, S.P.: Crochemore's partitioning on weighted strings and applications. Algorithmica **80**(2), 496–514 (2018). https://doi.org/10.1007/s00453-016-0266-0
11. Charalampopoulos, P., Iliopoulos, C.S., Liu, C., Pissis, S.P.: Property suffix array with applications. In: Bender, M.A., Farach-Colton, M., Mosteiro, M.A. (eds.) LATIN 2018. LNCS, vol. 10807, pp. 290–302. Springer, Cham (2018). https://doi.org/10.1007/978-3-319-77404-6_22
12. Charalampopoulos, P., Iliopoulos, C.S., Pissis, S.P., Radoszewski, J.: On-line weighted pattern matching. Inf. Comput. **266**, 49–59 (2019). https://doi.org/10.1016/j.ic.2019.01.001
13. Cormen, T.H., Leiserson, C.E., Rivest, R.L., Stein, C.: Introduction to Algorithms, 3rd edn. MIT Press (2009). https://mitpress.mit.edu/books/introduction-algorithms-third-edition
14. Cygan, M., Kubica, M., Radoszewski, J., Rytter, W., Waleń, T.: Polynomial-time approximation algorithms for weighted LCS problem. Discrete Appl. Math. **204**, 38–48 (2016). https://doi.org/10.1016/j.dam.2015.11.011
15. Horowitz, E., Sahni, S.: Computing partitions with applications to the knapsack problem. J. ACM **21**(2), 277–292 (1974). https://doi.org/10.1145/321812.321823
16. Impagliazzo, R., Paturi, R.: On the complexity of k-SAT. J. Comput. Syst. Sci. **62**(2), 367–375 (2001). https://doi.org/10.1006/jcss.2000.1727
17. Impagliazzo, R., Paturi, R., Zane, F.: Which problems have strongly exponential complexity? J. Comput. Syst. Sci. **63**(4), 512–530 (2001). https://doi.org/10.1006/jcss.2001.1774
18. Jiang, T., Li, M.: On the approximation of shortest common supersequences and longest common subsequences. SIAM J. Comput. **24**(5), 1122–1139 (1995). https://doi.org/10.1137/S009753979223842X
19. Karp, R.M.: Reducibility among combinatorial problems. In: Miller, R.E., Thatcher, J.W. (eds.) Symposium on the Complexity of Computer Computations. pp. 85–103. The IBM Research Symposia Series, Plenum Press, New York (1972). https://doi.org/10.1007/978-1-4684-2001-2_9
20. Kipouridis, E., Tsichlas, K.: Longest common subsequence on weighted sequences (2019). http://arxiv.org/abs/1901.04068
21. Kociumaka, T., Pissis, S.P., Radoszewski, J.: Pattern matching and consensus problems on weighted sequences and profiles. Theory Comput. Syst. **63**(3), 506–542 (2019). https://doi.org/10.1007/s00224-018-9881-2
22. Lokshtanov, D., Marx, D., Saurabh, S.: Lower bounds based on the Exponential Time Hypothesis. Bull. EATCS **105**, 41–72 (2011). http://eatcs.org/beatcs/index.php/beatcs/article/view/92
23. Maier, D.: The complexity of some problems on subsequences and supersequences. J. ACM **25**(2), 322–336 (1978). https://doi.org/10.1145/322063.322075

24. Radoszewski, J., Starikovskaya, T.: Streaming k-mismatch with error correcting and applications. In: Bilgin, A., Marcellin, M.W., Serra-Sagristà, J., Storer, J.A. (eds.) Data Compression Conference, DCC 2017, pp. 290–299. IEEE (2017). https://doi.org/10.1109/DCC.2017.14
25. Räihä, K., Ukkonen, E.: The shortest common supersequence problem over binary alphabet is NP-complete. Theor. Comput. Sci. **16**, 187–198 (1981). https://doi.org/10.1016/0304-3975(81)90075-X
26. Stormo, G.D., Schneider, T.D., Gold, L., Ehrenfeucht, A.: Use of the 'perceptron' algorithm to distinguish translational initiation sites in E. coli. Nucl. Acids Res. **10**(9), 2997–3011 (1982). https://doi.org/10.1093/nar/10.9.2997

Algorithms

Fast Identification of Heavy Hitters
by Cached and Packed Group Testing

Yusaku Kaneta[1]($\boxtimes$), Takeaki Uno[2], and Hiroki Arimura[3]

[1] Autonomous Networking Research and Innovation Department, Rakuten Mobile,
Inc. and Rakuten Institute of Technology, Rakuten, Inc., Tokyo, Japan
yusaku.kaneta@rakuten.com
[2] National Institute of Informatics, Tokyo, Japan
uno@nii.ac.jp
[3] IST, Hokkaido University, Sapporo, Japan
arim@ist.hokudai.ac.jp

Abstract. The ϵ-approximate ϕ-heavy hitters problem is, for any element from some universe $\mathbb{U} = [0..n)$, to maintain its frequency under an
arbitrary data stream of form $(x_i, \Delta_i) \in \mathbb{U} \times \mathbb{Z}$ that changes the frequency
of x_i by Δ_i, such that one can output every element with frequency
more than ϕN and no element with frequency no more than $(\phi - \epsilon)N$ for
$N = \sum_i \Delta_i$ and prespecified parameters $\epsilon, \phi \in \mathbb{R}$. To solve this problem
in small space, Cormode and Muthukrishnan (ACM TODS, 2005) have
proposed an $O(\rho\epsilon^{-1} \lg n)$-space probabilistic data structure with good
practical performance, where $\rho = \lg(1/(\delta\phi))$ for any failure probability
$\delta \in \mathbb{R}$. In this paper, we improve its output time from $O(\rho\epsilon^{-1}(\lg n + \rho))$
to $O(\rho^2\epsilon^{-1})$ for arbitrary updates ($\Delta_i \in \mathbb{Z}$) and its update time from
$O(\rho \lg n)$ to amortized $O(\rho)$ for constant updates ($\Delta_i \in O(1)$) with the
same space and output guarantee by removing application-specific $\lg n$
terms that are not tunable, unlike other parameters δ, ϵ, and ϕ.

1 Introduction

Identifying *heavy hitters* (also known as frequent items [7], hot items [10], or top-
k items [23]) is one of the most fundamental tasks in data stream models, where
the goal is, given some real $\phi \in \mathbb{R}$ and data stream $\mathcal{S}$ of form $\mathcal{S}[i] = (x_i, \Delta_i)$
for $i \in [1, |\mathcal{S}|]$ as input, to output every element x, called a *ϕ-heavy hitter*, with
frequency more than ϕN for $N = \sum_{i=1}^{|\mathcal{S}|} \Delta_i$. This task has attracted considerable
attention for a variety of its applications like search query analysis [27], network
anomaly detection [1,28,30], multidimensional data analysis [21], malicious event
detection [3], cache management [13], and so on. For more details on this topic, we
refer the reader to an excellent survey by Cormode and Hadjieleftheriou [8]. This
task is also related to the *α-majority problem* that has been extensively studied
by the string processing community in various settings [2,12,15,16,20,26], where
$\Delta_i = 1$ holds for any i and $\mathcal{S}$ is regarded as a string of symbols.

In this paper, we study the *ϵ-approximate ϕ-heavy hitters problem*, one of
the most well studied variants of this task, in the strict turnstile model, i.e.,

© Springer Nature Switzerland AG 2019
N. R. Brisaboa and S. J. Puglisi (Eds.): SPIRE 2019, LNCS 11811, pp. 241–257, 2019.
https://doi.org/10.1007/978-3-030-32686-9_17

Table 1. Summary of previous [10] and our results that solve the ϵ-approximate ϕ-heavy hitters problem with probability $1 - \delta$ based on the idea of combinatorial group testing (CGT), in terms of update/query time, where n is the universe size, $\Delta \in \mathbb{Z}$ is update weights supported, $\rho = \lg(1/(\delta\phi))$. All the results use $O(\rho\epsilon^{-1}\lg n)$ space.

Techniques	Update time	Query time	Remark
CGT [10]	$O(\rho\lg n)$	$O(\rho\epsilon^{-1}(\lg n + \rho))^{\text{a}}$	$\|\Delta\| \in \mathbb{N}$
This work (Sect. 4.1)	$O(\rho\lg n)$	$O(\rho^2\epsilon^{-1})$	$\|\Delta\| \in \mathbb{N}$
This work (Sect. 4.2)	$O(\rho)$ amortized	$O(\rho^2\epsilon^{-1})$	$\|\Delta\| \in O(1)$

[a]Note that the authors of [10] claim $O(\rho^2\epsilon^{-1}\lg n)$ query time. We argue that it is too pessimistic and $O(\rho\epsilon^{-1}(\lg n + \rho))$ query time is possible.

each Δ_i can be both positive and negative and any frequency never becomes negative. Hereafter, we give its formal definition and state our main contributions. Throughout this paper, $\mathbb{U} = [0..n)$ denotes the universe of size n from which each element x_i is chosen. As our model of computation, we assume the word RAM with word size $w \geq \max\{n, N\}$ for the universe size n and the total frequency N.

1.1 Heavy Hitters Problem

Let x be an element in the universe $\mathbb{U} = [0..n)$. For any integer $i \geq 0$, we denote by $n_x(i)$ the frequency of x at time i (i.e., after i updates complete) and, if i is clear from the context, simply denote $n_x(i)$ as n_x. In this paper, we assume the strict turnstile model [25] and an input data stream $\mathcal{S}$ of length $|\mathcal{S}|$ of form $\mathcal{S}[i] = (x_i, \Delta_i) \in \mathbb{U} \times \mathbb{Z}$ that changes the frequency of x_i by Δ_i, where $n_x(i) \geq 0$ holds for any element $x \in \mathbb{U}$ and time $i \geq 0$. The data stream $\mathcal{S}$ is said to be (a data stream with) *unitary updates* if $\Delta_i \in \{\pm 1\}$ and *constant updates* if $\Delta_i \in O(1)$ for any i. Given any Boolean expression E, we define $[E] = 1$ if E is true and $[E] = 0$ otherwise. Using this notation, $n_x = \sum_i [x = x_i] \cdot \Delta_i$ holds.

Given some real $\phi \in \mathbb{R}$, an element $x \in \mathbb{U}$ is said to be a ϕ-*heavy hitter* if it occurs more than ϕN times in $\mathcal{S}$, i.e., $n_x > \phi N$. The ϵ-approximate ϕ-heavy hitters problem (or simply the *heavy hitters problem*) is then defined as follows:

Definition 1 (The heavy hitters problem). Let ϵ and ϕ be some real numbers with $0 < \epsilon \leq \phi \leq 1/2$. The heavy hitters problem is to implement two operations:

update(x, Δ): updates n_x with $n_x + \Delta$ for $(x, \Delta) \in \mathbb{U} \times \mathbb{Z}$.
    query(): outputs every ϕ-heavy hitter and no other that occurs no more than $(\phi - \epsilon)N$ times.

1.2 Main Contributions

In this paper, we study the ϵ-approximate ϕ-heavy hitters problem with an emphasis on its practical aspects. Specifically, we revisit a randomized data structure of Cormode and Muthukrishnan [10], called *Combinatorial Group Testing*

(CGT for short), which is shown to be very competitive in practice by extensive experiments in [8].

As our main result, we develop two novel techniques that remove its dependence on the $\lg n$ term, which can be a bottleneck in practice as explained later, from its query and update times while keeping the same asymptotic space and output guarantee. Table 1 compares the update and query times of [10] and those of ours. Our main contributions are summarized as follows for any failure probability $\delta \in \mathbb{R}$ and $\rho = \lg(1/(\delta\phi))$:

1. We improve the query time of [10] from $O(\rho\epsilon^{-1}(\lg n + \rho))$ to $O(\rho^2\epsilon^{-1})$. Our idea is to identify candidates of heavy hitters at update time and to verify their frequency at query time, while [10] does both at query time. (Sect. 4.1)
2. We improve the update time of [10] from $O(\rho \lg n)$ to amortized $O(\rho)$ for constant updates. Our idea is to maintain an array of $\lg n$ bidirectional counters used by [10] efficiently by exploiting the word-level parallelism. (Sect. 4.2)
3. We conducted experiments on synthetic datasets and showed that, in various settings, our method for unitary updates outperformed previous competitive ones [9,10] in query time with comparable update time and space. (Sect. 5)

We argue that the $\lg n$ term in update and query times of [10] (and [9]) can be the main bottleneck in practice because it is application-specific and not flexibly tunable, unlike other parameters δ, ϵ, and ϕ, according to available computational resources. For example, $\lg n$ must be 32 and 128 if an element is an IPv4 and IPv6 address, respectively, and it must be doubled for tracking every pair of source and destination addresses. The authors of [10], in fact, attempted to reduce the computational cost of the $\lg n$ term in the update time and showed a tradeoff that replaces the $\lg n$ term in the update time by $\lg n/\lg b$ and that in the space and query time by $b \lg n/\lg b$ for any integer $b \in [2..n]$. In this study, we show that this dependence on the $\lg n$ term can be completely removed from the query time for arbitrary updates and (in amortized sense) from the update time for constant updates without increasing the space.

We note that our techniques can be used to deterministically solve the dynamic majority problem [10,14] (i.e., the exact $(1/2)$-heavy hitters problem under constant updates) in amortized constant update time and optimal constant query time using $O(\lg n)$ space. Our result improves Theorem 3.1 of [10], and, if we can scan the input twice, the linear space of [14] in case of $\lg n = o(N)$.

1.3 Related Work

The heavy hitters problem has a long history of research, dating back at least to the work of Boyer and Moore [5], where they showed an optimal deterministic algorithm to find the $(1/2)$-heavy hitter or the majority element from an input stream of size N in $O(N)$ time and constant space in the cash register model with $\Delta_i = +1$ for any i. Misra and Gries [24] extended the idea of [5] to any $0 < \phi < 1/2$ in amortized $O(\lg 1/\phi)$ update time and $O(1/\phi)$ query time using optimal $O(1/\phi)$ space. The same algorithm as [24] was rediscovered by

Demaine et al. [11], where they showed that the update time can be $O(1)$ in the worst case for $\Delta_i = +1$, and by Karp et al. [19] independently. All the aforementioned algorithms cannot work in the strict turnstile model, in contrast to ours.

For the ϵ-approximate ϕ-heavy hitters problem in the strict turnstile model, Cormode and Muthukrishnan [10] have proposed a randomized data structure, called Combinatorial Group Testing. In the same article [10], they also described a divide-and-conquer technique combined with Count sketch of Charikar et al. [7] for frequency estimation. In another article [9] of the same authors, they proposed Count-Min sketch for frequency estimation, which is similar to Count sketch, and solved the ϵ-approximate ϕ-heavy hitters problem by plugging it into their divide-and-conquer technique. This extension of Count (resp. Count-Min) sketch is called Hierarchical Count (resp. Count-Min) sketch for its implicit hierarchical structure of multiple Count (resp. Count-Min) sketch instances.

After seminal work of Cormode and Muthukrishnan [9,10] and Charikar et al. [7], there have been extensive studies on the heavy hitters problem in the turnstile model, most of which put on the emphasis on the theoretical aspects. A recent notable result is by Larsen et al. [22], where they introduced the cluster-preserving clustering technique and solved the ℓ_p heavy hitters problem in the general turnstile model for any $p \in (0..2]$. The master thesis of Hovmand and Nygaard [18] is a recent experimental study, where they have empirically compared Hierarchical Count-Min sketch [9] and its variant [22], called Hierarchical Constant-Min sketch, with Hierarchical Count sketch in the strict turnstile model. In contrast to our present study, their interest was not in Combinatorial Group Testing but in Hierarchical Count, Count-Min, and Constant-Min sketch. Most recently, Bender et al. [3] has introduced the online event detection problem to check if there is a ϕ-heavy hitter that occurs *exactly* $\lceil \phi N \rceil$ times at time $1 \leq i \leq N$ with no false positives and negatives in an online manner. Due to their requirements, they focused on exact solutions in the external memory model.

2   A Data Structure of Cormode and Muthukrishnan

In this section, we describe a randomized data structure of Cormode and Muthukrishnan [10] for the ϵ-approximate ϕ-heavy hitters problem, which we will improve in Sect. 4. Let $m = \lg n$ be the size of an element in $\mathbb{U}$ in bits, and ρ and c be some positive integers such that $\rho \geq \lg(1/(\delta\phi))$ and $c \geq 2\epsilon^{-1}$. Their data structure is composed of three components:

1. a counter N for the total frequency of elements,
2. a three-dimensional counter array count$[1..\rho][1..c][0..m]$, and
3. hash functions $h_1, \ldots, h_\rho : \mathbb{U} \mapsto [1..c]$.

Here, every h_i is chosen uniformly at random from a family of universal hash functions. Note that h_i can be stored and evaluated in constant space and time [6].

Let x be any element in $\mathbb{U}$ and bit(x, i) be its i-th least significant (or its i-th rightmost) bit. Given $x \in \mathbb{U}$ and $\Delta \in \mathbb{Z}$, update(x, Δ) increases both N and

count$[i][h_i(x)][0]$ by Δ and, for every $(i, k) \in [1..\rho] \times [1..m]$, count$[i][h_i(x)][k]$ by bit$(x, k) \cdot \Delta$. To describe how query() works, we introduce two key facts:

Fact 1 (Cormode and Muthukrishnan [10]). *Let x be any ϕ-heavy hitter. For any $i \in [1..\rho]$, with probability at least $1/2$, the total frequency of every other element y $(\neq x)$ with $h_i(x) = h_i(y)$ is no more than ϕN, that is,*

$$\sum_{y \in U} [(y \neq x) \wedge (h_i(x) = h_i(y))] \cdot n_y \leq \phi N.$$

Fact 2 (Corollary 3.8 of Cormode and Muthukrishnan [10]). *Let x be an element in U and n_x be its frequency. Then, $\hat{n}_x = \min_i$ count$[i][h_i(x)][0]$ satisfies $n_x \leq \hat{n}_x \leq n_x + \epsilon N$ with probability at least $1 - \phi \delta$.*

By Facts 1 and 2, query() is implemented as follows. We iterate over every $(i, j) \in [1..\rho] \times [1..c]$. Let $C = $ count$[i][j]$ be an array of $m + 1$ counters. Suppose that there exists some ϕ-heavy hitter x with $h_i(x) = j$. We then consider two inquiries: (i) $C[k] > \phi N$ and (ii) $C[0] - C[k] > \phi N$. Fact 1 then implies that, with probability at least $1/2$, either bit$(x, k) = 1$ iff (i) is true and (ii) is false or bit$(x, k) = 0$ iff (i) is false and (ii) is true for every $k \in [1..m]$. We can thus identify ϕ-heavy hitter x from C as follows:

$$\sum_{1 \leq k \leq m} [(C[k] > \phi N) \wedge (C[0] - C[k] \leq \phi N)] \cdot 2^{k-1}. \tag{1}$$

Note that we skip to the next iteration if (i) and (ii) are both true or both false for any $k \in [1..m]$ because there is no way to identify x from C, i.e., there can exist more than one heavy hitters and/or the total frequency of non-heavy hitters exceeds ϕN if (i) and (ii) are both true or there must exist no heavy hitter if (i) and (ii) are both false.

The failure probability of identifying a ϕ-heavy hitter on ρ independent tests can be bounded $(1/2)^\rho = \delta \phi$, and thus, from the union bound, that of identifying any ϕ-heavy hitter can be bounded by δ. After identified as a candidate of ϕ-heavy hitters from C, an element x can be verified in two ways: $h_i(x) = j$ and $\hat{n}_x > \phi N$. By Fact 2, any element $x \in U$ satisfies $n_x \leq \hat{n}_x \leq n_x + \epsilon N$, and thus, from $\hat{n}_x > \phi N$, it holds that $n_x > (\phi - \epsilon)N$ with probability at least $1 - \delta \phi$.

In summary, we can output every ϕ-heavy hitter and no other with frequency no more than ϕN with probability at least $1 - \delta$. The update time is obviously $O(\rho \lg n)$. The query time is $O(\rho \epsilon^{-1} (\lg n + \rho))$ because there are at most $O(\rho \epsilon^{-1})$ candidates, each of which is identified in $O(\lg n)$ time and verified in $O(\rho)$ time.

In the next section, we introduce a novel technique, called *packed bidirectional counters*, to improve the query and update times of [10], which we explain separately from our final data structure for the ϵ-approximate ϕ-heavy hitters problem since it may be of independent interest. In Sect. 4, we show how to improve the query and update times of [10] by another simple technique, called *cached candidates*, or packed bidirectional counters.

3 Packed Bidirectional Counters

As observed in Sect. 2, the $\lg n$ term in the update and query times of [10] arises from $O(\lg n)$-time operations of $\lg n$ bidirectional counters. Our goal here is thus to improve such $O(\lg n)$-time operations somehow.

In this section, we introduce an abstract data type, called *bidirectional counter arrays*, to support three basic operations on an array of counters defined as follows:

Definition 2 (The basic operations on bidirectional counter arrays). Let C be an array of m counters and x be any element in $\mathbb{U}$ for $m = \lg n = O(w)$. We define three basic operations on bidirectional counters arrays as follows:

increment(C, x): $C[i] \leftarrow C[i] + \mathsf{bit}(x, i)$ for $i \in [1..m]$.
decrement(C, x): $C[i] \leftarrow C[i] - \mathsf{bit}(x, i)$ for $i \in [1..m]$.
ispositive(C):   returns $\sum_{i=1}^{m} [C[i] > 0] \cdot 2^{i-1}$.

In the rest of this section, we present a novel technique, called *packed bidirectional counters*, to implement this abstract data type efficiently. Specifically, we show that increment/decrement can be supported in amortized constant time and ispositive in constant time. This technique will be used in Sect. 4.2 to improve the update and query times of [10] for constant updates.

The basic idea behind our technique is to maintain multiple counters in parallel by exploiting the word-level parallelism of the RAM. The main difficulty here is how to support bidirectional updates on them in a synchronized way, i.e., how to deal with such a situation that one counter needs no carry propagation while another needs it at all its digits. To handle such a situation, we represent each counter value using small subcounters or variables and update them in a regular manner independent of the value. Similar techniques have been proposed for string matching under Hamming distance [17] and regular expression matching with intervals [4], which can deal with unidirectional updates on multiple counters in amortized constant time.

For ease of explanation, we assume that C is a single bidirectional counter of w bits, i.e., $m = 1$. As explained later, we store the value of a counter in a set of variables whose value is in $\{0, \pm 1, \pm 2\}$. Thus, it is easy to extend our discussion to $m = O(w)$ by packing the corresponding variables of m counters into $O(1)$ words and exploiting the word-level parallelism in bitwise operations.

We represent C with two types of variables $D_i \in \{0, \pm 1, \pm 2\}$ and $P_i \in \{0, \pm 1\}$ for $i \in [0..w]$, all of which are initialized to 0. For any integer $x > 0$, we define $\mathsf{rmo}(x)$ to be the position of its rightmost one, i.e., $\mathsf{rmo}(x) = \min\{i \mid \mathsf{bit}(x, i) = 1\}$, and, for convenience, $\mathsf{rmo}(0) = w$. Let $C_k = \sum_{i=k}^{w} D_i \cdot 2^i$ be the weighted sum of $D_k, \ldots, D_w$. We then maintain four invariants at the t-th update for any $t \geq 0$:

(1) $C = C_0$.
(2) $D_i \in \{0, \pm 1\}$ for any $0 \leq i < \mathsf{rmo}(t)$ and $i = w$.
(3) $D_i \in \{0, \pm 1, \pm 2\}$ for any $\mathsf{rmo}(t) \leq i < w$.

$$(4) \ P_i = \begin{cases} +1 & C_i \geq +2^i \\ 0 & C_i = 0 \\ -1 & C_i \leq -2^i \end{cases} \quad \text{for any } 0 \leq i \leq w.$$

To achieve amortized constant update time, we maintain these invariants in such a way that we access $O(\text{rmo}(t))$ variables at the t-th update, as in previous techniques [4,17]. First, we explain how increment can maintain invariants (1)–(3). We start by updating D_0 with $D_0 + x$, then ensure that $D_i \in \{0, \pm 1\}$ by propagating the (redundant) carry bit in $\{0, \pm 1\}$ for every $i \in [0..\text{rmo}(t))$, and finally add the last carry bit to $D_{\text{rmo}(t)}$. We implement decrement similarly except that it subtracts x from D_0.

Lemma 3. *Invariants (1)–(3) can be maintained correctly.*

Proof. Invariants (1)–(3) hold after initialization because $D_i = 0$ for any $i \in [0..w]$ and $C = C_0 = 0$. Suppose that invariants (1)–(3) hold at the beginning of the t-th update. The invariant (1) is satisfied because increment (resp. decrement) does not change the counter value of C after adding (resp. subtracting) x to (resp. from) C. Invariants (2) and (3) can be maintained as follows. If t is odd, $\text{rmo}(t) = 0$ and $\text{rmo}(t-1) > 0$ hold. By inductive hypothesis, $D_0 \in \{0, \pm 1\}$ and thus $D_0 \pm x \in \{0, \pm 1, \pm 2\}$ hold, and no other variable changes. If t is even, $\text{rmo}(t) > 0$ and $\text{rmo}(t-1) = 0$ hold. Then, there must exist $t' = \max\{0, t - 2^{\text{rmo}}(t)\}$ such that $\text{rmo}(t') > \text{rmo}(t)$ and $D_{\text{rmo}(t)} \in \{0, \pm 1\}$ at the t'-th update. By inductive hypothesis, the (redundant) carry bit from variables $D_0, \ldots, D_{\text{rmo}(t)-1}$ must be in $\{0, \pm 1\}$, and thus, $D_{\text{rmo}(t)}$ must be in $\{0, \pm 1, \pm 2\}$. Invariants (1)–(3) can thus be maintained.

Next, we explain how increment and decrement can maintain invariant (4) after updating variables D_i, in the proof of the next lemma.

Lemma 4. *Invariant (4) can be maintained correctly.*

Proof. Invariant (4) can be maintained as follows. For any $i \in [0..w]$, we show that P_i can be properly updated from P_{i+1}, P_{i+2}, and D_i in constant time. It is enough to show it for $i \in [0..\text{rmo}(t)]$ because no variable changes for $i \in (\text{rmo}(t)..w]$ at time t. We focus on $P_i = +1$ because $P_i = -1$ is analogous. We show that we can determine if $C_i \geq 2^i$ in constant time for each value of P_{i+1}: (a) $P_{i+1} = 0$; (b) $P_{i+1} = +1$; and (c) $P_{i+1} = -1$. (a) If $P_{i+1} = 0$ and thus $C_{i+1} = 0$, then $C_i = D_i \cdot 2^i \geq 2^i$ holds for any $D_i \in \{+1, +2\}$, and can be examined in constant time. (b) If $P_{i+1} = +1$ and thus $C_{i+1} \geq 2^{i+1}$, then $C_i = D_i \cdot 2^i + C_{i+1} \geq 2^i$ holds except for $D_i = -2$ and $C_{i+1} = 2^{i+1}$, and can be examined in constant time. Note that $C_{i+1} = 2^{i+1}$ can be checked by $(P_{i+1} = +1) \wedge (P_{i+2} = 0)$. (c) If $P_{i+1} = -1$ and thus $C_{i+1} \leq -2^{i+1}$, $C_i \geq 2^i$ never holds because $C_i \leq 2^{i+1}$. Invariant (4) can thus be maintained.

Note again that both types of variables for each $i \in [0..w]$ can be maintained in constant time even for $m = O(w)$ using bitwise operations on packed variables. More specifically, we encode the k-th bits of variables $D_i[1], \ldots, D_i[m]$ to an

integer $\sum_{j=1}^{m} \text{bit}(D_i[j], k) \cdot 2^{j-1}$, and $P_i[1], \ldots, P_i[m]$ are stored similarly, where $D_i[j]$ and $P_i[j]$ are variables for $C[j]$. Using this "bit-split" representation of variables, we can maintain $D_i[j]$ and $P_i[j]$ in parallel for all $j \in [1..m]$ such that invariants (1)–(4) hold, and obtain the next lemma.

Lemma 5. increment(C, x) and decrement(C, x) can be implemented in amortized constant time for $m = O(w)$.

Proof. Note that $\text{rmo}(t) = i$ holds every 2^i updates and it can be computed in amortized constant time by scanning t from its least significant bit to its most significant bit. The amortized time for increment and decrement is thus bounded by $\sum_{i=0}^{w} 1/2^i = O(1)$.

Using invariant (4), ispositive can be implemented in constant time.

Lemma 6. ispositive(C) can be implemented in constant time.

Proof. Note that $P_0 = +1$ implies $C_0 = C \geq 2^0 = 1$ from invariant (4). Thus, $C > 0$ can be determined in constant time by examining P_0.

By Lemmas 3–6, we finally obtain the next lemma:

Lemma 7. There exists an $O(m)$-space data structure that implements increment and decrement in amortized constant time and ispositive in constant time on a bidirectional counter array of size $m = O(w)$.

4 Our Faster Data Structures

In this section, we describe our faster randomized data structures for the ϵ-approximate ϕ-heavy hitters problem in the strict turnstile model. First, we show how to improve the query time of [10] for arbitrary updates. Then, we show how to improve both the query and update times for constant updates by using packed bidirectional counters presented in Sect. 3.

4.1 Faster Query by Cached Candidates

First, we show how to improve the query time of Cormode and Muthukrishnan [10] from $O(\rho\epsilon^{-1}(\lg n + \rho))$ to $O(\rho^2\epsilon^{-1})$ in case of arbitrary updates while keeping the same update time, space, and output guarantee as [10]. To achieve this, we introduce a simple but effective technique, called the *cached candidates*.

Remember that query() described in Sect. 2 has two main steps of identifying candidates of ϕ-heavy hitters (identification step) and verifying them (verification step). Our basic idea is to identify and cache ρ candidates at update time and to verify all the cached candidates at query time. The main obstacle here is that the ϕN terms in (1) can change at every update. Because we can refresh at most ρ cached candidates at each update to ensure $O(\rho)$ update time, other candidates can be incoherent with the current ϕN. Our observation is that, as long as we keep all the "good" candidates originally identified at query time, we

can permit the presence of "bad" candidates because we can reject bad ones at the verification step, where we say that a candidate is good (resp. bad) if its estimated frequency is more (resp. no more) than ϕN.

Let cache$[1..\rho][1..c]$ be another two-dimensional array such that cache$[i][j]$ stores a candidate of ϕ-heavy hitters identified from count$[i][j]$. Our update(x, Δ) is implemented in the same manner described in Sect. 2 except that, after updating an array $C = $ count$[i][h_i(x)]$ of $m + 1$ counters, we identify a candidate y of ϕ-heavy hitters from C (in a modified way explained later) and store y in cache$[i][h_i(x)]$. The candidate y is identified as follows.

$$\sum_{1 \leq k \leq m} \left[C[k] > (1/2)C[0] \right] \cdot 2^{k-1}. \tag{2}$$

That is, we just replace the ϕN terms in (1) with $(1/2)C[0]$, where, for simplicity, we do not skip to the next iteration but rather continue with bit$(y, k) = 0$ in case of $C[k] = (1/2)C[0]$. Our query is implemented just by verifying each candidate stored in cache in the exactly same manner described in Sect. 2.

Lemma 8. update *can be implemented in* $O(\rho \lg n)$ *time.*

Proof. At update time, we modify $O(m)$ counters and identify one candidate both in $O(m)$ time for every $i \in [1..\rho]$. Thus, update can be implemented in $O(\rho \lg n)$ time for $m = \log n$.

Lemma 9. query *can be implemented in* $O(\rho^2 \epsilon^{-1})$ *time.*

Proof. At query time, we just read each cached candidate and then output it if its estimated frequency exceeds ϕN. By Fact 2, it takes $O(\rho)$ time. Because there are $O(\rho \epsilon^{-1})$ cached candidates, query can be implemented in $O(\rho^2 \epsilon^{-1})$ time.

Next, we show that our identification step based on (2) keeps any candidate identified at the original one based on (1).

Lemma 10. *Let R and R' be sets of the elements identified by (2) and (1), respectively. Then, $R \supseteq R'$ holds.*

Proof. Let C be an array of $m + 1$ counters and $x' \in R'$ be a candidate of ϕ-heavy hitters identified from C by (1) Note that bit$(x', k) = 1$ iff $(C[k] > \phi N) \wedge (C[0] - C[k] \leq \phi N)$, i.e., $C[k] > (1/2)C[0]$, for any $k \in [1..m]$. Let $x \in R$ be a candidate identified from C by (2). Because bit$(x, i) = 1$ iff $C[k] > (1/2)C[0]$, $x = x'$ must hold unless x' is null. We thus obtain $R \supseteq R'$.

By Lemmas 8, 9, and 10, we obtain the main result of this section:

Theorem 11. *There exists an $O(\rho \epsilon^{-1} \lg n)$-space randomized data structure that implements update in $O(\rho \lg n)$ and query in $O(\rho^2 \epsilon^{-1})$ time with probability at least $1 - \delta$ for arbitrary updates and $\rho = \lg (1/(\delta\phi))$.*

Algorithm 1. Our update(x, Δ) for unitary updates.

```
1: n ← n + Δ
2: for i ← {1,...,ρ} do
3:     count[i][hᵢ(x)][0] ← count[i][hᵢ(x)][0] + Δ
4:     if Δ = +1 then
5:         if t is odd then increment(count[i][hᵢ(x)], x)
6:         else decrement(count[i][hᵢ(x)], ~x)
7:     else
8:         if t is odd then decrement(count[i][hᵢ(x)], x)
9:         else increment(count[i][hᵢ(x)], ~x)
```

Algorithm 2. Our query() for unitary updates.

```
1: for i ← [1..ρ] do
2:     for j ← [1..c] do
3:         x ← ispositive(count[i][j])]
4:         n̂ₓ ← minₖ count[k][hₖ(x)][0]
5:         if (hᵢ(x) = j) ∧ (n̂ₓ > φN) then report x as a φ-heavy hitter
```

4.2 Faster Update by Packed Bidirectional Counters

Next, we show how to improve the update time of Cormode and Muthukrishnan [10] from $O(\rho \lg n)$ to amortized $O(\rho)$ and the query time of [10] from $O(\rho \epsilon^{-1}(\lg n + \rho))$ to $O(\rho^2 \epsilon^{-1})$ in case of constant updates. To obtain this result, we use the *packed bidirectional counters* presented in Sect. 3.

We focus on update(x, Δ) for unitary updates because it is easy to implement update(x, Δ) for constant updates by repeating update($x, \Delta/|\Delta|$) for $|\Delta|$ times. We implement update(x, Δ) for unitary updates as follows. We represent every array $C = \text{count}[i][j]$ of $m + 1$ bidirectional counters used by [10] (see Sect. 2 for details) by explicitly storing $C[0]$ and replacing $C[1..m]$ with packed bidirectional counters in Sect. 3. We then maintain $C[k]$ to hold the next value for $k \in [1..m]$:

$$\sum_{t=1}^{|\mathcal{S}|} [h_i(x_t) = j] \cdot \text{bit}(x_t, k) \cdot \Delta_t - \lceil (1/2) C[0] \rceil . \tag{3}$$

Note that $C[k]$ originally stores $\sum_{t=1}^{|\mathcal{S}|} [h_i(x_t) = j] \cdot \text{bit}(x_t, k) \cdot \Delta_t$. The above value (3) thus implies that our identification step described in Sect. 4.1 can be reduced to ispositive. Because every unitary update can change (3) by at most ± 1, we can still update C in amortized constant time using increment and decrement on packed bidirectional counters. Algorithms 1 and 2 describe our update and query for unitary updates, respectively. Note that cache used in Sect. 4.1 is no longer necessary because ispositive can return each candidate of heavy hitters in constant time. By Lemma 7 and Theorem 11, we obtain the main result of this section:

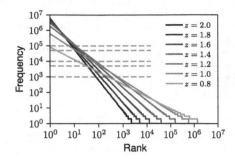

Fig. 1. The frequency distribution of our datasets of size $N = 10^7$ on $\mathbb{U} = [0..2^{64})$. The red dashed lines indicate thresholds ϕN for each $\phi \in \{0.0001, 0.0005, 0.001, 0.005, 0.01\}$. (Color figure online)

Theorem 12. *There exists an $O(\rho\epsilon^{-1}\lg n)$-space randomized data structure that implements* update *in amortized $O(\rho)$ time and* query *in $O(\rho^2\epsilon^{-1})$ time with probability at least $1 - \delta$ for unitary updates and $\rho = \lg(1/(\delta\phi))$.*

5   Experimental Results

In this section, we give an empirical comparison of three randomized approaches to the ϵ-approximate ϕ-heavy hitters problem in the strict turnstile model under unitary updates, in terms of space, recall/precision, and update/query time.

We implemented Hierarchical Count-Min sketch [9], Combinatorial Group Testing [10], and our method in Sect. 4.2 in C++ using multiply-shift hash functions described in [29], all of which were compiled by g++ 9.1.0 with -Ofast and -march=native options. All of them except for ours have a tradeoff parameter b, called *base*, to speed up their update time at the cost of their query time and space. We tested the method of [9] with base $b = 2$ (CMH2) and $b = 16$ (CMH16), that of [10] with base $b = 2$ (CGT2) and $b = 16$ (CGT16), and ours (Ours). As in an experimental study [8], we used 4 (resp. $2\epsilon^{-1}$) for parameter ρ (resp. c) of the Combinatorial Group Testing family (CGT2, CGT, and Ours) and for the depth (resp. width) of internal Count-Min sketches of the Hierarchical Count-Min sketch family (CMH2 and CMH16).

We generated 14 datasets of 10M integers chosen from $\mathbb{U} \in \{[0, 2^{32}), [0, 2^{64})\}$ according to the Zipf distribution of skewness $z \in \{0.8, 1.0, 1.2, 1.4, 1.6, 1.8, 2.0\}$, and then tested with threshold $\phi \in \{0.0001, 0.0005, 0.001, 0.005, 0.01\}$ and $\epsilon = \phi$. The frequency distribution of our datasets on $\mathbb{U} = [0, 2^{64})$ is shown in Fig. 1 as reference. Using each dataset, we conducted 5 trials with different random seeds, at each of which we executed 10M updates and then 100 queries on every method. We measured average update and query times at each trial and reported the median of the 5 trials. All the experiments were run on MacBook Pro with Intel(R) Core(TM) i7-8559U running at 2.7 GHz and 16 GB main memory. Because our main concern is practical scenarios where $\lg n$ is large, we mainly

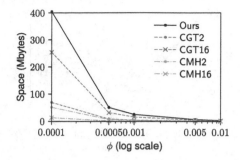

Fig. 2. Space for thresholds $\phi \in \{0.0001, 0.0005, 0.001, 0.005, 0.01\}$ on $\mathbb{U} = [0..2^{64})$.

discussed experimental results on $\mathbb{U} = [0, 2^{64})$. We showed those on $\mathbb{U} = [0, 2^{32})$ to verify that our update and query time were independent of the $\lg n$ term.

5.1 Space

Figure 2 shows the space usages in MB on $\mathbb{U} = [0..2^{64})$. In terms of space, CMH16 was the clear winner especially for $\phi = 0.0001$. Nevertheless, all the others required less than 100 MB for $\phi \geq 0.0005$ and were small enough to reside in main memory. As suggested in [8], both Hierarchical Count-Min sketch [9] and Combinatorial Group Testing [10] can have constant depth (4 in our experiments) to obtain reasonable recall and precision in practice. This means that CMH16 needs just $16/64 = 1/4$ of levels of sketches compared to CMH2 without changing their depth and width. The space used by CGT2 and CMH2 are same in theory. As in [8], however, our implementation of CMH2 tracked exact counts at lower levels, and thus it saved the space for some Count-Min sketches. Ours (resp. CGT16) required about 6 (resp. 4) times as much space as CGT2 because Ours (resp. CGT16) replaced each array of $m = 64$ words with our packed bidirectional counter array implemented using $6m$ words (resp. an array of $\lceil m/\lg b \rceil \cdot (b-1)$ words).

5.2 Recall and Precision

Figures 3(a) and 4(a) show the precision of all the methods on universes $\mathbb{U} = [0..2^{32})$ and $\mathbb{U} = [0..2^{64})$, respectively. We omitted the recall because they never missed any heavy hitter and thus their recall was 100%. As mentioned above, we focus on experimental results on $\mathbb{U} = [0..2^{64})$, i.e., Fig. 4. We observed that Ours showed much lower precision than others on datasets with $z \leq 1.2$. The easiest way to avoid its degradation of precision is to use more appropriate $\epsilon > \phi$. However, such an ϵ must increase its space and query time. Another and perhaps better approach is to add another Count-Min sketch with appropriate ϵ for frequency estimation. As z increased, Ours has improved its precision and finally reached the same level of precision as the others. It should be noted that for $\phi = 0.01$ and $z = 0.8$, Ours, CGT2, and CMH16 output only one false positive and their precision became 0 because there was no heavy hitter.

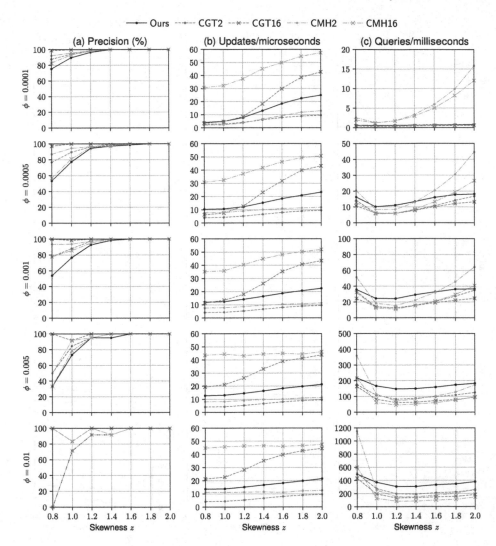

Fig. 3. (a) Precision, (b) Update speed, and (c) Query speed on $\mathbb{U} = [0..2^{32})$.

5.3 Update and Query Time

Figures 3(b) and 4(b) show the update speed (in updates per microsecond) on universes $\mathbb{U} = [0..2^{32})$ and $\mathbb{U} = [0..2^{64})$, respectively. We focus on CMH16, CGT16, and Ours because CMH2 and CGT2 were not competitive in terms of update time. Although their update time complexity is independent of both the threshold ϕ and skewness z in our experimental setup, their update times were much affected by ϕ and z, most likely due to cache effects. In fact, our experimental result is consistent with an intuition that the cache hit ratio can be improved as ϕ decreases (resp. z increases), because smaller ϕ (resp. larger z) reduces the

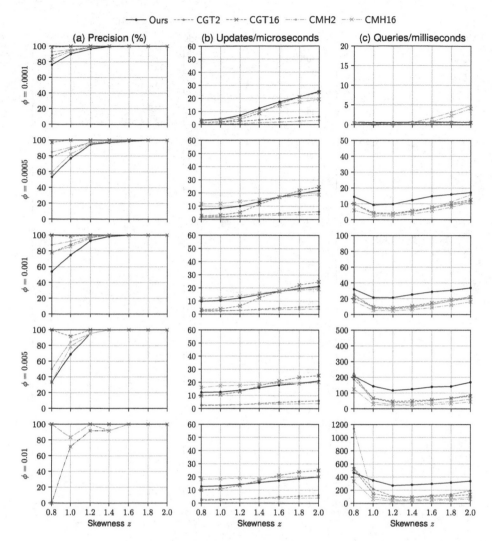

Fig. 4. (a) Precision, (b) Update speed, and (c) Query speed on $\mathbb{U} = [0..2^{64})$.

memory footprint (resp. the number of distinct elements) and thus can improve locality of memory access.

Figures 3(c) and 4(c) show the query speed (in queries per millisecond) on universes $\mathbb{U} = [0..2^{32})$ and $\mathbb{U} = [0..2^{64})$, respectively. First, we discuss experimental results for $\mathbb{U} = [0..2^{64})$ in Fig. 4. Overall, Ours outperformed all the others except for $\phi = 0.0001$ and a few points with $z = 0.8$. In case of $\phi = 0.0001$, Ours was not dominant because its query time complexity has a linear dependence on $2/\epsilon$ (and thus $2/\phi$ in our experiments). In addition, Ours, CGT2, and CGT16 have to examine all of $\rho \times c$ candidates, while CMH2 and CMH16 can be more output sensitive by pruning the search space of candidates using the estimated

frequency of their prefixes. Next, we verify that our update and query times are independent of $\lg n$. As shown in (a) and (b) of Figs. 3 and 4, our update and query times were less affected by the $\lg n$ term than those of the others.

6   Conclusion

In this paper, we presented two novel techniques for improving a randomized data structure of [10] for the ϵ-approximate ϕ-heavy hitters problem in the strict turnstile model. We showed that our first technique of caching candidates of heavy hitters improves the query time of [10] from $O(\rho\epsilon^{-1}(\lg n + \rho))$ to $O(\rho^2\epsilon^{-1})$ for arbitrary updates and that our second technique of packing bidirectional counters improves the update time from $O(\rho\lg n)$ to amortized $O(\rho)$ for constant updates. Thus, our improved randomized data structure for constant updates has no dependence on $\lg n$ in both update time (in amortized sense) and query time (in worst case sense) keeping the same space and output guarantee. Our experiments confirmed that our data structure for constant updates outperformed previous ones in query time for various combinations of thresholds and synthetic skewed datasets while keeping competitive update time and space.

Acknowledgements. The authors would like to thank anonymous referees for their comments that greatly improved the readability and structure of this paper.

References

1. Basat, R.B., Einziger, G., Friedman, R., Luizelli, M.C., Waisbard, E.: Constant time updates in hierarchical heavy hitters. In: Proceedings of the Conference of the ACM Special Interest Group on Data Communication, SIGCOMM 2017, pp. 127–140 (2017)
2. Belazzougui, D., Gagie, T., Navarro, G.: Better space bounds for parameterized range majority and minority. In: Dehne, F., Solis-Oba, R., Sack, J.-R. (eds.) WADS 2013. LNCS, vol. 8037, pp. 121–132. Springer, Heidelberg (2013). https://doi.org/10.1007/978-3-642-40104-6_11
3. Bender, M.A., et al.: The online event-detection problem. arXiv e-prints arXiv:1812.09824 (2018)
4. Bille, P., Thorup, M.: Regular expression matching with multi-strings and intervals. In: Proceedings of the 21 Annual ACM-SIAM Symposium on Discrete Algorithms, SODA 2010, pp. 1297–1308 (2010)
5. Boyer, R.S., Moore, J.S.: MJRTY: a fast majority vote algorithm. In: Boyer, R.S. (ed.) Automated Reasoning: Essays in Honor of Woody Bledsoe, pp. 105–118. Springer, Dordrecht (1991). https://doi.org/10.1007/978-94-011-3488-0_5
6. Carter, J.L., Wegman, M.N.: Universal classes of hash functions. J. Comput. Syst. Sci. **18**(2), 143–154 (1979)
7. Charikar, M., Chen, K.C., Farach-Colton, M.: Finding frequent items in data streams. Theor. Comput. Sci. **312**(1), 3–15 (2004)
8. Cormode, G., Hadjieleftheriou, M.: Methods for finding frequent items in data streams. VLDB J. **19**(1), 3–20 (2010)

9. Cormode, G., Muthukrishnan, S.: An improved data stream summary: the count-min sketch and its applications. J. Algorithms **55**(1), 58–75 (2005)

10. Cormode, G., Muthukrishnan, S.: What's hot and what's not: tracking most frequent items dynamically. ACM Trans. Database Syst. **30**(1), 249–278 (2005)

11. Demaine, E.D., López-Ortiz, A., Munro, J.I.: Frequency estimation of internet packet streams with limited space. In: Möhring, R., Raman, R. (eds.) ESA 2002. LNCS, vol. 2461, pp. 348–360. Springer, Heidelberg (2002). https://doi.org/10.1007/3-540-45749-6_33

12. Durocher, S., He, M., Munro, J.I., Nicholson, P.K., Skala, M.: Range majority in constant time and linear space. Inf. Comput. **222**, 169–179 (2013)

13. Feigenblat, G., Itzhaki, O., Porat, E.: The frequent items problem, under polynomial decay, in the streaming model. Theor. Comput. Sci. **411**(34–36), 3048–3054 (2010)

14. Frandsen, G.S., Skyum, S.: Dynamic maintenance of majority information in constant time per update. Inf. Process. Lett. **63**(2), 75–78 (1997)

15. Gagie, T., He, M., Munro, J.I., Nicholson, P.K.: Finding frequent elements in compressed 2D arrays and strings. In: Grossi, R., Sebastiani, F., Silvestri, F. (eds.) SPIRE 2011. LNCS, vol. 7024, pp. 295–300. Springer, Heidelberg (2011). https://doi.org/10.1007/978-3-642-24583-1_29

16. Gagie, T., He, M., Navarro, G.: Compressed dynamic range majority datastructures. In: 2017 Data Compression Conference, DCC 2017, pp. 260–269 (2017)

17. Grabowski, S., Fredriksson, K.: Bit-parallel string matching under Hamming distance in $O\left(nm/w\right]$) worst case time. Inf. Process. Lett. **105**(5), 182–187 (2008)

18. Hovmand, J.N., Nygaard, M.H.: Estimating frequencies and finding heavy hitters. Master's thesis, Aarhus University (2016)

19. Karp, R.M., Shenker, S., Papadimitriou, C.H.: A simple algorithm for finding frequent elements in streams and bags. ACM Trans. Database Syst. **28**(1), 51–55 (2003)

20. Karpinski, M., Nekrich, Y.: Searching for frequent colors in rectangles. In: Proceedings of the 20th Annual Canadian Conference on Computational Geometry, CCCG 2008 (2008)

21. Kveton, B., Muthukrishnan, S., Vu, H.T., Xian, Y.: Finding subcube heavy hitters in analytics data streams. In: Proceedings of the 2018 World Wide Web Conference WWW 2018, pp. 1705–1714 (2018)

22. Larsen, K.G., Nelson, J., Nguyen, H.L., Thorup, M.: Heavy hitters via cluster-preserving clustering. In: Proceedings of the IEEE 57th Annual Symposium on Foundations of Computer Science, FOCS 2016, pp. 61–70 (2016)

23. Metwally, A., Agrawal, D., Abbadi, A.E.: An integrated efficient solution for computing frequent and top-k elements in data streams. ACM Trans. Database Syst. **31**(3), 1095–1133 (2006)

24. Misra, J., Gries, D.: Finding repeated elements. Sci. Comput. Program. **2**(2), 143–152 (1982)

25. Muthukrishnan, S.: Data streams: algorithms and applications. Found. Trends Theor. Comput. Sci. **1**(2), 117–236 (2005)

26. Navarro, G., Thankachan, S.V.: Encodings for range majority queries. In: Kulikov, A.S., Kuznetsov, S.O., Pevzner, P. (eds.) CPM 2014. LNCS, vol. 8486, pp. 262–272. Springer, Cham (2014). https://doi.org/10.1007/978-3-319-07566-2_27

27. Pike, R., Dorward, S., Griesemer, R., Quinlan, S.: Interpreting the data: parallel analysis with Sawzall. Sci. Program. J. **13**, 277–298 (2005)

28. Sivaraman, V., Narayana, S., Rottenstreich, O., Muthukrishnan, S., Rexford, J.: Heavy-hitter detection entirely in the data plane. In: Proceedings of the Symposium on SDN Research, SOSR 2017, pp. 164–176 (2017)
29. Thorup, M.: High speed hashing for integers and strings. arXiv e-prints arXiv:1504.06804 (2015)
30. Tong, D., Prasanna, V.K.: Sketch acceleration on FPGA and its applications in network anomaly detection. IEEE Trans. Parallel Distrib. Syst. **29**(4), 929–942 (2018)

Range Shortest Unique Substring Queries

Paniz Abedin[1(✉)], Arnab Ganguly[2], Solon P. Pissis[3],
and Sharma V. Thankachan[1]

[1] Department of Computer Science, University of Central Florida, Orlando, USA
paniz@cs.ucf.edu, sharma.thankachan@ucf.edu
[2] Department of Computer Science, University of Wisconsin - Whitewater,
Whitewater, USA
gangulya@uww.edu
[3] CWI, Amsterdam, The Netherlands
solon.pissis@cwi.nl

Abstract. Let $T[1, n]$ be a string of length n and $T[i, j]$ be the sub-string of T starting at position i and ending at position j. A substring $T[i, j]$ of T is a repeat if it occurs more than once in T; otherwise, it is a unique substring of T. Repeats and unique substrings are of great interest in computational biology and in information retrieval. Given string T as input, the *Shortest Unique Substring* problem is to find a shortest substring of T that does not occur elsewhere in T. In this paper, we introduce the range variant of this problem, which we call the *Range Shortest Unique Substring* problem. The task is to construct a data structure over T answering the following type of online queries efficiently. Given a range $[\alpha, \beta]$, return a shortest substring $T[i, j]$ of T with exactly one occurrence in $[\alpha, \beta]$. We present an $\mathcal{O}(n \log n)$-word data structure with $\mathcal{O}(\log_w n)$ query time, where $w = \Omega(\log n)$ is the word size. Our construction is based on a non-trivial reduction allowing us to apply a recently introduced optimal geometric data structure [Chan et al. ICALP 2018].

Keywords: Shortest unique substring · Suffix tree · Heavy-light decomposition · Range queries · Geometric data structures

1 Introduction

Finding regularities in strings is one of the main topics of combinatorial pattern matching and its applications. Among the most well-studied types of string regularities is the notion of repeat. Let $T[1, n]$ be a string of length n. A substring $T[i, j]$ of T is called a repeat if it occurs more than once in T. The notion of unique substring is thus dual: it is a substring $T[i, j]$ of T that does not occur more than once in T. Computing repeats and unique substrings has applications in computational biology [14,23] and in information retrieval [19,22].

Supported in part by the U.S. National Science Foundation under CCF-1703489 and the Royal Society International Exchanges Scheme (IES\R1\180175).

N. R. Brisaboa and S. J. Puglisi (Eds.): SPIRE 2019, LNCS 11811, pp. 258–266, 2019.
https://doi.org/10.1007/978-3-030-32686-9_18

In this paper, we are interested in the notion of shortest unique substring. All shortest unique substrings of T can be computed in $\mathcal{O}(n)$ time using the suffix tree data structure [9,29]. Many different problems based on this notion have already been studied. Pei et al. [22] considered the following problem on the so-called position (or point) queries. Given a position i of T, return a shortest unique substring of T covering i. The authors gave an $\mathcal{O}(n^2)$-time and $\mathcal{O}(n)$-space algorithm, which finds the shortest unique substring covering every position of T. Since then, the problem has been revisited and optimal $\mathcal{O}(n)$-time algorithms have been presented by Ileri et al. [16] and by Tsuruta et al. [27]. Several other variants of this problem have been investigated [2,10,11,15,18,20,21,24,28].

We introduce a natural generalization of the shortest unique substring problem. Specifically, our focus is on the range version of the problem, which we call the *Range Shortest Unique Substring* (rSUS) problem. The task is to construct a data structure over T to be able to answer the following type of online queries efficiently. Given a range $[\alpha, \beta]$, return a shortest substring $T[k, k + h - 1]$ of T with exactly one occurrence in $[\alpha, \beta]$; i.e., $k \in [\alpha, \beta]$, there is no $k' \in [\alpha, \beta]$ such that $T[k, k + h - 1] = T[k', k' + h - 1]$, and h is minimal.

Range queries are a classic data structure topic [6,7,30]. A range query $q = f(A, i, j)$ on an array of n elements over some set S, denoted by $A[1, n]$, takes two indices $1 \le i \le j \le n$, a function f defined over arrays of elements of S, and outputs $f(A[i, j]) = f(A[i], \ldots, A[j])$. Range query data structures have also been considered specifically for strings [1,3,4,12]. For instance, in bioinformatics applications we are often interested in finding regularities in certain regions of a DNA sequence [5,17]. In the Range-LCP problem, defined by Amir et al. [3], the task is to construct a data structure over T to be able to answer the following type of online queries efficiently. Given a range $[\alpha, \beta]$, return $i, j \in [\alpha, \beta]$ such that the length of the longest common prefix of $T[i, n]$ and $T[j, n]$ is maximal among all pairs of suffixes within this range. The state of the art is an $\mathcal{O}(n)$-word data structure supporting $\mathcal{O}(\log^{\mathcal{O}(1)} n)$-time queries [1] (see also [12]).

Main Problem and Main Result

An *alphabet* Σ is a finite nonempty set of elements called *letters*. We fix a *string* $T[1, n] = T[1] \cdots T[n]$ over Σ. The *length* of T is denoted by $|T| = n$. By $T[i, j] = T[i] \cdots T[j]$, we denote the *substring* of T starting at position i and ending at position j of T. We say that another string P has an *occurrence* in T or, more simply, that P *occurs* in T if $P = T[i, i + |P| - 1]$, for some i. Thus, we characterize an occurrence of P by its starting position i in T. A *prefix* of T is a substring of T of the form $T[1, i]$ and a *suffix* of T is a substring of T of the form $T[i, n]$.

We next formally define the main problem considered in this paper.

Problem rSUS
Preprocess: String $T[1, n]$.
Query: Range $[\alpha, \beta]$, where $1 \le \alpha \le \beta \le n$.
Output: (p, ℓ) such that $T[p, p + \ell - 1]$ is a shortest string with exactly one occurrence in $[\alpha, \beta]$.

If $\alpha = \beta$ the answer $(\alpha, 1)$ is trivial. So, in the rest we assume that $\alpha < \beta$.

Example 1. Given $\mathsf{T} = $ $\overset{\text{1 2 3 4 5 6 7 8 9 10 11 12 13 14 15 16 17 18 19 20 21}}{\mathbf{caabcaddaacaddaaaabac}}$ and a query $[\alpha, \beta] = $ $[5, 16]$, we need to find a shortest substring of T with exactly one occurrence in $[5, 16]$. The output here is $(p, \ell) = (10, 2)$, because $\mathsf{T}[10, 11] = \mathbf{ac}$ is the shortest substring of T with exactly one occurrence in $[5, 16]$.

In what follows, we prove our main result (Theorem 1).

Theorem 1. *We can construct an $\mathcal{O}(n \log n)$-word data structure which answers rSUS queries in $\mathcal{O}(\log_w n)$ time per query in the word RAM model, where $w = \Omega(\log n)$ is the word size.*

Our construction is based on ingredients such as the suffix tree [29], heavy-light decomposition [25], and a geometric data structure for rectangle stabbing [8].

2   Our Data Structure

Let us start with some definitions.

Definition 1. *For a position $k \in [1, n]$ and $h \geq 1$, we define $\mathsf{Prev}(k, h)$ and $\mathsf{Next}(k, h)$ as follows:*

$$\mathsf{Prev}(k, h) = \max_j \{\{j < k \mid \mathsf{T}[k, k + h - 1] = \mathsf{T}[j, j + h - 1]\} \cup \{-\infty\}\}$$

$$\mathsf{Next}(k, h) = \min_j \{\{j > k \mid \mathsf{T}[k, k + h - 1] = \mathsf{T}[j, j + h - 1]\} \cup \{+\infty\}\}.$$

Intuitively, let x and y be the occurrences of $\mathsf{T}[k, k + h - 1]$ right before and right after the position k, respectively. Then, $\mathsf{Prev}(k, h) = x$ and $\mathsf{Next}(k, h) = y$. If x (resp., y) does not exist, then $\mathsf{Prev}(k, h) = -\infty$ (resp., $\mathsf{Next}(k, h) = +\infty$).

Definition 2. *Let $k \in [a, b]$. We define $\lambda(a, b, k)$ as follows:*

$$\lambda(a, b, k) = \min\{h \mid \mathsf{Prev}(k, h) < a \ \ \mathbf{and} \ \ \mathsf{Next}(k, h) > b\}.$$

Intuitively, $\lambda(a, b, k)$ denotes the length of the shortest substring that starts at position k with exactly one occurrence in $[a, b]$.

Definition 3. *For a position $k \in [1, n]$, we define C_k as follows:*

$$C_k = \{h \mid (\mathsf{Next}(k, h), \mathsf{Prev}(k, h)) \neq (\mathsf{Next}(k, h - 1), \mathsf{Prev}(k, h - 1))\}.$$

Example 2 (Running Example). Let $\mathsf{T} = $ $\overset{\text{1 2 3 4 5 6 7 8 9 10 11 12 13 14 15 16 17 18 19 20 21}}{\mathbf{caabcaddaacaddaaaabac}}$ and $k = 10$. We have that $(\mathsf{Next}(10, 1), \mathsf{Prev}(10, 1)) = (12, 9)$, $(\mathsf{Next}(10, 2), \mathsf{Prev}(10, 2)) = (20, -\infty)$, and $(\mathsf{Next}(10, 3), \mathsf{Prev}(10, 3)) = (+\infty, -\infty)$. Thus, $C_{10} = \{2, 3\}$.

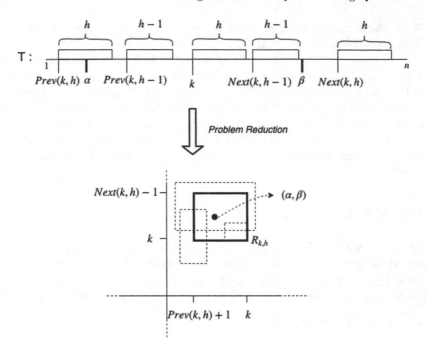

Fig. 1. Illustration of the problem reduction: (k, h) is the output of the rSUS problem with query range $[\alpha, \beta]$, where $h = \lambda(\alpha, \beta, k) \in C_k$. $R_{k,h}$ is the lowest weighted rectangle in $\mathcal{R}$ containing the point (α, β).

Intuitively, C_k stores the set of candidate lengths for shortest unique substrings starting at position k. We make the following observation.

Observation 1. $\lambda(a, b, k) \in C_k$, for any $1 \leq a \leq b \leq n$.

Example 3 (Running Example). Let $\mathsf{T} = \overset{\scriptscriptstyle 1\,2\,3\;\;4\,5\,6\;\,7\;\,8\;\;9\,10\,11\,12\,13\,14\,15\,16\,17\,18\,19\,20\,21}{\texttt{caabcaddaacaddaaaaabac}}$ and $k = 10$. We have that $C_{10} = \{2, 3\}$. For $a = 5$ and $b = 16$, $\lambda(5, 16, 10) = 2$, denoting substring $\texttt{ac}$. For $a = 5$ and $b = 20$, $\lambda(5, 20, 10) = 3$, denoting substring $\texttt{aca}$.

The following combinatorial lemma is crucial for efficiency.

Lemma 1. $\sum_k |C_k| = \mathcal{O}(n \log n)$.

The proof of Lemma 1 is deferred to Sect. 3.

We are now ready to present our construction. By Observation 1, for a given query range $[\alpha, \beta]$, the answer (p, ℓ) we are looking for is the pair (k, h) with the minimum h under the following conditions: $k \in [\alpha, \beta]$, $h \in C_k$, $\mathsf{Prev}(k, h) < \alpha$ and $\mathsf{Next}(k, h) > \beta$. Equivalently, (p, ℓ) is the pair (k, h) with the minimum h, such that $h \in C_k$, $\alpha \in (\mathsf{Prev}(k, h), k]$, and $\beta \in [k, \mathsf{Next}(k, h))$. We map each $h \in C_k$ into a weighted rectangle $R_{k,h}$ with weight h and defined as follows:

$$R_{k,h} = [\mathsf{Prev}(k, h) + 1, k] \times [k, \mathsf{Next}(k, h) - 1].$$

Let $\mathcal{R}$ be the set of all such rectangles, then the lowest weighted rectangle in $\mathcal{R}$ stabbed by the point (α, β) is $R_{p,\ell}$. In short, an rSUS query on $\mathsf{T}[1, n]$ with an input range $[\alpha, \beta]$ can be reduced to an equivalent top-1 rectangle stabbing query on a set $\mathcal{R}$ of rectangles with input point (α, β), where the task is to report the lowest weighted rectangle in $\mathcal{R}$ containing the point (α, β) (see Fig. 1 for an illustration). By Lemma 1, we have that $|\mathcal{R}| = \mathcal{O}(n \log n)$. Therefore, by employing the optimal data structure for top-1 rectangle stabbing presented by Chan et al. [8], which takes $\mathcal{O}(|\mathcal{R}|)$-word space supporting $\mathcal{O}(\log_w |\mathcal{R}|)$-time queries, we obtain the space-time trade-off in Theorem 1. This completes our construction.

3   Proof of Lemma 1

Let $\mathsf{lcp}(i, j)$ denote the length of the longest common prefix of the suffixes of T starting at positions i and j in T. Also, let S denote the set of all (x, y) pairs, such that $1 \le x < y \le n$ and $\mathsf{lcp}(x, y) > \mathsf{lcp}(x, z)$, for all $z \in [x + 1, y - 1]$.

The proof of Lemma 1 can be broken down into the following two lemmas.

Lemma 2. $\sum_k |C_k| = \mathcal{O}(|S|)$.

Lemma 3. $|S| = \mathcal{O}(n \log n)$.

3.1   Proof of Lemma 2

Let us fix a position k. Let

$$C_k' = \{h \mid \mathsf{Prev}(k, h) \ne \mathsf{Prev}(k, h - 1)\}$$
$$C_k'' = \{h \mid \mathsf{Next}(k, h) \ne \mathsf{Next}(k, h - 1)\}.$$

Clearly we have that $C_k = C_k' \cup C_k''$.

The following statements can be deduced by a simple contradiction argument:

1. Let $i = \mathsf{Prev}(k, h) \ne -\infty$, where $h \in C_k'$, then $i = \mathsf{Prev}(k, \mathsf{lcp}(i, k))$
2. Let $j = \mathsf{Next}(k, h) \ne \infty$, where $h \in C_k''$, then $j = \mathsf{Next}(k, \mathsf{lcp}(k, j))$.

Figure 2 illustrates the proof for the first statement. The second one can be proved in a similar fashion.

Clearly, $|C_k'|$ is proportional to the number of (i, k) pairs such that $\mathsf{lcp}(i, k) \ne 0$ and $i = \mathsf{Prev}(k, \mathsf{lcp}(i, k))$. Similarly, $|C_k''|$ is proportional to the number of (k, j) pairs such that $\mathsf{lcp}(k, j) \ne 0$ and $j = \mathsf{Next}(k, \mathsf{lcp}(k, j))$. Therefore, $\sum_k |C_k|$ is proportional to the number of (x, y) pairs, such that $\mathsf{lcp}(x, y) \ne 0$ and $\mathsf{lcp}(x, y) > \mathsf{lcp}(x, z)$, for all $z \in [x + 1, y - 1]$. This completes the proof of Lemma 2.

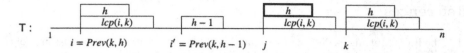

Fig. 2. Let $h \in C'_k$ and $i = \mathsf{Prev}(k, h)$. By contradiction, assume that there exists $j \in (i, k)$ such that $j = \mathsf{Prev}(k, \mathsf{lcp}(i, k))$. Since $h \le \mathsf{lcp}(i, k)$, $\mathsf{T}[j, j+h-1] = \mathsf{T}[k, k+h-1]$. This is a contradiction with $i = \mathsf{Prev}(k, h)$. Thus, $i = \mathsf{Prev}(k, \mathsf{lcp}(i, k))$.

3.2 Proof of Lemma 3

Consider the suffix tree data structure of string $\mathsf{T}[1, n]$, which is a compact trie of the n suffixes of T appended with a letter $\$ \notin \Sigma$ [29]. This suffix tree consists of n leaves (one for each suffix of T) and at most $n-1$ internal nodes. The edges are labeled with substrings of T. Let u be the lowest common ancestor of the leaves corresponding to the strings $\mathsf{T}[x, n]\$$ and $\mathsf{T}[y, n]\$$. Then, the concatenation of the edge labels on the path from the root to u is exactly the longest common prefix of $\mathsf{T}[x, n]\$$ and $\mathsf{T}[y, n]\$$. For any node u, we denote by $\mathsf{size}(u)$ the total number of leaf nodes of the subtree rooted at u.

We decompose the nodes in the suffix tree into *light* and *heavy* nodes. The root node is light and for any internal node, exactly one child is heavy. Specifically, the heavy child is the one having the largest number of leaves in its subtree (ties are broken arbitrarily). All other children are light. This tree decomposition is known as *heavy-light decomposition*. We have the following critical observation. Any path from the root to a leaf node contains many nodes, however, the number of light nodes is at most $\log n$ [13,25]. We have the following lemma.

Lemma 4 ([25]). *The sum of subtree sizes over all light nodes is $\mathcal{O}(n \log n)$.*

We are now ready to complete the proof. Let $S_u \subseteq S$ denote the set of pairs (x, y), such that the lowest common ancestor of the leaves corresponding to suffixes $\mathsf{T}[x, n]\$$ and $\mathsf{T}[y, n]\$$ is u. Clearly, the paths from the root to the leaves corresponding to suffixes $\mathsf{T}[x, n]\$$ and $\mathsf{T}[y, n]\$$ pass from two distinct children of node u and then at least one of the two must be a light node. Therefore, $|S_u|$ is at most twice the sum of $\mathsf{size}(\cdot)$ over all light children of u. Since $|S| = \sum_u |S_u|$, we can bound $|S|$ by the sum of $\mathsf{size}(\cdot)$ over all light nodes in the suffix tree, which is $\mathcal{O}(n \log n)$ by Lemma 4. This completes the proof of Lemma 3.

4 Open Questions

We leave the following related questions unanswered:

1. Can we design an efficient $\mathcal{O}(n)$-word data structure for the rSUS problem?
2. Can we design an efficient solution for the k mismatches/edits variation of the rSUS problem, perhaps using the framework of [26]?
3. Can our reduction be extended to other types of string regularities?

References

1. Abedin, P., et al.: A linear-space data structure for Range-LCP queries in poly-logarithmic time. In: Proceedings of Computing and Combinatorics - 24th International Conference, COCOON 2018, Qing Dao, China, 2–4 July 2018. pp. 615–625 (2018). https://doi.org/10.1007/978-3-319-94776-1_51
2. Allen, D.R., Thankachan, S.V., Xu, B.: A practical and efficient algorithm for the k-mismatch shortest unique substring finding problem. In: Shehu, A., Wu, C.H., Boucher, C., Li, J., Liu, H., Pop, M. (eds.) Proceedings of the 2018 ACM International Conference on Bioinformatics, Computational Biology, and Health Informatics, BCB 2018, Washington, DC, USA, 29 August–01 September 2018. pp. 428–437. ACM (2018). https://doi.org/10.1145/3233547.3233564
3. Amir, A., Apostolico, A., Landau, G.M., Levy, A., Lewenstein, M., Porat, E.: Range LCP. J. Comput. Syst. Sci. **80**(7), 1245–1253 (2014). https://doi.org/10.1016/j.jcss.2014.02.010
4. Amir, A., Lewenstein, M., Thankachan, S.V.: Range LCP queries revisited. In: Iliopoulos, C., Puglisi, S., Yilmaz, E. (eds.) SPIRE 2015. LNCS, vol. 9309, pp. 350–361. Springer, Cham (2015). https://doi.org/10.1007/978-3-319-23826-5_33
5. Ayad, L.A.K., Pissis, S.P., Polychronopoulos, D.: CNEFinder: finding conserved non-coding elements in genomes. Bioinformatics **34**(17), i743–i747 (2018). https://doi.org/10.1093/bioinformatics/bty601
6. Bender, M.A., Farach-Colton, M.: The LCA problem revisited. In: Gonnet, G.H., Viola, A. (eds.) LATIN 2000. LNCS, vol. 1776, pp. 88–94. Springer, Heidelberg (2000). https://doi.org/10.1007/10719839_9
7. Berkman, O., Vishkin, U.: Recursive star-tree parallel data structure. SIAM J. Comput. **22**(2), 221–242 (1993). https://doi.org/10.1137/0222017
8. Chan, T.M., Nekrich, Y., Rahul, S., Tsakalidis, K.: Orthogonal point location and rectangle stabbing queries in 3-D. In: 45th International Colloquium on Automata, Languages, and Programming, ICALP 2018, Prague, Czech Republic, 9–13 July 2018, pp. 31:1–31:14 (2018). https://doi.org/10.4230/LIPIcs.ICALP.2018.31
9. Farach, M.: Optimal suffix tree construction with large alphabets. In: 38th Annual Symposium on Foundations of Computer Science, FOCS 1997, Miami Beach, Florida, USA, 19–22 October 1997, pp. 137–143. IEEE Computer Society (1997). https://doi.org/10.1109/SFCS.1997.646102
10. Ganguly, A., Hon, W., Shah, R., Thankachan, S.V.: Space-time trade-offs for the shortest unique substring problem. In: 27th International Symposium on Algorithms and Computation, ISAAC 2016, Sydney, Australia, 12–14 December 2016, pp. 34:1–34:13 (2016). https://doi.org/10.4230/LIPIcs.ISAAC.2016.34
11. Ganguly, A., Hon, W., Shah, R., Thankachan, S.V.: Space-time trade-offs for finding shortest unique substrings and maximal unique matches. Theor. Comput. Sci. **700**, 75–88 (2017). https://doi.org/10.1016/j.tcs.2017.08.002
12. Ganguly, A., Patil, M., Shah, R., Thankachan, S.V.: A linear space data structure for range LCP queries. Fundam. Inform. **163**(3), 245–251 (2018). https://doi.org/10.3233/FI-2018-1741
13. Harel, D., Tarjan, R.E.: Fast algorithms for finding nearest common ancestors. SIAM J. Comput. **13**(2), 338–355 (1984). https://doi.org/10.1137/0213024
14. Haubold, B., Pierstorff, N., Möller, F., Wiehe, T.: Genome comparison without alignment using shortest unique substrings. BMC Bioinform. **6**, 123 (2005). https://doi.org/10.1186/1471-2105-6-123

15. Hon, W., Thankachan, S.V., Xu, B.: In-place algorithms for exact and approximate shortest unique substring problems. Theor. Comput. Sci. **690**, 12–25 (2017). https://doi.org/10.1016/j.tcs.2017.05.032

16. İleri, A.M., Külekci, M.O., Xu, B.: Shortest unique substring query revisited. In: Kulikov, A.S., Kuznetsov, S.O., Pevzner, P. (eds.) CPM 2014. LNCS, vol. 8486, pp. 172–181. Springer, Cham (2014). https://doi.org/10.1007/978-3-319-07566-2_18

17. Iliopoulos, C.S., Mohamed, M., Pissis, S.P., Vayani, F.: Maximal motif discovery in a sliding window. In: Gagie, T., Moffat, A., Navarro, G., Cuadros-Vargas, E. (eds.) SPIRE 2018. LNCS, vol. 11147, pp. 191–205. Springer, Cham (2018). https://doi.org/10.1007/978-3-030-00479-8_16

18. Inoue, H., Nakashima, Y., Mieno, T., Inenaga, S., Bannai, H., Takeda, M.: Algorithms and combinatorial properties on shortest unique palindromic substrings. J. Discrete Algorithms **52**, 122–132 (2018). https://doi.org/10.1016/j.jda.2018.11.009

19. Khmelev, D.V., Teahan, W.J.: A repetition based measure for verification of text collections and for text categorization. In: Proceedings of the 26th Annual International ACM SIGIR Conference on Research and Development in Information Retrieval, SIGIR 2003, pp. 104–110. ACM, New York (2003). https://doi.org/10.1145/860435.860456

20. Mieno, T., Inenaga, S., Bannai, H., Takeda, M.: Shortest unique substring queries on run-length encoded strings. In: Faliszewski, P., Muscholl, A., Niedermeier, R. (eds.) 41st International Symposium on Mathematical Foundations of Computer Science, MFCS 2016, Kraków, Poland, 22–26 August 2016. LIPIcs, vol. 58, pp. 69:1–69:11. Schloss Dagstuhl - Leibniz-Zentrum fuer Informatik (2016). https://doi.org/10.4230/LIPIcs.MFCS.2016.69

21. Mieno, T., Köppl, D., Nakashima, Y., Inenaga, S., Bannai, H., Takeda, M.: Compact data structures for shortest unique substring queries. CoRR abs/1905.12854 (2019), http://arxiv.org/abs/1905.12854

22. Pei, J., Wu, W.C.H., Yeh, M.Y.: On shortest unique substring queries. In: 2013 IEEE 29th International Conference on Data Engineering (ICDE), pp. 937–948. IEEE (2013)

23. Schleiermacher, C., Ohlebusch, E., Stoye, J., Choudhuri, J.V., Giegerich, R., Kurtz, S.: REPuter: the manifold applications of repeat analysis on a genomic scale. Nucleic Acids Res. **29**(22), 4633–4642 (2001). https://doi.org/10.1093/nar/29.22.4633

24. Schultz, D.W., Xu, B.: On k-mismatch shortest unique substring queries using GPU. In: Proceedings of Bioinformatics Research and Applications - 14th International Symposium, ISBRA 2018, Beijing, China, 8–11 June 2018, pp. 193–204 (2018). https://doi.org/10.1007/978-3-319-94968-0_18

25. Sleator, D.D., Tarjan, R.E.: A data structure for dynamic trees. In: Proceedings of the 13th Annual ACM Symposium on Theory of Computing, Milwaukee, Wisconsin, USA, 11–13 May 1981, pp. 114–122 (1981). https://doi.org/10.1145/800076.802464

26. Thankachan, S.V., Aluru, C., Chockalingam, S.P., Aluru, S.: Algorithmic framework for approximate matching under bounded edits with applications to sequence analysis. In: Proceedings of Research in Computational Molecular Biology - 22nd Annual International Conference, RECOMB 2018, Paris, France, 21–24 April 2018, pp. 211–224 (2018). https://doi.org/10.1007/978-3-319-89929-9_14

27. Tsuruta, K., Inenaga, S., Bannai, H., Takeda, M.: Shortest unique substrings queries in optimal time. In: Geffert, V., Preneel, B., Rovan, B., Štuller, J., Tjoa, A.M. (eds.) SOFSEM 2014. LNCS, vol. 8327, pp. 503–513. Springer, Cham (2014). https://doi.org/10.1007/978-3-319-04298-5_44
28. Watanabe, K., Nakashima, Y., Inenaga, S., Bannai, H., Takeda, M.: Shortest unique palindromic substring queries on run-length encoded strings. In: Proceedings of Combinatorial Algorithms - 30th International Workshop, IWOCA 2019, Pisa, Italy, 23–25 July 2019, pp. 430–441 (2019). https://doi.org/10.1007/978-3-030-25005-8_35
29. Weiner, P.: Linear pattern matching algorithms. In: Proceedings of the 14th Annual Symposium on Switching and Automata Theory (SWAT 1973), pp. 1–11. IEEE Computer Society, Washington, DC (1973). https://doi.org/10.1109/SWAT.1973.13
30. Yao, A.C.: Space-time tradeoff for answering range queries (extended abstract). In: Proceedings of the Fourteenth Annual ACM Symposium on Theory of Computing, STOC 1982, pp. 128–136. ACM, New York (1982). https://doi.org/10.1145/800070.802185

An Optimal Algorithm to Find
Champions of Tournament Graphs

Lorenzo Beretta[1]([✉]), Franco Maria Nardini[2]([✉]), Roberto Trani[2,3]([✉]),
and Rossano Venturini[2,3]([✉])

[1] Scuola Normale Superiore, Pisa, Italy
lorenzo.beretta@sns.it
[2] ISTI-CNR, Pisa, Italy
{francomaria.nardini,roberto.trani,rossano.venturini}@isti.cnr.it
[3] Department of Computer Science, University of Pisa, Pisa, Italy
{roberto.trani,rossano.venturini}@unipi.it

Abstract. A tournament graph $T = (V, E)$ is an oriented complete graph, which can be used to model a round-robin tournament between n players. In this short paper, we address the problem of finding a champion of the tournament, also known as Copeland winner, which is a player that wins the highest number of matches. Our goal is to solve the problem by minimizing the number of *arc lookups*, i.e., the number of matches played. We prove that finding a champion requires $\Omega(\ell n)$ comparisons, where ℓ is the number of matches lost by the champion, and we present a deterministic algorithm matching this lower bound without knowing ℓ. Solving this problem has important implications on several Information Retrieval applications including Web search, conversational AI, machine translation, question answering, recommender systems, etc.

Keywords: Tournament graph · Round-robin tournament · Copeland winner

1 Introduction

A *tournament graph* is an oriented complete graph $T = (V, E)$ [13]. This graph can be used to model a round-robin tournament between n players, where each player plays a match with any other player. The orientation of an arc in E tells the winner of the match, i.e., we have the arc $(u, v) \in E$ iff u beats v in their match. In this short paper, we address the problem of finding a champion of the tournament, also known as *Copeland winner* [6], which is a vertex in V with the maximum out-degree, i.e., a player that wins the highest number of matches. Our goal is to find a champion by minimizing the number of *arc lookups*, i.e.,

This paper is partially supported by the BIGDATAGRAPES (EU H2020 RIA, grant agreement Nº780751), the "Algorithms, Data Structures and Combinatorics for Machine Learning" (MIUR-PRIN 2017), the OK-INSAID (MIUR-PON 2018, grant agreement NºARS01_00917) projects, and Scuola Normale Superiore.

N. R. Brisaboa and S. J. Puglisi (Eds.): SPIRE 2019, LNCS 11811, pp. 267–273, 2019.
https://doi.org/10.1007/978-3-030-32686-9_19

the number of played matches. Note that a tournament graph may have more than one champion. In this case, we aim at finding any of them, even if all the algorithms are able to find all of them without increasing the complexity.

If the tournament is transitive (i.e., if u wins against v and v wins against w, then u wins against w), we can trivially identifying the (unique) champion with $\Theta(n)$ arc lookups. Indeed, the champion is the only vertex that wins all its matches and, thus, we can perform a *knock-out tournament* where the loser of any match is immediately eliminated. However, finding the champion of general tournament graphs requires $\Omega(n^2)$ arc lookups [7] and, thus, there is nothing better to do than playing all the matches. This means that the structure of the underlying tournament graph heavily impacts the complexity of the problem.

In this paper we parametrize the problem with the number ℓ of matches lost by the champion. We first show that $\Omega(\ell n)$ arc lookups are required, then we present an optimal deterministic (and non-trivial) algorithm for finding a champion and achieving this bound without knowing ℓ. This parametrization is motivated by applications in Information Retrieval and Machine Learning where we expect that a champion (e.g., the best item in a set) loses only few matches.

Motivations. The identification of the best candidate of a set of items is a crucial task in many Information Retrieval (IR) applications, also known as P@1, including Web search, conversational AI, machine translation, question answering, recommender systems, etc. [8,11]. The task can be solved in two different ways: (i) by piggybacking state-of-the-art Machine Learning (ML) techniques for ranking, and by selecting the candidate with the highest rank in the list; (ii) by employing a pairwise ML classifier, which is trained to identify the best among two candidates, and then by selecting the champion of the resulting tournament. While the former approach exploits only the information of a single candidate for computing the ranking, the latter approach is potentially more powerful because it exploits the information of two candidates while comparing pairs of items. However, the latter approach is more expensive than the former one due to the lack of efficient algorithms to reduce the number of time-consuming comparisons. This work aims at filling this gap since the P@1 problem can be solved by finding a champion of the tournament graph induced by the pairwise ML classifier. As it is possible to design very accurate ML classifiers for several tasks, we expect a very low number of matches ℓ lost by the best item, thus a quasi-linear number of arc lookups is required by our algorithm to find it.

2   Related Work

Tournament graphs are a well-known model that has been applied to several different areas such as sociology, psychology, statistics, and computer science. Examples of applications are round-robin tournaments, paired-comparison experiments, majority voting, communication networks, etc. [4,9,10,12,13].

For the purpose of this work, we identify two different research lines. The first one aims at defining different *notions of tournament winner*, while the second one aims at *efficiently ranking the list of candidates* using pairwise approaches.

Tournament Winner. According to previous works [4,10,12], there is no unique definition of the notion of a tournament winner. Nevertheless, all of them agree on defining the winner whenever there is a candidate, called *Condorcet winner*, which beats all the others. Different definitions of winner requires different complexities of the algorithms used to identify it.

The easiest cases to consider are the *transitive tournament graphs*, i.e., directed acyclic graphs, where it is trivial to find the Condorcet winner in linear time by performing a knock-out tournament. Instead, in the general case, the complexity of finding a winner is usually much higher and strictly depends on the definition of winner. The winner as defined by Slater [15], called *Slater solution*, is the Condorcet winner in a modified tournament graph T'. T' is obtained by reversing the minimum number of arcs of T that make it a transitive tournament. However, the computation of the Slater solution is $\mathcal{NP}$-hard. The $\mathcal{NP}$-hardness derives by reduction to the *Feedback Arc Set Problem* [5]. A winner as defined by Banks [3] is the Condorcet winners of a maximal transitive sub-tournament of T. As there may be several of these sub-tournaments, the *Banks solution* is the set of all these winners. The problem of finding just one winner can be computed in $\Theta(n^2)$ arc lookups, while finding all of them is a $\mathcal{NP}$-hard problem [9]. A third definition of winner is given by Copeland [6], called *Copeland solution*, where a winner is the candidate winning the majority of the matches. This is the definition used in this paper. As we already mentioned, the Copeland solution requires $\Omega(n^2)$ arc lookups and there is a trivial algorithm to match it [7].

There are several other notions of winner, and most of them can be computed in polynomial time. We refer to Hudry [9] for a complete survey on this topic.

Tournament Ranking. Several works deal with the problem of efficiently ranking vertices according to the structure of the tournament graph, see e.g., [1,2,9,13,14].

The result by Shen *et al.* [14] provides a ranking based on the definition of *king*. A vertex $u \in V$ is a king if for each $v \in V$ either (1) $(u,v) \in E$, or (2) $\exists w \in V$ such that $(u,w) \in E$ and $(w,v) \in E$. Equivalently, u is a king if for every vertex v there is a directed path from u to v of length at most 2 in T. Every tournament has at least a king and it can be easily computed in linear time. The ranking algorithm by Jian *et al.* [14] finds a sorted sequences of vertices $u_1, u_2, \ldots, u_n$ such that for every i (1) u_i beast u_{i+1}, and (2) u_i is a king in the sub-tournament induced by the items $u_{i+1}, u_{i+2}, \ldots, u_n$. The authors provide a $O(n^{3/2})$ algorithm to compute this sequence.

Ailon *et al.* [1,2] provide a bound to the error achieved by the Quicksort algorithm when used to sort vertices of the tournament graph. The error is defined as the number of mis-ordered pairs of vertices, i.e., $u, v \in V$ is such that u beats v, but v is ranked higher than u. Ailon *et al.* show that the expected error is at most two times the best possible error. It is apparent that the proposed algorithm requires $\Omega(n \log n)$ arc lookups with high probability.

We observe that the two results above are not suitable in our setting. Indeed, the definition of king is weaker than the one of Copeland winner, since the latter implies the former [13]. Moreover, it is easy to show that the algorithm by Ailon *et al.* fails in finding a winner w every time one of the Quicksort pivots beats w.

3   Algorithm

An adversarial argument is used by Gutin *et al.* [7] to prove that finding a champion requires $\Omega\left(n^2\right)$ arc lookups. Therefore, the trivial algorithm that finds a champion by performing all the possible matches is optimal in general. The problem is indeed much more interesting if we parameterize with ℓ, the number of matches lost by the champion. Note that ℓ is unknown to the algorithm.

The goal of this section is to prove the following theorem which states that $\Omega(\ell n)$ arc lookups are necessary to find a champion and to present an optimal algorithm requiring exactly that number of lookups.

Theorem 1. *Given a tournament graph T with n vertices and with ℓ matches lost by a champion, then*

- *finding a champion requires $\Omega(\ell n)$ arc lookups;*
- *there is an algorithm that finds every champion by requiring $\Theta(\ell n)$ arc lookups and time. The algorithm requires linear space.*

Lower Bound. The lower bound is proved by using a simple adversarial argument. Assume that there is an algorithm that claims that a vertex u, losing ℓ matches, is a champion by performing $\frac{1}{2}\ell(n-1)$ arc lookups. There must exist a node v such that the algorithm has unfolded less than ℓ arcs incident to v. We thus can make v win more matches than u by setting v's unfolded matches, then let the algorithm be incorrect. In other words any correct algorithm, claiming that a vertex u is a champion with ℓ matches lost, must be able to certificate its answer exhibiting (1) a list of $n-1-\ell$ matches won by u, and (2) a list of ℓ matches lost by v, for every vertex $v \in V \setminus \{u\}$.

Upper Bound. To understand the difficulty of the problem, let us think about it in terms of the (unknown) adjacency matrix of the tournament graph. A reasonable strategy to solve the problem is by applying a row-wise approach: we process all the vertices, one after the other, and we try to discard each of them, say v, by finding at least ℓ matches lost by v, i.e., we try to certificate that v cannot be a champion. Through an adversarial argument, we can see that the best possible way to process any vertex v in this row-wise approach is to select random opponents until we are able to discard v (or recognize v as a champion). We can exploit properties of the global structure of the tournament to prove that this approach requires $\Theta\left(\ell n \log n\right)$ arc lookups with high probability. Thus, this algorithm is randomized and is $\log n$ times worse than the lower bound.

We design a simple, deterministic, and optimal algorithm to find the champion (Algorithm 1). The number ℓ of matches lost by the champion is unknown to the algorithm. Thus, it performs an exponential search to find the suitable value of α such that $\alpha/2 \le \ell < \alpha$ (line 2) and tries to solve the problem by assuming that the champion loses less than α matches. At each iteration, the algorithm maintains a set A of alive vertices that is initially equal to V. Then, it performs an elimination tournament among the vertices in A by eliminating a player each time it loses α matches (line 12) until only 2α vertices remain alive (line 6).

Algorithm 1.

```
 1: procedure FINDCHAMPION(T = (V, E))
 2:     for (α = 1; true; α = 2α) do
 3:         A = V
 4:         S = {(u, u) | u ∈ V}
 5:         ∀u ∈ V  lost[u] = 0    ▷ Initialize the number of lost matches of each vertex
 6:         while |A| > 2α do
 7:             choose a pair of vertices u, v in A² \ S
 8:             S = S ∪ {(u, v), (v, u)}           ▷ Avoid selecting the same pair twice
 9:             loser = if (u, v) ∈ E then v else u
10:             ++lost[loser]
11:             if lost[loser] ≥ α then
12:                 A = A \ {loser}
13:         c, lostₒ = FINDCHAMPIONBRUTEFORCE(A, E)
14:         if lostₒ < α then return c
```

The matches are selected arbitrarily avoiding to play the same match multiple times (line 7). When the elimination tournament ends, a candidate champion is found via FINDCHAMPIONBRUTEFORCE procedure, which exhaustively finds the vertex c of A with the maximum out-degree in T. Whenever the candidate c loses at least α matches (line 14), the value of α is not the correct one and the champion may have been erroneously eliminated before. Thus, c could be not a champion and the algorithm continues with the next value of α (line 2).

Correctness. Let us first assume that the value of α is such that $\alpha/2 \leq \ell < \alpha$. We prove now that, under this assumption, the algorithm correctly identifies a champion. First, we observe that the algorithm cannot eliminate the champions as any of them loses less than α matches. Thus, if we prove that the algorithm terminates, the set A contains all the champions and the FINDCHAMPIONBRUTEFORCE procedure will identify any (potentially, all) of them. Note that a champion of T may not be a champion of the sub-tournament restricted to only the vertices in A. That is why FINDCHAMPIONBRUTEFORCE procedure computes the out-degrees of all vertices in A by looking at the original tournament T. We use the following lemma to prove that eventually the condition $|A| = 2\alpha$ is met and the algorithm terminates.

Lemma 1. *In any tournament T of n vertices there is at least one vertex having in-degree $(n-1)/2$.*

Proof. The sum of the in-degrees of all vertices of T is exactly $\binom{n}{2} = \frac{n(n-1)}{2}$. Since there are n vertices, there must be at least one vertex with in-degree $\frac{n-1}{2}$.

Thus, each tournament of $2\alpha + 1$ vertices, or more, has at least one vertex losing at least α matches. This means that the algorithm has always the opportunity to eliminate a vertex from A until there are 2α vertices left. Notice that the above discussion is valid for any value of α smaller than the target one. Thus,

any iterations of the exponential search will terminate and it eventually finds a suitable value of α, i.e., $\alpha/2 \leq \ell < \alpha$, where a champion will be identified.

Complexity. Let us now analyze the complexity of the algorithm. Let us first consider the cost of an iteration of the exponential search. Observe that each arc lookup increases one entry of *lost* by one and that none of these entries is ever greater than α. Thus, the elimination tournament takes no more than $n\alpha$ arc lookups. Moreover, the FINDCHAMPIONBRUTEFORCE procedure takes less than $2n\alpha$ arc lookups since it just unfolds every arc of the remaining 2α alive nodes. Thus, an iteration of the exponential search takes less than $3n\alpha$ arc lookups.

We get the complexity of the overall algorithm by summing up over all the possible values of α, which are all the powers of 2 from 1 up to 2ℓ. Thus, we have at most $3n \sum_{i=0}^{\lceil \log_2(2\ell) \rceil} 2^i = \Theta(\ell n)$ arc lookups.

Technical Details. We are left with the proof that the Algorithm 1 can be implemented in $\Theta(\ell n)$ time and linear space. We do this by exploiting the fact that Algorithm 1 allows us to choose any arc as soon as its vertices are alive and it has never looked up before. An efficient implementation is achieved by maintaining two arrays of n elements each: an array A storing alive vertices, and an array *lost* storing the number of matches lost by each vertex. A counter *numAlive* stores the number of alive vertices. Our implementation maintains the invariant that the prefix $A[1, numAlive]$ contains only alive vertices. We use two cursors p_1 and p_2 to iterate over the elements in A. At the beginning $p_1 = 1$, $p_2 = 2$ and $numAlive = n$. Our implementation performs a series of matches involving vertex $A[p_1]$ and all other vertices in $A[p_1 + 1, numAlive]$. Then, it moves p_1 to the next position. After every match between $A[p_1]$ and $A[p_2]$, we increment *lost* of the loser, say vertex v. Whenever $lost[v]$ equals α we eliminate v according to the following two cases. The first case occurs when v is $A[p_1]$. We swap $A[p_1]$ and $A[numAlive]$, we end the current series of matches, and we start a new one. The second case occurs when v is $A[p_2]$. Here, we swap $A[p_2]$ and $A[numAlive]$, and we continue the current series of matches. In both cases, we decrease $numAlive$ by 1 so that we preserve the invariant.

4   Conclusions and Future Work

We addressed the problem of finding champions in tournament graphs by minimizing the number of arc lookups. We showed that, given the number ℓ of matches lost by the champion, $\Omega(\ell n)$ arc lookups are required to find a champion. Then, we presented an optimal deterministic algorithm that solves the problem and matches the lower bound without knowing ℓ. As future work, we plan to experimentally evaluate the proposed algorithm to solve the Information Retrieval tasks outlined in the Introduction. We also plan to study the impact of randomization on this problem to design a Monte Carlo algorithm that lowers the complexity in charge of providing an incorrect output with small probability.

References

1. Ailon, N., Mohri, M.: An efficient reduction of ranking to classification. In: Servedio, R.A., Zhang, T. (eds.) 21st Annual Conference on Learning Theory - COLT 2008, Helsinki, Finland, 9–12 July 2008, pp. 87–98. Omnipress (2008). http://colt2008.cs.helsinki.fi/papers/32-Ailon.pdf
2. Ailon, N., Mohri, M.: Preference-based learning to rank. Mach. Learn. **80**(2–3), 189–211 (2010). https://doi.org/10.1007/s10994-010-5176-9
3. Banks, J.S.: Sophisticated voting outcomes and agenda control. Soc. Choice Welf. **1**(4), 295–306 (1985)
4. Brandt, F., Brill, M., Harrenstein, P.: Tournament solutions. In: Brandt, F., Conitzer, V., Endriss, U., Lang, J., Procaccia, A.D. (eds.) Handbook of Computational Social Choice, pp. 57–84. Cambridge University Press, Cambridge (2016). https://doi.org/10.1017/CBO9781107446984.004
5. Charbit, P., Thomassé, S., Yeo, A.: The minimum feedback arc set problem is NP-hard for tournaments. Comb. Probab. Comput. **16**, 1–4 (2007). https://doi.org/10.1017/S0963548306007887
6. Copeland, A.H.: A reasonable social welfare function. Technical report, Mimeo, University of Michigan (1951)
7. Gutin, G.Z., Mertzios, G.B., Reidl, F.: Searching for maximum out-degree vertices in tournaments. CoRR abs/1801.04702 (2018), http://arxiv.org/abs/1801.04702
8. Herlocker, J.L., Konstan, J.A., Terveen, L.G., Riedl, J.: Evaluating collaborative filtering recommender systems. ACM Trans. Inf. Syst. **22**(1), 5–53 (2004). https://doi.org/10.1145/963770.963772
9. Hudry, O.: A survey on the complexity of tournament solutions. Math. Soc. Sci. **57**(3), 292–303 (2009). https://doi.org/10.1016/j.mathsocsci.2008.12.002
10. Laslier, J.F.: Tournament Solutions and Majority Voting, vol. 7. Springer, Heidelberg (1997)
11. Maarek, Y.: Alexa and her shopping journey. In: Proceedings of the 27th ACM International Conference on Information and Knowledge Management, CIKM 2018, p. 1. ACM, New York (2018). https://doi.org/10.1145/3269206.3272923
12. Moon, J.W.: Topics on Tournaments. Holt, Rinehart, and Winston, New York (1968)
13. Reid, K.B.: Tournaments. In: Gross, J.L., Yellen, J., Zhang, P. (eds.) Handbook of Graph Theory, 2nd edn. Chapman and Hall/CRC, New York (2013)
14. Shen, J., Sheng, L., Wu, J.: Searching for sorted sequences of kings in tournaments. SIAM J. Comput. **32**, 1201–1209 (2003). https://doi.org/10.1137/S0097539702410053
15. Slater, P.: Inconsistencies in a schedule of paired comparisons. Biometrika **48**(3/4), 303–312 (1961)

A New Linear-Time Algorithm
for Centroid Decomposition

Davide Della Giustina[1], Nicola Prezza[2(✉)], and Rossano Venturini[2]

[1] Scuola Superiore, University of Udine, Udine, Italy
dellagiustina.davide@spes.uniud.it
[2] Department of Computer Science, University of Pisa, Pisa, Italy
nicola.prezza@di.unipi.it, rossano.venturini@unipi.it

Abstract. The centroid of a tree is a node that, when removed, breaks the tree in connected components of size at most half of that of the original tree. By recursing this procedure on the components, one obtains the centroid decomposition of the tree, also known as centroid tree. The centroid tree has logarithmic height and its construction is a powerful pre-processing step in several tree-processing algorithms. The folklore recursive algorithm for computing the centroid tree runs in $O(n \log n)$ time. To the best of our knowledge, the only result claiming $O(n)$ time is unpublished and relies on (dynamic) heavy path decomposition of the original tree. In this short paper, we describe a new simple and practical linear-time algorithm for the problem based on the idea of applying the folklore algorithm to a suitable decomposition of the original tree.

1 Introduction

The *centroid decomposition* of a tree $\mathcal{T}$ (also known as separator decomposition) is a popular and powerful technique to obtain a tree $\mathcal{T}_C$ of logarithmic height. The centroid tree in employed in several applications: cache-oblivious string B-trees [2,5,6], dynamic farthest point queries [1], balanced decomposition of simple polygons [9], jumbled pattern matching on trees [7], counting of square substrings in a tree [11], just to cite a few.

The decomposition is based on a theorem proved by Jordan in 1869 [10]: *Any tree $\mathcal{T}$ of n nodes has at least a node, called centroid, whose removal leaves connected components of size at most $n/2$.*

The centroid decomposition is defined recursively. Given $\mathcal{T}$, we identify a centroid node u, which is chosen to be the root of the new tree $\mathcal{T}_C$. Then, we remove u from $\mathcal{T}$ and recurse on each connected component to get u's subtrees in $\mathcal{T}_C$. The resulting decomposition is a new tree $\mathcal{T}_C$ on the same nodes whose height is $O(\log n)$. Tree $\mathcal{T}_C$ preserves some information about the topology of

Partially supported by the project MIUR-SIR CMACBioSeq ("Combinatorial methods for analysis and compression of biological sequences") grant n. RBSI146R5L and by the project MIUR-PRIN 2017 "Algorithms, Data Structures and Combinatorics for Machine Learning".

N. R. Brisaboa and S. J. Puglisi (Eds.): SPIRE 2019, LNCS 11811, pp. 274–282, 2019.
https://doi.org/10.1007/978-3-030-32686-9_20

the original tree $\mathcal{T}$. For example, for any pair of nodes u and v, the path from u to v in $\mathcal{T}$ can be decomposed in two subpaths of $\mathcal{T}_C$: the path from u to w and the path from w to v, where w is the lowest common ancestor of u and v in $\mathcal{T}_C$.

A folklore algorithm computes the centroid decomposition in $\Theta(n \log n)$ time as follows. We first observe that a centroid node of $\mathcal{T}$ can be easily identified in linear time. Indeed, we can arbitrary choose a root in $\mathcal{T}$ and visit the tree to compute the size of each subtree. Then, we start from the root and move to the largest subtree until we reach a node whose subtrees have size at most $n/2$. This node is a centroid of the tree. Thus, it easily follows that the decomposition of the tree can be computed in $\Theta(n \log n)$ time.

The first linear time algorithm to compute the centroid decomposition of a tree is due to Giubas et al. [9] but it assumes that $\mathcal{T}$ is a binary tree. The first linear time algorithm for arbitrary trees is by Brodal et al. [3]. Actually paper [3] claims the result which is described in its unpublished extension [4]. This algorithm is based on the *heavy path decomposition* [12] of $\mathcal{T}$ which is kept updated after subtrees removal. Let us say a node u is *heavy* if u is the children of its parent with the largest subtree (ties are broken arbitrarily). The heavy path decomposition is the set of paths, called heavy paths, that connect heavy nodes. Brodal et al. [3] show that the use of this decomposition leads to an alternative description of the folklore algorithm. First the algorithm computes the heavy path decomposition of $\mathcal{T}$, and then searches for the centroid node which must be a node in the heavy path that contains the root. The algorithm can now recur on each connected component. The main inefficiency of this algorithm is that it recomputes the sizes of all subtrees and the heavy paths for each recursive call. Brodal et al. [3] improves the algorithm by showing how to update the already computed heavy paths in $O(\log^2 k)$ time, where k is the number of nodes of the component processed by the current recursive call. This requires to keep a binary search tree for each heavy path supporting split, join and successor operations, and a priority queue for each node of $\mathcal{T}$.

In this short paper, we describe a new simple and practical linear-time algorithm for the problem based on the idea of applying the folklore algorithm to a suitable (static) decomposition of the original tree.

2   The Algorithm

The overall idea of our algorithm is to break the input tree in $\Theta(n/\log n)$ subtrees of size $O(\log n)$ and replace each group with a node to form a new meta-tree. The core property that we exploit is that a centroid can be identified by navigating this meta-tree of size $\Theta(n/\log n)$, plus $O(\log n)$ nodes of the original tree. The strategy is then applied recursively on the connected components obtained by removing the centroid. After some level of recursion, we obtain components that are small enough so that their centroid decomposition can be pre-computed in a small table and thus retrieved in linear time.

Tree Cover. Let $\mathcal{T}$ be a rooted tree of size n. The notation $\pi(x)$, where x is a node of $\mathcal{T}$, denotes the parent of x (or NULL for the root). When two nodes u and

v are connected, we assume that both the edges (u, v) and (v, u) are present (this simplifies the description). We use the tree covering procedure described in [8, Sec. 2.1] to decompose $\mathcal{T}$ in $\Theta(n/\log n)$ sub-trees containing $\Theta(\log n)$ nodes each (except, possibly, the root). Two subtrees are either disjoint or intersect only at their common root. We make all subtrees disjoint as follows, with a procedure that also colors nodes in red or black. At the beginning, all nodes of $\mathcal{T}$ are colored black. When $k > 1$ subtrees share a common root x, we (i) delete x, (ii) create k new red nodes $x_1, \ldots, x_k$ and make each of them be the root of each of the k subtrees, and (iii) create a new black node x' with parent $\pi(x') = \pi(x)$ and let $x_1, \ldots, x_k$ be its children. The new node x' belongs to a new subtree containing only x'. We denote as $\mathcal{T}'$ the tree obtained from $\mathcal{T}$ by performing these splitting and coloring operations, and keep a map β mapping black nodes of $\mathcal{T}'$ to the corresponding nodes of $\mathcal{T}$ (note that there is a bijection between black nodes of $\mathcal{T}'$ and nodes of $\mathcal{T}$). Let $\pi'(x)$ denote the parent of node x in $\mathcal{T}'$. We extend β to red nodes x as $\beta(x) = \beta(\pi'(x))$ (i.e., we take the black parent of x and apply β). Figures 1 and 2 illustrate our tree covering procedure. In the example, we have $\beta(\bar{0}) = 0, \ldots, \beta(\bar{7}) = 7, \beta(\bar{8}') = \beta(\bar{8}_1) = \beta(\bar{8}_2) = 8, \beta(\bar{9}) = 9, \ldots, \beta(\overline{19}) = 19$.

From now on, when we say *subtree of* $\mathcal{T}'$ we always refer to the subtrees obtained by our modified tree covering procedure. Note that $\mathcal{T}'$ is divided in $\Theta(n/\log n)$ subtrees, each containing $O(\log n)$ nodes (some of these subtrees may contain just one node, read above). We denote as $\mathcal{T}''$ the tree whose nodes are the subtrees (now disjoint) of $\mathcal{T}'$. Note that $\mathcal{T}''$ has $\Theta(n/\log n)$ nodes as well. We store explicitly $\mathcal{T}''$ and keep a map α mapping nodes of $\mathcal{T}''$ to the roots of the corresponding subtrees in $\mathcal{T}'$. Figure 3 illustrates the tree $\mathcal{T}''$ obtained from the previous example (in the next paragraph we describe the meaning of the weights associated with nodes and edges). In the example, we have $\alpha(A) = \bar{0}$, $\alpha(B) = \bar{3}$, $\alpha(C) = \bar{8}'$, $\alpha(D) = \bar{8}_1$, $\alpha(E) = \overline{13}$, and $\alpha(F) = \bar{8}_2$.

For each node u'' of $\mathcal{T}''$, we compute and store the number $\delta(u'')$ of black nodes contained in the subtree of $\mathcal{T}'$ rooted in $\alpha(u'')$. We call $\delta(u'')$ the *weight* of u''. After that, with a visit of $\mathcal{T}''$ we cumulate those weights and extend them to edges as follows. Let (u'', v'') be an edge of $\mathcal{T}''$. With $\delta(u'', v'')$ we denote the sum of all weights $\delta(w'')$ in the connected component rooted in v'' obtained after removing u'' from $\mathcal{T}''$. Said otherwise, $\delta(u'', v'')$ is the sum of the weights $\delta(w'')$ for all nodes w'' reached traversing (u'', v'') only once (and without counting $\delta(u'')$). We store $\delta(u'', v'')$ for each edge (u'', v'') of $\mathcal{T}''$ (remember also that, for each edge (u'', v''), also the reversed edge (v'', u'') exists therefore $\delta(v'', u'')$ is also defined). Intuitively, $\delta(u'', v'')$ corresponds to the number of black nodes in one of the connected components (the one containing node $\alpha(v'')$) obtained after removing the subtree rooted in $\alpha(u'')$ from $\mathcal{T}'$. In turn, this is exactly the number of nodes in the corresponding connected component of $\mathcal{T}$, and will be used to quickly compute a centroid. Figure 3 illustrates our construction.

Finding a Centroid. We now show that a centroid of $\mathcal{T}$ can be found in $O(n/\log n)$ time by visiting $\mathcal{T}''$ and a small (logarithmic) number of nodes of $\mathcal{T}'$. We prove the following lemma:

Lemma 1. *The following two properties hold:*

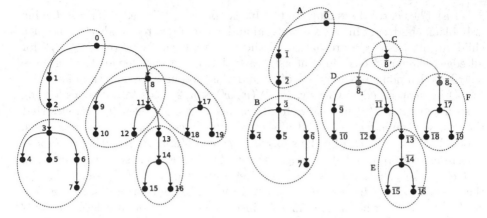

Fig. 1. Input tree $\mathcal{T}$, covered using the procedure described in [8, Sec. 2.1] with parameter $M = 4$.

Fig. 2. Modified tree $\mathcal{T}'$. We overline node names to distinguish them from those in $\mathcal{T}$. Note that sub-trees are disjoint.

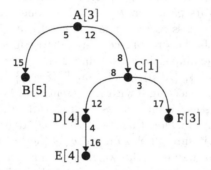

Fig. 3. Tree $\mathcal{T}''$, obtained by collapsing each sub-tree of $\mathcal{T}'$ into a node. Between square brackets, we show each node's weight (For example, $\delta(B) = 5$). We also show weights on the edges: for example, $\delta(C, F) = 3$ and $\delta(F, C) = 17$.

1. If c is a centroid of $\mathcal{T}$, then there exist a subtree $\mathcal{R} = (V_\mathcal{R}, E_\mathcal{R})$ of $\mathcal{T}'$ and a node $x' \in V_\mathcal{R}$ such that $\beta(x') = c$ and $\delta(x'', y'') \leq n/2$ for all (x'', y'') in $\mathcal{T}''$ such that $\alpha(x'')$ is the root of $\mathcal{R}$.
2. If u'' is a node of $\mathcal{T}''$ such that $\delta(u'', v'') \leq n/2$ for all (u'', v'') in $\mathcal{T}''$, then the subtree $\mathcal{R}$ rooted in $\alpha(u'')$ contains a node x' such that $\beta(x')$ is a centroid of $\mathcal{T}$.

Proof. Consider the function δ extended to edges of $\mathcal{T}'$ (we call it δ' to distinguish it from δ): $\delta'(u', v')$ is the number of black nodes in the tree containing v' obtained after removing node u'. Similarly, we will talk about the weights of edges of $\mathcal{T}$ (which are defined analogously). We start by proving claim (1), and consider two main cases.

(1.a) The centroid c is mapped to a black node c' of $\mathcal{T}'$ (i.e., $\beta(c') = c$) having only black children. Since c is a centroid and c is mapped to a node c' with black children, also its edges are preserved and thus we have that $\delta'(c', v') \leq n/2$ for all edges leaving c' (see, in contrast, case 1.b: in that case, edges leaving c are distributed among red children of c' and this property no longer holds). Let (c', v') be such an edge. Clearly, also $\delta'(v', w') \leq n/2$ holds for all edges leaving v' such that $w' \neq c'$: this follows from the fact that $\delta'(c', v') \leq n/2$ implies that the tree containing v' obtained after removing c' contains at most $n/2$ black nodes. Let $\mathcal{R} = (V_\mathcal{R}, E_\mathcal{R})$ be the subtree of $\mathcal{T}'$ containing c'. Iterating the above reasoning, we obtain that all edges (u', v') leaving $\mathcal{R}$ (i.e., such that $u' \in V_\mathcal{R}$ and $v' \notin V_\mathcal{R}$) satisfy $\delta'(u', v') \leq n/2$. Let z' be the root of $\mathcal{R}$, and let x'' be the node of $\mathcal{T}''$ such that $\alpha(x'') = z'$. Then, by definition of δ we have that $\delta(x'', y'') \leq n/2$ for all (x'', y'') in $\mathcal{T}''$, since δ' and δ coincide on edges leaving $\mathcal{R}$ and x'', respectively, and our claim holds for $x' = c'$.

(1.b) The centroid c is mapped to a black node c' of $\mathcal{T}'$ (i.e., c' is the only black node with $\beta(c') = c$) with only red children. Let $c'_1, \ldots, c'_k$ be the red children of c'. By construction of our tree decomposition, the edges leaving c have been partitioned. Each class of the partition contains either just the original edge connecting c with its parent in $\mathcal{T}$, or at least one edge connecting c with its children. The former class corresponds to the edge in $\mathcal{T}'$ connecting c' with its parent. Classes of the latter kind correspond to edges in $\mathcal{T}'$ connecting c' with red children. For example, in Figs. 1 and 2 the edges leaving node 8 have been partitioned in $\{(8,0)\}$, $\{(8,9), (8,11)\}$, and $\{(8,17)\}$. The first class $\{(8,0)\}$ contains just the edge connecting 8 with its parent, and becomes edge $(\bar{8}', \bar{0})$ in $\mathcal{T}'$. The latter two classes become edges $(\bar{8}', \bar{8}_1)$ and $(\bar{8}', \bar{8}_2)$ in $\mathcal{T}'$. Now, the fact that we collapse edges means that $\delta'(c', v') \leq n/2$ does not necessarily hold when v' is a red node (it surely holds only if v' is the black parent of c'), since $\delta'(c', v')$ corresponds to a sum of the weights of multiple edges. For example, in Fig. 2, $\delta'(\bar{8}', \bar{8}_1) = 8$ corresponds to the sum of the weights $\delta'(\bar{8}_1, \bar{9}) = 2$ and $\delta'(\bar{8}_1, \bar{11}) = 6$. While the weights of the latter two are surely at most $n/2$ (by definition of centroid), their sum could exceed $n/2$ (this is not the case of Fig. 2, where $n/2 = 10$). We consider two further sub-cases. (1.b.1) $\delta'(c', c'_k) \leq n/2$ for all edges leaving c' (this is the case of Fig. 2). Then, the same argument used in case (1.a) applies (it is actually simpler, since the subtree containing c' is a singleton subtree), and our claim holds with $x' = c' = \alpha(x'')$ and $\mathcal{R}$ being the singleton subtree containing c'. (1.b.2) $\delta'(c', c'_k) > n/2$ for at least one edge (c', c'_k) leaving c'. Then, our claim holds for $x' = c'_k = \alpha(x'')$ and $\mathcal{R}$ being the subtree containing c'_k: since $\delta'(c', c'_k) > n/2$, then $\delta'(c'_k, c') \leq n/2$. Moreover, the other edges (c'_k, w') leaving c'_k correspond to edges of the original tree $\mathcal{T}$ (i.e., not to group of edges), therefore $\delta'(c'_k, w') \leq n/2$. We can apply the argument used in case (1.a) and conclude that $\delta(x'', y'') \leq n/2$ for all edges leaving x''.

We now prove claim (2). Let u'' be a node of $\mathcal{T}''$ such that $\delta(u'', v'') \leq n/2$ for all (u'', v'') in $\mathcal{T}''$. Let moreover $\mathcal{R}$ be the subtree rooted in $\alpha(u'') = y'$. Then, $\delta'(y', w') \leq n/2$, where w' is the parent of y' in $\mathcal{T}'$. We have two cases. (2.a) $\delta'(y', w') \leq n/2$ for all children w' of y' in $\mathcal{T}'$. Then, clearly $\beta(y')$ is a centroid

of $\mathcal{T}$: if y' is a red node, then the weights of edges leaving $\beta(y')$ are at most $n/2$. If y' is a black node then the weights of edges leaving $\beta(y')$ correspond precisely to those leaving y'. (2.b) $\delta'(y', w') > n/2$ for some child w' of y'. Then, $\delta'(w', y') \leq n/2$ (i.e., the edge leading to the parent of w' weights at most $n/2$) and we can recurse the above reasoning to the children of w'. Clearly, sooner or later we will find a node q' in $\mathcal{R}$ such that $\delta'(q', w') \leq n/2$ for all edges leaving q' (including the edge leading to its parent, for which the property always holds true if we have moved to q'). Otherwise, it is easy to see that we obtain a contradiction. Suppose we reach a node q' of $\mathcal{R}$ such that some of its children lie outside $\mathcal{R}$. Denote $q'_1, \ldots, q'_k$ the children of q' leaving $\mathcal{R}$. Then, clearly $\delta'(q', q'_i) \leq n/2$ for all $i = 1, \ldots, k$ must hold since we are assuming that $\delta(u'', v'') \leq n/2$ for all edges (u'', v'') leaving u'' in $\mathcal{T}''$. This shows that we can recurse only on children internal to $\mathcal{R}$. However, at some point we will reach a node x' whose children (all of them) lie outside $\mathcal{R}$. Then, clearly all edges leaving x' must weight at most $n/2$.   □

By Lemma 1, this algorithm finds a centroid of $\mathcal{T}$ in $O(n/\log n + \log n)$ time: **(i)** visit $\mathcal{T}''$ and find a node u'' such that $\delta(u'', v'') \leq n/2$ for all edges (u'', v'') leaving u''. Let $\mathcal{R}$ be the subtree of $\mathcal{T}'$ associated with u'', i.e., the subtree rooted in $\alpha(u'')$. **(ii)** Visit $\mathcal{R}$ and find a node u' such that, if removed, it splits $\mathcal{T}'$ into connected components having at most $n/2$ black nodes each. **(iii)** Return $\beta(u')$.

Step **(i)** takes $O(n/\log n)$ time. Step **(ii)** can be implemented with a visit of $\mathcal{R}$. Note that the weights we store on $\mathcal{T}'''$'s edges are precisely the sizes of the connected components of $\mathcal{T}$ obtained after removing nodes u' in $\mathcal{R}$ whenever (u', v') is an edge that leaves $\mathcal{R}$. Since we can afford visiting the whole $\mathcal{R}$, those weights can be easily used to compute the sizes of the subtrees of $\mathcal{T}$ obtained after removing any node u' in $\mathcal{R}$. Steps **(i)**, **(ii)** run in $O(n/\log n + \log n)$ time.

In the above example, we have $n = 20$. The node of $\mathcal{T}''$ whose outgoing edges weight at most $n/2 = 10$ is C (its outgoing edges weight 8, 8, and 3). In this particular case, C corresponds to a unary subtree therefore step (ii) finds node $\bar{8}'$, and step (iii) returns $\beta(\bar{8}') = 8$.

Recursion. Note that Lemma 1 does not make any assumption on the subtree-decomposition of $\mathcal{T}'$. It follows that the above algorithm for finding a centroid can be iterated as follows. After finding a node u' of $\mathcal{T}'$ such that $\beta(u')$ is a centroid of $\mathcal{T}$, we remove from $\mathcal{T}'$ all nodes v' such that $\beta(v') = \beta(u')$ (i.e., all nodes that map to the centroid). We break every subtree $\mathcal{R}$ of $\mathcal{T}'$ containing one of the removed nodes into one singleton subtree (i.e., a subtree consisting of just one node) per remaining node of $\mathcal{R}$ (i.e., one subtree for each node that was not removed). The process of removing nodes breaks the original tree $\mathcal{T}'$ into q trees $\mathcal{T}'_1, \ldots, \mathcal{T}'_q$, for some $q \geq 2$, each of which contains at most $n/2$ black nodes. Crucially, note that each $\mathcal{T}'_i$ with n_i nodes is partitioned into at most $O(n_i/\log n) + O(\log n)$ subtrees: those of the original tree $\mathcal{T}'$ that have not been split into singleton subtrees, and at most $O(\log n)$ singleton subtrees. Similarly, we break $\mathcal{T}''$ into a forest. Some of the trees belonging to this forest will contain new nodes corresponding to newly-created singleton subtrees in $\mathcal{T}'$. Each such new node u'' gets a weight $\delta(u'') = 1$. The weight of the other nodes does not

change, since they correspond to subtrees of $\mathcal{T}'$ that have not been modified. At this point, we can re-compute the weights $\delta(u'', v'')$ on the edges of the forest in overall $O(n/\log n + \log n)$ time by using the stored weights $\delta(u'')$ for each node u'' of the forest. Figure 4 shows how the trees of Figs. 2 and 3 change after removing all the nodes u' such that $\beta(u') = 8$.

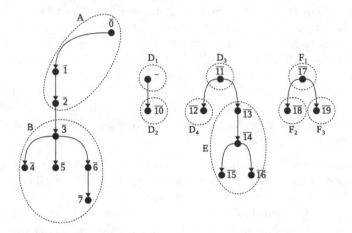

Fig. 4. The figure shows how $\mathcal{T}'$ changes after returning the centroid $\beta(\bar{8}') = 8$ and removing nodes $\bar{8}'$, $\bar{8}_1$, and $\bar{8}_2$. Each subtree of $\mathcal{T}'$ that contains a removed node (in particular, the subtrees D,F) has been split in singleton subtrees: D has been split in $D_1, \ldots, D_4$, and F has been split in $F_1, \ldots, F_3$. Note that node C disappears since it corresponds to a subtree of $\mathcal{T}'$ containing only a removed node ($\bar{8}'$). Tree $\mathcal{T}''$ changes similarly: it is broken into four trees whose nodes are the subtrees shown in the figure.

Lemma 1 can then be applied again recursively on $\mathcal{T}'_1, \ldots, \mathcal{T}'_q$. It is easy to see that each connected component at recursion depth j has at most $n/2^j$ black nodes and is divided into at most $n/(\log n \cdot 2^j) + O(j \cdot \log n)$ subtrees (note that each recursive iteration adds at most $O(\log n)$ singleton subtrees to each component). The complexity of finding a centroid in such a tree using Lemma 1 is $O(n/(\log n \cdot 2^j) + j \cdot \log n + \log n)$ (i.e., number of subtrees plus size of a subtree). We stop recursion as soon as we obtain components of size at most $\log^3 n$, i.e., at recursive depth $j = \log(n/\log^3 n)$. In this way, each component is a tree having at most $\log^3 n$ black nodes and divided into at most $n/(\log n \cdot 2^j) + O(j \cdot \log n) = O(\log^2 n)$ subtrees. Note that the base case of $\log^3 n$ for the tree size is the minimum (asymptotically) guaranteeing that the two components contributing to the number of subtrees (i.e., $n/(\log n \cdot 2^j)$ and $j \cdot \log n$) sum up to $O(n'/\log n)$, n' being the subtree's size. The total number of nodes contained in the trees at each recursion depth j is $O(n)$ and, by the above observation, applying Lemma 1 to one tree of size n' at any recursion depth takes time $O(n'/\log n)$. Overall, this adds up to $O(n/\log n)$ time per recursion level. Since the recursion depth is $O(\log n)$, the overall procedure terminates in $O(n)$ time. We obtain the following lemma.

Lemma 2. *In $O(n)$ time we can reduce the centroid decomposition problem to the same problem on a certain number of trees with at most $\log^3 n$ nodes each, whose union contains at most n nodes.*

We note that, in a practical implementation of the above algorithm, it is sufficient to apply the folklore algorithm to the trees of Lemma 2 (which are small enough to fit in cache). From the theoretical perspective, however, this solution runs in $O(n \log \log n)$ time. We now show how to reach linear time (though with a less practical solution).

Intuitively, we perform one more round of our recursive strategy on the trees of Lemma 2, obtaining trees of size $z = O(\log^3 \log n)$. Finally, we use tabulation: there are $o(n/\log n)$ possible trees with z nodes, so we can pre-compute their centroid decomposition in $O(n)$ time with the folklore algorithm. The following theorem states our final result.

Theorem 1. *The centroid decomposition of a tree with n nodes can be computed in $O(n)$ time and $O(n)$ space.*

Proof. By Lemma 2, we obtain trees of size at most $\log^3 n$. Our idea is to apply one more round of our recursive strategy to those trees. As a result, in $O(n)$ additional time we reduce the problem to that of computing the centroid decomposition of a certain number of trees of size at most $(\log \log^3 n)^3 = 27 \log^3 \log n$ whose union contains at most n nodes. The trees are now small enough to use tabulation. The number of distinct (rooted) trees with at most $z = 27 \log^3 \log n$ nodes is upper-bounded by $N = 2^{2z} = 2^{54 \log^3 \log n}$. We compute the centroid decomposition of each of them in total $O(N \cdot z \log z) = o(n)$ time using the folklore algorithm. We store the centroid tree of each of these trees in a table $U[k][p]$ indexed by the number k of nodes of the tree and a unique identifier p representing the rank of the tree among all trees with k nodes. This identifier can be, for example, the $2k$-bits integer corresponding to the balanced parentheses representation of the tree, which can be computed in linear time with a DFS visit. Table U takes $O(N \cdot z^2)$ words of space, which is again $o(n)$.

We use the table as follows. Given an unrooted tree $\mathcal{T}^*$ with $k' \leq 27 \log^3 \log n$ nodes, we root it arbitrarily[1] (storing the permutation associating nodes of the rooted and unrooted versions of $\mathcal{T}^*$), we compute its DFS-identifier p', and access $U[k'][p']$. This entry contains the centroid tree decomposition of the (rooted version) of $\mathcal{T}^*$. The process takes $O(k')$ (linear) time, therefore by applying the procedure to all those small trees we complete the centroid decomposition of our input tree $\mathcal{T}$ in additional $O(n)$ time. □

References

1. Aronov, B., et al.: Data structures for halfplane proximity queries and incremental voronoi diagrams. In: Proceedings of the 7th Latin American Symposium on Theoretical Informatics (LATIN), pp. 80–92 (2006). https://doi.org/10.1007/11682462_12

---

[1] Note that the centroid decomposition does not depend on how the tree is rooted.

2. Bender, M.A., Farach-Colton, M., Kuszmaul, B.C.: Cache-oblivious string b-trees. In: Proceedings of the Twenty-Fifth ACM SIGACT-SIGMOD-SIGART Symposium on Principles of Database Systems (PODS), pp. 233–242 (2006). https://doi.org/10.1145/1142351.1142385

3. Brodal, G.S., Fagerberg, R., Pedersen, C.N.S., Östlin, A.: The complexity of constructing evolutionary trees using experiments. In: Proceedings of the 28th International Colloquium on Automata, Languages and Programming (ICALP), pp. 140–151 (2001). https://doi.org/10.1007/3-540-48224-5_12

4. Brodal, G.S., Fagerberg, R., Pedersen, C.N.S., Östlin, A.: The complexity of constructing evolutionary trees using experiments. Technical Report BRICS-RS-01-1, BRICS, Department of Computer Science, University of Aarhus (2001). https://www.brics.dk/RS/01/1/BRICS-RS-01-1.pdf

5. Ferragina, P.: On the weak prefix-search problem. Theoret. Comput. Sci. **483**, 75–84 (2013). https://doi.org/10.1016/j.tcs.2012.06.011

6. Ferragina, P., Venturini, R.: Compressed cache-oblivious string b-tree. ACM Trans. Algorithms (TALG) **12**(4), 52:1–52:17 (2016). https://doi.org/10.1145/2903141

7. Gagie, T., Hermelin, D., Landau, G.M., Weimann, O.: Binary jumbled pattern matching on trees and tree-like structures. Algorithmica **73**(3), 571–588 (2015). https://doi.org/10.1007/s00453-014-9957-6

8. Geary, R.F., Raman, R., Raman, V.: Succinct ordinal trees with level-ancestor queries. ACM Trans. Algorithms (TALG) **2**(4), 510–534 (2006)

9. Guibas, L.J., Hershberger, J., Leven, D., Sharir, M., Tarjan, R.E.: Linear-time algorithms for visibility and shortest path problemsinside triangulated simple polygons. Algorithmica **2**, 209–233 (1987). https://doi.org/10.1007/BF01840360

10. Jordan, C.: Sur les assemblages de lignes. Journal für die reine und angewandte Mathematik **70**, 185–190 (1869)

11. Kociumaka, T., Pachocki, J., Radoszewski, J., Rytter, W., Walen, T.: Efficient counting of square substrings in a tree. Theoret. Comput. Sci. **544**, 60–73 (2014). https://doi.org/10.1016/j.tcs.2014.04.015

12. Sleator, D.D., Tarjan, R.E.: A data structure for dynamic trees. J. Comput. Syst. Sci. **26**(3), 362–391 (1983). https://doi.org/10.1016/0022-0000(83)90006-5

Computational Biology

COBS: A Compact Bit-Sliced
Signature Index

Timo Bingmann[1][✉], Phelim Bradley[2], Florian Gauger[1], and Zamin Iqbal[2]

[1] Institute of Theoretical Informatics,
Karlsruhe Institute of Technology, Karlsruhe, Germany
`bingmann@kit.edu`
[2] European Molecular Biology Laboratory,
European Bioinformatics Institute, Cambridge, UK

Abstract. We present COBS, a COmpact Bit-sliced Signature index, which is a cross-over between an inverted index and Bloom filters. Our target application is to index k-mers of DNA samples or q-grams from text documents and process approximate pattern matching queries on the corpus with a user-chosen coverage threshold. Query results may contain a number of false positives which decreases exponentially with the query length. We compare COBS to seven other index software packages on 100 000 microbial DNA samples. COBS' compact but simple data structure outperforms the other indexes in construction time and query performance with Mantis by Pandey et al. in second place. However, unlike Mantis and other previous work, COBS does not need the complete index in RAM and is thus designed to scale to larger document sets.

1 Introduction

In this paper we present an approximate q-gram index named COBS [13], short for COmpact Bit-sliced Signature index, which is a cross-over between an inverted index and Bloom filters. The current focus of COBS is to index DNA and protein k-mers from sequencing experiments, but the data structure can also be used for indexing q-grams from other domains such as English text.

In living cells, DNA exists as long contiguous molecules, typically textually encoded as strings of A, C, G, and T. Experimental methods for "reading" DNA have been developing rapidly; there are various approaches, but all involve breaking the DNA and "reading" (typically called "sequencing") those fragments (these short strings are typically called "reads"). Read lengths started out moderately long (500–1000 characters) in the late 1990s, dropped down to 30 characters in 2008 with the advent of massively parallel technologies, and in the recent past, bleeding edge technologies have enabled reading of fragments as long as 1 million characters, albeit with a higher error rate.

The output of sequencing experiments are stored both in raw format (text files of the read strings) and "assembled format" – semi-heuristic best approximations to the underlying genome, also in text format, but of very variable quality,

© Springer Nature Switzerland AG 2019
N. R. Brisaboa and S. J. Puglisi (Eds.): SPIRE 2019, LNCS 11811, pp. 285–303, 2019.
https://doi.org/10.1007/978-3-030-32686-9_21

in particular when based on short read data. Unambiguous reconstruction of the original string from the substrings is mathematically impossible unless the fragments are longer than the longest repeated substring. Another complication is that a great deal of data is generated by sequencing unknown mixtures of different genomes (e.g. mixtures of bacteria from within the human gut, or samples from humans infected by three different types of malaria parasite), making it very hard to reconstruct the underlying genomes.

As sequencing technology has advanced, it has also become much cheaper and more widespread, and its output has been stored in publicly available archives, e.g. the European Nucleotide Archive (ENA) and the Sequence Read Archive (SRA) which maintain mirrors of all the data. These archives now double in size every 18 months, and it is progressively more important to be able to search within the stored datasets, to find important genes or mutations, or combinations of mutations which are informative of function or ancestry. All of these search queries can be expressed in terms of exact or approximate matching of strings. In 2018, the ENA encompassed $1.5 \cdot 10^9$ microbial sequences and $8 \cdot 10^{15}$ base pairs (i.e. characters) of read data [17], while the European Bioinformatics Institute reached 160 PB of storage capacity [10].

Despite the obvious similarities to standard document retrieval problems, the properties of DNA k-mer data are very different from traditional text corpora. Google's index is reported to have in the order of 10^{13} documents containing 10^8 unique terms [6], whereas the small benchmark set of 100 000 microbial sequences used in our experiments already contain $2.2 \cdot 10^{10}$ distinct 31-mers, of which $1.8 \cdot 10^{10}$ occur only once. The frequency of terms in a natural language is power-law distributed, with underlying terms generated over hundreds of years, resulting in just a few new terms per document. Microbial genomes however encode many billions of years of evolution; each new genome generates thousands of novel k-mers. There are also two other aspects whereby searching biological data differs from standard text retrieval. The first is that the index must support *approximate queries* allowing detection of closely related DNA to the query. Approximate pattern matching however is a notoriously difficult subject for text indices [22,27]. The second is that users often want all hits, not just the top few as is typical in web search.

For COBS we chose the robust q-gram indexing approach [36] and combined it with Bloom filters to reduce the term space size. This can be considered a variant of *signature files*, which have a long history in information retrieval [12] but were pushed to the sidelines for text search by inverted indexes [41]. Recently, they have been reconsidered as acceleration filters for large text search corpora [15] by engineering them to adapt to the collection's characteristics. With COBS we venture to combine signature files with one-sided errors introduced by Bloom filters and inverted files to design an ultra fast and scalable q-gram index which supports approximate queries delivering a small reasonable number of expected false positives. Our contribution of making the signature files *compact* first enables the index to be applied to corpora with highly varying document sizes, such as microbial DNA samples.

After reviewing related work in the following subsection, we present the new COBS index design in Sect. 2. In Sect. 3 we then report on our experimental evaluation of COBS and seven other k-mer indexing software packages.

1.1 Related Work

Considering q-grams or k-mers of a sequence are a staple in bioinformatics [8].

The earliest use of Bloom filters as an index for a collection of independent documents we could find is called *Bloofi* by Crainiceanu and Lemire [11]. They propose to use a Bloom filter for each document and to arrange them either in a B-tree or as a *Flat-Bloofi*. The latter is similar to BIGSI and COBS without compaction.

The currently most cited line of work on DNA k-mer indices for approximate search are the *Sequence Bloom Trees* (SBTs) first proposed by Solomon and Kingsford [32]. In an SBT the k-mers of each document are indexed into individual Bloom filters, which are then arranged as the leaves of a binary tree. The inner nodes of the binary tree are union Bloom filters of their descendants. A query can then breadth-first traverse the tree, pruning search paths which no longer sufficiently cover a given threshold Θ of the query k-mers.

In the original SBT [32] a simple greedy clustering method is used, the bit union is stored in each inner node, and all nodes are RRR compressed [31] using SDSL [14]. The first improvement, the *Split Sequence Bloom Tree* (SSBT) [33], splits the inner nodes into two Bloom filters: a *similarity* filter and a *remainder* filter, where the first contains all bits in both child filters and the second those set in either child minus the *similarity* filter. This representation allows descendant nodes to omit storing the bits in the *similarity* filter explicitly, hence reducing space requirements while retaining the same information.

Simultaneously, Sun, Harris, Chikhi, and Medvedev proposed the *AllSome Sequence Bloom Tree* (AllSome-SBT) [34], which splits each inner node into an *all* and a *some* subfilter. The *all* filter contains bits in all leaves below the node, excluding those already set in the parent node, and the *some* filter all bits in some leaves but not all. Again, this representation allows exclusion of bits already known from the parent node's filters, and thus reducing space and enabling better compression. Furthermore, the AllSome-SBT also improves on the clustering methods by employing an agglomerative hierarchical technique and by constructing batch Bloom filters for large query sets.

The currently smallest SBT variant is called *HowDe Sequence Bloom Tree* (HowDe-SBT) by Harris and Medvedev [16]. It decomposes the Bloom filters in each inner node into two bit vectors: the *det* vector signals if a particular bit is *determined* at this inner node, meaning that it is equal in all descendant leaves, and the *how* vector signals if it is determined as zero or one. All determined bits can be omitted from any children. These two bit vectors are exactly the information needed to perform an efficient breadth-first search down the tree. Furthermore, the authors introduce a *culling* process to remove sparse inner nodes which don't reveal much information and thus speed up queries.

A completely different approach to indexing k-mers is taken by *Mantis* from Pandey et al. [28]. In Mantis, a counting quotient filter (CQF) [29] is used to construct a mapping from k-mers to *color classes*, wherein k-mers with identical occurrence vectors for all documents are mapped to the same color class. Incidence of color classes to documents can then be represented as a matrix, in which columns are associated with documents and each row corresponds to a color class. Hence, bits set in the rows signal occurrence of any k-mer mapping to the color in the corresponding document list. Mantis then compresses the bit vectors in the color matrix using RRR or with a spanning tree based approach. The k-mer mapping is built from CQFs constructed by *Squeakr* [30], a k-mer counting tool. Mantis differs from the other k-mer indexes referenced in this paper by being able to deliver *exact* approximate matching results without false positives.

SeqOthello [39] is another k-mer index software package. It contains an "ensemble" of encoding techniques for compressing the occurrence maps of k-mers in the document set. Occurrence maps are then grouped depending on their density and encoding into disjoint buckets. To locate the correct occurrence map for a k-mer, a hierarchy of *Othello*s is built inside each bucket and over all bucket *Othello*s. An *Othello* [38] is a minimum perfect hash function mapping, which is fast and scalable but can introduce false positive results due to mapping of alien k-mers to random results.

BIGSI (BItsliced Genomic Signature Index) by Bradley et al. [5] is the direct ancestor of COBS and also a combination of Bloom filters and inverted indexes. BIGSI however is a prototype programmed in Python and uses a key-value database such as BerkeleyDB or RocksDB as storage back-end. It also does not contain the compaction feature introduced in COBS.

Related to k-mer indexing are *colored de Bruijn graph* representation data structures, which often contain an exact k-mer index but do not support approximate k-mer pattern searches. The original implementation, *Cortex* [20,21], stored k-mers in a hash table, along with booleans for the four possible forward and backward edges in a single byte. This was then followed by *McCortex* [35], which added a second data structure to encode paths in the graph present in the original reads. By contrast, *VARI* [26], *Rainbowfish* [1], and *pufferfish* [2] explore use of succinct data structures, the Burrows-Wheeler transform, and minimal perfect hash functions to save space and possibly even accelerate operations. The *Bloom Filter Trie* by Holley et al. [19] is another colored de Bruijn graph representation based on the *burst trie* [18], wherein lookups for suffixes at compressed inner nodes are accelerated with Bloom filters.

2   A Compact Bit-Sliced Signature Index

In this section we present the index structure used in COBS. We first generally review Bloom filters as a q-gram index in Subsect. 2.1, then turn to COBS' more compact bit-sliced representation in Subsect. 2.2, and discuss implementation details and algorithm engineering aspects in Subsect. 2.3.

2.1   Approximate Matching with Bloom Filters of Signatures

Given are an ordered set of *documents* $\mathcal{D} = [d_0, \ldots, d_{|\mathcal{D}|-1}]$, where each document d is composed of a set of *strings* $\{t_0, \ldots, t_{|d|-1}\}$. The number of items in a set or array is denoted with $|\cdot|$. Each string t is a zero-based array of $|t|$ characters from a finite ordered *alphabet* Σ. In the context of indexing DNA, the alphabet is usually $\{A, C, G, T\}$, the documents are experiment samples, and the strings in each document can be reads or assembled genome sequences. When indexing web sites, the alphabet may be the ASCII characters or English words, the documents could be web pages, and the substrings may be words, sentences, or paragraphs.

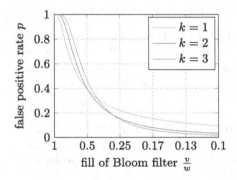

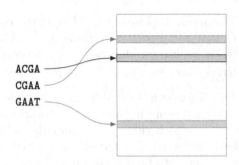

Fig. 1. Theoretical false positive rate of Bloom filters given fill and number of hash functions.

Fig. 2. Access pattern of the classical bit-sliced index.

To facilitate approximate pattern matching we consider *q-grams* of the strings [36], commonly called *k-mers* for DNA. For each string t with $|t| \geq q$ there are $|t| - q + 1$ consecutive substrings of length q. For a document d, we denote with $G_q(d)$ the union of all q-grams in the strings in d. Due to similarities with full-text indexing we also refer to the q-grams in a document as *terms*.

A COBS index is composed of $|\mathcal{D}|$ Bloom filters [4], each representing an approximate membership data structure with one-sided error. To construct a Bloom filter for a document d we assume k pairwise independent hash functions $h_0, \ldots, h_{k-1}$ with range $[0, w)$ and set the k bits $h_i(s)$ in an array f of w bits for each q-gram $s \in G_q(d)$. Testing for membership of a q-gram s is performed by checking if all k cells $h_i(s)$ are set, which can lead to false positives but never false negatives.

The entire document collection is thus represented by $|\mathcal{D}|$ bit arrays $[f_0, \ldots, f_{|\mathcal{D}|-1}]$, each a Bloom filter with possibly different parameters. From previous work, the false positive rate p of a Bloom filter of size w with k hash functions and v inserted elements is known to be at most $(1 - (1 - \frac{1}{w})^{kv})^k \leq (1 - e^{-kv/w})^k$. Given a desired false positive rate p and number of elements v, one

can calculate a partial derivative of the last bound to determine good approximate parameters $k = \frac{w}{v} \ln 2$ and $w = -\frac{v \ln p}{(\ln 2)^2}$ [7,24].

To perform approximate matching for a pattern P, we follow previous work [36] and determine the q-gram distance of P to all documents in the collection $\mathcal{D}$ by testing each of the query's q-grams $G_q(P)$ on all documents. In COBS we present this positively as the q-gram *score* of the query for each document. The score is used to rank and return all documents containing at least a given percentage K of the $|G_q(P)|$ terms in the query.

As Solomon and Kingsford already noticed for SBTs, in the case of approximate pattern search on Bloom filters, we are not interested in the false positive rate of a *single* Bloom filter lookup. Instead we are concerned with the false positive rate of a *query* P. More precisely, given $\ell = |G_q(P)|$ q-grams with the probability that more than $K\ell$ terms are false positives in the same filter.

Theorem 1 (False Positive Rate of a Query, Theorem 2 in [32]). *Let P be a query pattern containing $\ell = |G_q(P)|$ distinct terms. If we consider the terms as being independent, the probability that more than $\lfloor K\ell \rfloor$ false-positive terms occur in a filter f with false positive rate p is $1 - \sum_{i=0}^{\lfloor K\ell \rfloor} \binom{\ell}{i} p^i (1-p)^{\ell-i}$.*

This theorem is derived by considering lookups of terms as independent Bernoulli trials and summing over the probability of zero to $\lfloor K\ell \rfloor$ false positives among the ℓ trials, which yields a binomial distribution. Given $K \geq p$, Solomon and Kingsford also apply a Chernoff bound and show that the false positive probability for a query to be detected in a document is $\leq \exp(-\ell(K-p)^2/(2(1-p)))$.

These repeated trials into the Bloom filter allow us to push the false positive rate p up higher than commonly used. Figure 1 shows the false positive rate $(1 - e^{-kv/w})^k$ of Bloom filters depending on its fill $\frac{v}{w}$ and the number of hash functions k. Traditional uses of Bloom filters for approximate membership queries consider an error rate of 0.01 or less and multiple hash functions as desirable. Due to the inverse exponential relationship of a query's false positive rate with its length, coupled with the fact that more hash functions cost more cache faults or I/Os, the minimum $k = 1$ and a high false positive rate around 0.3 are desirable for our q-gram index application.

For example, if we consider a query of length 100 containing $\ell = 70$ distinct 31-grams, a false positive rate of $p = 0.3$, and threshold $K = 0.5$, then Theorem 1 yields a false positive rate of about 0.000143. Which means there will be about 143 false positive results in one million documents on average.

2.2 Bit-Slicing and Compaction

Provided all Bloom filters are of the same size w, one can store them as a $w \times |\mathcal{D}|$ bit matrix such that a row contains all bit cells at one index in the $|\mathcal{D}|$ filters (see left side of Fig. 3). This is also called a "bit-sliced" layout [37] and was chosen for BIGSI and COBS to reduce the number of random accesses needed to evaluate a query. Each row of a term can be scanned sequentially, as shown in Fig. 2. This is particularly important if the index is read from external memory, where

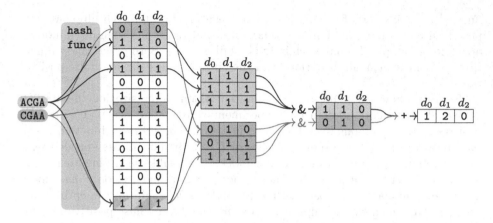

Fig. 3. Architecture of the bit-sliced signature index and query processing steps.

scanning is much more efficient than random accesses. The approach however requires all Bloom filters to use the same hash functions and be the same size.

Figure 3 also illustrates how a query P is performed using the bit-sliced Bloom filter matrix. The q-grams of the query are hashed to determine the corresponding rows. These $k|G_q(P)|$ rows are then scanned and an *AND* join of k rows is performed to determine which q-grams occur in which document. This yields an indicator bit vector ordered by document number. All indicator vectors are then added together to calculate the score for each document. Only those documents reaching the query threshold $K|G_q(P)|$ are then reported as approximate matches. Due to the one-sided error of the Bloom filters, only *more* documents may be reported due to hash collisions; false negatives, i.e. missed hits, cannot occur.

One can also view the Bloom filter bit matrix as an inverted index: each row simply lists the document numbers containing the corresponding q-gram as indexes in a bit vector. Unlike a traditional inverted index however, *multiple* q-gram terms are superimposed in one row. This leads to false positive matches. In theory, one could apply all the methods developed by the information retrieval community [40] to these bit vectors or posting lists.

The current version of a bit-sliced index however relies on all documents and resulting Bloom filters having the same size. But larger documents result in denser bit vectors and smaller documents in sparser, as the number of bits set depends on the number of q-gram terms in the document. Depending on the dataset, this creates vastly different false positive rates in the bit matrix. Hence, we propose to *adapt* the size of each Bloom filter bit array to the document it indexes and aim to keep the false positive rate *constant*. We call this a *compact bit-sliced signature index* (the CO in COBS).

In theory one could adapt the Bloom filter size and hash function for each document. In practice we want to store bits of rows as blocks of size $\Theta(B)$ in external memory, thus keep the parameters constant for B consecutive documents.

Furthermore, instead of calculating a new hash function for each filter, we propose to use only one function with a larger output range and then use a modulo operation to map it down to each individual filter's size. Both practical optimizations only incur a small deviation from the optimal index size and false positive rates.

Figure 4 shows in light blue the desired Bloom filter size for the 100 000 microbial documents used in our experiments ordered by size and with false positive rate 0.3 and one hash function. The dark blue staircase function above the upward sloping curve shows batches of $B = 8\,192$ documents encoded with the maximum Bloom filter size of that block. The visible dark blue area is the minor overhead for encoding documents block-wise. If one uses only one Bloom filter size (the classic approach), then the index size would be the entire filled orange area, which extends upward to ensure the desired false positive rate for the largest document.

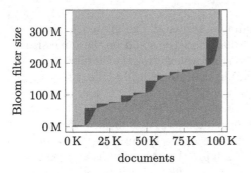

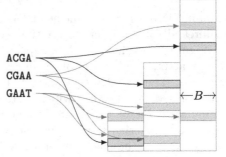

Fig. 4. Compact index composed of small sub-indexes containing B documents.

Fig. 5. Access pattern of our compact bit-sliced index.

Due to the variance in size of microbial and other real-world documents, the *compact* representation in COBS is essential. In designing COBS, we also considered that today's SSD and NVMe storage technology now has orders of magnitude faster random access speeds [3] compared to rotational disks. Thus with these new storage devices, the batched random access for many smaller blocks of size B, as used in the compact layout and illustrated in Fig. 5, first becomes viable.

2.3   Implementation and Engineering

We implemented COBS as a command line search engine tool using C++ and plan to provide a Python interface to the underlying algorithm library. The tool is open source and available from https://panthema.net/cobs/. It can read DNA FASTA files, multi-document protein FASTA files, McCortex, or text files as documents and extract q-grams from them. Depending on the format, the input data is broken into different q-gram sets: DNA reads are for example

hashed independently, while English text is processed continuously. We used xxHash [9] for hashing the q-gram strings. The q-grams or k-mers can optionally be canonicalized if their reverse complement are considered equivalent.

Classic and Compact. The COBS program can currently construct two index variants: *classic* (ClaBS) and *compact* (COBS). In the classic index all documents are hashed using the same Bloom filter size, which depends on the desired false positive rate and the number of q-grams in the *largest* document. This is the non-compact version, which is similar to BIGSI, but was written for performance in C++ and with direct file accesses. We will refer to it as **ClaBS** in the experimental results.

When constructing a compact index, the size of all documents are determined and the document set is reordered by size. Then a subindex is constructed for every B documents, as described in Subsect. 2.2. Each subindex is actually a ClaBS index. The subindices are simply concatenated into one large file.

While classic indexes with the same parameters can be concatenated straightforwardly, compact indexes are more difficult to merge. We may implement this in future versions of COBS by keeping some slack in the $\Theta(B)$ blocks and packing new documents into the best free block or by storing the subindices as separate files. This would allow incremental augmentation of COBS compact indices.

Parallelization. Due to the massive amount of data to process, we parallelized construction and query for shared-memory systems. ClaBS index construction we parallelized by building temporary indexes over batches of the documents and then merging them into larger indexes. For compact index construction we parallelized construction of the subindices.

Pattern search in COBS can optionally be parallelized by processing disjoint partitions of the document scores in parallel and then selecting the top scores sequentially using a partial sort operation.

Memory Mapped I/O. For querying an index, we map the file into virtual address space using mmap. The necessary rows of the inverted Bloom filter index are then read using simple memory transfers. We experimented with directly issuing asynchronous I/O commands, but found only a negligible performance advantage that did not outweigh the higher code complexity.

Alternatively, COBS can also read the complete index into RAM and then run all queries. This was added to compare performance against other indexing software which only work in RAM, e.g. Mantis, in Sect. 3.

Single-Instruction Multiple-Data (SIMD). Besides the I/O bottleneck, extracting the bits from the index rows and adding them together required a considerable amount of running time in the query.

In the *ADD* step of the query process (Fig. 3), the rows are summed up to create the query result. In this illustration we hid the fact that the rows that are output from the *AND* step are bit-packed: each cell is represented by one bit. In the output of the *ADD* step, however, each document's score is represented by an integer specifying the number of matched query terms. This poses a problem

3.1 Results

In this section we present and discuss the results of our experiments with the eight index software packages. The machine we selected for the experiments is a large server-class platform with 80 cores and large amounts of RAM. While these properties are always good, we primarily chose it due to the 8 TB of fast SSD storage, which is many times faster than traditional rotational disks. For rapidly performing the experiments, this storage speed was crucial.

On the other hand the fast storage speed and massive multi-core processing power in our machine may highlight different aspects in the indexing software than previous comparisons. Most prominently, algorithms which previously only had to process data rates known from rotational disks (100s of MiB/s) may become a bottleneck when dealing with SSD speeds (currently around 10 GiB/s). Furthermore, most of the index software packages had no built-in provisioning for utilizing multi-core parallelism. While we were able to accelerate embarrassingly parallel parts of the construction using bash (like creating Bloom filters for each file), in some software the main index build was still sequential. On the other hand, one can argue that index construction time is not as important as query performance, but it still limits scalability.

Table 2 shows our results from all eight software packages for only 1 000 microbial DNA documents. The steps in the construction of each index are shown as separate rows if it was possible to measure these independently.

Table 2. Construction wall-clock time, CPU time, memory usage, and resulting index size for 1 000 microbial documents and all k-mer index software in our experiment

Phase	SBT	SSBT	AllSome-SBT	HowDe-SBT	Seq-Othello	Mantis	BIGSI	ClaBS	COBS
	Construction wall-clock time in seconds								
Count	2 018	1 974	1 954	1 959					
Bloom	114	117	140	144	295	232	1 881		
Build	3 097	21 378	1 401	68 034	2 225	987	2 574	99	43
Compress	1 768	5 187	80	3 802		45			
Total	6 996	28 657	3 576	73 939	2 520	1 264	4 455	99	43
	Construction CPU (User) time in seconds								
Count	4 574	4 511	4 475	4 488					
Bloom	11 133	10 967	10 234	10 278	28 123	19 162	169 345		
Build	855	5 178	449	66 872	2 198	943	1 767	1 604	1 430
Compress	1 569	4 832	1 663	2 857		3 423			
Total	18 131	25 489	16 821	84 495	30 320	23 527	171 113	1 604	1 430
	Construction maximum RSS memory usage in MiB								
Count	518	518	518	518					
Bloom	641	640	640	640	634	1 756	4 244		
Build	11 028	1 523	7 140	108 147	12 137	88 357	246 806	16 245	2 616
Compress	10 953	992	560	963		16 613			
Maximum	11 028	1 523	7 140	108 147	12 137	88 357	246 806	16 245	2 616
	Index size in MiB								
Size	19 844	3 254	21 335	1 911	4 410	16 486	27 794	16 236	3 022

We show both wall-clock time and CPU user time such that parallelized construction can highlight its speedup without obscuring the actual amount of computation. Table 3 considers the time to run the query sets. We only show wall-clock for queries due to space; all query computations are performed with a single thread such that this is a fair comparison. Furthermore, for ClaBS and COBS the index is *completely loaded* into RAM such that the comparison with the others is fair. In future, it will become important to measure how many bytes were read from the disks per query, but in the current comparison we assume all index data is resident in RAM.

Considering construction wall-clock time, COBS is clearly the fastest index taking only 43 s on 1 000 documents. ClaBS is a factor 2.3 slower, Mantis a factor 30 slower, SeqOthello a 59 factor, and AllSome-SBT a factor 83 slower than COBS. The same is reflected in construction CPU time, with COBS being fastest and taking 1430 s. ClaBS is a factor 1.12 slower, AllSome-SBT a factor 11.8 slower, and Mantis a factor 16.5.

Table 3. Query wall-clock time for 1 000 microbial documents and all k-mer index software in our experiment

Phase	SBT	SSBT	AllSome-SBT	HowDe-SBT	Seq-Othello	Mantis	BIGSI	ClaBS	COBS
ℓ	Query wall-clock time in seconds								
31 bp r0	31	80	20	34	62	12	281	10	8
31 bp r2	26	76	19	33	62	13	289	9	8
100 bp r0	663	3183	100	600	73	22	783	14	9
100 bp r2	649	3153	95	588	73	23	455	14	9
1000 bp r0	794	3466	112	670	63	21	660	15	10
1000 bp r2	781	3435	108	659	64	27	310	13	10
10000 bp r0	802	3273	112	622	62	23	699	16	11
10000 bp r2	790	3243	111	613	62	22	316	15	11
Total r0–r2	6 775	29 833	1 007	5 710	783	252	5 177	154	114
	Document false positive rate for 31 bp queries								
Rate	0.004	0.004	0.004	0.004	0.001	0.000	0.027	0.024	0.227

One can also see that we parallelized the Bloom filter construction (the "bloom" row) effectively for all indexes, while the build steps are usually only partially parallelized. COBS has a CPU/wall-clock speedup of 33, while BIGSI has 38, Mantis has 18, and SeqOthello 12. However, since COBS performs *the least amount* of computation and has among the highest speedups, the combination of these two factors really diminishes wall-clock construction time. Considering CPU user time, the index requiring most work for construction is BIGSI, probably due to the Python implementation. It however is parallelized, such that the wall-clock time is on par with the SBTs.

The amount of RAM required by the indexing software also limits their applicability, especially if the complete index itself needs to be constructed in RAM. BIGSI, HowDe-SBT, and Mantis have the highest main memory usage

in the experiment. For BIGSI and Mantis memory was the limiting scalability factor, while for HowDe-SBT the construction time grew too long.

The index sizes of all packages for the 1 000 microbial documents was smaller than the input in McCortex format (41 GiB input size). The software with the smallest index was the HowDe-SBT with only 1.9 GiB, followed by COBS with around 3.0 GiB and SSBT with 3.3 GiB.

In terms of query performance, the fastest index was COBS with 114 s to run all query sets three times, followed by ClaBS with 154 s. Mantis was a factor 2.2 slower, SeqOthello a factor 6.9 slower, and the fastest SBT version, AllSome-SBT, was a factor 8.8 slower.

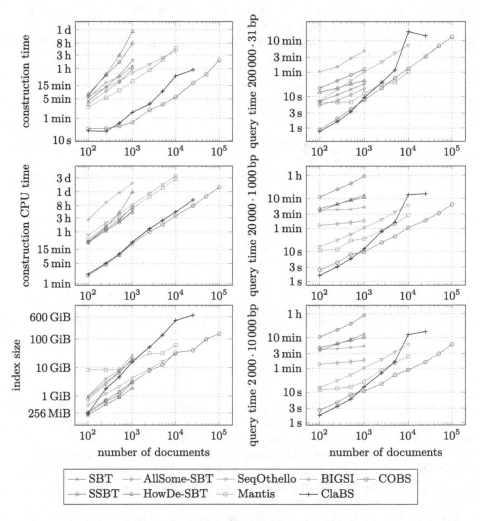

Fig. 6. Construction time, index size, query time for $200\,000 \cdot 31$ bp, $20\,000 \cdot 1\,000$ bp, and for $2\,000 \cdot 10\,000$ bp round 2 after disk cache warm-up.

Using result checkers we verified that all software packages calculated correct results and counted the false positives contained in the returned list of the single k-mer query set ($\ell = 31$). The notable exception was SeqOthello, which produced false positives consistently for each multi-k-mer query and started returning false negative (missing) results when run on the 10 000 dataset. We could not investigate this issue further. The SBT variants and SeqOthello showed a very low false positive rate less than 0.5 %. Mantis produced zero false positives as expected. BIGSI and ClaBS are nearly identical in underlying data structure design, and deliver around 2.6 % false positives on single k-mer queries. COBS is designed to deliver about the prescribed error rate of 0.3, hence the 22.7 %

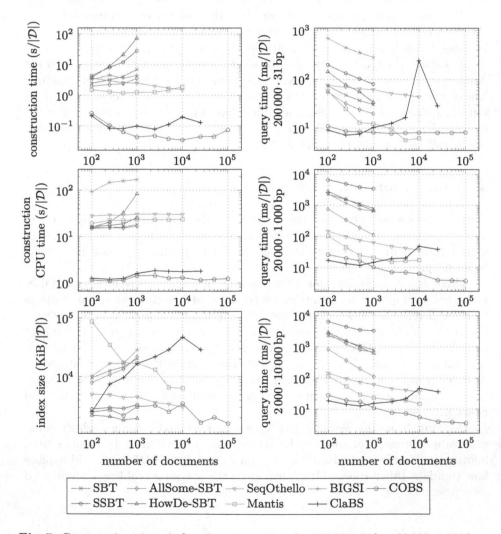

Fig. 7. Construction time, index size, query time for $200\,000 \cdot 31\,\mathrm{bp}$, $20\,000 \cdot 1\,000\,\mathrm{bp}$, and for $2\,000 \cdot 10\,000\,\mathrm{bp}$ in the first round divided by the number of documents $|\mathcal{D}|$.

false positives, which enables us to construct a more compact index. We also calculated the number of false positives in larger multi-k-mer query sets, and found all indexes except SeqOthello but including COBS to return *zero* false positives for all queries with $\ell \geq 100$ in the experiment.

Figure 6 shows scaling results for all software packages on increasing subsets of the indexed microbial document set. We skipped running the SBT variants for data sets larger than 10 000 because their construction time was growing super-linearly. SeqOthello and Mantis scaled much better in terms of construction time per document. Figure 7 shows construction time per document. These plots show that COBS scales well, with an order of magnitude faster construction time per document than Mantis and SeqOthello, both in wall-clock and CPU time. While ClaBS's index size appears to increase with the number of documents (due to the maximum document size), the size per document of COBS actually decreases because it can better pack documents into blocks.

As expected COBS' query time for single k-mers increases linearly with the number of documents in the index, due to the scoring method without pruning. The query time of all other indexes also increases with the number of documents, but not quite linearly. The best index in terms of query time increase per document is the AllSome-SBT followed by HowDe-SBT, but only COBS index scales to our full 100 000 microbial dataset.

4   Conclusion and Future Work

With COBS we presented a signature index based on Bloom filters which enables approximate pattern matching on large q-gram datasets. It outperforms all other q-gram indexes in both construction and query time for multi-q-gram queries due to its simple data structure.

There are many avenues for future work on possible improvements to COBS' ideas. For example, dynamic operations on the index such as insertion, replacement, and removal of documents are very important for practical applications. We already provide a merge operation for classic indexes, but not for compact ones. Our current COBS implementation also already supports querying of multiple index files, such that a frontend may select different datasets or categories. Another important topic is better support for batch or bulk queries. And for further scalability it is important to explore distributed index construction and query processing.

Deriving from the simplicity of COBS are research avenues which could explore compression of rows in the Bloom filter matrix using techniques from information retrieval. And similar to Mantis' use of the CQF one could explore how to adapt other Bloom filter variants to the indexing problem with allowed false positives.

References

1. Almodaresi, F., Pandey, P., Patro, R.: Rainbowfish: a succinct colored de Bruijn graph representation. In: 17th International Workshop on Algorithms in Bioinformatics (WABI). LIPIcs, vol. 88, pp. 18:1–18:15. Schloss Dagstuhl, August 2017. preprint bioRxiv:138016

2. Almodaresi, F., Sarkar, H., Srivastava, A., Patro, R.: A space and time-efficient index for the compacted colored de Bruijn graph. Bioinformatics **34**(13), i169–i177 (2018)

3. Bingmann, T.: NVMe "disk" bandwidth and latency for batched block requests, March 2019. Online Article, http://panthema.net/2019/0322-nvme-batched-block-access-speed

4. Bloom, B.H.: Space/time trade-offs in hash coding with allowable errors. Commun. ACM **13**(7), 422–426 (1970)

5. Bradley, P., den Bakker, H.C., Rocha, E.P.C., McVean, G., Iqbal, Z.: Ultrafast search of all deposited bacterial and viral genomic data. Nat. Biotechnol. **37**, 152–159 (2019)

6. Brin, S., Page, L.: The anatomy of a large-scale hypertextual web search engine. Comput. Networks ISDN Syst. **30**(1–7), 107–117 (1998)

7. Broder, A.Z., Mitzenmacher, M.: Network applications of Bloom filters: a survey. Internet Math. **1**(4), 485–509 (2003)

8. Chikhi, R., Holub, J., Medvedev, P.: Data structures to represent sets of k-long DNA sequences. Computing Research Repository (CoRR), arXiv:1903.12312:1–16, March 2019

9. Collet, Y.: xxHash: extremely fast non-cryptographic hash algorithm, 2014. Git repository. https://github.com/Cyan4973/xxHash. Accessed July 2019

10. Cook, C.E., Lopez, R., Stroe, O., Cochrane, G., Brooksbank, C., Birney, E., Apweiler, R.: The European Bioinformatics Institute in 2018: tools, infrastructure and training. Nucleic Acids Res. **47**(D1), D15–D22 (2019)

11. Crainiceanu, A., Lemire, D.: Bloofi: multidimensional bloom filters. Inf. Syst. **54**, 311–324 (2015)

12. Faloutsos, C., Christodoulakis, S.: Signature files: an access method for documents and its analytical performance evaluation. ACM Trans. Inf. Syst. (TOIS) **2**(4), 267–288 (1984)

13. Gauger, F.: Engineering a compact bit-sliced signature index for approximate search on genomic data. Master Thesis. Karlsruhe Institute of Technology, Germany, February 2018

14. Gog, S., Beller, T., Moffat, A., Petri, M.: From theory to practice: plug and play with succinct data structures. In: Gudmundsson, J., Katajainen, J. (eds.) SEA 2014. LNCS, vol. 8504, pp. 326–337. Springer, Cham (2014). https://doi.org/10.1007/978-3-319-07959-2_28

15. Goodwin, B., et al.: BitFunnel: revisiting signatures for search. In: 40th ACM SIGIR Conference on Research and Development in Information Retrieval, pp. 605–614. ACM, August 2017

16. Harris, R.S., Medvedev, P.: Improved representation of sequence Bloom trees. bioRxiv, pp. 501452, December 2018

17. Harrison, P.W., et al.: The european nucleotide archive in 2018. Nucleic Acids Res. **D47**(1), D84–D88 (2019)

18. Heinz, S., Zobel, J., Williams, H.E.: Burst tries: a fast, efficient data structure for string keys. ACM Trans. Inf. Syst. (TOIS) **20**(2), 192–223 (2002)

19. Holley, G., Wittler, R., Stoye, J.: Bloom filter trie: an alignment-free and reference-free data structure for pan-genome storage. Algorithms Mol. Biol. **11**(1), 3 (2016)
20. Iqbal, Z., Caccamo, M., Turner, I., Flicek, P., McVean, G.: De novo assembly and genotyping of variants using colored de Bruijn graphs. Nat. Genet. **44**(2), 226 (2012)
21. Iqbal, Z., Turner, I., McVean, G.: High-throughput microbial population genomics using the cortex variation assembler. Bioinformatics **29**(2), 275–276 (2012)
22. Krugel, J.: Approximate Pattern Matching with Index Structures. Ph.D. thesis, Technische Universität München, Germany, February 2016
23. Marçais, G., Kingsford, C.: A fast, lock-free approach for efficient parallel counting of occurrences of k-mers. Bioinformatics **27**(6), 764–770 (2011)
24. Mitzenmacher, M., Upfal, E.: Probability and Computing: Randomized Algorithms and Probabilistic Analysis. Cambridge University Press, Cambridge (2005)
25. Mohamadi, H., Khan, H., Birol, I.: ntCard: a streaming algorithm for cardinality estimation in genomics data. Bioinformatics **33**(9), 1324–1330 (2017)
26. Muggli, M.D., et al.: Succinct colored de Bruijn graphs. Bioinformatics **33**(20), 3181–3187 (2017). preprint bioRxiv:040071
27. Navarro, G., Baeza-Yates, R.A., Sutinen, E., Tarhio, J.: Indexing methods for approximate string matching. IEEE Bull. Tech. Committee Data Eng. **24**(4), 19–27 (2001)
28. Pandey, P., Almodaresi, F., Bender, M.A., Ferdman, M., Johnson, R., Patro, R.: Mantis: a fast, small, and exact large-scale sequence-search index. Cell Systems, June 2018. preprint bioRxiv:217372
29. Pandey, P., Bender, M.A., Johnson, R., Patro, R.: A general-purpose counting filter: making every bit count. In: ACM International Conference on Management of Data, pp. 775–787. ACM (2017)
30. Pandey, P., Bender, M.A., Johnson, R., Patro, R.: Squeakr: an exact and approximate k-mer counting system. Bioinformatics **34**(4), 568–575 (2018). preprint bioRxiv:122077
31. Raman, R., Raman, V., Srinivasa Rao, S.: Succinct indexable dictionaries with applications to encoding k-ary trees and multisets. In: 13th ACM-SIAM Symposium on Discrete Algorithms (SODA), pp. 233–242. SIAM, January 2002
32. Solomon, B., Kingsford, C.: Fast search of thousands of short-read sequencing experiments. Nat. Biotechnol. **34**(3), 300–312 (2016)
33. Solomon, B., Kingsford, C.: Improved search of large transcriptomic sequencing databases using split sequence Bloom trees. J. Comput. Biol. **25**(7), 755–765 (2018)
34. Sun, C., Harris, R.S., Chikhi, R., Medvedev, P.: AllSome sequence Bloom trees. J. Computat. Biol. **25**(5), 467–479 (2018)
35. Turner, I., Garimella, K.V., Iqbal, Z., McVean, G.: Integrating long-range connectivity information into de Bruijn graphs. Bioinformatics **34**(15), 2556–2565 (2018)
36. Ukkonen, E.: Approximate string-matching with q-grams and maximal matches. Theoret. Comput. Sci. **92**(1), 191–211 (1992)
37. Wong, H.K.T., Liu, H.-F., Olken, F., Rotem, D., Wong, L.: Bit transposed files. In 11th International Conference on Very Large Data Bases (VLDB), pp. 448–457. VLDB Endowment, August 1985
38. Ye, Y., Belazzougui, D., Qian, C., Zhang, Q.: Memory-efficient and ultra-fast network lookup and forwarding using othello hashing. IEEE/ACM Trans. Networking **26**(3), 1151–1164 (2018)
39. Ye, Y., et al.: SeqOthello: querying RNA-seq experiments at scale. Genome Biol. **19**(1), 167 (2018). preprint bioRxiv:258772

40. Zobel, J., Moffat, A.: Inverted files for text search engines. ACM Comput. Surveys (CSUR) **38**(2), 6 (2006)
41. Zobel, J., Moffat, A., Ramamohanarao, K.: Inverted files versus signature files for text indexing. ACM Trans. Database Syst. (TODS) **23**(4), 453–490 (1998)

An Index for Sequencing Reads Based on the Colored de Bruijn Graph

Diego Díaz-Domínguez[1,2(✉)]

[1] Department of Computer Science, University of Chile, Santiago, Chile
diediaz@dcc.uchile.cl
[2] CeBiB — Center for Biotechnology and Bioengineering,
University of Chile, Santiago, Chile

Abstract. In this article, we show how to transform a colored de Bruijn graph (dBG) into a practical index for processing massive sets of sequencing reads. Similar to previous works, we encode an instance of a colored dBG of the set using *BOSS* and a color matrix C. To reduce the space requirements, we devise an algorithm that produces a smaller and more sparse version of C. The novelties in this algorithm are (i) an incomplete coloring of the graph and (ii) a greedy coloring approach that tries to reuse the same colors for different strings when possible. We also propose two algorithms that work on top of the index; one is for reconstructing reads, and the other is for contig assembly. Experimental results show that our data structure uses about half the space of the plain representation of the set (1 Byte per DNA symbol) and that more than 99% of the reads can be reconstructed just from the index.

Keywords: de Bruijn graphs · DNA sequencing · Compact data structures

1 Introduction

A set of *sequencing reads* is a massive collection $R = \{R_1, \ldots, R_n\}$ of n overlapping short strings that together encode the sequence of a DNA sample. Analyzing this kind of data allows scientists to uncover complex biological processes that otherwise could not be studied. There are many ways for extracting information from a set of reads (see [27] for review). However, in most of the cases, the process can be reduced to build a *de Bruijn graph* (dBG) of the collection and then search for graph paths that spell segments of the source DNA (see [6,15,28] for some examples).

Briefly, a dBG is a directed labeled graph that stores the transitions of the substrings of size k, or *kmers*, in R. Constructing it is relatively simple, and

Partially supported by Basal Funds FB0001, Conicyt, Chile; by a Conicyt Ph.D. Scholarship; by Fondecyt Grants 1-171058 and 1-170048, Chile; and by the European Union's Horizon 2020 research and innovation programme under the Marie Sklodowska-Curie [grant agreement No. 690941].

© Springer Nature Switzerland AG 2019
N. R. Brisaboa and S. J. Puglisi (Eds.): SPIRE 2019, LNCS 11811, pp. 304–321, 2019.
https://doi.org/10.1007/978-3-030-32686-9_22

the resulting graph usually uses less space than the input text. Nevertheless, this data structure is lossy, so it is not always possible to know if the label of a path matches a substring of the source DNA. The only paths that fulfill this property are those in which all nodes, except the first and last, have indegree and outdegree one [16]. Still, they represent just a fraction of the complete dBG.

More branched parts of the graph are also informative, but traverse them requires extra information to avoid spelling incorrect sequences. A simple solution is to augment the dBG with colors, in other words, we assign a particular color c_i to every string $R_i \in R$, and then we store the same c_i in every edge that represents a kmer of R_i. In this way, we can walk over the graph always following the successor node colored with the same color of the current node.

The idea of coloring dBGs was first proposed by Iqbal et al. [15]. Their data structure, however, contemplated a union dBG built from several string collections, with colors assigned to the collections rather than particular strings. Considering the potential applications of colored dBGs, Boucher et al. [4] proposed a succinct version of the data structure of Iqbal et al. In their index, called VARI, the topology of the graph is encoded using $BOSS$ [5], and the colors are stored separately from the dBG in a binary matrix C, in which the rows represent the kmers and the columns represent the colors. Since the work of Boucher et al., some authors have tried to compress and manipulate C even further; including that of [2,13,25], while others, such as [21] and [22] have proposed methods to store compressed and dynamic versions C.

An instance of a colored dBG for a single set R can also be encoded using a color matrix. The only difference though is that the number of columns is proportional to the number of sequences in R. Assigning a particular color to every sequence is not a problem if the collection is of small or moderated size. However, massive datasets are rather usual in Bioinformatics, so even using a succinct representation of C might not be enough. One way to reduce the number of columns is to reuse colors for those sequences that do not share any kmer in the dBG. Alpanahi et al. [1] addressed this problem, and showed that it is unlikely that the minimum-size coloring can be approximated in polynomial time.

Alpanahi et al. also proposed a heuristic for recoloring the colored dBG of a set of sequences that, in practice, dramatically reduces the number of colors when R is a set of sequencing reads. Their coloring idea, however, might still produce incorrect sequences, so the applications of their version of the colored dBG are still limited.

Our Contributions. In this article, we show how to use a colored dBG to store and analyze a collection of sequencing reads succinctly. Similarly to VARI, we use $BOSS$ and the color matrix C to encode the data. However, we reduce the space requirements by partially coloring the dBG and greedily reusing the same colors for different reads when possible. We also propose two algorithms that work on top of the data structure, one for reconstructing the reads directly from the dBG and other for assembling contigs. We believe that these two algorithms can serve as a base to perform Bioinformatics analyses in compressed space.

Our experimental results show that on average, the percentage of nodes in $BOSS$ that need to be colored is about 12.4%, the space usage of the whole index is about half the space of the plain representation of R (1 Byte/DNA symbol), and that more than 99% of the original reads can be reconstructed from the index.

2   Preliminaries

DNA Strings. A DNA sequence R is a string over the alphabet $\Sigma = \{\mathtt{a}, \mathtt{c}, \mathtt{g}, \mathtt{t}\}$ (which we map to $[2..5]$), where every symbol represents a particular nucleotide in a DNA molecule. The DNA *complement* is a permutation $\pi[2..\sigma]$ that reorders the symbols in Σ exchanging $\mathtt{a}$ with $\mathtt{t}$ and $\mathtt{c}$ with $\mathtt{g}$. The *reverse complement* of R, denoted R^{rc}, is a string transformation that reverses R and then replaces every symbol $R[i]$ by its complement $\pi(R[i])$. For technical convenience we add to Σ the so-called *dummy* symbol \$, which is always mapped to 1.

de Bruijn Graph. A de Bruijn graph (dBG) [7] of the string collection $R = \{R_1, \ldots, R_n\}$ is a labeled directed graph $G = (V, E)$ that encodes the transitions between the substrings of size k of R, where k is a parameter. Every node $v \in V$ is labeled with a unique $k - 1$ substring of R. Two nodes v and u are connected by a directed edge $(v, u) \in E$ if the $k - 2$ suffix of v overlaps the $k - 2$ prefix of u and the k-string resulted from the overlap exists as substring in R. The label of the edge is the last symbol of the label of node u.

Rank and Select Data Structures. Given a sequence $B[1..n]$ of elements over the alphabet $\Sigma = [1..\sigma]$, $\mathtt{rank}_b(B, i)$ with $i \in [1..n]$ and $b \in \Sigma$, returns the number of times the element b occurs in $B[1..i]$, while $\mathtt{select}_b(B, i)$ returns the position of the ith occurrence of b in B. For binary alphabets, B can be represented in $n + o(n)$ bits so that $\mathtt{rank}$ and $\mathtt{select}$ are solved in constant time [9]. When B has $m \ll n$ 1s, a compressed representation using $m \lg \frac{n}{m} + \mathcal{O}(m) + o(n)$ bits, still solving the operations in constant time, is of interest [26].

BOSS Index. The $BOSS$ data structure [5] is a succinct representation for dBGs based on the *Burrows-Wheeler Transform* (BWT) [8]. In this index, the labels of the nodes are regarded as rows in a matrix and sorted in reverse lexicographical order, i.e., strings are read from right to left. Suffixes and prefixes in R of size below $k - 1$ are also included in the matrix by padding them with \$ symbols in the right size (for suffixes) or the left side (for prefixes). These padded nodes are also called *dummy*. The last column of the matrix is stored as an array $K[1..\sigma]$, with $K[i]$ being the number of labels lexicographically smaller than any other label ending with character i. Additionally, the symbols of the outgoing edges of every node are sorted and then stored together in a single array E. A bit vector $B[1..|E|]$ is also set to mark the position in E of the first outgoing symbol of each node. The complete index is thus composed of the vectors E, K, and B. It can be stored in $|E|(\mathcal{H}_0(E) + \mathcal{H}_0(B))(1 + o(1)) + \mathcal{O}(\sigma \log n)$ bits,

where $\mathcal{H}_0$ is the zero-order empirical entropy [23, Sec 2.3]. This space is reached with a Huffman-shaped Wavelet Tree [18] for E, a compressed bitmap [26] for B (as it is usually very dense), and a plain array for K. Bowe et al. [5] defined the following operations over $BOSS$ to navigate the graph:

- outdegree(v): number of outgoing edges of v.
- forward(v, a): node reached by following an edge from v labeled with a.
- indegree(v): number of incoming edges of v.
- backward(v): list of the nodes with an outgoing edge to v.
- nodeLabel(v): label of node v.

The first four operations can be answered in $\mathcal{O}(\log \sigma)$ time while the last one takes $\mathcal{O}(k \log \sigma)$ time. For our purposes, we also define the following operations:

- forward_r(v, r): node reached by following the r-th outgoing edge of v in lexicographical order.
- label2Node(S): identifier in $BOSS$ of the node labeled with the $(k - 1)$-string S.

The function forward_r is a small variation forward, and it maintains the original time, while the function label2Node is the opposite of nodeLabel, but it also maintain its complexity in $\mathcal{O}(k \log \sigma)$ time.

Graph Coloring. The problem of coloring a graph $G = (V, E)$ consists of assigning an integer $c(v) \in [1..\omega]$ to each node $v \in V$ such that (i) no adjacent nodes have the same color and (ii) ω is minimal. The coloring is *complete* if all the nodes of the graph are assigned with one color, and it is *proper* if constraint (i) is met for each node. The chromatic number of a graph G, denoted by $\chi(G)$, is the minimum number of colors required to generate a coloring that is complete and proper. A coloring using exactly $\chi(G)$ colors is considered to be optimal. Determining if there is a feasible ω-coloring for G is well known to be NP-complete, while the problem of inferring $\chi(G)$ is NP-hard [17].

Colored dBG. The first version of the colored dBG [15] was described as a union graph G built from several dBGs of different string collections. The edges in G that encode the kmers of the *i-th* collection are assigned the color i. The compacted version of this graph [4] represents the topology of G using the $BOSS$ index and the colors using a binary matrix C, where the position $C[i, j]$ is set to true if the kmer represented by the *i-th* edge in the ordering of $BOSS$ is assigned color j. The rows of C are then stored using the compressed representation for bit vectors of [26], or using Elias-Fano encoding [10, 11, 24] if the rows are very sparse. In the single set version of the colored dBG, the colors are assigned to every string. Therefore, the number of columns in C grows with the size of the collection. Alipanahi et al. [1] noticed that we could reduce the space of C by using the same colors in those strings that have no common kmers. This new problem was named the *CDBG-recoloring*, and formally stated

as follows; given a set R of strings and its dBG G, find the minimum number of colors such that (i) every string $R_i \in R$ is assigned one color and (ii) strings having two or more kmers in common in G cannot have the same color. Alipanahi et al. [1] showed that an instance $I(G')$ of the *Graph-Coloring* problem can be reduced in polynomial time to another instance $I'(G)$ of the *CDBG-recoloring* problem. Thus, any algorithm that finds $\chi(G')$, also finds the minimum number of colors for dBG G. However, they also proved that the decision version of *CDBG-Recoloring* is NP-complete.

3   Definitions

Let $R = \{R_1, R_2, ..., R_n\}$ be a collection of n DNA sequencing *reads*, and let $R' = \{R_1, R_1^{rc}..R_n, R_n^{rc}\}$ be a collection of size $2n$ that contains the strings in R along with their DNA reverse complements. The dBG of order k constructed from R' is defined as $G_{R'}^k = (V, E)$, and an instance of *BOSS* for $G_{R'}^k$ is denoted as $BOSS(G_{R'}^k) = (V', E')$, where V' and E' include the dummy nodes and their edges. For simplicity, we will refer to $BOSS(G_{R'}^k)$ just as $BOSS(G)$. A node in V' is considered a *starting* node if its $k - 1$ label is of the form $\$A$, where $\$$ is a dummy symbol and A is a $k - 2$ prefix of one or more sequences in R'. Equivalently, a node is considered an *ending* node if its $k - 1$ label is of the form $A\$$, with $\$$ being a dummy and A being a $k - 2$ suffix of one or more sequences in R'. Nodes whose labels do not contain dummy symbols are considered *solid*, and solid nodes with at least one predecessor node with outdegree more than one are considered *critical*. For practical reasons, we define two extra functions, isStarting and isEnding that are used to check if a node is starting or ending respectively.

A *walk* P over the dBG of $BOSS(G)$ is a sequence $(v_0, e_0, v_1...v_{t-1}, e_t, v_t)$ where $v_0, v_1, ...v_{t-1}, v_t$ are nodes and $e_1..e_t$ are edges, e_i connecting v_{i-1} with v_i. P is a *path* if all the nodes are different, except possibly the first and the last. In such case, P is said to be a *cycle*. A sequence $R_i \in R$ is *unambiguous* if there is a path in $BOSS(G)$ whose label matches the sequence of R_i and if no pair of colored nodes in $(u, v) \in P$ share a predecessor node $v' \in P$. In any other case, R_i is *ambiguous*. Finally, the path P_i that spells the sequence of R_i is said to be *safe* if every one of its branching nodes has only one successor colored with the color of R_i.

We assume that R is a *factor-free* set, i.e., no $R_i \in R$ is also a substring of another sequence R_j, with $i \neq j$.

4   Coloring a dBG of Reads

In this section, we define a coloring scheme for $BOSS(G)$ that generates a more succinct color matrix, and that allows us to reconstruct and assemble unambiguous sequences of R'. We use the dBG of R' because most of the Bioinformatic analyses require the inspection of the reverse complements of the reads. Unlike previous works, the rows in C represent the nodes in $BOSS(G)$ instead of the edges.

A Partial Coloring. We make C more sparse by coloring only those nodes in the graph that are *strictly* necessary for reconstructing the sequences. We formalize this idea with the following lemma:

Lemma 1. *For the path of an unambiguous sequence $R_i \in R'$ to be safe we have to color the starting node s_i that encodes the $k - 2$ prefix of R_i, the ending node e_i that encodes the $k - 2$ suffix of R_i and the critical nodes in the path.*

Proof. We start a walk from s_i using the following rules: (i) if the current node v in the walk has outdegree one, then we follow its only outgoing edge, (ii) if v is a branching node, i.e., it has outdegree more than one, then we inspect its successor nodes and follow the one colored with the same color of s_i and (iii) if v is equal to e_i, then we stop the traversal.   □

Note that the successor nodes of a branching node are critical by definition, so they are always colored. On the other hand, nodes with outdegree one do not require a color inspection because they have only one possible way out.

Coloring the nodes s_i and e_i for every R_i is necessary; otherwise, it would be difficult to know when a path starts or ends. Consider, for example, using the solid nodes that represent the $k - 1$ prefix and the $k - 1$ suffix of R_i as starting and ending points respectively. It might happen that the starting point of R_i can also be a critical point of another sequence R_j. If we start a reconstruction from s_i and pick the color of R_j, then we will generate an incomplete sequence. A similar argument can be used for ending nodes. The concepts associated with our coloring idea are depicted in Figs. 2A and B.

Unsafe Coloring. As explained in Sect. 2, we can use the recoloring idea of [1] to reduce the number of columns in C. Still, using the same colors for unrelated strings is not safe for reconstructing unambiguous sequences.

Lemma 2. *Using the same color c for two unambiguous sequences $R_i, R_j \in R'$ that do not share any $k - 1$ substring might produce an unsafe path for R_i or R_j.*

Proof. Assume there is another pair of sequences $R_x, R_y \in R'$ that do not share any $k - 1$ subsequence either, to which we assign them color c'. Suppose that the paths of R_x and R_j crosses the paths of R_i and R_j such that the resulting dBG topology resembles a grid. In other words, if R_i has the edge (u, u') and R_j has the edge (v, v'), then R_x has the edge (u, v) and R_y has the edge (u', v'). In this scenario, v will have two successors, node v' from the path of R_j and some other node v'' from the path of R_x. Both v' and v'' are critical by definition so they will be colored with c and c' respectively. The problem is that node v' is also a critical node for R_y, so it will also have color c'. The reason is that u', a node that precedes v', appears in R_i and R_y. As a consequence, the path of R_x is no longer safe because one of its nodes (v in this example) has to successors colored with c'. A similar argument can be made for R_i and color c. Figure 1 depicts the idea of this proof.   □

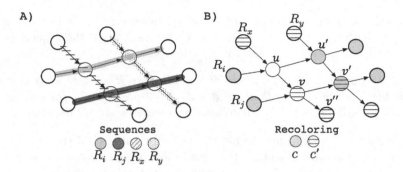

Fig. 1. Example of unsafe paths produced by a graph recoloring. (A) The dBG generated from the unambiguous sequences R_i, R_j, R_x and R_y. Every texture represents the path of a specific string. (B) Recolored dBG. Sequences R_i and R_j are assigned the same color c (light gray) as they do not share any $k-1$ substring. Similarly, sequences R_x and R_y are assigned another color c' (horizontal lines) as they do not share any $k-1$ sequence neither. Nodes u, u', v, v' and v'' are those mentioned in the Proof of Lemma 2. The sequences of R_i and R_x cannot be reconstructed as their paths become unsafe after the recoloring.

When spurious edges connect paths of unrelated sequences that are assigned the same color (as in the proof of Lemma 2), we can generate chimeric strings if, by error, we follow one of those edges. In the algorithm, we solve this problem by assigning different colors to those strings with sporadic edges, even if they do not share any $k-1$ substring.

Safer and Greedy Coloring. Our greedy coloring algorithm starts by marking in a bitmap $N = [1..|V'|]$ the p nodes of $BOSS(G)$ that need to be colored (starting, ending and critical). After that, we create an array M of p entries. Every $M[j]$ with $j \in [1..p]$ will contain a dynamic vector that stores the colors of the j-th colored node in the $BOSS$ ordering. We also add $\mathtt{rank}_1$ support to N to map a node $v \in V$ to its array of colors in M. Thus, its position can be inferred as $\mathtt{rank}_1(N, v)$.

The only inputs we need for the algorithm are N, R' and $BOSS(G)$. For every $R_i \in R'$ we proceed as follows; we append a dummy symbol at the ends of the string, and then use the function $\mathtt{label2Node}$ to find the node v labeled with the $k-1$ prefix of R_i. Note that this prefix will map a starting node as we append dummies to R_i. From v, we begin a walk on the graph and follow the edges whose symbols match the characters in the suffix $R_i[k..|R_i|]$. Note now that the last node v' we visit in this walk is an ending node that maps the $k-1$ suffix of R_i. As we move through the edges, we store in an array W_i the starting, ending, and critical nodes associated with R_i. Additionally, we push into another array I_i the neighbor nodes of the walk that need to be inspected to assign a color to R_i. The rules for pushing elements into I_i are as follows; (i) if v is a node in the path of R_i with outdegree more than one, then we push all its successor

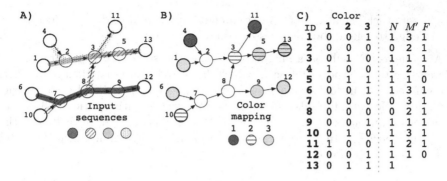

ID	Color 1	2	3	N	M'	F
1	0	0	1	1	3	1
2	0	0	0	0	2	1
3	0	1	0	1	1	1
4	1	0	0	1	2	1
5	0	1	1	1	1	0
6	0	0	1	1	3	1
7	0	0	0	0	3	1
8	0	0	0	0	2	1
9	0	0	1	1	1	1
10	0	1	0	1	3	1
11	1	0	0	1	2	1
12	0	0	1	1	1	0
13	0	1	1	1		

Fig. 2. Succinct colored dBG. (A) The topology of the graph. Colors and textures represent the paths that spell the input sequences of the dBG. Numbers over the nodes are their identifiers. Nodes 4,1,6 and 10 are starting nodes (darker borders). Nodes 11,13 and 12 are ending nodes and nodes 3,9,11 and 5 are critical. (B) Our greedy coloring algorithm. (C) The binary matrix C that encodes the colors of Fig. B. The left side is C in its uncompressed form and the right side is our succinct version of C using the arrays N, M', and F.

nodes into I_i, (ii) if v is a node in the path of R_i with indegree more than one, then we visit every predecessor node v' of v, and if v' has outdegree more than one, then we push into I_i the successor nodes of v'. Once we finish the traversal, we create a hash map H_i and fill it with the colors that were previously assigned to the nodes in I_i and W_i. After that, we pick the smallest color c' that is not in the keys of H_i, and push it to every array $M[\mathtt{rank}_1(N, j)]$ with $j \in W_i$. After we process all the sequences in R', the final set of colors is represented by the values in M. The whole processing of coloring a R_i is described in detail by the procedure $\mathtt{greedyCol}$ in Pseudocode (Algorithm 1).

The construction of the sets W_i and I_i is independent for every string in R', so it can be done in parallel. However, the construction of the hash map H_i and the assignment of the color c' to the elements of W_i has to be performed sequentially as all the sequences in R' need concurrent access to M.

Ambiguous Sequences. Our scheme, however, cannot safely retrieve sequences that are ambiguous.

Lemma 3. *Ambiguous sequences of R' cannot be reconstructed safely from the color matrix C and $BOSS(G)$.*

Proof. Assume that collection R is composed just by one string $R_1 = XbXc$, where X is a repeated substring and b, c are two different symbols in Σ. Consider also that the kmer size for $BOSS(G)$ is $k = |X| + 1$. The instance of $BOSS(G)$ will have a node v labeled with X, with two outgoing edges, whose symbols are b and c. Given our coloring scheme, the successor nodes of v will be both colored with the same color. As a consequence, if during a walk we reach node v, then

we will get stuck because there is not enough information to decide which is the correct edge to follow (both successor nodes have the same color).   □

A sequence R_i will be ambiguous if it has the same $k - 1$ pattern in two different contexts. Another case in which R_i is ambiguous is when a spurious edge connects an uncolored node of R_i with two or more critical nodes in the same path. Note that unlike unambiguous sequences with spurious edges, an ambiguous sequence will always be encoded by an unsafe path, regardless of the recoloring algorithm. In general, the number of ambiguous sequences will depend on the value we use for k.

5   Compressing the Colored dBG

The pair (M, N) can be regarded as a compact representation of C, where the empty rows were discarded. Every $M[i]$, with $i \in [1..|M|]$, is a row with at least one value, and every color $M[i][j]$, with $j \in [1..|M[i]|]$, is a column. However, M is not succinct enough to make it practical. We are still using a computer word for every color of M. Besides, we need $|M|$ extra words to store the pointers for the lists in M.

We compress M by using an idea similar to the one implemented in $BOSS$ to store the edges of the dBG. The first step is to sort the colors of every list $M[i]$. Because the greedy coloring generates a set of unique colors for every node, each $M[i]$ becomes an array of strictly increasing elements after the sorting. Thus, instead of storing the values explicitly, we encode them as deltas, i.e., $M[i][j] = M[i][j] - M[i][j-1]$. After transforming M, we concatenate all its values into one single list M' and create a bit map $F = [1..|M'|]$ to mark the first element of every $M[i]$ in M'. We store M' using Elias-Fano encoding [10,11] and F using the compressed representation for bit maps of [26]. Finally, we add $select_1$ support to F to map a range of elements in M' to an array in M. The complete representation of the color matrix now becomes $C = N + F + M'$ (see Fig. 2C). The complete index of the colored dBG is thus composed of our version of C and $BOSS(G)$. We now formalize the idea of retrieving the colors of a node from the succinct representation of C.

– getColors(v): list of colors assigned to node v.

Theorem 1. *the function* getColors(v) *computes in* $\mathcal{O}(c)$ *time the* c *colors assigned to node* v.

Proof. We first compute the rank r of node v within the colored nodes. This operation is carried out with $r = rank_1(N, v)$. After retrieving r, we obtain the range $[i..j]$ in M' where the values of v lie. For this purpose, we perform two $select_1$ operations over F, $[i, j] = (select_1(F, r), select_1(F, r + 1) - 1)$. Finally, we scan the range $[i..j]$ in M', and as we read the values, we incrementally reconstruct the colors from the deltas. All the rank and select operations takes $\mathcal{O}(1)$, and reading the $c = j - i + 1$ entries from M' takes $\mathcal{O}(c)$, because retrieving an element from an Elias-Fano-encoded array also takes $\mathcal{O}(1)$. In conclusion, computing the colors of v takes $\mathcal{O}(c)$.   □

6   Algorithms for the Colored dBG

Reconstructing Unambiguous Sequences. We describe now an online algorithm that works on top of our index and that reconstructs all the unambiguous sequences in R'. We cannot tell, however, if a reconstructed string R_i was present in the original set R or if it was its reverse complement R_i^{rc}. This is not really a problem, because a sequence and its reverse complement are equivalent in most of the Bioinformatic analyses.

The algorithm receives as input a starting node v. It first computes an array A with the colors assigned to v using the function `getColors` (see Sect. 5), and then sets a string $S = $ `nodeLabel`(v). For every color $a \in A$, the algorithm performs the following steps; initializes two temporary variables, an integer $v' = v$ and string $S' = S$, and then begins a graph walk from v'. If the outdegree of v' is one, then the next node in the walk is the successor node $v' = $ `outgoing_r`$(v', 1)$. On the other hand, if the outdegree of v' is more than one, then the algorithm inspects all the successor nodes of v' to check which one of them is the node v'' colored with a. If there is only one such v'', then it sets $v' = v''$. This procedure continues until v' becomes an ending node. During the walk, the edge symbols are pushed into S'. When an ending node is reached, the algorithm reports $S'[1..|S'| - 1]$ as the reconstructed sequence.

If at some point during a walk, the algorithm reaches a node with outdegree more than one, and with more than one successor colored with a, then aborts the reconstruction of the string as path is unsafe for color a. Then, it returns to v and continues with the next sequence. The complete procedure is detailed in the function `buildSeqs` of Algorithm 2.

Assembling Contigs. Our coloring scheme for the dBG allows us to report sequences that represent the overlap of two or more strings of R'. There are several ways in which a set of sequences can be arranged such that they form valid overlaps, but in practice, we are not interested in all such combinations. What we want is to compute only those union strings that describe real segments of the underlying genome of R', a.k.a *contig* sequences. In this work we do not go deep into the complexities of contig assembly (see [14,16,19,20] for some review). Instead, we propose a simple heuristic, that work on top of our index, and that it is aimed to produce contigs that are longer than those produced by uncolored dBGs.

Similar to `buildSeqs`, this method traverses the graph to reconstruct the contigs. During the process, it uses the color information to weight the outgoing edges *on the fly*, and thus, inferring which is the most probable path that matches a real segment of the source DNA.

The algorithm receives as input a starting node v and initializes a set L and hash map Q. Both data structures are used to store information about the strings that belong to the contig of v. A read $R_i \in R'$ is identified in the index as a pair (c, v'), where c is a color assigned to R_i and v' is the starting node of its path. L contains the reads already traversed while Q contains the active reads.

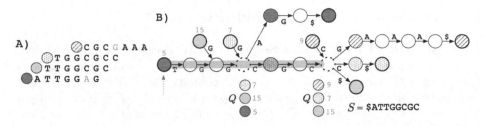

Fig. 3. Example of the assembly of a contig using our index. (A) Inexact overlap of four sequences. The circle to the left of every string represents its color in the dBG. Light gray symbols are mismatches in the overlap. (B) The colored dBG of the sequences. Circles with darker borders are starting and ending nodes. Light gray values over the starting nodes are their identifiers. The contig assembly begins in node 5 (denoted with a dashed arrow) and the threshold x to continue the extension is set to 0.5. The state of the hash map Q when the walk reaches a branching node (dashed circles) is depicted below the graph. The assembly ends in the right-most branching node as it has not a successor node that contains at least 50% of the colors in Q. The final contig is shown as a light grey path over the graph, and its sequence is stored in S.

The algorithm also initializes a string $S = \texttt{nodeLabel}(v)$ and pushes every pair $Q[c_i] = v$ with $c_i \in \texttt{getColor}(v)$. After that, it begins a walk from v and pushes into S the symbols of the edges it visits. For every new node v' reached during the walk, the algorithm checks if one of its predecessor nodes, say u, is a starting node. If so, then for every $c_i \in \texttt{getColors}(u)$ sets $Q[c_i] = u$ if (c_i, u) does not exist in L. On the other hand, if one of the successors of v', say u', is an ending node, then for every $c_i \in \texttt{getColors}(u')$ sets $L[(c, Q[c_i])]$ and then removes the entry $Q[c_i]$. After updating Q and L, it selects one of the outgoing edges of v' to continue the walk. For this purpose, the algorithm uses the following rules; (i) if v' has outdegree one, then it takes its only outgoing edge, (ii) if v' has outdegree more than one, then it inspects how the colors in Q distribute among the successors of v'. If there is only one successor node of v', say v'', colored with at least x fraction of the colors of Q, where x is a parameter, then the algorithm follows v'', and removes from Q the colors of the other successor nodes of v'.

The algorithm will stop if; (i) there is no such v'' that meet the x threshold, (ii) there is more than one successor of v' with the same color or (iii) v' has outdegree one, but the successor node is an ending node. After finishing the walk, the substring $S[2..|S|]$ is reported as the contig. The procedure $\texttt{contigAssm}$ in Pseudocode (Algorithm 3) describes in detail the contig assembly algorithm, and a graphical example is shown in Fig. 3.

7   Experiments

We use a set of reads generated from the E.coli genome[1] to test the ideas described in this article. The raw file was in **FASTQ** format and contained

---

[1] http://spades.bioinf.spbau.ru/spades_test_datasets/ecoli_mc.

14,214,324 reads of 100 characters long each. We preprocessed the file by removing sequencing errors using the tool of [3], and discarding reads with N symbols. The preprocessing yielded a data set of 8,655,214 reads (a FASTQ file of 2 GB). Additionally, we discarded sequencing qualities and the identifiers of the reads as they are not considered in our data structure. From the resulting set R (a text file of 833.67 MB), we created another set R' that considers the elements in R and their reverse complements.

Our version of the colored dBG, the algorithm for greedy coloring and the algorithms for reconstructing and assembly reads were implemented in C++[2], on top of the SDSL-lite library [12]. In our implementation, arrays M' and F are precomputed beforehand to store the colors directly to them, because using the dynamic list M is not cache-friendly. Additionally, all our code, except the algorithm for contig assembly, can be executed using multiple threads.

We built six instances of our index using R' as input. We choose different values for k, from 25 to 50 in steps of five. The coloring of every one of these instances was carried out using eight threads. Statistics about the graph topologies are shown in Table 1, and statistics about the coloring process are shown in Table 2. In every instance, we reconstructed the unambiguous reads (see Table 2). Additionally, we generated an FM-index of R' to locate the reconstructed reads and check that they were real sequences. All the tests were carried out on a machine with Debian 4.9, 252 GB of RAM and processor Intel(R) Xeon(R) Silver @ 2.10 GHz, with 32 cores.

8 Results

The average compression rate achieved by our index is 1.89, meaning that, in all the cases, the data structure used about half the space of the plain representation of the reads (see Table 1). We also note that the smaller the value for k, the greater the size of the index. This behavior is expected as the dBG becomes denser when we decrease k. Thus, we have to store a higher number of colors per node.

The number of colors of every instance is several orders of magnitude smaller than the number of reads, being $k = 25$ the instance with more colors (6552) and $k = 50$ the instance with the fewest (1689). Even though the fraction of colored nodes in every instance is small, the percentage of the index space used by the color matrix is still high (73% on average). Regarding the time for coloring the graph, it seems to be reasonable for practical purposes if we use several threads. In fact, building, filling and compacting C took 5,015 s on average, and the value decreases if we increment k. The working space, however, is still considerable. We had memory peaks ranging from 3.03 GB to 4.3 GB, depending on the value for k (see Table 2).

The process of reconstructing the reads yielded a small number of ambiguous sequences in all the instances (2,760 sequences on average), and decreases with higher values of k, especially for values above 40 (see Table 2).

---

[2] https://bitbucket.org/DiegoDiazDominguez/colored_bos/src/master.

Table 1. Statistics about the different colored dBGs generated in the experiments. The index size is expressed in MB and considers the space of $BOSS(G)$ plus the space of our succinct version of C. The compression rate was calculated as the space of the plain representation of the reads (833.67 MB) divided by the index size.

k	Total number of nodes	Number of solid nodes	Number of edges	Index size	Compression rate
25	106,028,714	11,257,781	120,610,151	446.38	1.86
30	142,591,410	11,425,646	157,186,548	443.82	1.87
35	179,167,289	11,561,630	193,773,251	441.18	1.88
40	215,751,326	11,667,364	230,365,635	438.23	1.90
45	252,337,929	11,743,320	266,958,709	435.30	1.91
50	288,925,674	11,791,640	303,552,318	432.13	1.92

Table 2. Statistics about our greedy coloring algorithm. The column "Color space usage" refers to the percentage of the index space used by our succinct version of C. Elapsed time and memory peak are expressed in seconds and MB, respectively, and both consider only the process of building, filling, and compacting the color matrix.

k	Number of colored nodes	Number of colors	Color space usage	Ambiguous sequences	Elapsed time	Memory peak
25	21,882,874	6,552	83.03	1904	5,835	4,391
30	21,907,324	4,944	79.14	1502	5,551	4,119
35	21,926,687	2,924	75.27	1224	5,131	3,847
40	21,942,083	2,064	71.40	1054	4,872	3,575
45	21,954,138	1,888	67.51	714	4,507	3,303
50	21,964,947	1,689	63.58	176	4,199	3,030

9   Conclusions and Further Work

Experimental results shows our data structure is succinct, and that has a practical use. Still, we believe that a more careful algorithm for constructing the index is still necessary to reduce the memory peaks during the coloring. Further compaction of the color matrix can be achieved by using more elaborated compression techniques. However, this extra compression can increase the construction time of the colored dBG and produce a considerable slow down in the algorithms that work on top of it for extracting information from the reads. Comparison of our results with other similar data structures is difficult for the moment. Most of the indexes based on colored dBGs were not designed to handle huge sets of colors like ours and the greedy recoloring of [1] does not scale well

and needs extra information for reconstructing the reads. Still, it is a promising approach that, with further work, can be used in the future as a base for performing Bioinformatics analyses in compressed space.

A   Appendix

A.1   Pseudocodes

Algorithm 1. Function greedyCol

1: **procedure** greedyCol(G,N,R_i,M)   ▷ G is a dBG, N is a bitmap, R_i is a string and M is array of lists
2:      $R_i \leftarrow \$R_i\$$ ▷ append dummy symbols at the ends of R_i
3:      $v \leftarrow \text{string2node}(R_i[1..k-1])$
4:      $W_i \leftarrow \emptyset$
5:      $I_i \leftarrow I_i \cup \text{rank}_1(N, v)$
6:      **for each** $r \in R_i[k-1..|R_i|]$ **do**   ▷ traverse the dBG path of R_i
7:          $o \leftarrow \text{outdegree}(G, v)$
8:          **if** $o > 1$ **then**
9:              **for** $j \leftarrow 1$ **to** o **do**
10:                  $I_i \leftarrow I_i \cup \text{rank}_1(N, \text{forward\_r}(G, v, j))$
11:          $i \leftarrow \text{indegree}(G, v)$
12:          **if** $i > 1$ **then**
13:              **for** $j \leftarrow 1$ **to** i **do**
14:                  $v' \leftarrow \text{incomming\_r}(G, v, j)$
15:                  $o' \leftarrow \text{outdegree}(G, v'))$
16:                  **if** $o' > 1$ **then**
17:                      **for** $j \leftarrow 1$ **to** o' **do**
18:                          $I_i \leftarrow I_i \cup \text{rank}_1(N, \text{forward\_r}(G, v', j))$
19:          **if** $N[v]$ is **true then**
20:              $W_i \leftarrow W_i \cup \text{rank}_1(N, v)$
21:          $v \leftarrow \text{forward}(G, v, r)$
22:      $W_i \leftarrow W_i \cup \text{rank}_1(N, v)$
23:      $I_i \leftarrow I_i \cup \text{rank}_1(N, v)$
24:      **for each** $n \in I_i$ **do**   ▷ compute the colors already used
25:          **for each** $c \in M[n]$ **do**
26:              $H_i[c] \leftarrow \textbf{true}$
27:      $c' \leftarrow$ minimum color not in H_i
28:      **for each** $n \in W_i$ **do**   ▷ color the nodes
29:          $M[n] \leftarrow M[n] \cup c'$

Algorithm 2. Function buildSeqs

1: **procedure** buildSeqs(G,v) ▷ G is a colored dBG and v is a starting node
2:     $L \leftarrow \emptyset$ ▷ list of rebuilt sequences
3:     $A \leftarrow \text{getColors}(G,v)$
4:     $S \leftarrow \text{nodeLabel}(G,v)$ ▷ initialize an string with the label of v
5:     **for each** $a \in C$ **do**
6:         $v' \leftarrow v$ ▷ temporal dBG node
7:         $S' \leftarrow S, amb \leftarrow \textbf{false}$
8:         **while** isEnding(G,v') is **false** and amb is **false do**
9:             $o \leftarrow \text{outdegree}(G,v')$
10:             **if** o is 1 **then**
11:                 $S' \leftarrow S' \cup \text{edgeSymbol}(G,v',1)$ ▷ push the new symbol into S'
12:                 $v' \leftarrow \text{forward\_r}(G,1)$
13:             **else**
14:                 $m \leftarrow 0$
15:                 **for** $u \leftarrow 1$ **to** o **do** ▷ check which successors of v' has color a
16:                     **if** $a \in \text{getColors}(\text{forward\_r}(G,v',u))$ **then**
17:                         $v' \leftarrow \text{forward\_r}(G,v',u)$
18:                         $m \leftarrow m + 1$
19:                 **if** $m > 1$ **then**   ▷ more than one successor v' has color a
20:                     $amb \leftarrow \textbf{true}$
21:             **if** amb **not true then**
22:                 $L \leftarrow L \cup S[2..|S|-1]$
23:     **return** L

Algorithm 3. Function `contigAssm`

1: **procedure** contigAssm(G,v,x)   ▷ v is a starting node and x is a threshold
2:     $L \leftarrow \emptyset$
3:     $S \leftarrow \text{nodeLabel}(G, v)$
4:     **for each** $c_i \in \text{getColors}(v)$ **do**
5:        $Q[c_i] \leftarrow v$
6:     **while true do**
7:        **if** $\text{indegree}(G, v) > 1$ **then**
8:           $v' \leftarrow \text{backward\_r}(G, v, 1))$
9:           **if** $\text{isStarting}(v')$ **then**   ▷ add new reads to the contig
10:              **for each** $c_i \in \text{getColors}(v')$ **do**
11:                 **if** $L[(c_i, v')]$ is not **true then**
12:                    $Q[c_i] \leftarrow v'$
13:        **if** $o \leftarrow \text{outdegree}(G, v) > 1$ **then**
14:           $t \leftarrow v, v \leftarrow 0$
15:           **for** $i \leftarrow 1$ **to** o **do**   ▷ compute the most probable successor node
16:              $v' \leftarrow \text{forward\_r}(G, t, i)$
17:              **if** $\text{isEnding}(v')$ **then**   ▷ discard reads ending at v
18:                 **for each** $c_i \in \text{getColors}(v')$ **do**
19:                    $L[(c_i, Q[c_i])] \leftarrow \text{true}$
20:                 $Q \leftarrow Q \setminus A$
21:              **else**
22:                 $A \leftarrow \text{getColors}(v')$
23:                 $w \leftarrow (Q \cap A)/|Q|$ ▷ weight the successor node
24:                 **if** $w \geq x$ **then**
25:                    $v \leftarrow v'$
26:                    $Q \leftarrow A$
27:                    $S \leftarrow S \cup \text{edgeSymbol}(G, t, i)$
28:                    **break**
29:           **if** v is 0 **then break**   ▷ no successor has the minimum weight x
30:        **else**
31:           $v \leftarrow \text{forward\_r}(G, v, 1)$
32:           **if** $\text{isEnding}(v)$ **then break**
33:           $S \leftarrow S \cup \text{edgeSymbol}(G, v, 1)$
34:     **return** $S[2..|S|]$

References

1. Alipanahi, B., Kuhnle, A., Boucher, C.: Recoloring the colored de Bruijn graph. In: Proceedings of 25th International Symposium on String Processing and Information Retrieval (SPIRE), pp. 1–11 (2018). https://doi.org/10.1007/978-3-030-00479-8_1

2. Almodaresi, F., Pandey, P., Patro, R.: Rainbowfish: a succinct colored de Bruijn graph representation. In: Proceedings of 17th International Workshop on Algorithms in Bioinformatics (WABI). Article 18 (2017). https://doi.org/10.4230/LIPIcs.WABI.2017.18

3. Bankevich, A., et al.: SPAdes: a new genome assembly algorithm and its applications to single-cell sequencing. J. Comput. Biol. **19**(5), 455–477 (2012). https://doi.org/10.1089/cmb.2012.0021

4. Boucher, C., Bowe, A., Gagie, T., Puglisi, S.J., Sadakane, K.: Variable-order de Bruijn graphs. In: Proceedings of 25th Data Compression Conference (DCC), pp. 383–392 (2015). https://doi.org/10.1109/DCC.2015.70

5. Bowe, A., Onodera, T., Sadakane, K., Shibuya, T.: Succinct de Bruijn graphs. In: Proceedings of 12th International Workshop on Algorithms in Bioinformatics (WABI), pp. 225–235 (2012). https://doi.org/10.1007/978-3-642-33122-0_18

6. Bray, N., Pimentel, H., Melsted, P., Pachter, L.: Near-optimal probabilistic RNA-seq quantification. Nat. Biotechnol. **34**(5), 525–527 (2016). https://doi.org/10.1038/nbt.3519

7. de Bruijn, N.G.: A combinatorial problem. Koninklijke Nederlandse Akademie v. Wetenschappen **49**(49), 758–764 (1946)

8. Burrows, M., Wheeler, D.: A block sorting lossless data compression algorithm. Technical report 124, Digital Equipment Corporation (1994)

9. Clark, D.: Compact PAT trees. Ph.D. thesis, University of Waterloo, Canada (1996)

10. Elias, P.: Efficient storage and retrieval by content and address of static files. J. ACM **21**(2), 246–260 (1974). https://doi.org/10.1145/321812.321820

11. Fano, R.M.: On the number of bits required to implement an associative memory. Massachusetts Institute of Technology (1971)

12. Gog, S., Beller, T., Moffat, A., Petri, M.: From theory to practice: plug and play with succinct data structures. In: Proceedings of 13th International Symposium on Experimental Algorithms (SEA), pp. 326–337 (2014). https://doi.org/10.1007/978-3-319-07959-2_28

13. Holley, G., Wittler, R., Stoye, J.: Bloom filter trie - a data structure for pan-genome storage. In: Proceedings of 15th International Workshop on Algorithms in Bioinformatics (WABI), pp. 217–230 (2015). https://doi.org/10.1007/978-3-662-48221-6_16

14. Idury, R.M., Waterman, M.S.: A new algorithm for DNA sequence assembly. J. Comput. Biol. **2**(2), 291–306 (1995). https://doi.org/10.1089/cmb.1995.2.291

15. Iqbal, Z., Caccamo, M., Turner, I., Flicek, P., McVean, G.: De novo assembly and genotyping of variants using colored de Bruijn graphs. Nat. Genet. **44**(2), 226–232 (2012). https://doi.org/10.1038/ng.1028

16. Kececioglu, J.D., Myers, E.W.: Combinatorial algorithms for DNA sequence assembly. Algorithmica **13**(1), 7–51 (1995). https://doi.org/10.1007/BF01188580

17. Lewis, R.: A Guide to Graph Colouring. Springer, Cham (2015). https://doi.org/10.1007/978-3-319-25730-3

18. Mäkinen, V., Navarro, G.: Succinct suffix arrays based on run-length encoding. Nordic J. Comput. **12**(1), 40–66 (2005). https://doi.org/10.1007/11496656_5

19. Medvedev, Paul, Georgiou, Konstantinos, Myers, Gene, Brudno, Michael: Computability of Models for Sequence Assembly. In: Giancarlo, Raffaele, Hannenhalli, Sridhar (eds.) WABI 2007. LNCS, vol. 4645, pp. 289–301. Springer, Heidelberg (2007). https://doi.org/10.1007/978-3-540-74126-8_27

20. Medvedev, P., Pham, S., Chaisson, M., Tesler, G., Pevzner, P.: Paired de bruijn graphs: a novel approach for incorporating mate pair information into genome assemblers. J. Comput. Biol. **18**(11), 1625–1634 (2011). https://doi.org/10.1089/cmb.2011.0151

21. Mustafa, H., Kahles, A., Karasikov, M., Raetsch, G.: Metannot: a succinct data structure for compression of colors in dynamic de Bruijn graphs. bioRxiv, Article 236711 (2017). https://doi.org/10.3929/ethz-b-000236153

22. Mustafa, H., Schilken, I., Karasikov, M., Eickhoff, C., Rätsch, G., Kahles, A.: Dynamic compression schemes for graph coloring. Bioinformatics **35**(3), 407–414 (2018). https://doi.org/10.1093/bioinformatics/bty632

23. Navarro, G.: Compact Data Structures: A Practical Approach. Cambridge University Press, Cambridge (2016). https://doi.org/10.1017/CBO9781316588284

24. Okanohara, D., Sadakane, K.: Practical entropy-compressed rank/select dictionary. In: Proceedings of 9th Workshop on Algorithm Engineering and Experiments (ALENEX), pp. 60–70 (2007). https://doi.org/10.1137/1.9781611972870.6

25. Pandey, P., Almodaresi, F., Bender, M.A., Ferdman, M., Johnson, R., Patro, R.: Mantis: a fast, small, and exact large-scale sequence-search index. Cell Syst. **7**(2), 201–207 (2018). https://doi.org/10.1016/j.cels.2018.05.021

26. Raman, R., Raman, V., Satti, S.R.: Succinct indexable dictionaries with applications to encoding k-ary trees, prefix sums and multisets. ACM Trans. Algorithms **3**(4), Article 43 (2007). https://doi.org/10.1145/1290672.1290680

27. Reuter, J., Spacek, D., Snyder, M.: High-throughput sequencing technologies. Mol. Cell **58**(4), 586–597 (2015). https://doi.org/10.1016/j.molcel.2015.05.004

28. Salmela, L., Walve, R., Rivals, E., Ukkonen, E.: Accurate self-correction of errors in long reads using de Bruijn graphs. Bioinformatics **33**(6), 799–806 (2016). https://doi.org/10.1093/bioinformatics/btw321

Linear Time Maximum Segmentation Problems in Column Stream Model

Bastien Cazaux[1]([⊠]) [iD], Dmitry Kosolobov[2] [iD], Veli Mäkinen[1] [iD],
and Tuukka Norri[1] [iD]

[1] Department of Computer Science, University of Helsinki, Helsinki, Finland
{bastien.cazaux,veli.makinen,tuukka.norri}@helsinki.fi
[2] Ural Federal University, Ekaterinburg, Russia
dkosolobov@mail.ru

Abstract. We study a lossy compression scheme linked to the biological problem of founder reconstruction: The goal in founder reconstruction is to replace a set of strings with a smaller set of founders such that the original connections are maintained as well as possible. A general formulation of this problem is NP-hard, but when limiting to reconstructions that form a segmentation of the input strings, polynomial time solutions exist. We proposed in our earlier work (WABI 2018) a linear time solution to a formulation where minimum segment length was bounded, but it was left open if the same running time can be obtained when the targeted compression level (number of founders) is bounded and lossyness is minimized. This optimization is captured by the Maximum Segmentation problem: Given a threshold M and a set $\mathcal{R} = \{\mathcal{R}_1, \ldots, \mathcal{R}_m\}$ of strings of the same length n, find a minimum cost partition P where for each segment $[i, j] \in P$, the *compression level* $|\{\mathcal{R}_k[i, j] : 1 \leq k \leq m\}|$ is bounded from above by M. We give linear time algorithms to solve the problem for two different (compression quality) measures on P: the average length of the intervals of the partition and the length of the minimal interval of the partition. These algorithms make use of positional Burrows–Wheeler transform and the range maximum queue, an extension of range maximum queries to the case where the input string can be operated as a queue. For the latter, we present a new solution that may be of independent interest. The solutions work in a streaming model where one column of the input strings is introduced at a time.

Keywords: Pan-genome indexing · Founder reconstruction · Dynamic programming · Positional Burrows–Wheeler transform · Range maximum queue

1 Introduction

Given a *set of recombinants* $\mathcal{R} = \{\mathcal{R}_1, \ldots, \mathcal{R}_m\}$, i.e. a set of strings of the same length, a *set of founders* is a set of strings of the same length where all the

This work was partially supported by the Academy of Finland (grant 309048).

N. R. Brisaboa and S. J. Puglisi (Eds.): SPIRE 2019, LNCS 11811, pp. 322–336, 2019.
https://doi.org/10.1007/978-3-030-32686-9_23

recombinants can be mapped on the founders for the common positions, i.e. for each position k all the characters at the position k of the recombinants are included in all the characters at the position k of the founders.

Minimizing the number of *crossovers*, i.e. positions where recombinants need to change between mapped founders, corresponds to the problem *founder sequence reconstruction* [2,8,9]. As this problem is NP-hard [2,8], Ukkonen suggested taking a polynomial variant of this problem through construction of a *segmentation* [9]. Here a segmentation is a decomposition of the set of recombinants into blocks. For a partition P, the corresponding segmentation corresponds to a set of *segments* where for $[i,j] \in P$, the segment of the interval $[i,j]$ is the set of strings $\{\mathcal{R}_k[i,j] : 1 \le k \le m\}$.

Ukkonen proposed three different measures for segmentations of recombinants: λ_{min} (the minimum size of the intervals), λ_{ave} (the ratio between the length of a string of $\mathcal{R}$ and the number of segments) and λ_{max} (the maximum of segment sizes) and gave an optimal solution in $O(mn^2)$ time where λ_{min} is bounded by a given user-defined value M and λ_{max} is minimized (*Minimum Segmentation problem*) and an optimal solution in $O(mn)$ time where λ_{ave} is bounded and λ_{max} is minimized [9]. Norri et al. improved the first result to $O(mn)$ time by using the *positional Burrows–Wheeler transform* (pBWT) [3]. This problem corresponds to finding the segmentation that minimizes the maximal segment size where each interval of the corresponding partition have length bigger than a threshold K.

Ukkonen [9] proved that we can find in $O(n(m+M^3))$ time a set of founders which minimizes the number of crossovers in the case where all the segments of the segmentation have the same size M. Norri et al. [7] apply the algorithm of Ukkonen to the case where the segments have different sizes; experimental results show that this approach works well in practice, although optimality is not guaranteed in this case.

Instead of minimizing λ_{max} where λ_{min} is bounded as in the Minimum Segmentation problem, in this paper we study the problem where λ_{max} is bounded and we want to maximize λ_{min} or λ_{ave}. In other words, we take the dual problem of Minimum Segmentation problem where we bound the maximum size of segments, i.e. the number of founders, and we want to optimize the partition corresponding to this segmentation (either maximize the size of the minimal interval of the partition or minimize the number of intervals). This formulation is motivated by the ability to control the size of the *pan-genome index* proposed in [10]: Multiple alignment of thousands of human genomes can be replaced with a multiple alignment of founders. We demonstrated in [7] that reduction from 5009 sequences to 130 founders gives average distance 9358 bases between two crossovers. Such preservation of continuity is sufficient for the approach in [10]. With our new formulation, we can directly control the target number of founders without needing to try out different bounds for the segment length. In more general terms, the approach can be seen as a lossy compression scheme, where the targeted compression level (number of founders) is fixed and the *compression quality* (preservation of continuities) is optimized. Such general formulation might find other applications beyond genome research.

We propose two different algorithms to solve maximization of λ_{min} and λ_{ave}, respectively. In Sect. 3, we give a greedy algorithm which finds an optimal solution in $O(mn)$ for the first problem and in Sect. 4, we present a dynamic programming algorithm using a *Range Maximum Queue* to solve the second problem in linear time. The Range Maximum Queue is an extension of *min-queue* [5]. A min-queue supports minimum value queries on queue in constant time, with the update operations taking also constant time. Our structure supports querying maximum value over a range of items in the queue in constant time, with the update operations taking amortized constant time.

We consider all of the Maximum Segmentation problems in a specific streaming data model. As every input of the Maximum Segmentation problems consists of a bound and a set of strings of the same length, we define the *Column Stream Model* such that we are given the bound and a stream that yields the set of strings of the same length, column-by-column. In essence, at the kth step we have the kth character of each input string. A justification of this model can be found in Appendix.

2 Preliminaries

In this section we present the problems of Maximum Segmentation and some terminologies we will need.

To begin we define some notations for strings and sets. Given a *string* $w = a_1 \ldots a_n$, the *length* of w, denoted by $|w|$, is n, the ith element of w, denoted by $w[i]$, is a_i and the *substring* denoted by $w[i, j]$ is $a_i \ldots a_j$. We use an analogous notation for the arrays. Given two integers i and j with $i \leq j$, we denote by $[i, j]$ the set of integers between i and j, i.e. $\{k \in \mathbb{Z} : i \leq k \leq j\}$. Given a finite set S, a *partition* $P = \{S_1, \ldots, S_k\}$ is a set of subsets of S such that $\cup_{S_i \in P} S_i = S$ and $i \neq j \Rightarrow S_i \cap S_j = \emptyset$. The *cardinality* of S is denoted by $|S|$.

The input of our problems is a set of *recombinants* $\mathcal{R} = \{\mathcal{R}_1, \ldots, \mathcal{R}_m\}$ which is a set of m strings of the same length n ($|\mathcal{R}_1| = \ldots = |\mathcal{R}_m| = n$). In what follows, we use m as the number of strings of $\mathcal{R}$ and n as the length of each string of $\mathcal{R}$.

For an integer interval $[i, j]$ with $1 \leq i \leq j \leq n$, we denote by $\mathcal{R}[i, j]$ the set of all the substrings $\mathcal{R}_k[i, j]$ with $k \in [1, m]$, i.e. $\mathcal{R}[i, j] = \{\mathcal{R}_k[i, j] : k \in [1, m]\}$. Given a partition P of $[1, n]$, we define the following three measures: $\lambda_{min}(\mathcal{R}, P) = \min_{[i,j] \in P} |[i, j]|$, $\lambda_{ave}(\mathcal{R}, P) = \frac{n}{|P|}$ and $\lambda_{max}(\mathcal{R}, P) = \max_{[i,j] \in P} |\mathcal{R}[i, j]|$. When there is no confusion, we just use the notations λ_{min}, λ_{ave} and λ_{max}.

Example 1. The set $\mathcal{R}$ of 6 recombinants of size 10:

$$
\begin{array}{cccccccccc}
1 & 2 & 3 & 4 & 5 & 6 & 7 & 8 & 9 & 10 \\
0 & 1 & 1 & 2 & 2 & 1 & 0 & 2 & 2 & 1 \\
0 & 1 & 1 & 2 & 1 & 2 & 0 & 1 & 0 & 1 \\
2 & 1 & 0 & 2 & 1 & 2 & 0 & 2 & 1 & 0 \\
0 & 2 & 1 & 2 & 2 & 1 & 0 & 2 & 2 & 1 \\
2 & 1 & 0 & 2 & 2 & 1 & 0 & 2 & 2 & 1 \\
0 & 2 & 1 & 2 & 1 & 2 & 0 & 1 & 0 & 1
\end{array}
$$

By taking the partition $P_1 = \{[1,3], [4,7], [8,10]\}$,

one has $\lambda_{min}(\mathcal{R}, P_1) = \min\{|[1,3]|, |[4,7]|, |[8,10]|\} = \min\{3, 4, 3\} = 3$, $\lambda_{ave}(\mathcal{R}, P_1) = \frac{10}{3}$ and $\lambda_{max}(\mathcal{R}, P_1) = \max\{|\mathcal{R}[1,3]|, |\mathcal{R}[4,7]|, |\mathcal{R}[8,10]|\} = \max\{3, 2, 3\} = 3$.

By taking the partition $P_2 = \{[1,2], [3,6], [7,8], [9,10]\}$,

one has $\lambda_{min}(\mathcal{R}, P_2) = \min\{|[1,2]|, |[3,6]|, |[7,8]|, |[9,10]|\} = \min\{2, 4, 2, 2\} = 2$, $\lambda_{ave}(\mathcal{R}, P_2) = \frac{10}{4} = 2.5$ and $\lambda_{max}(\mathcal{R}, P_2) = \max\{|\mathcal{R}[1,2]|, |\mathcal{R}[3,6]|, |\mathcal{R}[7,8]|, |\mathcal{R}[9,10]|\} = \max\{3, 4, 2, 2\} = 4$.

Definition 1. *The problem of λ_{min}-Maximum Segmentation Partition (or λ_{min}-MSP) is, given a bound M and a set of recombinants $\mathcal{R}$, to find a partition of $[1,n]$ which maximizes $\lambda_{min}(\mathcal{R}, P)$ subject to $\lambda_{max}(\mathcal{R}, P) \leq M$. The problem of λ_{min}-Maximum Segmentation Length (or λ_{min}-MSL) is, given a bound M and a set of recombinants $\mathcal{R}$, to find the measure λ_{min} for an optimal partition P of λ_{min}-MSP.*

We denote analogously by λ_{ave}-*Maximum Segmentation Partition* (or λ_{ave}-MSP) and λ_{ave}-*Maximum Segmentation Length* (or λ_{ave}-MSL) the similar problems in which $\lambda_{min}(\mathcal{R}, P)$ is substituted with $\lambda_{ave}(\mathcal{R}, P)$.

Hereafter we assume that M is at least $\max\{|\mathcal{R}[k, k]| : k \in [1, n]\}$, otherwise all the Maximum Segmentation problems admit no solution.

Remark 1. The notation of $\lambda_{min}(\mathcal{R}, P)$ and $\lambda_{ave}(\mathcal{R}, P)$ corresponds to the "λ_{min}" and "λ_{ave}" of [9]. The Maximum Segmentation Length problem corresponds to the segmented version of *Maximum fragment length* of [9]. The *Minimum Segmentation problem* of [7] corresponds to finding the measure λ_{max} of a partition P which minimizes λ_{max} where λ_{min} is bounded from below by an integer in input.

In the two following sections we present different algorithms to find in linear time (in the size of the input) an optimal solution of λ_{ave}-MSP and λ_{ave}-MSL (see Sect. 3) and λ_{min}-MSP and λ_{min}-MSL (see Sect. 4).

3 λ_{ave}-Maximum Segmentation Problems

As $\lambda_{ave} = \frac{n}{|P|}$, maximizing λ_{ave} corresponds to minimizing $|P|$, i.e. the number of intervals of P. The idea of our algorithm solving the λ_{ave}-Maximum Segmentation problems is to use a greedy algorithm from left to right depending on values given by a special data structure, the *positional Burrows–Wheeler Transform*. We begin by explaining which values we want to compute, what is the positional Burrows–Wheeler Transform and how we can use it to build our values and finally we give a proof of the correctness for our greedy algorithm.

3.1   Optimal Solution of λ_{ave}-MSP

Lemma 1. *Let i, j, i' and j' be four integers such that $i \leq i' \leq j' \leq j$. We have $|\mathcal{R}[i', j']| \leq |\mathcal{R}[i, j]|$.*

Proof. The property is due to the fact that each string of $\mathcal{R}[i', j']$ is a substring of $\mathcal{R}[i, j]$ on the same interval.   □

Hence, for a fixed j, the function $|\mathcal{R}[i, j]|$ is decreasing in i. For a bound M and an integer k, we define c_k as the value such that $|\mathcal{R}[c_k, k]| > M$ and $|\mathcal{R}[c_k + 1, k]| \leq M$; in the case $|\mathcal{R}[1, k]| \leq M$, we take $c_k = 0$.

Remark 2. We can equivalently define c_k as follows:

$$c_k = \min\{j \in [1, k] : |\mathcal{R}[j, k]| \leq M\} - 1$$
$$= \max\{j \in [1, k] : |\mathcal{R}[j, k]| > M\}.$$

With all the values of c_k for all $k \in [1, n]$, we can build an optimal solution of λ_{ave}-MSP.

Lemma 2. *The solution* $P = \{[1, b_p], \ldots, [b_3 + 1, b_2], [b_2 + 1, b_1]\}$ *is an optimal solution of the* λ_{ave}-*Maximum Segmentation Partition problem where* $b_1 = n$, $b_{k+1} = c_{b_k}$ *for* $k \geq 1$, *and* $c_{b_p} = 0$.

Proof. The set $P = \{[1, b_p], \ldots, [b_3 + 1, b_2], [b_2 + 1, b_1]\}$ is a partition of $[1, n]$ because $b_1 = n$ and $b_{k+1} = c_{b_k}$. We are to prove that P is an optimal solution. Let $P_{opt} = \{[1, o_{p'}], \ldots, [o_2 + 1, o_1]\}$ be an optimal partition with $o_1 = n$. We denote $P_1 = P_{opt}$ and $P_k = P_{k-1} \setminus ([o_{k+1} + 1, o_k] \cup [o_k + 1, b_{k-1}]) \bigcup ([o_{k+1} + 1, b_k] \cup [b_k + 1, b_{k-1}])$. We are going to prove by induction on k that each P_k is optimal (for $k \in [1, p']$). The base of induction $k = 1$ holds by definition ($P_1 = P_{opt}$). We assume that P_{k-1} is an optimal solution. As $b_k = c_{b_{k-1}}$, we have $o_k \geq b_k$.

If $b_k \geq o_{k+1} + 1$, by Lemma 1, as $o_{k+1} + 1 \leq b_k \leq o_k$, we have $|\mathcal{R}[o_{k+1} + 1, b_k]| \leq |\mathcal{R}[o_{k+1} + 1, o_k]| \leq M$ and thus P_k is a solution and $|P_k| = |P_{k-1}|$. Hence by induction, P_k is an optimal solution.

If $b_k < o_{k+1} + 1$, $P_k^\star = P_{k-1} \setminus ([o_{k+2} + 1, o_{k+1}] \cup [o_{k+1} + 1, o_k] \cup [o_k + 1, b_{k-1}]) \bigcup ([o_{k+2} + 1, b_k] \cup [b_k + 1, b_{k-1}])$ is a solution and $|P_k^\star| = |P_{k-1}| - 1$ and thus P_{k-1} is not optimal which is impossible by induction.

As $P_{p'}$ is optimal and $P_{p'} = P$, the partition P is an optimal solution.   □

3.2   pBWT and Linear Algorithm for λ_{ave}-MSP

Given a string $T[1, m] = t_1 \ldots t_m$, we denote by $\overleftarrow{T}$ the string corresponding to the reverse of T, i.e. $\overleftarrow{T} = t_m \ldots t_1$. The *positional Burrows–Wheeler Transform* [3] (or pBWT) of a set of recombinants $\mathcal{R}$ is two sets of n arrays of size m – an array a_k and an array d_k for all $k \in [1, n]$ – where for $k \in [1, n]$, $a_k[1, m]$ is a permutation of $[1, m]$ such that $\overleftarrow{\mathcal{R}_{a_k[1]}}[1, k] \leq \ldots \leq \overleftarrow{\mathcal{R}_{a_k[m]}}[1, k]$ lexicographically and $d_k[i] = 1 + \max\{j \in [1, k] : \overleftarrow{\mathcal{R}_{a_k[i]}}[j] \neq \overleftarrow{\mathcal{R}_{a_k[i-1]}}[j]\}$, for $i \in [2, m]$ and $d_k[1] = k + 1$.

Durbin [3] showed that we can compute recursively a_k and d_k from a_{k-1} and d_{k-1} in $O(m)$ time for a binary alphabet and Mäkinen and Norri [6] further generalized the construction for integer alphabets of size $O(m)$.

Lemma 3 ([6]). *The arrays* a_k *and* d_k *can be computed from* a_{k-1} *and* d_{k-1} *in* $O(m)$ *time, assuming the input alphabet is* $[0, |\Sigma| - 1]$ *with* $|\Sigma| = O(m)$.

Norri et al. [7] use three different arrays s_k, t_k, e_k to store the array d_k in increasing sorted order where s_k contains all distinct elements from d_k in the increasing sorted order (so that the length of s_k might be less than m), e_k is the normalized array d_k where $s_k[e_k[j]] = d_k[j]$ for all $j \in [1, m]$ and t_k is an array of the same length as s_k such that, for any j, $t_k[j]$ indicates the number of times the value $s_k[j]$ occurs in d_k.

Lemma 4 ([7]). *The arrays* a_k, s_k, e_k *and* t_k *can be computed from* a_{k-1}, s_{k-1}, e_{k-1} *and* t_{k-1} *in* $O(m)$ *time, assuming the input alphabet is* $[0, |\Sigma| - 1]$ *with* $|\Sigma| = O(m)$.

With s_k, e_k and t_k we can redefine c_k. As $|\mathcal{R}[s_k[j] - 1, k]| = \sum_{i \in [j, |t_k|]} t_k[i]$, one has

$$c_k = \max\{j \in [1, |s_k|] : \sum_{i \in [j, |t_k|]} t_k[i] > M\}. \tag{1}$$

With this new definition of c_k we obtain the following theorem.

Theorem 1. *Given a bound M and a set of recombinants $\mathcal{R}$, there is an algorithm that computes an optimal solution of the λ_{ave}-Maximum Segmentation Partition problem in a streaming fashion in $O(mn)$ time and $O(m + n)$ space.*

Proof. By Lemma 4 and Eq. 1, we can build c_k in $O(m)$ time for each value $k \in [1, n]$. Lemma 2 gives us an optimal solution by using the values c_k. Finally we build and store all the values c_k and make a backtracking from n to 0.   □

3.3   Right Greedy and Linear Algorithm for λ_{ave}-MSL

Lemma 2 gives us a greedy solution from right to left, a "Left greedy" version. Here we present a "Right greedy" version working from left to right.

Lemma 5. *The solution $P = \{[b_1, b_2 - 1], [b_2, b_3 - 1], \ldots, [b_p, n]\}$ is an optimal solution of the λ_{ave}-MSP problem where $b_1 = 1$ and $b_{k+1} = \min\{j \in [b_k, n] : c_j \geq b_k\}$.*

Proof. By Lemma 1, we know that for all $k \in [1, n]$ and for all $j \in [k, n]$, $c_k \leq c_j$. Hence, we can adapt the proof of Lemma 2 by extending the optimal solution of the right to prove this result.   □

By using the solution of Lemma 5 instead of Lemma 2, we do not need to store all the array of c_k to build the solutions of λ_{ave}-MSL and of λ_{ave}-MSP and we can extend the result of Theorem 1.

Theorem 2. *Given a bound M and a set of recombinants $\mathcal{R}$, there is an algorithm that computes an optimal solution of the λ_{ave}-Maximum Segmentation Length problem in a streaming fashion in $O(mn)$ time and $O(m)$ space. One can also find in $O(mn)$ time and $O(m + |P|)$ space the corresponding partition P, thus solving the λ_{ave}-Maximum Segmentation Partition problem.*

4 λ_{min}-Maximum Segmentation Problems

In this section, we give an $O(mn)$ time algorithm building an optimal solution of λ_{min}-MSP and λ_{min}-MSL. We focus on solving the problem λ_{min}-MSL by using dynamic programming algorithm; the corresponding partition (solution of λ_{min}-MSP) can be reconstructed by "backtracking" in a standard way (see [9]).

4.1   Dynamic Programming Algorithm

Given an integer M and a set of recombinants $\mathcal{R}$, the λ_{min}-Maximum Segmentation Length problem seeks to maximize the smallest cardinality of the intervals of a partition P subject to $\lambda_{max} \leq M$. In other words, this problem is to compute

$$\max_{P \in \mathcal{P}_{M,\mathcal{R}}} \min\{j - i + 1 : [i,j] \in P\} \tag{2}$$

where $\mathcal{P}_{M,\mathcal{R}}$ is the set of all partitions P of $[1,n]$ such that for all $[i,j] \in P$, $|\mathcal{R}[i,j]| \leq M$.

To solve λ_{min}-MSL, we define the following recursion which solves (2):

$$N(k) = \begin{cases} \infty & \text{If } k = 0, \\ \max_{c_k \leq j < k} \min\{N(j), k - j\} & \text{Otherwise.} \end{cases} \tag{3}$$

Given k between 1 and n, we denote by $\text{Previous}(k)$ the set of previous values of k by (3), i.e. $\text{Previous}(k) = \text{Argmax}_{c_k \leq j < k} \min\{N(j), k - j\}$
$= \{j \in [c_k, k-1] : N(k) = \min\{N(j), k - j\}\}$.

We can exhibit two recursive properties, one on c_k and one on $\text{Previous}(k)$ (with Algorithm 1).

Algorithm 1. The algorithm $Next(x, k)$.

1: $z \leftarrow k - N(x)$;
2: $w \leftarrow \text{Argmax}\{N(u) : x \leq u < z\}$;
3: **if** $x < z$ and $N(w) > N(x)$ **then**
4:     **return** $Next(x + 1, k)$;
5: **else**
6:     **return** x;

Lemma 6. *Given $k \in [1, n-1]$, we have*

1. $c_k \leq c_{k+1}$,
2. For all $j \in \text{Previous}(k)$, $Next(\max\{j, c_{k+1}\}, k+1) \in \text{Previous}(k+1)$.

Proof. For 1, we straightforwardly obtain $c_k \leq c_{k+1}$ due to Lemma 1.

For 2, we begin by proving that for all $j_k \in \text{Previous}(k)$, there exists $j_{k+1} \in \text{Previous}(k+1)$ with $j_k \leq j_{k+1}$. Assume that it is not the case. Let be $j_k \in \text{Previous}(k)$ such that $\forall j_{k+1} \in \text{Previous}(k+1)$, $j_{k+1} < j_k$. In this case, we have $c_{k+1} < j_k$ and for all $j' \in [c_{k+1}, j_k - 1]$, $N(j') \leq N(k) = \min\{N(j_k), k - j_k\}$. If $N(j_k) \leq k - j_k$, we have that $N(j_k) < k - j_k + 1$ and thus $N(k+1) \leq N(k)$. As $j_k \in [c_{k+1}, k+1]$, j_k is an element of $\text{Previous}(k+1)$ which is impossible. Otherwise, we have $N(j_k) > k - j_k$, $N(j_k) \geq k - j_k + 1$ and thus $j_k \in \text{Previous}(k+1)$ which is also impossible.

Now, we know that we can search j_{k+1} in $[\max\{j_k, c_{k+1}\}, k+1]$. In Algorithm 1, we decrease the size of the interval by the left until finding an element of Previous($k + 1$). Indeed for $Next(x, k)$, if $N(w) > N(x)$, there exists $u \in [x, k - N(x) - 1]$ such that $N(u) > N(x)$ and thus $\min\{N(x), k - x\} < \min\{N(u), k - u\}$ and $x \notin$ Previous(k). Otherwise, we know that for all $u \in [x, k - N(x) - 1]$, $\min\{N(x), k - x\} \geq N(u) \geq \min\{N(u), k - u\}$ and for all $u \in [k - N(x), k]$, $\min\{N(x), k - x\} \geq k - u \geq \min\{N(u), k - u\}$. Hence, we have for $x = Next(\max\{j_k, c_{k+1}\}, k + 1)$, for all $u \in [\max\{j_k, c_{k+1}\}, k + 1]$, $\min\{N(x), k + 1 - x\} \geq \min\{N(u), k + 1 - u\}$ and thus $x \in$ Previous($k + 1$). $\square$

Theorem 3. *Given a bound M and a set of recombinants $\mathcal{R}$, there is an algorithm that computes an optimal solution of the λ_{min}-Maximum Segmentation Length problem in a streaming fashion in $O(nm)$ time and $O(m + \mathcal{C})$ space where $\mathcal{C} = \max\{k - c_k + 1 : k \in [1, n]\}$. Using an additional array of length n, one can also find in $O(n)$ time the corresponding partition, thus solving the λ_{min}-Maximum Segmentation Partition problem.*

Proof. By using a Range Maximum Queue (see Lemma 8) on the table of $N(.)$ initialized in size $\mathcal{C}$, we can build one recursive step of $Next$ (Algorithm 1) in $O(1)$. By using the pBWT, we can precompute to find $\mathcal{C}$ and build all the c_k in $O(nm)$ time (see Lemma 4 and Eq. 1).

By Lemma 6, we can call $O(k)$ times the algorithm $Next$ to build $N(k)$. Hence, we can solve the λ_{min}-Maximum Segmentation Length in the optimal $O(nm)$ time. $\square$

4.2   Range Maximum Queue

Our algorithm for solving λ_{min}-MSL (Theorem 3) requires a Range Maximum Queue data structure. We begin by presenting a semi-dynamic RMQ data structure that can answer RMQ queries on the array Q in constant time and can "extend" Q to the right (see Lemma 7).

Lemma 7. *There exists a data structure that maintains an integer array $Q[1, n]$ and supports the append query, which adds a new element to the end of Q and increments n, in $O(1)$ amortized time and the Range Maximum Query, which, for given $i \in [1, n]$ and $j \in [i, n]$, computes a position $h \in [i, j]$ such that $Q[h] = \max\{Q[\ell] : i \leq \ell \leq j\}$, in $O(1)$ time.*

Proof. Our solution is a straightforward modification of the classical static Range Minimum Queury approach used in, for instance, [4] and [1].

Let n be the current length of Q. Denote $b = \lceil \frac{\log n}{4} \rceil$. We split $Q[1, n]$ into blocks of length b. As is standard, a Range Maximum Query on $Q[i, j]$ is reduced to two queries inside blocks and one "block-aligned" query: provided i and j belong to different blocks (i.e., $\lfloor (i - 1)/b \rfloor < \lfloor (j - 1)/b \rfloor$), the new three query ranges are $Q[i, i']$, $Q[i'+1, i'']$, $Q[i''+1, j]$ (each might be empty) such that i' and i'' are multiples of b, $i' - i < b$, and $j - i'' < b$.

To process the query on $Q[i'+1, i'']$, we maintain, for each $k \in [0, \log n]$, an array $P_k[1, \lfloor \frac{n}{b} \rfloor]$ storing positions of maximums in ranges of 2^k blocks; more precisely, for $h \in [1, \lfloor \frac{n}{b} \rfloor]$, we have $(h - 2^k)b < P_k[h] \le hb$ and $Q[P_k[h]] = \max\{Q[\ell] : (h - 2^k)b < \ell \le hb\}$, assuming $Q[\ell] = +\infty$ for $\ell \le 0$. Then, putting $k = \lfloor \log((i'' - i')/b) \rfloor$, the maximum in $Q[i'+1, i'']$ obviously is $\max\{Q[P_k[i''/b]], Q[P_k[i'/b + 2^k]]\}$ and we return either $P_k[i''/b]$ or $P_k[i'/b + 2^k]$ accordingly. To calculate k in $O(1)$ time, we use either a special processor instruction or a precomputed table $L[1, 2^{\lceil \frac{\log n}{2} \rceil}]$ such that $L[x] = \lfloor \log x \rfloor$ for $x \in [1, 2^{\lceil \frac{\log n}{2} \rceil}]$ (hence, $\lfloor \log x \rfloor = L[x/2^{\lceil \frac{\log n}{2} \rceil}] + \lceil \frac{\log n}{2} \rceil$ for $x \in [2^{\lceil \frac{\log n}{2} \rceil + 1}, n]$). Note that the length of L is $O(\sqrt{n})$.

For "in-block" queries, we maintain an array $C[1, \lceil \frac{n}{b} \rceil]$ succinctly encoding Cartesian trees for all blocks (see below). The *Cartesian tree*[1] for an array $A[h, h']$ is a binary tree with vertices $[h, h']$ whose root is the smallest $r \in [h, h']$ such that $A[r] = \max\{A[\ell] : h \le \ell \le h'\}$, and the left (resp., right) child of r (if any) is the root of the Cartesian tree for $A[h, r-1]$ (resp., $A[r+1, h']$). For each $h \in [1, \lceil \frac{n}{b} \rceil - 1]$, we encode the Cartesian tree for the block $Q[(h-1)b+1, hb]$ as a sequence of $2b$ balanced parentheses and store it as a $2b$-bit integer in $C[h]$ (zero/one bits correspond to opening/closing parentheses); $C[\lceil \frac{n}{b} \rceil]$ stores the Cartesian tree for $Q[(\lceil \frac{n}{b} \rceil - 1)b+1, n]$. It is well known that if a is the lowest common ancestor of two vertices p and q ($p \le q$) in the Cartesian tree for $A[h, h']$, then $A[a] = \max\{A[\ell] : p \le \ell \le q\}$. We precalculate a table $T[0, 2^{2b}-1][1, b][1, b]$ such that, given numbers $p, q \in [1, b]$ and a $2b$-bit integer x encoding the Cartesian tree for an integer array $A[1, b]$, $T[x][p][q]$ stores the lowest common ancestor of p and q in the tree (if x does not encode any such tree, the value of $T[x][p][q]$ is undefined). Now, using the table T and the array C, one can straightforwardly answer in-block queries in $O(1)$ time. Note that the size of T is $O(2^{2b}b^2) = O(\sqrt{n} \log^{O(1)} n)$.

It remains to describe how the defined structures are modified. Suppose that a new value is appended to the end of Q and n is incremented. If the new n is a multiple of b, a new element $P_k[\lfloor \frac{n}{b} \rfloor]$ is added to each array P_k: $P_0[\lfloor \frac{n}{b} \rfloor]$ is computed naively in $O(b)$ time and, for $k > 0$, $P_k[\lfloor \frac{n}{b} \rfloor]$ is set to either $\ell = P_{k-1}[\lfloor \frac{n}{b} \rfloor]$ or $\ell' = P_{k-1}[\lfloor \frac{n}{b} \rfloor - 2^{k-1}]$, depending on whether $Q[\ell] < Q[\ell']$. Thus, we spend $\Theta(b + \log n) = \Theta(b)$ time to update all P_k, which is amortized among the previous $b - 1$ appends to Q in which n was not a multiple of b.

To maintain C, we utilize a well-known fact that the Cartesian tree for any array $A[h, h']$ can be constructed in $O(h' - h)$ time online, i.e., we read $A[h, h']$ from left to right and, after processing each prefix $A[h, h'']$, have the Cartesian tree for $A[h, h'']$ (e.g., see [4]). The Cartesian tree for the last block $Q[(\lceil \frac{n}{b} \rceil - 1)b+1, n]$ of Q is maintained using this online algorithm and, thus, when n is incremented, we have a new tree and have to update $C[\lceil \frac{n}{b} \rceil]$. To this end, we construct $b - 1$ tables $T_j[0, 2^{2j}-1][1, 2j]$, for $j \in [1, b-1]$, such that, given $h \in [1, 2j]$ and a $2j$-bit number x that encodes a tree F with j vertices $[1, j]$ in a balanced parentheses form, $T_j[x][h]$ contains a $2(j+1)$-bit integer encoding a tree obtained from F by attaching the new vertex $j + 1$ as a leaf to h if $h \le j$,

---

[1] The original of cartesian tree is for Range Minimum Query.

or by making $j + 1$ the new parent of $h - j$ (so that the old parent of $h - j$, if any, is the parent of $j + 1$) if $h > j$; $T_j[x][h]$ is undefined if x does not encode any tree. Using the tables T_j and the online algorithm, one can maintain C in $O(n)$ total time. Note that the size of all T_j is $O(2^{2b}b^2) = O(\sqrt{n}\log^{O(1)} n)$.

Finally, if the new value of $\lceil\log n\rceil$ differs from the old one, we rebuild all structures in a straightforward way: the tables T, T_j, and L are precomputed in $O(\sqrt{n}\log^{O(1)} n) = o(n)$ time, P_0 is constructed from Q in one pass, each array P_k with $k \in [1, \log n]$ is computed using P_{k-1} in $\Theta(\frac{n}{b}) = \Theta(\frac{n}{\log n})$ time, and the Cartesian trees for all blocks are built in $O(n)$ total time and encoded in the array C. Overall, the rebuilding takes $O(n)$ time. Since this process is initiated only when n is a power of two, the total time is $O(n)$ in the end.   □

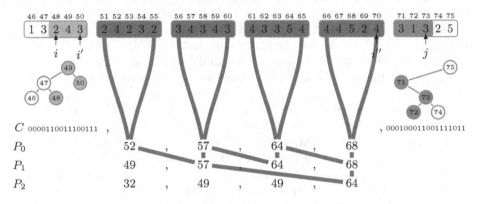

Fig. 1. Example of the construction of the semi-dynamic Range Maximum Query of Lemma 7. For the interval $[48, 73]$ with $b = 5$, we split this interval in three intervals: $[48, 50]$, $[51, 70]$ and $[71, 73]$. We build the positions of the maximum for these three intervals which are $T[0001001111][2][5] = 49$, $P_2[15] = 64$ and $T[0010010111][1][3] = 71$ and we take this one with the maximum value in the array which is 64.

In our algorithm, all the RMQ queries of the semi-dynamic RMQ data structure are made on a window from left to right. We can see this data structure as a *queue*, i.e. a data structure where we can make insertion at the end and deletion at the beginning in constant time. A *Range Maximum Queue* (or RMQe) is the data structure Q that supports queue operations and range maximum query, which, for given $i \in [1, n]$ and $j \in [i, n]$, computes a position $h \in [i, j]$ such that $Q[h] = \max\{Q[\ell] : i \le \ell \le j\}$, in $O(1)$ time. If we know the maximum number of elements of the RMQe data structure, we can improve the space complexity of our algorithm by removing the first elements of the array that will no longer query (Fig. 1).

Lemma 8. *Let N be an integer. There exists a queue data structure that maintains an integer array $Q[1, n]$ with $n \le N$ and supports the Range Maximum query, which adds a new element to the end of Q and increments n, in $O(1)$*

amortized time, remove elements to the beginning of Q in $O(1)$ amortized time and the Range Maximum Query, which, for given $i \in [1, n]$ and $j \in [i, n]$, computes a position $h \in [i, j]$ such that $Q[h] = \max\{Q[\ell] : i \leq \ell \leq j\}$, in $O(1)$ time.

Proof. We use the data structure that we explain in Lemma 7 but we initialize all the arrays in function of N instead of n. Hence we initialize $b = \lceil \frac{\log N}{4} \rceil$, $\log N + 1$ arrays $P_k[1, \lceil \frac{N}{b} \rceil]$ (with $k \in [0, \log N]$), $L[1, 2^{\lceil \frac{\log N}{2} \rceil}]$ and $C[1, \lceil \frac{N}{b} \rceil]$. We add also an integers *begin* initialized to 1 to store the beginning of the arrays P_k and the array C and another integer *start* initiated to 0 to store the shift in the first block (during all our algorithm $start \in [0, b-1]$). We define by $+_A$ the modular addition (plus one) in $[1, A]$, i.e. for all x and y in $[1, A]$, $x +_A y$ is equal to $x + y$ if $x + y \leq A$ and $x + y - A$ otherwise ($x +_A y \in [1, A]$).

To remember, in Lemma 7, a Range Maximum Query on $Q[i, j]$ is reduced to two queries inside blocks $Q[i, i']$ and $Q[i''+1, j]$ and one "block-aligned" query $Q[i'+1, i'']$ with i' and i'' are multiples of b, $i' - i < b$, and $j - i'' < b$. As we shift of *start* elements on the right, we take i' and i'' two multiples of b such that $i + start \leq i' \leq i'' \leq j + start$, $i' - (i + start) < b$, and $(j + start) - i'' < b$. To build $Q[i'+1, i'']$ we put $k = \lfloor \log((i'' - i')/b) \rfloor$ and return $P_k[i''/b +_{\lceil \frac{N}{b} \rceil} begin]$ or $P_k[i'/b + 2^k +_{\lceil \frac{N}{b} \rceil} begin]$. To build the queries inside blocks $Q[i, i']$ and $Q[i''+1, j]$ we need to compute $T[C[i'/b +_{\lceil \frac{N}{b} \rceil} begin]][i + start][i']$ and $T[C[i''/b +_{\lceil \frac{N}{b} \rceil} begin]][i''][j + start]$.

To remove an element at the beginning of our RMQe we only need to increase *start* by 1: *start* becomes $start + 1$ if $start < b - 1$ and otherwise *start* becomes 0 and we update *begin* to $begin +_{\lceil \frac{N}{b} \rceil} 1$.

To add an element at the end, we use the same online algorithm of Lemma 7 (see [4]) to maintain the P_k arrays and C by updating the elements of index *end* except the fact that if n is a multiple of b, we do not create a new element, we just update *end* to $end +_{\lceil \frac{N}{b} \rceil} 1$.   □

The lemma can be further strenghtened by removing the requirement of knowing the bound N: In that case, one needs to consider the case when $\log n$ changes. Unlike in Lemma 7, series of alternating insertions and deletions can now cause $\lceil \log n \rceil$ to change at each operation, so the amortization argument cannot be used directly. However, this case can be handled by maintaining all structures for two consective $\lceil \log n \rceil$ values: Consider $x = \lceil \log n \rceil$ to change into $x+1$ due to insertion. We build all structures for $x+1$ as in the proof of Lemma 7, but also keep the structures for x. All insertions and deletions are applied on both structures until $\lceil \log n \rceil$ becomes $x-1$ or $x+2$. We then build structures for $x - 1$ and keep structures for $x - 1$ and x, or build structures for $x + 2$ and keep structures for $x + 1$ and $x + 2$. Now the $O(n)$ rebuilding cost can be amortized to the $O(n)$ work done before it takes place.

5   Conclusion

In this article, we described linear algorithms for Maximum Segmentation problems (see Table 1). In our Column Stream Model, we assume that we see our data column by column and thus the time complexity is $\Omega(mn)$ and the space complexity is $\Omega(m + \mathcal{X})$ where $\mathcal{X}$ is the size of the output ($\mathcal{X}$ is equal to 1 for λ_{ave}-MSL and λ_{min}-MSL and the cardinality of the optimal partition for λ_{ave}-MSP and λ_{min}-MSP). All of these algorithms can be applied in the random access data model (without data streams): they give exactly the same time complexities.

Table 1. Summary of Maximum Segmentation problems in the column stream model.

Problems	Time complexity	Space complexity	Source
λ_{ave}-MSP	$O(mn)$	$O(m + |P|)$ where P is an optimal partition	Theorem 1
λ_{ave}-MSL	$O(mn)$	$O(m)$	Theorem 2
λ_{min}-MSP	$O(mn)$	$O(m + n)$	Theorem 3
λ_{min}-MSL	$O(mn)$	$O(m + \max\{k - c_k : k \in [1, n]\})$	Theorem 3

As future work, we plan to implement the algorithms and offer them as new features of our founder reconstruction toolbox.[2]

Appendix

About the Column Stream Model

Given an algorithm for a problem with an input $\mathcal{I}$ and an output $\mathcal{O}$, the space complexity of this algorithm corresponds to the space used by $\mathcal{I}$ and by $\mathcal{O}$ and the *auxiliary space* which is the temporary space used by this algorithm. Therefore the space complexity is in $\Omega(|\mathcal{I}| + |\mathcal{O}|)$. In the case of problems of Maximum Segmentation, all algorithms have a space complexity of $\Omega(nm)$ where the input is a set of m strings of size n. As we want to avoid an auxiliary space of $\Theta(nm)$ (this could be too big for a computer), we cannot use the random access model. Indeed the random access model corresponds to open all the file in input in the temporary memory. We suggest a specific streaming data model where the set of strings of the same length is seen column by column: the *Column Stream Model*. In this model, the size of the input is in $\Theta(m)$ which is acceptable.

   To prove the realism of this model, we implemented a streaming way to read a file and we tested this implementation with files of different sizes (see Fig. 2). The experiments were run on a machine with an Intel Xeon E5-2680 v4 2.4GHz CPU, which has a 35 MB Intel SmartCache. The machine has 256 gigabytes of memory at a speed of 2400MT/s. The code was compiled with g++ using the -Ofast optimization flag.

---

[2] https://github.com/tsnorri/founder-sequences.

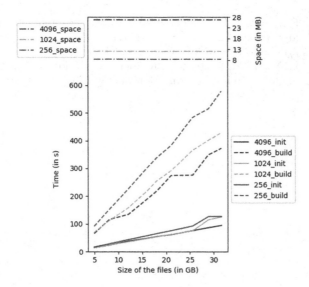

Fig. 2. Time and space complexity to read a set of recombinants depending of the buffer size (256, 1024 and 4096).

References

1. Bender, M.A., Farach-Colton, M.: The LCA problem revisited. In: Gonnet, G.H., Viola, A. (eds.) LATIN 2000. LNCS, vol. 1776, pp. 88–94. Springer, Heidelberg (2000). https://doi.org/10.1007/10719839_9
2. Blin, G., Rizzi, R., Sikora, F., Vialette, S.: Minimum mosaic inference of a set of recombinants. Int. J. Found. Comput. Sci. **24**(1), 51–66 (2013)
3. Durbin, R.: Efficient haplotype matching and storage using the positional Burrows-Wheeler transform (PBWT). Bioinformatics **30**(9), 1266–1272 (2014)
4. Fischer, J., Heun, V.: Theoretical and practical improvements on the RMQ-problem, with applications to LCA and LCE. In: Lewenstein, M., Valiente, G. (eds.) CPM 2006. LNCS, vol. 4009, pp. 36–48. Springer, Heidelberg (2006). https://doi.org/10.1007/11780441_5
5. Gajewska, H., Tarjan, R.E.: Deques with heap order. Inf. Process. Lett. **22**(4), 197–200 (1986)
6. Mäkinen, V., Norri, T.: Applying the positional Burrows-Wheeler transform to all-pairs hamming distance. Submitted manuscript (2018)
7. Norri, T., Cazaux, B., Kosolobov, D., Mäkinen, V.: Minimum segmentation for pan-genomic founder reconstruction in linear time. In 18th International Workshop on Algorithms in Bioinformatics, WABI 2018, Helsinki, Finland, 20–22 August 2018, pp. 15:1–15:15 (2018). https://doi.org/10.4230/LIPIcs.WABI.2018.15
8. Rastas, P., Ukkonen, E.: Haplotype inference via hierarchical genotype parsing. In: Giancarlo, R., Hannenhalli, S. (eds.) WABI 2007. LNCS, vol. 4645, pp. 85–97. Springer, Heidelberg (2007). https://doi.org/10.1007/978-3-540-74126-8_9

9. Ukkonen, E.: Finding founder sequences from a set of recombinants. In: Guigó, R., Gusfield, D. (eds.) WABI 2002. LNCS, vol. 2452, pp. 277–286. Springer, Heidelberg (2002). https://doi.org/10.1007/3-540-45784-4_21

10. Valenzuela, D., Norri, T., Niko, V., Pitkänen, E., Mäkinen, V.: Towards pangenome read alignment to improve variation calling. BMC Genomics 19(Suppl. 2), 87 (2018)

Space-Efficient Merging of Succinct de Bruijn Graphs

Lavinia Egidi[1] [iD], Felipe A. Louza[2(✉)] [iD], and Giovanni Manzini[1,3] [iD]

[1] University of Eastern Piedmont, Alessandria, Italy
{lavinia.egidi,giovanni.manzini}@uniupo.it
[2] Faculty of Electrical Engineering, Universidade Federal de Uberlândia,
Uberlândia, Brazil
louza@ufu.br
[3] IIT CNR, Pisa, Italy

Abstract. We propose a new algorithm for merging succinct representations of *de Bruijn* graphs introduced in [Bowe *et al.* WABI 2012]. Our algorithm is based on the lightweight BWT merging approach by Holt and McMillan [Bionformatics 2014, ACM-BCB 2014]. Our algorithm has the same asymptotic cost of the state of the art tool for the same problem presented by Muggli *et al.* [bioRxiv 2017, Bioinformatics 2019], but it uses less than half of its working space. A novel important feature of our algorithm, not found in any of the existing tools, is that it can compute the *Variable Order* succinct representation of the union graph within the same asymptotic time/space bounds.

Keywords: de Bruijn graphs · Succinct data structures · Merging · Variable-order · Colored graphs · External memory algorithms

1 Introduction

The *de Bruijn* graph for a collection of strings is a key data structure in genome assembly [19]. After the seminal work of Bowe *et al.* [5], many succinct representations of this data structure have been proposed in the literature [2–4,18] offering more and more functionalities still using a fraction of the space required to store the input collection uncompressed. In this paper we consider the problem of merging two existing succinct representations of de Bruijn graphs built for different collections. Since the de Bruijn graph is a lossy representation and from it we cannot recover the original input collection, the alternative to merging is storing a copy of each collection to be used for building new de Bruijn graphs from scratch.

Recently, Muggli *et al.* [16,17] have proposed a merging algorithm for colored de Bruijn graphs and have shown the effectiveness of the merging approach for the construction of de Bruijn graphs for very large datasets. The algorithm in [16] is based on an MSD Radix Sort procedure of the graph edges and its running

© Springer Nature Switzerland AG 2019
N. R. Brisaboa and S. J. Puglisi (Eds.): SPIRE 2019, LNCS 11811, pp. 337–351, 2019.
https://doi.org/10.1007/978-3-030-32686-9_24

time is $\mathcal{O}(mk)$, where m is the total number of edges and k is the order of the de Bruijn graph.

A fundamental parameter of any construction algorithm for succinct data structures is its *space usage* since this parameter determines the size of the largest dataset that can be handled by a machine with a given amount of memory. For a graph with m edges and n nodes the merging algorithm by Muggli *et al.* uses, in addition to the input and the output, $2(m \log \sigma + m + n)$ bits plus $\mathcal{O}(\sigma)$ words of working space, where σ is the alphabet size. This value represents a three fold improvement over previous results, but it is still larger than the size of the resulting de Bruijn graph which is upper bounded by $2(m \log \sigma + m) + o(m)$ bits.

In this paper, we present a new merging algorithm that still runs in $\mathcal{O}(mk)$ time, but only uses $4n$ bits plus $\mathcal{O}(\sigma)$ words of working space. For genome collections ($\sigma = 5$) our algorithm uses less than half the space of Muggli *et al.*'s: our advantage grows with the size of the alphabet and with the average outdegree m/n. Notice that the working space of our algorithm is always less than the space of the resulting de Bruijn graph. In Sect. 4 we will discuss the practical significance of this space reduction.

Our new merging algorithm is based on a mixed LSD/MSD Radix Sort algorithm which is inspired by the lightweight BWT merging algorithm introduced by Holt and McMillan [11,12] and later improved in [8,9]. In addition to its small working space, our algorithm has the remarkable feature that it can compute as a by-product, with no additional cost, the LCS (Longest Common Suffix) between the node labels, thus making it possible to construct succinct Variable Order de Bruijn graph representations [4], a feature not shared by any other merging algorithm.

The rest of the paper is organized as follows. After reviewing succinct de Bruijn graphs in Sect. 2, we describe our algorithm in Sect. 3. In Sect. 4 we describe the implementation details and compare our result to the state of the art. In Sect. 5 we discuss the case of colored or variable order de Bruijn graphs. In Sect. 6 we show that combining an external memory version of our merging algorithm with recent results on external memory de Bruijn graph construction [6,7] we get a space efficient external memory procedure for building succinct representations of de Bruijn graphs for very large collections.

2   Notation and Background

Given the alphabet $\Sigma = \{1, 2, \ldots, \sigma\}$ and a collection of strings $\mathcal{C} = s_1, \ldots, s_d$ over Σ, we prepend to each string s_i k copies of a symbol $\$ \notin \Sigma$ which is lexicographically smaller than any other symbol. The order-k *de Bruijn graph* $G(V, E)$ for the collection $\mathcal{C}$ is a directed edge-labeled graph containing a node v for every **unique k-mer** appearing in one of the strings of $\mathcal{C}$. For each node v we denote by $\overrightarrow{v} = v[1, k]$ its associated k-mer, where $v[1] \ldots v[k]$ are symbols. The graph G contains an edge (u, v), with label $v[k]$, iff one of the strings in $\mathcal{C}$ contains a **$(k + 1)$-mer** with prefix $\overrightarrow{u}$ and suffix $\overrightarrow{v}$. The edge (u, v) therefore represents the $(k+1)$-mer $u[1, k]v[k]$. Note that each node has at most $|\Sigma|$ outgoing edges and all edges incoming to node v have label $v[k]$.

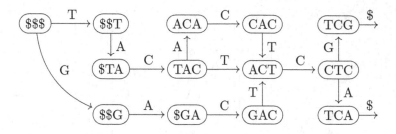

Fig. 1. de Bruijn graph for $\mathcal{C} = \{\text{TACACT, TACTCG, GACTCA}\}$.

BOSS Succinct Representation. In 2012, Bowe *et al.* [5] introduced a succinct representation for the de Bruijn graph, usually referred to as BOSS representation, for the authors' initials. The authors showed how to represent the graph in small space supporting fast navigation operations. The BOSS representation of the graph $G(V, E)$ is defined by considering the set of nodes $v_1, v_2, \ldots v_n$ sorted according to the colexicographic order of their associated k-mer. Hence, if $\overleftarrow{v} = v[k] \ldots v[1]$ denotes the string $\overrightarrow{v}$ reversed, the nodes are ordered so that

$$\overleftarrow{v_1} \prec \overleftarrow{v_2} \prec \cdots \prec \overleftarrow{v_n} \tag{1}$$

By construction the first node is $\overleftarrow{v_1} = \$^k$ and all $\overleftarrow{v_i}$ are distinct. For each node v_i, $i = 1, \ldots, n$, we define W_i as the sorted sequence of symbols on the edges leaving from node v_i; if v_i has out-degree zero we set $W_i = \$$. Let $\mathsf{Node}[i]$ denote the node label for W_i. Finally, we define

1. $W[1, m]$ as the concatenation $W_1 W_2 \cdots W_n$;
2. $W^-[1, m]$ as the bitvector such that $W^-[i] = \mathbf{1}$ iff $W[i]$ corresponds to the label of the edge (u, v) such that $\overleftarrow{u}$ has the smallest rank among the nodes that have an edge going to node v;
3. $\mathsf{last}[1, m]$ as the bitvector such that $\mathsf{last}[i] = 1$ iff $i = m$ or the outgoing edges corresponding to $W[i]$ and $W[i + 1]$ have different source nodes.
4. $\mathsf{C}[1, \sigma]$ as the integer array, such that $\mathsf{C}[c]$ stores the number of symbols smaller than $c \in \Sigma \cup \{\$\}$ in the last symbol of Node.

The length m of the arrays W, W^-, and last is equal to the number of edges plus the number of nodes with out-degree 0. In addition, the number of $\mathbf{1}$'s in last is equal to the number of nodes n, and the number of $\mathbf{1}$'s in W^- is equal to the number of nodes with positive in-degree, which is $n - 1$ since $v_1 = \$^k$ is the only node with in-degree 0. Array C can be obtained by scanning W, W^- and last, therefore, array $\mathsf{Node}[1, m]$ is not stored explicitly.

Note that there is a natural one-to-one correspondence, called LF for historical reasons, between the indices i such that $W^-[i] = \mathbf{1}$ and the the set $\{2, \ldots, n\}$: in this correspondence $LF(i) = j$ iff v_j is the destination node of the edge associated to $W[i]$. See example in Figs. 1 and 2.

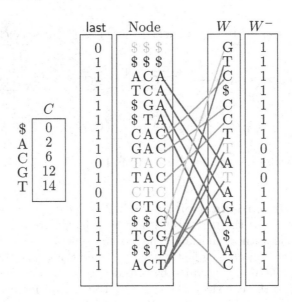

Fig. 2. BOSS representation of the graph in Fig. 1. The colored lines connect each label in W to its destination node; edges of the same color have the same label. Note that edges of the same color do not cross because of Property 1. (Color figure online)

Property 1. The LF map is order preserving in the following sense: if $W^-[i] = W^-[j] = 1$ then

$$W[i] < W[j] \implies LF(i) < LF(j),$$
$$(W[i] = W[j]) \wedge (i < j) \implies LF(i) < LF(j). \tag{2}$$

$\square$

In [5] it is shown that given array C, enriching the arrays W, W^-, and last with the data structures from [10,20] supporting constant time rank and select operations, we can efficiently navigate the graph G. The cost to store array C is $\mathcal{O}(\sigma \log n)$ bits. The overall cost of encoding the three arrays and the auxiliary data structures is bounded by $m \log \sigma + 2m + o(m)$ bits, with the usual time/space tradeoffs available for rank/select data structures.

Colored BOSS. The colored de Bruijn graph [13] is an extension of the de Bruijn graphs for a multiset of individual graphs, where each edge is associated with a set of "colors" that indicates which graphs contain that edge.

The BOSS representation for a set of graphs $\mathcal{G} = \{G_1, \dots, G_t\}$ contains the union of all individual graphs. In its simplest representation, the colors of all edges $W[i]$ are stored in a two-dimensional binary array $\mathcal{M}$, such that $\mathcal{M}[i, j] = 1$ iff the i-th edge is present in graph G_j. There are different compression alternatives for the color matrix $\mathcal{M}$ that support fast operations [2,15,18]. Recently, Alipanah *et al.* [1] presented a different approach to reduce the size of $\mathcal{M}$ by recoloring.

Variable-Order BOSS. The order k (dimension) of a de Bruijn graph is an important parameter for genome assembling algorithms. The graph can be very small and uninformative when k is small, whereas it can become too large or disconnected when k is large. To add flexibility to the BOSS representation, Boucher *et al.* [4] suggest to enrich the BOSS representation of an order-k de Bruijn graph with the length of the longest common suffix (LCS) between the k-mers of consecutive nodes $v_1, v_2, \ldots, v_n$ sorted according to (1). These lengths are stored in a wavelet tree using $O(n \log k)$ additional bits. The authors show that this enriched representation supports navigation on all de Bruijn graphs of order $k' \le k$ and that it is even possible to vary the order k' of the graph on the fly during the navigation up to the maximum value k.

The LCS between $\overrightarrow{v_i}$ and $\overrightarrow{v_{i+1}}$ is equivalent to the length of the longest common prefix (LCP) between their reverses $\overleftarrow{v_i}$ and $\overleftarrow{v_{i+1}}$. The LCP (or LCS) between the nodes $v_1, v_2, \cdots, v_n$ can be computed during the k-mer sorting phase. In the following we denote by VO-BOSS the **variable order** succinct de Bruijn graph consisting of the BOSS representations enriched with the LCS/LCP information.

3   Merging Plain BOSS Representations

Suppose we are given the BOSS representations of two de Bruijn graphs $\langle W_0, W_0^-, \mathsf{last}_0 \rangle$ and $\langle W_1, W_1^-, \mathsf{last}_1 \rangle$ obtained respectively from the collections of strings $\mathcal{C}_0$ and $\mathcal{C}_1$. In this section we show how to compute the BOSS representation for the union collection $\mathcal{C}_{01} = \mathcal{C}_0 \cup \mathcal{C}_1$. The procedure does not change in the general case when we are merging an arbitrary number of graphs. Let G_0 and G_1 denote respectively the (uncompressed) de Bruijn graphs for $\mathcal{C}_0$ and $\mathcal{C}_1$, and let

$$v_1, \ldots, v_{n_0} \quad \text{and} \quad w_1, \ldots, w_{n_1}$$

denote their respective set of nodes sorted in colexicographic order. Hence, with the notation of the previous section we have

$$\overleftarrow{v_1} \prec \cdots \prec \overleftarrow{v_{n_0}} \quad \text{and} \quad \overleftarrow{w_1} \prec \cdots \prec \overleftarrow{w_{n_1}} \tag{3}$$

We observe that the k-mers in the collection $\mathcal{C}_{01}$ are simply the union of the k-mers in $\mathcal{C}_0$ and $\mathcal{C}_1$. To build the de Bruijn graph for $\mathcal{C}_{01}$ we need therefore to: (1) merge the nodes in G_0 and G_1 according to the colexicographic order of their associated k-mers, (2) recognize when two nodes in G_0 and G_1 refer to the same k-mer, and (3) properly merge and update the bitvectors W_0^-, last_0 and W_1^-, last_1.

3.1   Phase 1: Merging k-mers

The main technical difficulty is that in the BOSS representation the k-mers associated to each node $\overrightarrow{v} = v[1, k]$ are not directly available. Our algorithm

will reconstruct them using the symbols associated to the graph edges; to this end the algorithm will consider only the edges such that the corresponding entries in W_0^- or W_1^- are equal to $\mathbf{1}$. Following these edges, first we recover the last symbol of each k-mer, following them a second time we recover the last two symbols of each k-mer and so on. However, to save space we do not explicitly maintain the k-mers; instead, using the ideas from [11, 12] our algorithm computes a bitvector $Z^{(k)}$ representing how the k-mers in G_0 and G_1 should be merged according to the colexicographic order.

To this end, our algorithm executes $k-1$ iterations of the code shown in Fig. 3 (note that lines 8–10 and 17–22 of the algorithm are related to the computation of the B array that is used in the following section). For $h = 2, 3, \ldots, k$, during iteration h, we compute the bitvector $Z^{(h)}[1, n_0 + n_1]$ containing n_0 $\mathbf{0}$'s and n_1 $\mathbf{1}$'s such that $Z^{(h)}$ satisfies the following property

Property 2. For $i = 1, \ldots, n_0$ and $j = 1, \ldots n_1$ the i-th $\mathbf{0}$ precedes the j-th $\mathbf{1}$ in $Z^{(h)}$ if and only if $\overleftarrow{v_i}[1, h] \preceq \overleftarrow{w_j}[1, h]$.   □

Property 2 states that if we merge the nodes from G_0 and G_1 according to the bitvector $Z^{(h)}$ the corresponding k-mers will be sorted according to the lexicographic order restricted to the first h symbols of each reversed k-mer. As a consequence, $Z^{(k)}$ will provide us the colexicographic order of all the nodes in G_0 and G_1. To prove that Property 2 holds, we first define $Z^{(1)}$ and show that it satisfies the property, then we prove that for $h = 2, \ldots, k$ the code in Fig. 3 computes $Z^{(h)}$ that still satisfies Property 2.

For $c \in \Sigma$ let $\ell_0(c)$ and $\ell_1(c)$ denote respectively the number of nodes in G_0 and G_1 whose associated k-mers end with symbol c. These values can be computed with a single scan of W_0 (resp. W_1) considering only the symbols $W_0[i]$ (resp. $W_1[i]$) such that $W_0^-[i] = 1$ (resp. $W_1^-[i] = 1$). By construction, it is

$$n_0 = 1 + \sum_{c \in \Sigma} \ell_0(c), \qquad n_1 = 1 + \sum_{c \in \Sigma} \ell_1(c)$$

where the two 1's account for the nodes v_1 and w_1 whose associated k-mer is $\$^k$. We define

$$Z^{(1)} = \underline{\mathbf{01}}\ \underline{\mathbf{0}^{\ell_0(1)}\mathbf{1}^{\ell_1(1)}}\ \underline{\mathbf{0}^{\ell_0(2)}\mathbf{1}^{\ell_1(2)}} \cdots \underline{\mathbf{0}^{\ell_0(\sigma)}\mathbf{1}^{\ell_1(\sigma)}}\ . \qquad (4)$$

The first pair $\mathbf{01}$ in $Z^{(1)}$ accounts for v_1 and w_1; for each $c \in \Sigma$ group $\mathbf{0}^{\ell_0(c)}\mathbf{1}^{\ell_1(c)}$ accounts for the nodes ending with symbol c. Note that, apart from the first two symbols, $Z^{(1)}$ can be logically partitioned into σ subarrays one for each alphabet symbol. For $c \in \Sigma$ let

$$\mathsf{start}(c) = 3 + \sum_{i < c} (\ell_0(i) + \ell_1(i))$$

then the subarray corresponding to c starts at position $\mathsf{start}(c)$ and has size $\ell_0(c) + \ell_1(c)$. As a consequence of (3), the i-th $\mathbf{0}$ (resp. j-th $\mathbf{1}$) belongs to the subarray associated to symbol c iff $\overleftarrow{v_i}[1] = c$ (resp. $\overleftarrow{w_j}[1] = c$).

To see that $Z^{(1)}$ satisfies Property 2, observe that the i-th $\mathbf{0}$ precedes j-th $\mathbf{1}$ iff the i-th $\mathbf{0}$ belongs to a subarray corresponding to a symbol not larger than the symbol corresponding to the subarray containing the j-th $\mathbf{1}$; this implies $\overleftarrow{v_i}[1,1] \preceq \overleftarrow{w_j}[1,1]$.

The bitvectors $Z^{(h)}$ computed by the algorithm in Fig. 3 can be logically divided into the same subarrays we defined for $Z^{(1)}$. In the algorithm we use an array $F[1,\sigma]$ to keep track of the next available position of each subarray. Because of how the array F is initialized and updated, we see that every time we read a symbol c at line 14 the corresponding bit $b = Z^{(h-1)}[k]$, which gives us the graph containing c, is written in the portion of $Z^{(h)}$ corresponding to c (line 16). The only exception are the first two entries of $Z^{(h)}$ which are written at line 6 which corresponds to the nodes v_1 and w_1. We treat these nodes differently since they are the only ones with in-degree zero. For all other nodes, we implicitly use the one-to-one correspondence (2) between entries $W[i]$ with $W^-[i] = 1$ and nodes v_j with positive in-degree.

The following Lemma proves the correctness of the algorithm in Fig. 3.

Lemma 1. *For $h = 2, \ldots, k$, the array $Z^{(h)}$ computed by the algorithm in Fig. 3 satisfies Property 2.*

Proof. To prove the "if" part of Property 2 let $1 \le f < g \le n_0 + n_1$ denote two indexes such that $Z^{(h)}[f]$ is the i-th $\mathbf{0}$ and $Z^{(h)}[g]$ is the j-th $\mathbf{1}$ in $Z^{(h)}$ for some $1 \le i \le n_0$ and $1 \le j \le n_1$. We need to show that $\overleftarrow{v_i}[1,h] \preceq \overleftarrow{w_j}[1,h]$.

Assume first $\overleftarrow{v_i}[1] \neq \overleftarrow{w_j}[1]$. The hypothesis $f < g$ implies $\overleftarrow{v_i}[1] < \overleftarrow{w_j}[1]$, since otherwise during iteration h the j-th $\mathbf{1}$ would have been written in a subarray of $Z^{(h)}$ preceding the one where the i-th $\mathbf{0}$ is written. Hence $\overleftarrow{v_i}[1,h] \preceq \overleftarrow{w_j}[1,h]$ as claimed.

Assume now $\overleftarrow{v_i}[1] = \overleftarrow{w_j}[1] = c$. In this case during iteration h the i-th $\mathbf{0}$ and the j-th $\mathbf{1}$ are both written to the subarray of $Z^{(h)}$ associated to symbol c. Let f', g' denote respectively the value of the main loop variable p in the procedure of Fig. 3 when the entries $Z^{(h)}[f]$ and $Z^{(h)}[g]$ are written. Since each subarray in $Z^{(h)}$ is filled sequentially, the hypothesis $f < g$ implies $f' < g'$. By construction $Z^{(h-1)}[f'] = 0$ and $Z^{(h-1)}[g'] = 1$. Say f' is the i'-th $\mathbf{0}$ in $Z^{(h-1)}$ and g' is the j'-th $\mathbf{1}$ in $Z^{(h-1)}$. By the inductive hypothesis on $Z^{(h-1)}$ it is

$$\overleftarrow{v_{i'}}[1, h-1] \preceq \overleftarrow{w_{j'}}[1, h-1]. \tag{5}$$

By construction there is an edge labeled c from $v_{i'}$ to v_i and from $w_{j'}$ to w_j hence

$$\overrightarrow{v_i}[1,h] = \overrightarrow{v_{i'}}[1,h-1]c, \qquad \overrightarrow{w_j}[1,h] = \overrightarrow{w_{j'}}[1,h-1]c;$$

therefore

$$\overleftarrow{v_i}[1,h] = c\overleftarrow{v_{i'}}[1,h-1], \qquad \overleftarrow{w_j}[1,h] = c\overleftarrow{w_{j'}}[1,h-1];$$

using (5) we conclude that $\overleftarrow{v_i}[1,h] \preceq \overleftarrow{w_j}[1,h]$ as claimed.

For the "only if" part of Property 2, assume $\overleftarrow{v_i}[1,h] \preceq \overleftarrow{w_j}[1,h]$ for some $i \geq 1$ and $j \geq 1$. We need to prove that in $Z^{(h)}$ the i-th $\mathbf{0}$ precedes the j-th $\mathbf{1}$. If $\overleftarrow{v_i}[1] \neq \overleftarrow{w_j}[1]$ the proof is immediate. If $c = \overleftarrow{v_i}[1] = \overleftarrow{w_j}[1]$ then

$$\overleftarrow{v_i}[2,h] \preceq \overleftarrow{w_j}[2,h].$$

Let i' and j' be such that $\overleftarrow{v_{i'}}[1,h-1] = \overleftarrow{v_i}[2,h]$ and $\overleftarrow{w_{j'}}[1,h-1] = \overleftarrow{w_j}[2,h]$. By induction hypothesis, in $Z^{(h-1)}$ the i'-th $\mathbf{0}$ precedes the j'-th $\mathbf{1}$.

During phase h, the i-th $\mathbf{0}$ in $Z^{(h)}$ is written to position f when processing the i'-th $\mathbf{0}$ of $Z^{(h-1)}$, and the j-th $\mathbf{1}$ in $Z^{(h)}$ is written to position g when processing the j'-th $\mathbf{1}$ of $Z^{(h-1)}$. Since in $Z^{(h-1)}$ the i'-th $\mathbf{0}$ precedes the j'-th $\mathbf{1}$ and since f and g both belong to the subarray of $Z^{(h)}$ corresponding to the symbol c, their relative order does not change and the i-th $\mathbf{0}$ precedes the j-th $\mathbf{1}$ as claimed.   □

```
 1: for c ← 1 to σ do
 2:     F[c] ← start(c)                                      ▷ Init F array
 3:     Block_id[c] ← −1                                     ▷ Init Block_id array
 4: end for
 5: i₀ ← i₁ ← 1                                              ▷ Init counters for W₀ and W₁
 6: Z^(h) ← 01                                               ▷ First two entries correspond to v₁ and w₁
 7: for p ← 1 to n₀ + n₁ do
 8:     if B[p] ≠ 0 and B[p] ≠ h then
 9:         id ← p                                           ▷ A new block of Z^(h−1) is starting
10:     end if
11:     b ← Z^(h−1)[p]                                       ▷ Get bit b from Z^(h−1)
12:     repeat                                               ▷ Current node is from graph G_b
13:         if W_b^−[i_b] = 1 then
14:             c ← W_b[i_b]                                 ▷ Get symbol from outgoing edges
15:             q ← F[c]++                                   ▷ Get destination for b according to symbol c
16:             Z^(h)[q] ← b                                 ▷ Copy bit b to Z^(h)
17:             if Block_id[c] ≠ id then
18:                 Block_id[c] ← id                         ▷ Update block id for symbol c
19:                 if B[q] = 0 then                         ▷ Check if already marked
20:                     B[q] ← h                             ▷ A new block of Z^(h) will start here
21:                 end if
22:             end if
23:         end if
24:     until last_b[i_b++] ≠ 1                              ▷ Exit if c was last edge
25: end for
```

Fig. 3. Main procedure for merging succinct de Bruijn graphs. Lines 8–10 and 17–22 are related to the computation of the B array introduced in Sect. 3.2.

3.2   Phase 2: Recognizing Identical k-mers

Once we have determined, via the bitvector $Z^{(h)}[1, n_0 + n_1]$, the colexicographic order of the k-mers, we need to determine when two k-mers are identical since in this case we have to merge their outgoing and incoming edges. Note that two identical k-mers will be consecutive in the colexicographic order and they will necessarily belong one to G_0 and the other to G_1.

Following Property 2, and a technique introduced in [8], we identify the i-th 0 in $Z^{(h)}$ with $\overleftarrow{v_i}$ and the j-th 1 in $Z^{(h)}$ with $\overleftarrow{w_j}$. Property 2 is equivalent to state that we can logically partition $Z^{(h)}$ into $b(h) + 1$ h-blocks

$$Z^{(h)}[1, \ell_1], \ Z^{(h)}[\ell_1 + 1, \ell_2], \ \ldots, \ Z^{(h)}[\ell_{b(h)} + 1, n_0 + n_1] \tag{6}$$

such that each block corresponds to a set of k-mers which are prefixed by the same length-h substring. Note that during iterations $h = 2, 3, \ldots, k$ the k-mers within an h-block will be rearranged, and sorted according to longer and longer prefixes, but they will stay within the same block.

In the algorithm of Fig. 3, in addition to $Z^{(h)}$, we maintain an integer array $B[1, n_0 + n_1]$, such that at the end of iteration h it is $B[i] \neq 0$ if and only if a block of $Z^{(h)}$ starts at position i. Initially, for $h = 1$, since we have one block per symbol, we set

$$B = \underline{10} \, \underline{10^{\ell_0(1) + \ell_1(1) - 1}} \, \underline{10^{\ell_0(2) + \ell_1(2) - 1}} \cdots \underline{10^{\ell_0(\sigma) + \ell_1(\sigma) - 1}}.$$

During iteration h, new block boundaries are established as follows. At line 9 we identify each existing block with its starting position. Then, at lines 17–22, if the entry $Z^{(h)}[q]$ has the form $c\alpha$, while $Z^{(h)}[q - 1]$ has the form $c\beta$, with α and β belonging to different blocks, then we know that q is the starting position of an h-block. Note that we write h to $B[q]$ only if no other value has been previously written there. This ensures that $B[q]$ is the smallest position in which the strings corresponding to $Z^{(h)}[q - 1]$ and $Z^{(h)}[q]$ differ, or equivalently, $B[q] - 1$ is the LCP between the strings corresponding to $Z^{(h)}[q - 1]$ and $Z^{(h)}[q]$. The above observations are summarized in the following Lemma, which is a generalization to de Bruijn graphs of an analogous result for BWT merging established in Corollary 4 in [8].

Lemma 2. *After iteration k of the merging algorithm for $q = 2, \ldots, n_0 + n_1$ if $B[q] \neq 0$ then $B[q] - 1$ is the LCP between the reverse k-mers corresponding to $Z^{(k)}[q - 1]$ and $Z^{(k)}[q]$, while if $B[q] = 0$ their LCP is equal to k, hence such k-mers are equal.*   □

The above lemma shows that using array B we can establish when two k-mers are equal and consequently the associated graph nodes should be merged.

3.3   Phase 3: Building BOSS Representation for the Union Graph

We now show how to compute the succinct representation of the union graph $G_0 \cup G_1$, consisting of the arrays $\langle W_{01}, W_{01}^-, \mathrm{last}_{01} \rangle$, given the succinct representations of G_0 and G_1 and the arrays $Z^{(k)}$ and B.

The arrays W_{01}, W_{01}^-, last_{01} are initially empty and we fill them in a single sequential pass. For $q = 1, \ldots, n_0 + n_1$ we consider the values $Z^{(k)}[q]$ and $B[q]$. If $B[q] = 0$ then the k-mer associated to $Z^{(k)}[q-1]$, say $\overleftarrow{v_i}$ is identical to the k-mer associated to $Z^{(k)}[q]$, say $\overleftarrow{w_j}$. In this case we recover from W_0 and W_1 the labels of the edges outgoing from v_i and w_j, we compute their union and write them to W_{01} (we assume the edges are in the lexicographic order), writing at the same time the representation of the out-degree of the new node to last_{01}. If instead $B[q] \neq 0$, then the k-mer associated to $Z^{(k)}[q-1]$ is unique and we copy the information of its outgoing edges and out-degree directly to W_{01} and last_{01}.

When we write the symbol $W_{01}[i]$ we simultaneously write the bit $W_{01}^-[i]$ according to the following strategy. If the symbol $c = W_{01}[i]$ is the first occurrence of c after a value $B[q]$, with $0 < B[q] < k$, then we set $W_{01}^-[i] = \mathbf{1}$, otherwise we set $W_{01}^-[i] = \mathbf{0}$. The rationale is that if no values $B[q]$ with $0 < B[q] < k$ occur between two nodes, then the associated (reversed) k-mers have a common LCP of length $k-1$ and therefore if they both have an outgoing edge labelled with c they reach the same node and only the first one should have $W_{01}^-[i] = \mathbf{1}$.

4   Implementation Details and Analysis

Let $n = n_1 + n_0$ denote the sum of number of nodes in G_0 and G_1, and let $m = |W_0| + |W_1|$ denote the sum of the number of edges. The k-mer merging algorithm as described executes in $\mathcal{O}(m)$ time a first pass over the arrays W_0, W_0^-, and W_1, W_1^- to compute the values $\ell_0(c) + \ell_1(c)$ for $c \in \Sigma$ and initialize the arrays $F[1,\sigma]$, $\mathsf{start}[1,\sigma]$, $\mathsf{Block\_id}[1,\sigma]$ and $Z^{(1)}[1,n]$ (Phase 1). Then, the algorithm executes $k-1$ iterations of the code in Fig. 3 each iteration taking $\mathcal{O}(m)$ time. Finally, still in $\mathcal{O}(m)$ time the algorithm computes the succinct representation of the union graph (Phases 2 and 3). The overall running time is therefore $\mathcal{O}(m\,k)$.

We now analyze the space usage of the algorithm. In addition to the input and the output, our algorithm uses $2n$ bits for two instances of the $Z^{(\cdot)}$ array (for the current $Z^{(h)}$ and for the previous $Z^{(h-1)}$), plus $n\lceil \log k \rceil$ bits for the B array. Note, however, that during iteration h we only need to check whether $B[i]$ is equal to 0, h, or some value within 0 and h. Similarly, for the computation of W_{01}^- we only need to distinguish between the cases where $B[i]$ is equal to 0, k or some value $0 < B[i] < k$. Therefore, we can save space replacing $B[1,n]$ with an array $B_2[1,n]$ containing two bits per entry representing the four possible states $\{0, 1, 2, 3\}$. During iteration h, the values in B_2 are used instead of the ones in B as follows: An entry $B_2[i] = 0$ corresponds to $B[i] = 0$, an entry $B_2[i] = 3$ corresponds to an entry $0 < B[i] < h - 1$. In addition, if h is even, an entry $B_2[i] = 2$ corresponds to $B[i] = h$ and an entry $B_2[i] = 1$ corresponds to $B[i] = h - 1$; while if h is odd the correspondence is $2 \to h - 1$, $1 \to h$. The reason for this apparently involved scheme, first introduced in [6], is that during phase h, an entry in B_2 can be modified either before or after we have read it at Line 9. Using this technique, the working space of the algorithm, i.e., the space in addition to the input and the output, is $4n$ bits plus $3\sigma + \mathcal{O}(1)$ words of RAM for the arrays start, F, and $\mathsf{Block\_id}$.

Theorem 1. *The merging of two succinct representations of two order-k de Bruijn graphs can be done in $\mathcal{O}(m\,k)$ time using $4n$ bits plus $\mathcal{O}(\sigma)$ words of working space.*   □

We stated the above theorem in terms of working space, since the total space depends on how we store the input and output, and for such storage there are several possible alternatives. The usual assumption is that the input de Bruijn graphs, i.e. the arrays $\langle W_0, W_0^-, \mathsf{last}_0 \rangle$ and $\langle W_1, W_1^-, \mathsf{last}_1 \rangle$, are stored in RAM using overall $m \log \sigma + 2m$ bits. Since the three arrays representing the output de Bruijn graph are generated sequentially in one pass, they are usually written directly to disk without being stored in RAM, so they do not contribute to the total space usage. Also note that during each iteration of the algorithm in Fig. 3, the input arrays are all accessed sequentially. Thus we could keep them on disk reducing the overall RAM usage to just $4n$ bits plus $\mathcal{O}(\sigma)$ words; the resulting algorithm would perform additional $\mathcal{O}(k(m \log \sigma + 2m)/D)$ I/Os where D denotes the disk page size in bits.

Comparison with the State of the Art. The de Bruijn graph merging algorithm by Muggli *et al.* [16,17] is similar to ours in that it has a *planning phase* consisting of the colexicographic sorting of the $(k+1)$-mers associated to the edges of G_0 and G_1. To this end, the algorithm uses a standard MSD radix sort. However only the most significant symbol of each $(k+1)$-mer is readily available in W_0 and W_1. Thus, during each iteration the algorithm computes also the next symbol of each $(k+1)$-mer that will be used as a sorting key in the next iteration. The overall space for such symbols is $2m\lceil \log \sigma \rceil$ bits, since for each edge we need the symbol for the current and next iteration. In addition, the algorithm uses up to $2(n+m)$ bits to maintain the set of intervals consisting in edges whose associated reversed $(k+1)$-mer have a common prefix; these intervals correspond to the blocks we implicitly maintain in the array B_2 using only $2n$ bits.

Summing up, the algorithm by Muggli *et al.* runs in $\mathcal{O}(mk)$ time, and uses $2(m\lceil \log \sigma \rceil + m + n)$ bits plus $\mathcal{O}(\sigma)$ words of working space. Our algorithm has the same time complexity but uses less space: even for $\sigma = 5$ as in bioinformatics applications, our algorithm uses less than half the space ($4n$ bits vs. $6.64m + 2n$ bits). This space reduction significantly influences the size of the largest de Bruijn graph that can be built with a given amount of RAM. For example, in the setting in which the input graphs are stored on disk and all the RAM is used for the working space, our algorithm can build a de Bruijn graph whose size is twice the size of the largest de Bruijn graph that can be built with the algorithm of Muggli *et al.*.

We stress that the space reduction was obtained by substantially changing the sorting procedure. Although both algorithms are based on radix sorting they differ substantially in their execution. The algorithm by Muggli *et al.* follows the traditional MSD radix sort strategy; hence it establishes, for example, that $ACG \prec ACT$ when it compares the third 'digits' and finds that $G < T$. In our algorithm we use a mixed LSD/MSD strategy: in the above example we also find that $ACG \prec ACT$ during the third iteration, but this is established without comparing directly G and T, which are not explicitly available. Instead, during

the second iteration the algorithm finds that $CG \prec CT$ and during the third iteration it uses this fact to infer that $ACG \prec ACT$: this is indeed a remarkable sorting trick first introduced in [12] and adapted here to de Bruijn graphs.

5   Merging Colored and VO-BOSS Representations

Our algorithm can be easily generalized to merge colored and VO (variable-order) BOSS representations. Note that the algorithm by Muggli *et al.* can also merge colored BOSS representations, but in its original formulation, it cannot merge VO representations.

Given the colored BOSS representation of two de Bruijn graphs G_0 and G_1, the corresponding color matrices $\mathcal{M}_0$ and $\mathcal{M}_1$ have size $m_0 \times c_0$ and $m_1 \times c_1$. We initially create a new color matrix $\mathcal{M}_{01}$ of size $(m_0 + m_1) \times (c_0 + c_1)$ with all entries empty. During the merging of the union graph (Phase 3), for $q = 1, \ldots, n$, we write the colors of the edges associated to $Z^{(h)}[q]$ to the corresponding line in $\mathcal{M}_{01}$ possibly merging the colors when we find nodes with identical k-mers in $\mathcal{O}(c_{01})$ time, with $c_{01} = c_0 + c_1$. To make sure that color ids from $\mathcal{M}_0$ are different from those in $\mathcal{M}_1$ in the new graph we add the constant c_0 (the number of distinct colors in G_0) to any color id coming from the matrix $\mathcal{M}_1$.

Theorem 2. *The merging of two succinct representations of colored de Bruijn graphs takes $\mathcal{O}(m \max(k, c_{01}))$ time and $4n$ bits plus $\mathcal{O}(\sigma)$ words of working space, where $c_{01} = c_0 + c_1$.* □

We now show that we can compute the variable order VO-BOSS representation of the union of two de Bruijn graphs G_0 and G_1 given their *plain*, eg. non variable order, BOSS representations. For the VO-BOSS representation we need the LCS array for the nodes in the union graph $\langle W_{01}, W_{01}^-, \text{last}_{01}\rangle$. Notice that after merging the k-mers of G_0 and G_1 with the algorithm in Fig. 3 (Phase 1) the values in $B[1, n]$ already provide the LCP information between the reverse labels of all consecutive nodes (Lemma 2). When building the union graph (Phase 3), for $q = 1, \ldots, n$, the LCS between two consecutive nodes, say v_i and w_j, is equal to the LCP of their reverses $\overleftarrow{v_i}$ and $\overleftarrow{w_j}$, which is given by $B[q] - 1$ whenever $B[q] > 0$ (if $B[q] = 0$ then $\overleftarrow{v_i} = \overleftarrow{w_j}$ and nodes v_i and v_j should be merged). Hence, our algorithm for computing the VO representation of the union graph consists exactly of the algorithm in Fig. 3 in which we store the array B in $n \log k$ bits instead of using the 2-bit representation described in Sect. 4. Hence the running time is still $\mathcal{O}(mk)$ and the working space becomes the space for the bitvectors $Z^{(h-1)}$ and $Z^{(h)}$ (recall we define the working space as the space used in addition to the space for the input and the output).

Theorem 3. *Merging two succinct representations of variable order de Bruijn graphs takes $\mathcal{O}(mk)$ time and $2n$ bits plus $\mathcal{O}(\sigma)$ words of working space.* □

6   External Memory Construction

In this section we show that using our merging algorithm we can design a complete external memory algorithm to construct succinct de Bruijn graphs.

We preliminary observe that at each iteration of the algorithm in Fig. 3 not only the arrays $\langle W_0, W_0^-, \mathsf{last}_0 \rangle$ and $\langle W_1, W_1^-, \mathsf{last}_1 \rangle$ but also $Z^{(h-1)}$ and B_2 are read sequentially from beginning to end. At the same time, the arrays $Z^{(h)}$ and B_2 are written sequentially but into σ different partitions whose starting positions are the values in $\mathsf{start}[1, \sigma]$ which are the same for each iteration. Thus, if we split $Z^{(\cdot)}$ and B_2 into σ different files, all accesses are sequential and our algorithm runs in external memory in $\mathcal{O}(mk)$ time, doing $\mathcal{O}(mk)$ sequential I/Os and using only $\mathcal{O}(\sigma)$ words of RAM.

Assume now we are given a string collection $\mathcal{C} = s_1, \ldots, s_d$ of total length N, the desired order k, and the amount of available RAM M. First, we split $\mathcal{C}$ into smaller subcollections $r_i = s_j, \ldots, s_{j'}$, such that we can compute the BWT and LCP array of each subcollection in linear time in RAM using M bytes, using *e.g.* the suffix sorting algorithm gSACA-K [14]. For each subcollection we then compute, and write to disk, the BOSS representation of its de Bruijn graph using the algorithm described in [6, Section 5.3]. Since these are linear algorithms the overall cost of this phase is $\mathcal{O}(N)$ time and $\mathcal{O}(N)$ sequential I/Os.

Finally, we merge all de Bruijn graphs into a single BOSS representation of the union graph with the external memory variant just described. Since the number of subcollections is $\mathcal{O}(N/M)$, a total of $\log(N/M)$ merging rounds will suffice to get the BOSS representation of the union graph.

Theorem 4. *Given a strings collection* $\mathcal{C} = s_1, \ldots, s_d$ *of total length* N, *we can build the corresponding order-k succinct de Bruijn graph in* $\mathcal{O}(N\,k \log(N/M))$ *time and* $\mathcal{O}(N\,k \log(N/M))$ *sequential I/Os using* $\mathcal{O}(M)$ *words of RAM.* □

Note that our construction algorithm can be easily extended to generate the colored/variable order variants of the de Bruijn graph. For the colored variant it suffices to use gSACA-K to generate also the document array [14] and then use the colored merging variant. For the variable order representation, it suffices to store the LCP/LCS values during the very last merging phase, using the techniques described in [6, Section 3] to handle them in external memory.

Acknowledgments. *Funding.* L.E. and G.M. were partially supported by PRIN grant 2017WR7SHH. L.E. was partially supported by the University of Eastern Piedmont project *Behavioural Types for Dependability Analysis with Bayesian Networks.* F.A.L. was supported by the grants #2017/09105-0 and #2018/21509-2 from the São Paulo Research Foundation (FAPESP). G.M. was partially supported by INdAM-GNCS Project 2019 *Innovative methods for the solution of medical and biological big data* and by the LSBC_19-21 Project from the University of Eastern Piedmont.

References

1. Alipanahi, B., Kuhnle, A., Boucher, C.: Recoloring the colored de Bruijn graph. In: Gagie, T., Moffat, A., Navarro, G., Cuadros-Vargas, E. (eds.) SPIRE 2018. LNCS, vol. 11147, pp. 1–11. Springer, Cham (2018). https://doi.org/10.1007/978-3-030-00479-8_1
2. Almodaresi, F., Pandey, P., Patro, R.: Rainbowfish: a succinct colored de Bruijn graph representation. In: WABI. LIPIcs, vol. 88, pp. 18:1–18:15. Schloss Dagstuhl - Leibniz-Zentrum fuer Informatik (2017)
3. Belazzougui, D., Gagie, T., Mäkinen, V., Previtali, M., Puglisi, S.J.: Bidirectional variable-order de Bruijn graphs. Int. J. Found. Comput. Sci. **29**(08), 1279–1295 (2018)
4. Boucher, C., Bowe, A., Gagie, T., Puglisi, S.J., Sadakane, K.: Variable-order de Bruijn graphs. In: DCC, pp. 383–392. IEEE (2015)
5. Bowe, A., Onodera, T., Sadakane, K., Shibuya, T.: Succinct de Bruijn graphs. In: Raphael, B., Tang, J. (eds.) WABI 2012. LNCS, vol. 7534, pp. 225–235. Springer, Heidelberg (2012). https://doi.org/10.1007/978-3-642-33122-0_18
6. Egidi, L., Louza, F.A., Manzini, G., Telles, G.P.: External memory BWT and LCP computation for sequence collections with applications. In: WABI. LIPIcs, vol. 113, pp. 10:1–10:14. Schloss Dagstuhl - Leibniz-Zentrum fuer Informatik (2018)
7. Egidi, L., Louza, F.A., Manzini, G., Telles, G.P.: External memory BWT and LCP computation for sequence collections with applications. Algorithms Mol. Biol. **14**(1), 6:1–6:15 (2019)
8. Egidi, L., Manzini, G.: Lightweight BWT and LCP merging via the gap algorithm. In: Fici, G., Sciortino, M., Venturini, R. (eds.) SPIRE 2017. LNCS, vol. 10508, pp. 176–190. Springer, Cham (2017). https://doi.org/10.1007/978-3-319-67428-5_15
9. Egidi, L., Manzini, G.: Lightweight merging of compressed indices based on BWT variants. CoRR (2019). http://arxiv.org/abs/1903.01465
10. Ferragina, P., Manzini, G., Mäkinen, V., Navarro, G.: Compressed representations of sequences and full-text indexes. ACM Trans. Algorithms **3**(2) (2007)
11. Holt, J., McMillan, L.: Constructing Burrows-Wheeler transforms of large string collections via merging. In: BCB, pp. 464–471. ACM (2014)
12. Holt, J., McMillan, L.: Merging of multi-string BWTs with applications. Bioinformatics **30**(24), 3524–3531 (2014)
13. Iqbal, Z., Caccamo, M., Turner, I., Flicek, P., McVean, G.: De novo assembly and genotyping of variants using colored de Bruijn graphs. Nat. Genet. **44**(2), 226–232 (2012)
14. Louza, F.A., Gog, S., Telles, G.P.: Inducing enhanced suffix arrays for string collections. Theor. Comput. Sci. **678**, 22–39 (2017)
15. Marcus, S., Lee, H., Schatz, M.C.: Splitmem: a graphical algorithm for pan-genome analysis with suffix skips. Bioinformatics **30**(24), 3476–3483 (2014)
16. Muggli, M.D., Alipanahi, B., Boucher, C.: Building large updatable colored de Bruijn graphs via merging. Bioinformatics **35**(14), i51–i60 (2019). https://doi.org/10.1093/bioinformatics/btz350
17. Muggli, M.D., Boucher, C.: Succinct de Bruijn graph construction for massive populations through space-efficient merging. bioRxiv (2017). https://doi.org/10.1101/229641
18. Muggli, M.D., et al.: Succinct colored de Bruijn graphs. Bioinformatics **33**(20), 3181–3187 (2017)

19. Pevzner, P.A., Tang, H., Waterman, M.S.: An Eulerian path approach to DNA fragment assembly. Proc. Natl. Acad. Sci. **98**(17), 9748–9753 (2001)
20. Raman, R., Raman, V., Rao, S.: Succinct indexable dictionaries with applications to encoding k-ary trees, prefix sums and multisets. ACM Trans. Algorithms **3**(4) (2007)

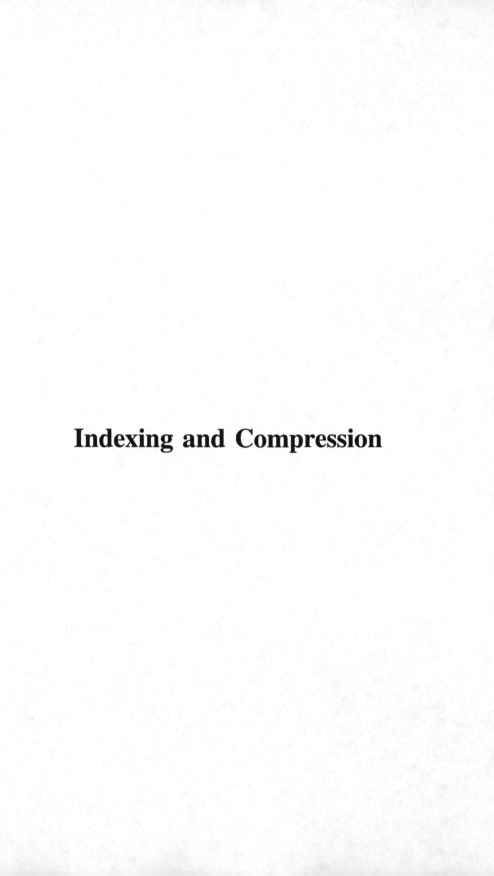

Indexing and Compression

Run-Length Encoding in a Finite Universe

N. Jesper Larsson[✉]

Faculty of Technology and Society, Malmö University, Malmö, Sweden
jesper.larsson@mau.se

Abstract. Text compression schemes and compact data structures usually combine sophisticated probability models with basic coding methods whose average codeword length closely match the entropy of known distributions. In the frequent case where basic coding represents run-lengths of outcomes that have probability p, i.e. the geometric distribution $\Pr(i) = p^i(1-p)$, a *Golomb code* is an optimal instantaneous code, which has the additional advantage that codewords can be computed using only an integer parameter calculated from p, without need for a large or sophisticated data structure. Golomb coding does not, however, gracefully handle the case where run-lengths are bounded by a known integer n. In this case, codewords allocated for the case $i > n$ are wasted. While negligible for large n, this makes Golomb coding unattractive in situations where n is recurrently small, e.g., when representing many short lists of integers drawn from limited ranges, or when the range of n is narrowed down by a recursive algorithm.

We address the problem of choosing a code for this case, considering efficiency from both information-theoretic and computational perspectives, and arrive at a simple code that allows computing a codeword using only $O(1)$ simple computer operations and $O(1)$ machine words. We demonstrate experimentally that the resulting representation length is very close (equal in a majority of tested cases) to the optimal Huffman code, to the extent that the expected difference is practically negligible. We describe efficient branch-free implementation of encoding and decoding.

1 Introduction

We are concerned with efficiently computing short codewords for values that follow a geometric distribution up to a bounding integer. Before discussing application areas and outlining the nature of our result, we begin with a scenario that illustrates the precise nature of our problem.

A Context for the Problem. In the original publication of a coding method for a geometric distribution, Golomb [6] presents an introductory example involving Agent 00111 at the Casino, "playing a game of chance, while the fate of mankind hangs in the balance". During 00111's game, the Secret Service wishes to continually receive status updates, and has enrolled the bartender to communicate

© Springer Nature Switzerland AG 2019
N. R. Brisaboa and S. J. Puglisi (Eds.): SPIRE 2019, LNCS 11811, pp. 355–371, 2019.
https://doi.org/10.1007/978-3-030-32686-9_25

the length of every run of consecutive *wins*, where a win is the more probable of two possible outcomes. Golomb presents his code as the solution to the problem of using the minimum number of bits in the transmission of run lengths.

To continue Golomb's scenario, we place ourselves in the situation of a competing agency that wishes to transmit the same information more efficiently, using the observation that there is always a finite upper bound n on the run length of consecutive wins, due to the need for 00111 to leave the game at predictable intervals to renew his cocktail. Since this means that only a finite number of possible run lengths exists, a *Huffman code* for the different possible values would seem to be an option. But since n varies, a new code would have to be computed for every codeword transmitted, and unlike Golomb coding, Huffman coding requires construction of a size n data structure. To circumvent the Casino's scanning visitors for cheating devices, codewords must be computed by a computing unit woven into our agent's dress, with only a few registers of storage and minimal power consumption. Thus, in algorithmic terms, our mission, should we decide to accept it, is to find a method to compute short codewords for this scenario in $O(1)$ simple operations using $O(1)$ storage.

Application Areas. Because of the prevalence of the geometric distribution in data, Golomb coding, and the computationally simplified version *Rice coding* [16] are – along with Huffman coding [9] – among the most common methods for low-level entropy coding in data compression. Although arithmetic coding [18] can often produce better compression, it is inconvenient in many applications, e.g. due to the necessity to decode previous parts of the bit sequence in order to access an arbitrary encoded value, which makes it less attractive for use in compact data structures [15].

A common context for Golomb coding is representing lengths of gaps between events that are estimated to occur with some known small probability $1 - p$. One common application is compressed representations of *inverted index* data structures [23,24], but many others exist. The significance of Golomb coding is demonstrated by a steady flow of new works citing Golomb [6] as well as the variant devised and proven optimal by Gallager and van Voorhis [4], which has become the standard formulation.[1] To take a small selection of examples, recent publications have included work on Golomb parameter estimation using machine learning [10], use for biomedical data [1], lossless video compression [2], encoding phasor measurement to monitor the power grid [21], and replacements for Golomb and Rice coding for random access [12].

In the situation addressed in this work – values to be encoded bounded by a relatively small integer n – Golomb coding is unattractive (and may therefore fail to be considered a possibility), because the part that the Golomb code dedicates to values above n would be wasted, rendering redundancy unnecessarily large.

In the compressed-inverted-index context, this could arise when document references in many (presumably small) lists are drawn from a limited collection. More typically, the upper bound on run length appears because integers are

---

[1] Golomb's formulation partitions values into groups of equal codeword length, which is a slightly less convenient take on the same family of codes.

encoded recursively, contained in intervals that shrink with recursion depth, similarly to schemes such as interpolative coding [14], tournament coding [22], wavelet trees [7], or recursive subset coding [13].

An available efficient entropy coding method for the bounded geometric distribution can contribute to development of effective compression schemes that do not yet exist.

Outline. We propose a method that computes a codeword for the run length of outcomes that have a known probability $p \geq \frac{1}{2}$, given an upper bound n, and two integer parameters m and m'' chosen on the basis of p. The method uses only a constant number of machine words, and a constant number of simple computational operations per computed codeword. The expected codeword length is close to minimum.

We generally assume (although our coding does not strictly depend on it) that only a small number of different p – and hence a small collection of m and m'' – appear in processing a single file or data structure. Therefore, if m and m'' need to be included in the compressed representation (because p is not available to the decoder), we assume that the encoding length as well as encoding time is negligible. Furthermore, we assume that n is available from the context to both encoder and decoder, and does not need to be explicitly represented. See Sect. 5 for a more detailed account of time complexity, along with a description of an efficient branch-free implementation.

Section 2 sets the scene by relating theory and relevant previous methods, and Sect. 3 presents our suggested code. Section 4 evaluates compression performance, Sect. 5 addresses time complexity and efficient implementation, and we conclude with some short comments in Sect. 6.

2   Entropy Codes and Code Trees

Let X be a random variable that takes on non-negative integer values $i \in 0, \ldots, n$ with probabilities $\Pr(X = i) = p_i$. We are concerned with finding codes that map the possible values of X to binary codewords of minimum expected length. A lower bound is the *entropy* $H(X) = -\sum_{i=0}^{n} p_i \lg p_i$ bits, where $\lg$ denotes logarithm in base 2. The bound can be matched only if all probabilities are *dyadic*, i.e., $\lg p_i \in \mathbb{Z}$. The *redundancy* of a code is the difference between its expected codeword length and the entropy. Any *instantaneous* code, where no codeword is a prefix of another, allows us to equate each code with a *code tree*: a binary tree with all possible values as leaves, and codewords identified by their paths from the root (0 for left branching, 1 for right branching). We refer to the number of values in a code tree (i.e. the number of leaves) as the *size* of the tree. It is well known that a *Huffman tree*, constructed bottom up by repeatedly joining the two subtrees of lowest probability, is optimal, i.e., no code that maps an individual value to each codeword can have shorter expected codeword length [3,23].

When probabilities are roughly equal, or more specifically, the highest probability is no greater than the sum of the two lowest probabilities, codeword lengths in an optimal code differ by at most one bit. Therefore, an optimal code

is achieved by using codeword length $h - 1$ for the s lowest-probability values and h for the other $n + 1 - s$, where $h = \lceil \lg(n + 1) \rceil$ and $s = 2^h - (n + 1)$ (see Fig. 1). We refer to this as a *balanced code*.[2]

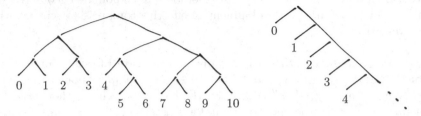

Fig. 1. Balanced code tree for $n = 10$ (left), and unlimited unary code tree (right).

On the other extreme, if the p_i decrease at least as quickly as proportionally to a reverse Fibonacci sequence, an optimal code is achieved by a code tree that is as skewed as possible, e.g. by encoding i as i one-bits followed by a single zero bit. This is referred to as a *unary* code (Fig. 1). With an upper bound n for i, the final zero in the codeword for n can be excluded.

When $p_i = p^i(1 - p)$ (the geometric distribution) for some $\frac{1}{2} \leq p < 1$, $i \geq 0$, and $p^m = \frac{1}{2}$ for some integer m, an optimal code is achieved by partitioning values into groups of size m, which we shall refer to as *bunches* in the code tree, and composing the code from two parts: a unary code that numbers the bunch of each value counting from the root, and a balanced code that determines the offset of each value inside its bunch. This is referred to as a *Golomb code* (Fig. 2). More generally, let m be the integer such that p^m is as close as possible to $\frac{1}{2}$. If $p^m \geq \frac{1}{2}$, then $p^{m+1} < \frac{1}{2}$, and m is the smallest integer such that $p^m - \frac{1}{2} \leq \frac{1}{2} - p^{m+1} \Leftrightarrow p^m + p^{m+1} \leq 1$. If $p^m \leq \frac{1}{2}$, m is again the smallest integer for which $p^m + p^{m+1} \leq 1$. Consequently,

$$m = \min \left\{ \ell \in \mathbb{Z} \mid p^\ell + p^{\ell+1} \leq 1 \right\} = \left\lceil \frac{\lg(1 + p)}{-\lg p} \right\rceil \tag{1}$$

Note that if $p = \frac{1}{2}$, then $m = 1$, and the Golomb code is equal to a unary code.

Gallager and van Voorhis [4] showed, using a bottom-up Huffman tree argument, that choosing m according to Eq. 1 always produces an optimal code tree. An alternative extended treatment was given by Golin [5].

3   Finite Code Trees for the Geometric Distribution

We now turn to the problem of forming code trees for a geometric distributions up to a maximum value n:

$$\Pr(X = i) = p^i(1 - p) \text{ for } i \in \mathbb{Z} \text{ and } 0 \leq i < n,$$

---

[2] Many authors use the term *binary code*. However, this clashes with the fact that all codes we address are binary in a wider sense.

$$\Pr(X = n) = 1 - \sum_{i=0}^{n-1} p^i(1-p) = p^n, \text{ and}$$

$\Pr(X = x) = 0$ for all other x.

We write $H(p,n) = -\sum_{i=0}^{n} p^i(1-p)\lg(p^i(1-p))$ for the entropy under this distribution, and denote the expected code length for a corresponding optimal code tree (e.g. Huffman) $L_H(p,n) \geq H(p,n)$

For convenience in encoding and decoding, we require values $0, \ldots, n$ to be in left to right order in the code tree, which does not impose any penalty on codeword lengths. To see this, consider starting from the *canonical* code tree [19,23], where codeword lengths are optimal, and leaves have monotonically non-decreasing depth from left to right. This differs from our desired code tree only in the placement of leaf n, which does not generally have the smallest codeword length, and is therefore not generally the rightmost leaf in the canonical tree. For every internal node in the path from the root to n, exchange the left and right subtrees wherever n is in the left subtree. This places n as the rightmost leaf, while maintaining the depths of all leaves. Hence, exchanging subtrees can displace the order only among leaves with the same depth, and since all leaves other than n have strictly decreasing probability order, they can be set in the right order without compromising the optimality of the tree.

3.1 Approach

When $n \to \infty$, a Golomb code tree is optimal [4], and a fair guess is that the code for finite n has some similarity to a Golomb code. Indeed, since the wasted probability p^n is negligible for large n, a hybrid scheme that uses Huffman coding if n is smaller than some acceptable constant, and Golomb coding otherwise, would theoretically solve our problem. However, in the interest of practical efficiency and simplicity, we seek a coding method that operates uniformly for all n.

A first attempt might be to use a Golomb code with minimal modification: Let m be defined by Eq. 1, and define $d_1 = \lfloor \frac{n-1}{m} \rfloor$ and $m_1 = n - d_1 m \leq m$. Assign the first $d_1 m$ values the same codewords as in the corresponding Golomb code. Let the rightmost internal node at depth d_1 have leaf n as its right child, and a balanced code tree for values $n - m_1, \ldots, n-1$, as its left subtree (Fig. 2). Consider the nodes along the rightmost path of the resulting tree. In general, the internal node at depth k along this path ($0 \leq k < d_1$) separates values into two subtrees with probability weights $\sum_{i=km}^{(k+1)m-1} p^i(1-p) = p^{km} - p^{(k+1)m}$ and $p^{(k+1)m}$. Dividing by p^{km}, we see that the relative weights of left and right subtree are $1 - p^m$ and p^m, which given our definition of m implies (unsurprisingly) that the subtrees are as weight-balanced as possible. The final internal node on the rightmost path has a left subtree of size m_1 (with probability weight $p^{n-m_1} - p^n$), and leaf n for right child (with weight p^n). This implies that at least locally, this node also balances the weights among its descendents as equally as possible.

Consequently, this tree is similar to what would be obtained by the top-down method known as Shannon–Fano coding, or more precisely (since our leaf probabilities are not necessarily monotonically non-increasing), a weight-balanced

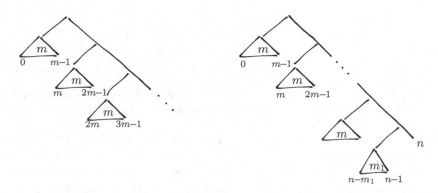

Fig. 2. Golomb code tree (left), and top-down weight balanced tree (right). Triangles are balanced code subtrees with sizes (number of leaves) shown inside and value ranges below.

code tree as studied by Rissanen [17] and Horibe [8]. The expected codeword length is therefore guaranteed to be asymptotically close to $H(p,n)$, but not necessarily optimal. In particular, note the striking anomaly when m_1 is much smaller than m, that a shorter codeword may be assigned to $n - m_1$ (and possibly others) than to $n - m_1 - 1$. This causes substantial redundancy for some p and n. For instance, when $p = 0.88$ and $n = 6$, the expected codeword length is 2.63, ten percent more than the optimal $L_H(0.88, 6) = 2.38$.

The reason is that greedy top-down partitioning can leave us with a severe unbalance between leaf n and the size-m_1 subtree, while a Huffman tree construction, which is bottom-up, can distribute weight among bunches of lower depth for better overall balance. For example, with $p = 0.9$ and $n = 43$, the optimal tree has four bunches of size $m = 7$ and one of size 8. With the same p and $n = 44$, only two size m bunches emerge, three of size 6, and one of size 5.

It is unclear whether a rule can be found that assigns bunch sizes optimally while allowing any codeword to be computed without explicitly processing $\omega(1)$ nodes in the code tree. Therefore, rather than computing the optimal code, we propose the following coding, which renders codewords almost as easy to compute as the top-down weight balanced code trees.

3.2   Suggested Code

Our suggested code trees are shaped similarly to the corresponding Golomb trees, with the exception of a subtree at the bottom that we refer to as the *tail*. Unless $n < m$, the tail has at least $m+1$ and at most $2m$ leaves, whose rightmost leaf is n, and whose other leaves have as equal depths as possible (similarly to a balanced code). If $n < 2m$, the whole tree is comprised of the tail.

Encoding and decoding takes two parameters m and m'', where m should be chosen according to Eq. 1, and the choice of integer m'', $m < m'' \leq 2m$, is discussed at the end of this section and defined in Eq. 2. Define

$$m' = \min\{m + (n \bmod m), n\}, \text{ and}$$

$$d_t = \frac{n - m'}{m}.$$

Note that $\min\{n, m\} \leq m' < 2m$, and that d_t, the *tail depth*, is an integer.

The codeword for i where $0 \leq i < d_t m$ consists of $\lfloor i/m \rfloor$ one-bits followed by the binary representation for $i \bmod m$ left-padded with zeros up to $h = \lceil \lg m \rceil$ bits if $i \bmod m < 2^h - m$, and up to $h + 1$ bits otherwise. This is equivalent to a unary code for the bunch number followed by a balanced code for the offset inside the bunch, i.e., the same as in the corresponding Golomb code.

The encoding of the remaining $m' + 1$ values (the tail) depends on three parameters: the height of the tail subtree, which we denote h'; the depth of n within the tail subtree, which is either 1 or 2 and denoted e_n; and the number of shorter codewords among the values $n - m', \ldots, n - 1$, denoted s' (cf. s for the m-sized bunches). The values are assigned as follows:

- If $m' < m''$, we let $e_n = 1$, $h' = \lceil \lg m' \rceil + 1$, and $s' = 2^{h'-1} - m'$ (depicted top right in Fig. 3).
- Otherwise, $e_n = 2$, $h' = \left\lceil \lg \frac{4m'}{3} \right\rceil$, and $s' = 3 \cdot 2^{h'-2} - m'$ (depicted bottom right in Fig. 3). Our choice of m'' will ensure that this case arises only when $m' \geq 3$.

The codeword for i when $n - m' \leq i < n$ is d_t one-bits followed by the binary representation of $j = i - m \cdot d_t$, left-padded with zeros up to $h' - 1$ bits when $j < s'$, and up to h' bits when $j \geq s'$. Finally, the codeword of n is $d_t + e_n$ one-bits. A demonstration implementation of encoding and decoding in Python can be found in Appendix 1.

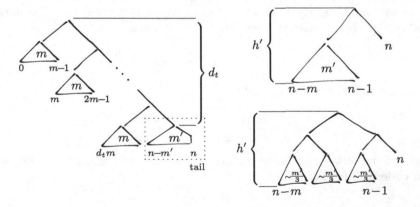

Fig. 3. General code tree for suggested coding. Left, the overall code tree. Top right, the tail for case $e_n = 1$, and bottom right for case $e_n = 2$.

Correctness and Decoding. Our encoding corresponds to the general code tree shown in Fig. 3. Therefore, given m, m'' and n, it yields a well-defined instantaneous code for integers i from 0 to n.

Decoding reads one-bits from the input, stopping after a zero-bit has appeared or d_t bits have been read. The number of one-bits found, which we denote d_i, is the number of an m-sized bunch if $d_i < d_t$, in which case decoding procedes with reading the offset in the bunch as an ordinary balanced code. Otherwise, we interpret the following h' bits as a number j that identifies a leaf in the tail. (If less than h' bits remain in the input, we right-pad with any bits.) If the e_n most significant bits of j are one, the decoded value is n. Otherwise, if $j < 2s$ we are in the range of the shorter codewords, which implies that $h' - 1$ bits should be consumed and the decoded value of i is $m \cdot d_t + \lfloor \frac{j}{2} \rfloor$; and if $j \geq 2s$, h' bits should be consumed and $i = m \cdot d_t + j - s'$

Justification and Choice of m''. If m divides n and $n \geq m$, then $m' = m$. Since we stipulated that $m'' > m$, the shape of our code tree in this case is the same as the optimal tree for unbounded values [4] except that the subtree for values above $n - 1$, with total probability p^n, is contracted into a single leaf. Since the code tree of Gallager and van Voorhis is optimal, our tree must also be optimal for $m' = m$, since otherwise the subtree that our code contracts would have had to be placed differently in the optimal unbounded tree.

Consequently, it is clear that the ideal value of e_n (the local depth of n in the tail subtree) is 1 when $m' = m$. On the other hand, we can see that as m' approaches $2m$, the ideal value of e_n should ultimately be 2. To decide where in the range between m and $2m$ to place the limit m'' between cases $e_n = 1$ and $e_n = 2$, we compare the expected codeword lengths for the two tail variants in Fig. 3. The codeword lengths for values below n are between $\lfloor \lg m' \rfloor$ and $\lceil \lg m' \rceil$ in the $e_n = 1$ case, and between $2 + \lfloor \lg \frac{4m'}{3} \rfloor$ and $2 + \lceil \lg \frac{4m'}{3} \rceil$ in the $e_n = 2$ case. Using the respective approximation $\lg m'$ for the codeword lengths in the $e_n = 1$ case and $2 + \lg \frac{4m'}{3}$ in the $e_n = 2$ case, we estimate that $e_n = 1$ results in a smaller expected codeword length than $e_n = 2$ when

$$(p^{n-m'} - p^n)(1 + \lg m') + 1 \cdot p^{m'} < p^n (2 + \lg \frac{m'}{3}) + 2 \cdot p^{m'}.$$

Simplifying and solving for m' yields

$$m' < \frac{\lg(\frac{1}{\lg 3 - 1} + 1)}{- \lg p} \approx \frac{1.4380}{- \lg p}.$$

Hence, it is reasonable to choose m'', the value that determines the value of e_n in our code specification, as

$$m'' = \left\lceil \frac{1.4380}{- \lg p} \right\rceil. \tag{2}$$

We note that this implies that $m'' = 2$ only when $m = 1 \Rightarrow m' = 1$, which avoids the degenerate case where $e_n = 2$ coincides with $m' < 3$, necessary for our code tree to be well-defined.

The choice of m'' is a heuristic that is not guaranteed to result in an optimal code tree. The same is true for our choice to assign codewords to $n-m', \ldots, n-1$ whose lengths differ by at most one: a difference of two (but never more than two, by the choice of m') may result in a slightly superior code. Also, as noted in Sect. 3.1, our choice of placing all but the last m' values in size-m bunches is not always optimal. Rather, these choices were made in order to obtain easily computable codewords without deviating much from the optimum. The next section evaluates to what extent this is successful.

4   Evaluation of Code Efficiency

Let $L(p, n)$ be the expected codeword length of our code for specified p and n. Using a bound derived by Horibe [8], we can conclude that $L(p, n) \leq H(p, n) + 2 - (n + 1)p^{n-1}(1 - p)$. However, as worst-case bounds on codeword length tend to be, this is pessimistic, and not a useful realistic estimate of code performance.

For a more practically valid assessment, we first note that our codes are proven optimal when $m = m'$, i.e., for every mth value of n (as we saw in Sect. 3.2). For other values of n, our codes may deviate slightly from the optimal, but we have made heuristic choices so as to stay as close as possible, without sacrificing the possibility of efficiently computing codewords. In particular, we note that our choice to assign values $n - m', \ldots, n - 1$ codewords whose lengths are monotonically nondecreasing and differing by at most one bit, eliminates the anomaly of the top-down weight-balanced tree that a less probable value may be assigned a shorter codeword (see Sect. 3.1). Furthermore, we note that both our code and an optimal Huffman code approach a Golomb code for relatively large n. For instance, when $p = 0.9$, the redundancy of our code, a Huffman code, and a Golomb code are all below one percent when $n \geq 54$.

Therefore, our main concern is to evaluate the redundancy of our code for smaller n. We experimentally compare our code to the calculated entropy (which can be matched in practice only by high-precision arithmetic coding) as well as to the optimal Huffman code, when p varies across its probability range and $n < 3m$. This range of n is the most interesting, since it results in at most one m-sized bunch plus the tail. For reference, we also include a comparison with Golomb coding for the same test cases.

Tables 1, 2 and 3 summarize the results of tests for 10^7 values of p evenly spread over the interval $\frac{1}{2} \leq p < 1$. For each value of p, 10 random values of n were selected uniformly at random in the range $[2, 3m)$ (the case $n = 1$ is uninteresting) for the comparison with Huffman (Table 2) and Golomb codes (Table 3). For the comparison with the entropy (Table 1), the range $\left[\max\{2, \lceil \frac{m}{2} \rceil\}, 3m\right)$ was used, because the entropy for smaller n tend to be impossible to achieve with any code tree, and therefore less interesting for our evaluation. Python source code used for measurements can be found at http://fgcode.avadeaux.net/.

Overall, summing up average lengths over the ranges of the experiments, the number of bits generated by our was about

Table 1. Evaluation in relation to entropy. The first column is the high endpoint of the redundancy range, e.g., we have 1–2% redundancy in 33.0% of the test cases. The p and n columns are examples that produce results in the respective range.

High $(L - H)/H$	%	Sample p, n
0	0.0	0.5, 2
10^{-5}	0.4	0.501, 2
10^{-4}	0.8	0.502, 2
0.001	2.5	0.506, 2
0.005	7.1	0.974, 52
0.01	26.7	0.984, 81
0.02	33.0	0.987, 87
0.03	13.0	0.919, 12
0.05	14.1	0.988, 34
0.1	1.9	0.983, 20
0.5	0.6	0.994, 45
>0.5	0	

Table 2. Evaluation in relation to Huffman coding. The first column is the high endpoint of the length increase range, e.g., our code yields output 1–2% larger than optimal in 0.3% of our test cases, and optimal in 86.2% of the cases. The p and n columns are examples that produce results in the respective range.

High $(L - L_H)/L_H$	%	Sample p, n
0	86.2	0.985, 93
10^{-5}	0.1	0.979, 68
10^{-4}	0.6	0.992, 175
0.001	4.1	0.972, 62
0.005	7.1	0.971, 67
0.01	1.6	0.938, 21
0.02	0.3	0.904, 12
>0.02	0	

Table 3. Comparison to Golomb coding. The first column is the high endpoint of the length decrease range, e.g., our code yields output at least 5% shorter than Golomb codes for all test cases, and 10–50% reduction in 84.2% of the cases. The p and n columns are examples that produce results in the respective range.

High $(L_{\text{Golomb}} - L)/L_{\text{Golomb}}$	%	Sample p, n
0.05	0	
0.1	7.5	0.862, 14
0.5	84.2	0.972, 62
1.0	8.3	0.994, 88
>1.0	0	

– 1.015 times the entropy (1.5% redundancy), and
– 1.0005 times the corresponding optimal Huffman code (an increase of five hundredths of a percent)

This indicates that all but a negligible part of the redundancy is due to the rounding effect of a code tree for a non-dyadic probability distribution, and not to inferior behavior in relation to Huffman coding.

Finally, the total number of bits was 0.737 times the corresponding Golomb codes, i.e., a 26% reduction. This may appear to be a surprisingly large improvement, but consider that the experiment was done for relatively small n in order to evaluate the redundancy of our code. For $n \gg m$, the difference between our code and a Golomb code is negligible.

5   Time Complexity and Efficient Implementation

Computation in encoding and decoding can be separated into three phases. The first computes m and m'' when only p is known. This involves floating point operations (logarithm and division), integer ceiling-log, and some integer bitwise and arithmetic operations. The second phase can take place when n is known, and computes m'. For the third phase, encoding or decoding of a value i can then take place. Disregarding input/output operations, phases two and three involve division, conditional branching (*if-else*) ceiling-log, bitwise, and arithmetic operations, but no floating point calculations.

We assume that arithmetic and bitwise operations (including shift, used for calculating powers of two) are constant-time machine operations. Ceiling-log, $\lceil \lg x \rceil$, cannot be directly expressed in most programming languages, but if not available as a special operation, it can either be computed by converting to floating point and extracting the exponent bits, or by locating the highest one-bit in the binary representation of $x - 1$ using $\Theta(\lg \lg N)$ operations, where N is an upper bound for x. Assuming a fixed word size of at most 64 bits, $\lg \lg N$ cannot exceed 6, and for practical purposes can be regarded as taking constant time. The floating-point logarithm involved in the first phase, however, is less easily dismissed, as it would typically involve a Taylor series computation.

We have generally made the assumption (see Sect. 1) that computing (and representing) m and m'' is rare enough to be considered negligible, while efficiency is critical for phase 2 computation and encoding/decoding.

Consequently, depending on machine model assumptions, time complexity of the critical parts of computation are bounded by either $O(1)$ (arguably true in most practical situations) or $O(\lg \lg N)$ where N is an upper bound for n. The time for the first phase, executed a small number of times, takes additional time for two floating point logarithm calculations, typically bounded by $O(w)$, where w is the number of bits of accuracy for the floating-point calculations.

Branch-Free Implementation. Classical time complexity analysis assumes that CPU instructions are executed more or less in sequence, where each instruction takes constant time. The truth for modern processors, where *pipelining* and

out-of-order execution are crucial to efficiency, is more complex [20]. Conditional branches, which the processor may or may not be able to predict, potentially forces execution out of *streaming mode*, and performance suffers. Consequently, when writing program code without detailed knowledge of code generation and target machine architecture, it is generally a good idea to write programs with as few conditional branching constructs (e.g., *if* and *while* statements) as possible.

With a straightforward implementation of our algorithms, a single encode or decode runs through at least three *if-else* statements. Although constant-time in the sense of classical algorithm analysis, they can be detrimental to practical performance. Our algorithms do however lend themselves to be implemented *without* any conditional statements, by using some bitwise tricks that take advantage of the two's complement integer representation of modern computers. The following tricks can partly be regarded as folklore in programming communities. Knuth [11] has examined some related techniques more closely. Our examples use the C programming language.

In two's complement representation, the leftmost bit is one for any negative number, and zero for any non-negative number. At the core of the tricks is smearing the sign bit across the whole word, producing -1 (all one-bits, which we take as our representation of *true*) if the number is negative, and zero (representing *false*) otherwise. Using *arithmetic shift right*, this is done simply by shifting the sign bit $w - 1$ positions to the right, where w is the number of bits in a word. For simplicity, we henceforth assume that $w = 32$. (Transformation to other word sizes is straightforward.) Hence with, >> denoting right shift, the operation is "x>>31". On the other hand, if shifting is *logical*, bits shifted in from the left are zero rather than copies of the sign bit, producing 1 instead of -1, we must negate the result, i.e., "-(x>>31)". As it is not always possible to know whether shifting is arithmetic or logical (the C standard leaves it undefined) we use a conditional expression (see `is_neg` in Fig. 4), trusting that the compiler will not translate it to a conditional branch, since the value of the condition is known in compile time.

```
inline int32_t is_neg(int32_t x) { return (-1 >> 1 < 0) ? x>>31 : -(x>>31); }
inline int32_t is_gt(int32_t x, int32_t y) { return is_neg(y-x); }
inline int32_t if_gt_then(int32_t x, int32_t y, int32_t z) { return z & is_gt(x, y); }
inline int32_t if_then(int32_t cond, int32_t x) { return cond & x; }
inline int32_t if_then_else(int32_t cond, int32_t x, int32_t y) { return cond & x | ~cond & y; }
```

```
#define MSB_TEST(x, y, b)           \        inline int32_t ceil_lg(int32_t x) {
  j = if_gt_then(x, (1<<b)-1, b);   \          int y = 0, j;
  y += j;                           \          x -= 1;
  x >>= j;                                     MSB_TEST(x, y, 16);
                                               MSB_TEST(x, y, 8);
                                               MSB_TEST(x, y, 4);
                                               MSB_TEST(x, y, 2);
                                               MSB_TEST(x, y, 1);
                                               return y+x;
                                             }
```

Fig. 4. Components for branch-free implementation.

```
int32_t not_tail = is_gt(n - m', i);
int32_t not_n = is_gt(n, i);
int32_t d_i = if_then_else(not_tail, i/m, d_t);
int32_t j = i - d_i * m;
int32_t s_i = if_then_else(not_tail, s, s');
int32_t not_short = ~is_gt(s_i, j);
int32_t h_i = if_then_else(not_tail, h, h'-1) - not_short;
int32_t v = j + if_then(not_short, s_i);
write_onebits(d_i + if_then(~not_n, e_n));
write_bits(if_then(not_n, v), if_then(not_n, h_i));
```

Fig. 5. Branch-free implementation of encoding. Values m, m', d_t, s, s', h, h', and e_n are assumed computed in the previous phases (not shown).

For positive numbers x and y, $x > y \Leftrightarrow y - x < 0$. Hence, we immediately have a method to compare the numbers that produces -1 if $x > y$ and zero otherwise (`is_gt` in Fig. 4). Since -1 is represented as all one-bits, a bitwise *and* "x&c" preserves x if c$= -1$ ("true") and is zero if c is zero ("false"). This is useful in itself (see `is_gt_then` in Fig. 4), and can be expanded to a full conditional expression by the additional use of bitwise *or* and negation (`if_then` and `if_then_else` in Fig. 4). As noted in in the first part of this section, $\lceil \lg x \rceil$ can be computed in $O(\lg w)$ tests for whether the most significant bit of $x - 1$ is in the left or right half of a range of bits (reminiscent of binary search). Since $\lg 32 = 5$, we can implement integer ceiling-log simply using five tests. (`ceil_lg` in Fig. 4 implements this with the help of the macro `MSB_TEST`.)

Figure 5 shows an example implementation of encoding (without the two phases of parameter computation). The full branch-free implementation can be found in Appendix 2, as well as via http://fgcode.avadeaux.net/.

It should be noted that the branch-free version is not guaranteed to be the most efficient for all combinations of compilers and target machines. An obvious potential inefficiency is due to branch-free techniques often computing values that are then discarded, since it effectively follows all execution paths of the algorithm.

6 Conclusion

Adding an efficient coding for the geometric distribution bounded by a potentially small constant to the repertoire of entropy codings, can plausibly contribute to the design of efficient text compression algorithms and compact data structures.

Given that our code is quite simple and addresses a natural situation that arises in practice (from experience by the author of this work), it could be argued that the problem might as well have been solved at least as early as in the 1970s. However, as no code with these properties is previously established in the data compression community, this research fills in a quite literal gap.

Appendix 1: Straightforward Python Implementation

```python
from collections import namedtuple
from math import ceil, log

Pparams = namedtuple('Pparams', ['m', 'm"', 'h', 's'])
Nparams = namedtuple('Nparams', ['m'', 'h'', 'h"', 's'', 'd_t', 'e_n'])

def phase1(p):
    neglgp = -log(p, 2)
    m = ceil(log(1+p, 2) / neglgp)
    m" = ceil(1.438 / neglgp)
    h = (m-1).bit_length()
    s = (1 << h) - m                    # short cws in m-sized bunch
    return Pparams(m, m", h, s)

def phase2(n, m, m"):
    m' = min(m + n%m, n)
    d_t = (n-m')//m                     # tail depth
    if m' < m":
        e_n = 1                         # in-tail depth of n
        h' = (m'-1).bit_length() + 1    # tail height
        s' = (1 << h'-1) - m'           # short cws in tail
    else:
        e_n = 2                         # in-tail depth of n
        h' = (m' + m'//3 - 1).bit_length() # tail height
        s' = 3*(1 << h'-2) - m'         # short cws in tail
    h" = h'-e_n
    return Nparams(m', h', h", s', d_t, e_n)

def enc(i, n, m, h, s, m', h', s', d_t, e_n, bit_out):
    if i < n - m':                      # not in tail
        d_i = i // m
        bit_out.write_ones(d_i)
        j = i - d_i*m;                  # offset in bunch
        if j < s:                       # short cw in balanced code
            bit_out.write_bits(j, h)
        else:                           # long cw
            bit_out.write_bits(j+s, h+1)
    elif i < n:                         # in tail, but not n
        bit_out.write_ones(d_t)
        j = i - (n - m')                # offset in tail
        if (j < s'):                    # short cw in tail
            bit_out.write_bits(j, h'-1)
        else:                           # long cw
            bit_out.write_bits(j+s', h')
    else:
        bit_out.write_ones(d_t+e_n)          # n

def dec(n, m, h, s, m', h', h", s', d_t, e_n, bit_in):
    d_i = bit_in.read_unary(d_t)
    if d_i < d_t:                       # not in tail
        j = bit_in.next_bits(h)
        if j < 2*s:                     # short cw in balanced code
            bit_in.skip_bits(h-1)
            return d_i*m + (j >> 1)
        else:                           # long cw
            bit_in.skip_bits(h)
            return d_i*m + j - s
    else:
        j = bit_in.next_bits(h')
        if j >> h" < (e_n1l):           # not starting with e_n 1-bits
            if j < 2*s':                # short cw in tail
                bit_in.skip_bits(h'-1)
                return n - m' + (j >> 1)
            else:                       # long cw
                bit_in.skip_bits(h')
                return n - m' + j - s'
        else:                           # n
            bit_in.skip_bits(e_n)
            return n
```

Appendix 2: Branch-Free C Implementation

```c
extern void write_onebits(int nbits);
extern void write_bits(int32_t bval, int nbits);
extern int32_t read_unary();
extern int32_t next_bits(int nbits);
extern void skip_bits(int nbits);

inline int32_t is_neg(int32_t x) { return (-1 >> 1 < 0) ? x>>31 : -(x>>31); }
inline int32_t is_gt(int32_t x, int32_t y) { return is_neg(y-x); }
inline int32_t if_gt_then(int32_t x, int32_t y, int32_t z) { return z & is_gt(x, y); }
inline int32_t if_then(int32_t cond, int32_t x) { return cond & x; }
inline int32_t if_then_else(int32_t cond, int32_t x, int32_t y) { return cond & x | ~cond & y; }

#define MSB_TEST(x, y, b)              \
  j = if_gt_then(x, (1<<b)-1, b);      \
  y += j;                              \
  x >>= j;

inline int32_t ceil_lg(int32_t x) {
  int y = 0, j;
  x -= 1;
  MSB_TEST(x, y, 16);
  MSB_TEST(x, y, 8);
  MSB_TEST(x, y, 4);
  MSB_TEST(x, y, 2);
  MSB_TEST(x, y, 1);
  return y+x;
}

// Phase 1 computation of m, m", h and s, given p
double neg_log_p = -log(p);
m = ceil(log(1+p)/neg_log_p);
m" = ceil(0.996768283334579/neg_log_p);   // 1.438*log(2) = 0.9967...
h = ceil_lg(m);
s = (1 << h) - m;

// Phase 2 computation of m', h', h", s', d_t, e_n, given n
int32_t n_ge_m = ~is_gt(m, n);
int n_div_m = n/m;
m' = n + if_then(n_ge_m, (1-n_div_m)*m);
d_t = if_then(n_ge_m, n_div_m - 1);
int32_t case2 = ~is_gt(m", m');
e_n = 1 - case2;
h' = ceil_lg(m' + if_then_else(case2, m'/3, m'));
h" = h' - e_n;
s' = 1 << h"; s' += if_then(case2, s' << 1) - m';

// Encoding value i
int32_t not_tail = is_gt(n - m', i);
int32_t not_n = is_gt(n, i);
int32_t d_i = if_then_else(not_tail, i/m, d_t);
int32_t j = i - d_i * m;
int32_t s_i = if_then_else(not_tail, s, s');
int32_t not_short = ~is_gt(s_i, j);
int32_t h_i = if_then_else(not_tail, h, h'-1) - not_short;
int32_t v = j + if_then(not_short, s_i);
write_onebits(d_i + if_then(~not_n, e_n));
write_bits(if_then(not_n, v), if_then(not_n, h_i));

// Decoding
int32_t d_i = read_unary(d_t);
int32_t not_tail = is_gt(d_t, d_i);
int32_t h_j = if_then_else(not_tail, h, h');
int32_t s_i = if_then_else(not_tail, s, s');
int32_t j = next_bits(h_j);
int32_t short_words = if_then_else(not_tail, s, s');
int32_t is_short = is_gt(short_words << 1, j);
int32_t not_n = not_tail | is_gt(e_n|1, j >> h");
int32_t h_i = h_j - is_short;
int32_t o_i = if_then_else(is_short, j >> 1, j - s_i);
skip_bits(if_then_else(not_n, h_i, e_n));
i = d_i*m + if_then_else(not_n, o_i, m');
```

References

1. Capurro, I., Lecumberry, F., Martín, Á., Ramírez, I., Rovira, E., Seroussi, G.: Efficient sequential compression of multichannel biomedical signals. IEEE J. Biomed. Health Inform. **21**(4), 904–916 (2017)
2. Choi, J.A., Ho, Y.S.: Efficient residual data coding in CABAC for HEVC lossless video compression. Signal Image and Video Process. **9**(5), 1055–1066 (2015)
3. Cover, T.M., Thomas, J.A.: Elements of Information Theory, 2nd edn. Wiley-Blackwell, Hoboken (2006)
4. Gallager, R.G., van Voorhis, D.C.: Optimal source codes for geometrically distributed integer alphabets. IEEE Trans. Inf. Theory **21**(2), 228–230 (1975)
5. Golin, M.J.: A combinatorial approach to Golomb forests. Theor. Comput. Sci. **263**(1–2), 283–304 (2001)
6. Golomb, S.W.: Run-length encodings. IEEE Trans. Inf. Theory **12**(3), 399–401 (1966)
7. Grossi, R., Gupta, A., Vitter, J.S.: High-order entropy-compressed text indexes. In: Proceedings of the Fourteenth Annual ACM-SIAM Symposium on Discrete Algorithms, pp. 841–850, January 2003
8. Horibe, Y.: An improved bound for weight-balanced tree. Inf. Control **34**(2), 148–151 (1977)
9. Huffman, D.A.: A method for the construction of minimum-redundancy codes. Proc. Inst. Electr. Radio Eng. **40**(9), 1098–1101 (1952)
10. Jiang, Z., Pan, W.D., Shen, H.: Universal Golomb-Rice coding parameter estimation using deep belief networks for hyperspectral image compression. IEEE J. Sel. Top. Appl. Earth Obs. Remote. Sens. **11**(10), 3830–3840 (2018)
11. Knuth, D.E.: Combinatorial Algorithms, Part 1, The Art of Computer Programming, vol. 4A. Addison-Wesley, Boston (2011)
12. Külekci, M.O.: Enhanced variable-length codes: improved compression with efficient random access. In: Proceedings of the Data Compression Conference, pp. 362–371, March 2014
13. Larsson, N.J.: Integer set compression and statistical modeling. Preprint, February 2014. arXiv:1402.1936 [cs.IT]
14. Moffat, A., Stuiver, L.: Exploiting clustering in inverted file compression. In: Proceedings of the IEEE Data Compression Conference, pp. 82–91, April 1996
15. Navarro, G.: Compact Data Structures. Cambridge University Press, Cambridge (2016)
16. Rice, R.F.: Some practical universal noiseless coding techniques. Technical report, Jet Propulsion Laboratory, California Institute of Technology Pasadena, CA, United States, March 1979
17. Rissanen, J.: Bounds for weight balanced trees. IBM J. Res. Dev. **17**(2), 101–105 (1973)
18. Rissanen, J.J.: Generalized kraft inequality and arithmetic coding. IBM J. Res. Dev. **20**(3), 198–203 (1976)
19. Schwartz, E.S., Kallick, B.: Generating a canonical prefix encoding. Commun. ACM **7**(3), 166–169 (1964)
20. Shen, J.P., Lipasti, M.H.: Modern Processor Design. McGraw-Hill, New York (2005)
21. Tate, J.E.: Preprocessing and Golomb-Rice encoding for lossless compression of phasor angle data. IEEE Trans. Smart Grid **7**(2), 718–729 (2016)

22. Teuhola, J.: Tournament coding of integer sequences. Comput. J. **52**(3), 368–377 (2009)
23. Witten, I.H., Moffat, A., Bell, T.C.: Managing Gigabytes: Compressing and Indexing Documents and Images, 2nd edn. Morgan Kaufmann, San Francisco (1999)
24. Zobel, J., Moffat, A.: Inverted files for text search engines. ACM Comput. Surv. **38**(2) (2006)

On the Computation of Longest Previous Non-overlapping Factors

Enno Ohlebusch$^{(\boxtimes)}$ and Pascal Weber

Institute of Theoretical Computer Science, Ulm University, 89069 Ulm, Germany
{Enno.Ohlebusch,Pascal-1.Weber}@uni-ulm.de

Abstract. The f-factorization of a string is similar to the well-known Lempel-Ziv (LZ) factorization, but differs from it in that the factors must be non-overlapping. There are two linear time algorithms that compute the f-factorization. Both of them compute the array of longest previous non-overlapping factors (LPnF-array), from which the f-factorization can easily be derived. In this paper, we present a simple algorithm that computes the LPnF-array from the LPF-array and an array prevOcc that stores positions of previous occurrences of LZ-factors. The algorithm has a linear worst-case time complexity if prevOcc contains leftmost positions. Moreover, we provide an algorithm that computes the f-factorization directly. Experiments show that our first method (combined with efficient LPF-algorithms) is the fastest and our second method is the most space efficient way to compute the f-factorization.

1 Introduction

The Lempel-Ziv (LZ) factorization [20] of a string has played an important role in data compression for more than 40 years and it is also the basis of important algorithms on strings, such as the detection of all maximal repetitions (runs) in a string [16] in linear time. Because of its importance in data compression, there is extensive literature on algorithms that compute the LZ-factorization and [1–4, 10, 12, 13, 18, 19] is an incomplete list.

A variant of the LZ-factorization is the f-factorization, which played an important role in solving a long standing open problem: it enabled the development of the first linear time algorithm for seeds computation by Kociumaka et al. [15].

Definition 1. *Let $S = S[0..n-1]$ be a string of length n on an alphabet Σ. The f-factorization $s_1 s_2 \cdots s_m$ of S can be defined as follows. Given $s_1 s_2 \cdots s_{j-1}$, the next factor s_j is obtained by a case distinction on the character $c = S[i]$, where $i = |s_1 s_2 \cdots s_{j-1}|$:*

(a) if c does not occur in $s_1 s_2 \cdots s_{j-1}$ then $s_j = c$
(b) else s_j is the longest prefix of $S[i..n-1]$ that is a substring of $s_1 s_2 \cdots s_{j-1}$.

© Springer Nature Switzerland AG 2019
N. R. Brisaboa and S. J. Puglisi (Eds.): SPIRE 2019, LNCS 11811, pp. 372–381, 2019.
https://doi.org/10.1007/978-3-030-32686-9_26

The difference to the LZ-factorization is that the factors must be non-overlapping. There are two linear time algorithms that compute the f-factorization [5,6]. Both of them compute the LPnF-array (defined below), from which the f-factorization can be derived (in case (b), the factor s_j is the length LPnF[i] prefix of $S[i..n-1]$).

i	0	1	2	3	4	5	6	7	8	9	10	11	12	13	14	15
$S[i]$	a	a	a	a	a	a	a	a	a	a	a	a	a	a	a	a
LPnF[i]	0	1	2	3	4	5	6	7	8	7	6	5	4	3	2	1
LPF[i]	0	15	14	13	12	11	10	9	8	7	6	5	4	3	2	1
(rm)prevOcc[i]	⊥	0	1	2	3	4	5	6	7	8	9	10	11	12	13	14
(lm)prevOcc[i]	⊥	0	0	0	0	0	0	0	0	0	0	0	0	0	0	0

Fig. 1. The LPnF, LPF, and prevOcc arrays of the string $S = aaaaaaaaaaaaaaaa$.

Definition 2. *For a string S of length n, the* longest previous non-overlapping factor (LPnF) *array of size n is defined for $0 \le i < n$ by*

$$\mathsf{LPnF}[i] = \max\{\ell \mid 0 \le \ell \le n - i; S[i..i + \ell - 1] \text{ is a substring of } S[0..i - 1]\}$$

In the following, we will give a simple algorithm that directly bases the computation of the LPnF-array on the LPF-array, which is used in several algorithms that compute the LZ-factorization. The LPF-array is defined by ($0 \le i < n$)

$$\mathsf{LPF}[i] = \max\{\ell \mid 0 \le \ell \le n - i; S[i..i + \ell - 1] \text{ is a substring of } S[0..i + \ell - 2]\}$$

In data compression, we are not only interested in the length of the longest previous factor but also in a previous position at which it occurred (because otherwise decompression would be impossible). For an LPF-array, the positions of previous occurrences are stored in an array prevOcc. If LPF[i] = 0, we set prevOcc[i] = ⊥ (for decompression, one can use the definition prevOcc[i] = $S[i]$). Figure 1 depicts the LPF-array of $S = a^{16}$ and two of many possible instances of the prevOcc-array: one that stores the rightmost (rm) positions of occurrences of longest previous factors and one that stores the leftmost (lm) positions.

2   Computing LPnF from LPF

Algorithm 1 computes the LPnF-array by a right-to-left scan of the LPF-array and its prevOcc-array. The computation of an entry $\ell = \mathsf{LPnF}[i]$ is solely based on entries LPF[j] and prevOcc[j] with $j \le i$. Consequently, after the calculation of ℓ, it can be stored in LPF[i]. Since Algorithm 1 overwrites the LPF-array with the LPnF-array (and the prevOcc-array of LPF with the prevOcc-array of LPnF), no extra space is needed. Algorithm 1 is based on the following simple idea:

1. If the factor starting at position i and its previous occurrence starting at position $j = \mathsf{prevOcc}[i]$ do not overlap, then clearly LPnF[i] = LPF[i].

Algorithm 1. Given LPF and its prevOcc-array, the algorithm computes LPnF and stores it in LPF.

```
 1: function COMPUTELPNF(LPF,prevOcc)
 2:     for i ← n − 1 downto 0 do
 3:         if LPF[i] > 0 then                          ▷ hence prevOcc[i] ≠ ⊥
 4:             j ← prevOcc[i]
 5:             if j + LPF[i] > i then                  ▷ overlapping case
 6:                 ℓ ← i − j
 7:                 while LPF[j] > ℓ do                 ▷ hence prevOcc[j] ≠ ⊥
 8:                     ℓ ← min{LPF[i], LPF[j]}
 9:                     j ← prevOcc[j]
10:                     if j + ℓ ≤ i then               ▷ non-overlapping case
11:                         break
12:                     else                           ▷ overlapping case
13:                         ℓ ← i − j
14:             prevOcc[i] ← j
15:             LPF[i] ← ℓ
```

2. Otherwise, the length of the (currently best) previous non-overlapping factor is $\ell = i-j$. A longer previous non-overlapping factor exists if $\mathsf{LPF}[j] > \ell$ (note that $\mathsf{LPF}[i] > \ell$ holds): the prefix of $S[i..n-1]$ of length $\min\{\mathsf{LPF}[i], \mathsf{LPF}[j]\}$ also occurs at position $\mathsf{prevOcc}[j]$ and even if the two occurrences (starting at i and $\mathsf{prevOcc}[j]$) overlap, their non-overlapping part must be greater than ℓ because $\mathsf{prevOcc}[j] < j$.

3. Step 2 is repeated until there is no further candidate (condition of the while-loop in line 7) or the two occurrences under consideration do not overlap (line 11 of Algorithm 1).

On the one hand, the example in Fig. 1 shows that Algorithm 1 may have a quadratic run-time if it uses the prevOcc-array that stores the rightmost positions of previous occurrences. On the other hand, the next lemma proves that Algorithm 1 has a linear worst-case time complexity if it uses the prevOcc-array that stores the leftmost positions of previous occurrences. Its proof is based on the following notion: An integer p with $0 < p \le |\omega|$ is called a period of $\omega \in \Sigma^+$ if $\omega[i] = \omega[i+p]$ for all $i \in \{0, 1, \ldots, |\omega| - p - 1\}$.

Lemma 1. *If* prevOcc *stores the leftmost positions of previous occurrences, then the else-case on line 12 in Algorithm 1 cannot occur.*

Proof. For a proof by contradiction, suppose that the else-case on line 12 in Algorithm 1 occurs for some i. We have $\mathsf{LPF}[i] > 0$, $j = \mathsf{prevOcc}[i]$ is the leftmost occurrence of the longest previous factor ω_i starting at i, and $j + \mathsf{LPF}[i] > i$. Suppose $\mathsf{LPF}[j] > i - j$, i.e., the while-loop is executed. Let $m = \min\{\mathsf{LPF}[i], \mathsf{LPF}[j]\}$ and $k = \mathsf{prevOcc}[j]$. If $m = \mathsf{LPF}[i]$, then it would follow that an occurrence of ω_i starts at k. This contradicts the fact that j is the leftmost occurrence of ω_i. Consequently, $m = \mathsf{LPF}[j] < \mathsf{LPF}[i]$. The else-case on line 12 occurs when $k + m > i$. This implies $k + m > j$ because $i > j$. Let ω_j be the longest previous

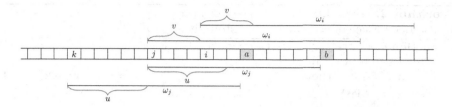

Fig. 2. Proof of Lemma 1: i, j, and k are positions, while a and b are characters.

factor starting at j. Let $a = S[k + m]$ (the character following the occurrence of ω_j starting at k) and $b = S[j + m]$ (the character following the occurrence of ω_j starting at j); see Fig. 2. By definition, $a \neq b$. We will derive a contradiction by showing that $a = b$ must hold in the else-case on line 12.

Since $k + m > j$, the occurrence of ω_j starting at k overlaps with the occurrence of ω_j starting at j. Let u be the non-overlapping part of the occurrence of ω_j starting at k, i.e., $u = S[k..j - 1]$. Because the occurrence of ω_j starting at j has u as a prefix and overlaps with the occurrence of ω_j starting at k, it follows that $|u|$ is a period of ω_j; see Fig. 2. By a similar reasoning, one can see that $|v|$ is a period of ω_i, where $v = S[j..i - 1]$. Since ω_j is a length m prefix of $S[j..n-1]$ and ω_i is a length $\mathsf{LPF}[i]$ prefix of $S[j..n - 1]$, where $m = \mathsf{LPF}[j] < \mathsf{LPF}[i]$, it follows that ω_j is a prefix of ω_i. Hence $|v|$ is also a period of ω_j. In summary, both $|u|$ and $|v|$ are periods of ω_j. Fine and Wilf's theorem [8] states that if $|\omega_j| \geq |u| + |v| - gcd(|u|, |v|)$, then the greatest common divisor $gcd(|u|, |v|)$ of $|u|$ and $|v|$ is also a period of ω_j. Since $m = |\omega_j| \geq |u| + |v|$, the theorem is applicable. Let γ be the length $gcd(|u|, |v|)$ prefix of ω_j. It follows that $v = \gamma^q$ for some integer $q > 0$, hence $|\gamma|$ is a period of ω_i. Recall that $a = S[k + m]$ is the character $\omega_j[m - |u|] = \omega_i[m - |u|]$ and $b = S[j + m] = \omega_i[m]$. We derive $a = \omega_i[m - |u|] = \omega_i[m] = b$ because $|\gamma|$ is a period of ω_i and $|u|$ is a multiple of $|\gamma|$. This contradiction proves the lemma.

To the best of our knowledge, Abouelhoda et al. [1] first computed the LZ-factorization based on the suffix array (and the LCP-array) of S. Their algorithm computes the LPF-array and the prevOcc-array that stores leftmost positions of previous occurrences of longest factors. So the combination of their algorithm and Algorithm 1 gives a linear-time algorithm that computes the LPnF-array. Subsequent work (e.g. [2–4,12,13,18]) concentrated on LZ-factorization algorithms that are faster in practice or more space-efficient (or both). Some of them also first compute the arrays LPF and prevOcc, but their prevOcc-arrays neither store leftmost nor rightmost occurrences (in fact, these algorithms are faster because they use lexicographically nearby suffixes–a *local* property–while being the leftmost occurrence is a *global* property). However, leftmost occurrences can easily be obtained by Algorithm 2. The algorithm is based on the following simple observation: If $\mathsf{LPF}[i] > 0$, $j = \mathsf{prevOcc}[i]$, and $\mathsf{LPF}[j] \geq \mathsf{LPF}[i]$, then $\mathsf{prevOcc}[j]$ is also the starting position of an occurrence of the factor starting at i. Since $\mathsf{prevOcc}[j] < j$, an occurrence left of j has been found.

Algorithm 2. Given LPF and its prevOcc-array, the algorithm computes the leftmost occurrence of each factor and stores it in prevOcc.

```
1: function COMPUTE-LEFTMOST-OCCURRENCE(LPF,prevOcc)
2:     for i ← 0 to n − 1 do
3:         if LPF[i] > 0 then                         ▷ hence prevOcc[i] ≠ ⊥
4:             j ← prevOcc[i]
5:             while LPF[j] ≥ LPF[i] do
6:                 j ← prevOcc[j]
7:         prevOcc[i] ← j
```

i	0	1	2	3	4	5	6	7	8	9	10	11	12	13
$S[i]$	a	$\#_1$	a	a	$\#_2$	a	a	a	$\#_3$	a	a	a	a	$\#_4$
$LPF[i]$	0	0	1	1	0	2	2	1	0	3	3	2	1	0
$prevOcc[i]$	⊥	⊥	0	2	⊥	2	5	5	⊥	5	9	9	9	⊥
$(lm)prevOcc[i]$	⊥	⊥	0	0	⊥	2	2	0	⊥	5	5	2	0	⊥
$iterations$	0		0	1		0	1	2		0	1	2	3	

Fig. 3. LPF and prevOcc arrays of the string $S = a^1\#_1a^2\#_2a^3\#_3a^4\#_4$.

The while-loop in Algorithm 2 repeats this procedure until the leftmost occurrence is found. Note that the algorithm overwrites the prevOcc-array. Consequently, if its for-loop is executed for i, then for every $0 \leq j < i$, prevOcc[j] stores a leftmost position. The next example shows that Algorithm 2 is not linear in the worst-case. Consider the string $S = a^1\#_1a^2\#_2a^3\#_3a^4\#_4 \cdots a^m\#_m$, where $m > 0$ and $\#_k$ are pairwise distinct separator symbols. Clearly, the length of S is $n = m + \sum_{k=1}^{m} k = m + m(m+1)/2 = m(m+3)/2$. If Algorithm 2 is applied to the arrays LPF and prevOcc in Fig. 3, it computes the leftmost (lm) prevOcc array and the number of iterations of its while-loop (last row in Fig. 3) is $\sum_{j=1}^{m-1} \sum_{k=1}^{j} k = (\sum_{j=1}^{m-1} j^2 + \sum_{j=1}^{m-1} j)/2 = (m-1)m(m+1)/6$.

3   Direct Computation of the f-Factorization

Algorithm 3 computes the f-factorization of S based on backward search on $T = S^{rev}$ and range maximum queries (RMQs) on the suffix array of T.[1] It uses ideas of [2, Algorithm CPS2] and [18, Algorithm LZ_bwd]. In fact, Algorithm 3 computes the right-to-left f-factorization of the reverse string S^{rev} of S. It is not difficult to see that $s_1s_2 \cdots s_m$ is the (left-to-right) f-factorization of S if and only if $s_m^{rev} \cdots s_2^{rev} s_1^{rev}$ is the right-to-left f-factorization of S^{rev}. In this subsection, we assume a basic knowledge of suffix arrays (SA), the Burrows-Wheeler transform (BWT), and wavelet trees; see e.g. [7,18]. Given a substring ω of T, there is a suffix array interval $[sp..ep]$— called the ω-interval— so that ω is a prefix of every suffix $T[SA[k]..n]$ if and only if $sp \leq k \leq ep$. For a character c, the $c\omega$-interval can be computed by one backward search

---

[1] In the implementation, T is terminated by a special (EOF) symbol.

Algorithm 3. f-factorization of S based on backward search on $T = S^{rev}$

1: **function** COMPUTE-F-FACTORIZATION(SA, wavelet tree of BWT)
2:     $i \leftarrow n - 1$ $\triangleright\, T = S^{rev}[0..n-1]$
3:     **while** $i \geq 0$ **do**
4:         $sp \leftarrow 0;\ ep \leftarrow n;\ pos \leftarrow i;\ m \leftarrow \perp$
5:         **repeat**
6:             $[sp..ep] \leftarrow backwardSearch(T[i], [sp..ep])$
7:             $max \leftarrow$ SA[RMQ(sp, ep)]
8:             **if** $max \leq pos$ **then**
9:                 break
10:             $m \leftarrow max$
11:             $i \leftarrow i - 1$
12:         **until** $i < 0$
13:         output $pos - i$

---

step $backwardSearch(c, [sp..ep])$; this takes $O(\log|\Sigma|)$ time if backward search is based on the wavelet tree of the BWT of T. A linear time preprocessing is sufficient to obtain a space-efficient data structure that supports RMQs in constant time; see [9] and the references therein. RMQ(sp, ep) returns the index of the maximum value among SA[sp], SA[$sp+1$], ..., SA[ep]; hence SA[RMQ(sp, ep)] is the maximum of these SA-values. Suppose Algorithm 3 has already computed $s_{j-1}^{rev} \cdots s_2^{rev} s_1^{rev}$ and let $i = n - (|s_{j-1}^{rev} \cdots s_2^{rev} s_1^{rev}| + 1)$. It computes the next factor s_j^{rev} as follows. First, it stores the starting position i in a variable pos. In line 6, $backwardSearch(T[i], [0..n])$ returns the c-interval $[sp..ep]$, where $c = T[i]$. In line 7, the maximum max of SA[sp], SA[$sp + 1$], ..., SA[ep] is determined. If $max = pos$ ($max < pos$ is impossible because $c = T[pos]$), then there is no occurrence of c in $T[pos + 1..n]$, so that $s_j^{rev} = c$ (the algorithm outputs 0, meaning that the next factor is the next character). Otherwise, there is a previous occurrence of c at position $max > pos$ and the process is iterated, i.e., i is decremented by one and the new $T[i..pos]$-interval is computed etc. Consider an iteration of the repeat-loop, where $[sp..ep]$ is the $T[i..pos]$-interval for some $i < pos$. The repeat-loop must be terminated early (line 9) if $max \leq pos$ because then the rightmost occurrence of $T[i..pos]$ starts left of $pos + 1$. In other words, $T[i..pos]$ is not a substring of $T[pos + 1..n]$. Since the repeat-loop did not terminate in the previous iteration, $T[i+1..pos]$ is a substring of $T[pos+1..n]$ that has a previous occurrence at position $m > pos$, where m is the maximum SA-value of the previous iteration. So $s_j^{rev} = T[i + 1..pos]$ and the algorithm outputs its length $|s_j^{rev}| = pos - (i+1) + 1 = pos - i$, which coincides with $|s_j|$. Note that the algorithm can easily be extended so that it also computes positions of previous occurrences. Algorithm 3 has run-time $O(n \log|\Sigma|)$ because one backward search step takes $O(\log|\Sigma|)$ time.

Kolpakov and Kucherov [17] used the reversed f-factorization (they call it reversed LZ-factorization) for searching for gapped palindromes. The reversed f-factorization is defined by replacing case (b) in Definition 1 with: (b) else s_j is the longest prefix of $S[i..n - 1]$ that is a substring of $(s_1 s_2 \cdots s_{j-1})^{rev}$. It is not

difficult so see that Algorithm 3 can be modified in such a way that it computes the reversed f-factorization of S in $O(n \log |\Sigma|)$ time (to find the next factor s_j, match prefixes of $S[i..n-1]$ against $T = S^{rev}$).

4   Experimental Results

Our implementation is based on the sdsl-lite library [11] and we experimentally compared it with the LPnF construction algorithm of Crochemore and Tischler [6], called CT-algorithm henceforth. Another LPnF construction algorithm is described in [5], but we could not find an implementation (this algorithm is most likely slower than the CT-algorithm because it uses *two* kinds of range minimum queries—one on the suffix array and one on the LCP-array—and range minimum queries are slow; see below). The experiments were conducted on a 64 bit Ubuntu 16.04.4 LTS system equipped with two 16-core Intel Xeon E5-2698v3 processors and 256 GB of RAM. All programs were compiled with the O3 option using g++ (version 5.4.1). Our programs are publically available.[2] The test data—the files dblp.xml, dna, english, and proteins—originate from the Pizza & Chili corpus.[3] In our first experiment, we computed the LPnF-array from the LPF-array. Three algorithms that compute the LPF-array were considered:

- AKO: algorithm by Abouelhoda et al. [1]
- LZ_OG: algorithm by Ohlebusch and Gog [18]
- KKP3: algorithm by Kärkkäinen et al. [14]

It is known that AKO is slower than the others, but in contrast to the other algorithms it calculates leftmost prevOcc-arrays. Thus, there was a slight chance that AKO in combination with Algorithm 1 is faster than LZ_OG or KKP3 in combination with Algorithm 1. However, our experiments showed that this is not the case. AKO is missing in Fig. 4 because the differences between the run-times of the other algorithms become more apparent without it. For the same reason, we did not take the suffix array construction time into account (note that each of the algorithms needs the suffix array). To find out whether or not it is advantageous to compute a leftmost prevOcc-array by Algorithm 2 before Algorithm 1 is applied, we also considered the combinations of LZ_OG and KKP3 with both algorithms. Figures 4 and 5 show the results of the first experiment. As one can see in Fig. 4, for real world data it seems disadvantageous to apply Algorithm 2 before Algorithm 1 because the overall run-time becomes slightly worse. However, for 'problematic' strings such as a^n and $a^n b$ it is advisable to use Algorithm 2: *With* it both LZ_OG and KKP3 outperformed the CT-algorithm (data not shown), but *without* it both did not terminate after 20 min. All in all, KKP3 in combination with Algorithms 1 and 2 is the best choice for the construction of the LPnF-array. In particular, it clearly outperforms the CT-algorithm in terms of run-time and memory usage.

---

[2] https://www.uni-ulm.de/in/theo/research/seqana/.
[3] http://pizzachili.dcc.uchile.cl.

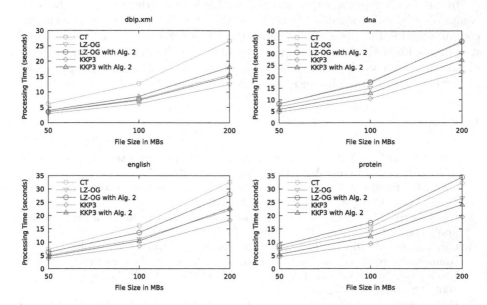

Fig. 4. Run-time comparison of LPnF-array construction (without suffix array construction, which on average takes 50% of the overall run-time)

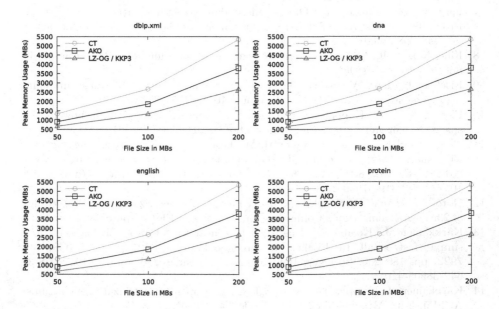

Fig. 5. Peak memory comparison of LPnF-array construction (with suffix array construction)

In the second experiment, we compared Algorithm 3—the only algorithm that computes the f-factorization directly—with the other algorithms (which first compute the LPnF-array and then derive the f-factorization from it). Algorithm 3 uses only 44% of the memory required by KKP3, but its run-time is by an order of magnitude worse (data not shown). We blame the range maximum queries for the rather bad run-time because these are slow in practice.

References

1. Abouelhoda, M.I., Kurtz, S., Ohlebusch, E.: Replacing suffix trees with enhanced suffix arrays. J. Discrete Algorithms **2**(1), 53–86 (2004)
2. Chen, G., Puglisi, S.J., Smyth, W.F.: Lempel-Ziv factorization using less time & space. Math. Comput. Sci. **1**(4), 605–623 (2008)
3. Crochemore, M., Ilie, L.: Computing longest previous factor in linear time and applications. Inf. Process. Lett. **106**(2), 75–80 (2008)
4. Crochemore, M., Ilie, L., Iliopoulos, C.S., Kubica, M., Rytter, W., Waleń, T.: LPF computation revisited. In: Fiala, J., Kratochvíl, J., Miller, M. (eds.) IWOCA 2009. LNCS, vol. 5874, pp. 158–169. Springer, Heidelberg (2009). https://doi.org/10.1007/978-3-642-10217-2_18
5. Crochemore, M., Kubica, M., Iliopoulos, C.S., Rytter, W., Waleń, T.: Efficient algorithms for three variants of the LPF table. J. Discrete Algorithms **11**, 51–61 (2012)
6. Crochemore, M., Tischler, G.: Computing longest previous non-overlapping factors. Inf. Process. Lett. **111**, 291–295 (2011)
7. Ferragina, P., Manzini, G.: Opportunistic data structures with applications. In: Proceedings of 41st Annual IEEE Symposium on Foundations of Computer Science, pp. 390–398 (2000)
8. Fine, N.J., Wilf, H.S.: Uniqueness theorem for periodic functions. Proc. Am. Math. Soc. **16**, 109–114 (1965)
9. Fischer, J., Heun, V.: Space-efficient preprocessing schemes for range minimum queries on static arrays. SIAM J. Comput. **40**(2), 465–492 (2011)
10. Fischer, J., I, T., Köppl, D., Sadakane, K.: Lempel-Ziv factorization powered by space efficient suffix trees. Algorithmica **80**(7), 2048–2081 (2018)
11. Gog, S., Beller, T., Moffat, A., Petri, M.: From theory to practice: plug and play with succinct data structures. In: Gudmundsson, J., Katajainen, J. (eds.) SEA 2014. LNCS, vol. 8504, pp. 326–337. Springer, Cham (2014). https://doi.org/10.1007/978-3-319-07959-2_28
12. Goto, K., Bannai, H.: Simpler and faster Lempel Ziv factorization. In: Proceedings of 23rd Data Compression Conference, pp. 133–142. IEEE Computer Society (2013)
13. Kärkkäinen, J., Kempa, D., Puglisi, S.J.: Linear time Lempel-Ziv factorization: simple, fast, small. In: Fischer, J., Sanders, P. (eds.) CPM 2013. LNCS, vol. 7922, pp. 189–200. Springer, Heidelberg (2013). https://doi.org/10.1007/978-3-642-38905-4_19
14. Kärkkäinen, J., Kempa, D., Puglisi, S.J.: Lazy Lempel-Ziv factorization algorithms. ACM J. Exp. Algorithmics **21**(2), Article 2.4 (2016)
15. Kociumaka, T., Kubica, M., Radoszewski, J., Rytter, W., Waleń, T.: A linear time algorithm for seeds computation. In: Proceedings of 23rd Symposium on Discrete Algorithms, pp. 1095–1112 (2012)

16. Kolpakov, R., Kucherov, G.: Finding maximal repetitions in a word in linear time. In: Proceedings of 40th Annual IEEE Symposium on Foundations of Computer Science, pp. 596–604 (1999)
17. Kolpakov, R., Kucherov, G.: Searching for gapped palindromes. Theor. Comput. Sci. **410**(51), 5365–5373 (2009)
18. Ohlebusch, E., Gog, S.: Lempel-Ziv factorization revisited. In: Giancarlo, R., Manzini, G. (eds.) CPM 2011. LNCS, vol. 6661, pp. 15–26. Springer, Heidelberg (2011). https://doi.org/10.1007/978-3-642-21458-5_4
19. Policriti, A., Prezza, N.: LZ77 computation based on the run-length encoded BWT. Algorithmica **80**(7), 1986–2011 (2018)
20. Ziv, J., Lempel, A.: A universal algorithm for sequential data compression. IEEE Trans. Inf. Theory **23**(3), 337–343 (1977)

Direct Linear Time Construction of Parameterized Suffix and LCP Arrays for Constant Alphabets

Noriki Fujisato$^{(\boxtimes)}$, Yuto Nakashima, Shunsuke Inenaga, Hideo Bannai[ID], and Masayuki Takeda

Kyushu University, 744 Motooka, Nishi-ku, Fukuoka 819-0395, Japan
{noriki.fujisato,yuto.nakashima,inenaga,bannai,takeda}@inf.kyushu-u.ac.jp

Abstract. We present the first worst-case linear time algorithm that directly computes the parameterized suffix and LCP arrays for constant sized alphabets. Previous algorithms either required quadratic time or the parameterized suffix tree to be built first. More formally, for a string over static alphabet Σ and parameterized alphabet Π, our algorithm runs in $O(n\pi)$ time and $O(n)$ words of space, where π is the number of distinct symbols of Π in the string.

Keywords: Parameterized pattern matching · Paramterized suffix array paramterized LCP array

1 Introduction

Parameterized pattern matching is one of the well studied "non-standard" pattern matching problems which was initiated by Baker [1], in an application to find duplicated code where variable names may be renamed. In the parameterized matching problem, we consider strings over an alphabet partitioned into two sets: the parameterized alphabet Π and the static alphabet Σ. Two strings $x, y \in (\Pi \cup \Sigma)^*$ of length n are said to parameterized match (p-match), if one can be obtained from the other with a bijective mapping over symbols of Π, i.e., there exists a bijection $\phi : \Pi \to \Pi$ such that for all $1 \leq i \leq n$, $x[i] = y[i]$ if $x[i] \in \Sigma$, and $\phi(x[i]) = y[i]$ if $x[i] \in \Pi$. For example, if $\Pi = \{x, y, z\}$ and $\Sigma = \{A, B, C\}$, strings xxAzxByzBCzy and yyAxyBzxBCxz p-match, since we can choose $\phi(x) = y, \phi(y) = z$, and $\phi(z) = x$, while strings xyAzzByxBCz and yyAzxByxBCy do not p-match, since there is no such bijection on Π. As parameterized matching captures the "structure" of the string, it has also been extended to RNA structural matching [20].

Baker introduced the so-called *prev encoding* of a p-string which maps each symbol of the p-string that is in Π to the distance to its previous occurrence (or 0 if it is the first occurrence), and showed that two p-strings p-match if and only if their prev encodings are equivalent. For example, the prev encodings for p-strings xxAzxByzBCzy and yyAxyBzxBCxz are both $(0, 1, A, 0, 3, B, 0, 4, B, C, 3, 5)$.

© Springer Nature Switzerland AG 2019
N. R. Brisaboa and S. J. Puglisi (Eds.): SPIRE 2019, LNCS 11811, pp. 382–391, 2019.
https://doi.org/10.1007/978-3-030-32686-9_27

Thus, the parameterized matching problem amounts to efficiently comparing the prev encodings of the p-strings.

Using the prev encoding allows for the development of data structures that mimic those of standard strings. The central difficulty, in contrast with standard strings, is in coping with the following property of prev encodings; a substring of a prev encoding is not necessarily equivalent to the prev encoding of the corresponding substring.

Nevertheless, several data structures and algorithms have so far successfully been developed. Baker proposed the parameterized suffix tree (PST), an analogue of the suffix tree for standard strings [21], and showed that for a string of length n, it could be built in $O(n|\Pi|)$ time and $O(n)$ words of space [2]. Using the PST for T, all occurrences of a substring in T which parameterized match a given pattern P can be computed in $O(|P|(\log(|\Pi| + |\Sigma|)) + occ)$ time, where occ is the number of occurrences of the pattern in the text. Kosaruju [19] further improved the running time of construction to $O(n\log(|\Pi| + |\Sigma|))$. Furthermore, Shibuya [20] proposed an on-line algorithm for constructing the PST that runs in the same time bounds.

Deguchi et al. [7] proposed the parameterized suffix array (PSA). Given the PST of a string, the PSA can be constructed in linear time, but as in the case for standard strings, the direct construction of PSAs has been a topic of interest. Deguchi et al. [7] showed a linear time algorithm for the special case of $|\Pi| = 2$ and $\Sigma = \emptyset$. I et al. [14] proposed a lightweight and practically efficient algorithm for larger Π, but the worst-case time was still quadratic in n. Beal and Adjeroh [5] proposed an algorithm based on arithmetic coding that runs in $O(n)$ time on average. Furthermore, they claimed a worst-case running time of $o(n^2)$. However, the proved upperbound is $O(n^2(\frac{\log(n - \log^{1+\varepsilon} n)}{\log^{1+\varepsilon} n}))$ for a very small $\varepsilon > 0$ (Corollary 27 of [5]), so it is only slightly better than quadratic.

In this paper, we break the worst-case quadratic time barrier considerably, and present the first worst-case linear time algorithm for constructing the parameterized suffix and LCP arrays of a given p-string, when the number of distinct parameterized symbols in the string is constant. Namely, when assuming that Π and Σ are linearly sortable integer alphabets, our algorithm runs in $O(n\pi)$ time and $O(n)$ words of space, where π is the number of distinct symbols of Π in the string. Furthermore, when $|\Pi|$ is constant and Σ is an integer alphabet, our algorithm runs in $O(n)$ time, which is faster than first building the PST which takes $O(n\log|\Sigma|)$ time. Since the PST can be constructed from the parameterized suffix and LCP arrays in linear time using essentially the same algorithm for standard strings by Kasai et al. [16], our algorithm is the fastest algorithm for constructing PST in such cases.

Several other indices for parameterized pattern matching have been proposed. Diptarama et al. [8] and Fujisato et al. [11] proposed the parameterized position heaps (PPH), an analogue of the position heap for standard strings [9], and showed that it could be built in $O(n\log(|\Sigma| + |\Pi|))$ time and $O(n)$ words of space. Using the PPH for T, all occurrences of a substring in T which parameterized

match a given pattern P can be computed in $O(|P|(|\Pi| + \log(|\Pi| + |\Sigma|)) + occ)$ time, where occ is the number of occurrences of the pattern in the text. Parameterized BWT's have been proposed in [13]. Also, parameterized text index with one wildcard was proposed in [12].

2 Preliminaries

For any set A of symbols, A^* denotes the set of strings over the alphabet A. Let $|x|$ denote the length of a string x. The empty string is denoted by ε. For any string $w \in A^*$, if $w = xyz$ for some (possibly empty) $x, y, z \in A^*$, x, y, z are respectively called a *prefix, substring, suffix* of w. When $x, y, z \neq w$, they are respectively called a *proper* prefix, substring, and suffix of w. For any integer $1 \leq i \leq |x|$, $x[i]$ denotes the ith symbol in x, and for any $1 \leq i \leq j \leq |x|$, $x[i..j] = x[i] \cdots x[j]$. For convenience, let $x[i..j] = \varepsilon$ when $j < i$. Let $\prec$ denote a total order on A, as well as the lexicographic order it induces. For two strings $x, y \in A^*$, $x \prec y$ if and only if x is a proper prefix of y, or there is some position $1 \leq k \leq \min\{|x|, |y|\}$ such that $x[1..k-1] = y[1..k-1]$ and $x[k] \prec y[k]$.

Let Π and Σ denote disjoint sets of symbols. Π is called the parameterized alphabet, and Σ is called the static alphabet. A string in $(\Pi \cup \Sigma)^*$ is sometimes called a p-string. Two p-strings $x, y \in (\Pi \cup \Sigma)^*$ of equal length are said to *parameterized match*, denoted $x \approx y$, if there exists a bijection $\phi : \Pi \to \Pi$, such that for all $1 \leq i \leq |x|$, $x[i] = y[i]$ if $x[i] \in \Sigma$, and $\phi(x[i]) = y[i]$ if $x[i] \in \Pi$.

The *prev encoding* of a p-string x of length n is the string $prev(x)$ over the alphabet $\Sigma \cup \{0, \ldots, n-1\}$ defined as follows:

$$prev(x)[i] = \begin{cases} x[i] & \text{if } x[i] \in \Sigma, \\ 0 & \text{if } x[i] \in \Pi \text{ and } x[i] \neq x[j] \text{ for any } 1 \leq j < i, \\ i - j & \text{if } x[i] \in \Pi, x[i] = x[j] \text{ and } x[i] \neq x[k] \text{ for any } j < k < i. \end{cases}$$

For example, if $\Pi = \{\mathtt{s}, \mathtt{t}, \mathtt{u}\}$, $\Sigma = \{\mathtt{A}\}$ and p-string $x = \mathtt{ssuAAstuAst}$, then $prev(x) = (0, 1, 0, \mathtt{A}, \mathtt{A}, 4, 0, 5, \mathtt{A}, 4, 4)$. Baker showed that $x \approx y$ if and only if $prev(x) = prev(y)$ [3]. We assume that Π and Σ are disjoint integer alphabets, where $\Pi = \{0, \ldots, n^{c_1}\}$ for some constant $c_1 \geq 1$ and $\Sigma = \{n^{c_1} + 1, \ldots, n^{c_2}\}$ for some constant $c_2 \geq 1$. This way, we can distinguish whether a symbol of a given prev encoding belongs to Σ or not. Also, given p-string x of length n, we can compute $prev(x)$ in $O(n)$ time and space (in words), by sorting the pairs $(x[i], i)$ using radix sort, followed by a simple scan of the result.

The following are the data structures that we consider in this paper.

Definition 1 (Parameterized Suffix Array [7]). *The parameterized suffix array of a p-string x of length n, is an array $PSA[1..n]$ of integers such that $PSA[i] = j$ if and only if $prev(x[j..n])$ is the ith lexicographically smallest string in $\{prev(x[i..n]) \mid i = 1, \ldots, n\}$.*

Definition 2 (Parameterized LCP Array [7]). *The parameterized LCP array of a p-string x of length n, is an array $pLCP[1..n]$ of integers such that*

$pLCP[1] = 0$, and $pLCP[i]$, for any $i \in \{2, \ldots, n\}$, is the longest common prefix between $prev(x[PSA[i-1]..n])$ and $prev(x[PSA[i]..n])$.

The difficulty when dealing with the prev encoding of suffixes of a string, is that they are not necessarily the suffixes of the prev encoding of the string. It is important to notice however, that, given the prev encoding $prev(x)$ of the whole string x, any specific value of the prev encoding of an arbitrary suffix of x can be retrieved in constant time, i.e., for any $1 \le i \le n$ and $1 \le k \le n - i + 1$,

$$prev(x[i..n])[k] = \begin{cases} 0 & \text{if } x[k'] \in \Pi \text{ and } prev(x)[k'] \ge k, \\ prev(x)[k'] & \text{otherwise,} \end{cases}$$

where $k' = i + k - 1$. The critical problem for suffix sorting is that even if two prev encodings $prev(x[i..n])$ and $prev(x[j..n])$ share a common prefix and satisfies $prev(x[i..n]) \prec prev(x[j..n])$, it may still be that $prev(x[j+1..n]) \prec prev(x[i+1..n])$.

Figure 1 shows an example of PSA and $pLCP$ for the string stssAtssAs. For example, we have that $prev(x[6..10]) \prec prev(x[1..10])$, which share a common prefix of length 2, yet $prev(x[2..10]) \prec prev(x[7..10])$.

i	$PSA[i]$	$prev(x[PSA[i]..|r|])$	$pLCP[i]$
1	10	0	0
2	6	0 0 1 A 2	1
3	2	0 0 1 A 4 3 1 A 2	4
4	1	0 0 2 1 A 4 3 1 A 2	2
5	3	0 1 A 0 3 1 A 2	1
6	7	0 1 A 2	3
7	4	0 A 0 3 1 A 2	1
8	8	0 A 2	2
9	9	A 0	0
10	5	A 0 0 1 A 2	2

Fig. 1. An example of the parameterized suffix and LCP arrays for a p-string $x =$ stssAtssAs, where $\Sigma = \{A\}, \Pi = \{s, t\}$.

3   Algorithms

In this section we describe our algorithms for constructing the parameterized suffix and LCP arrays. First, we mention a simple observation below.

From the definition of $prev(x)$, we have that $prev(x)[i] = 0$ for some position i if and only if i is the first occurrence of symbol $x[i] \in \Pi$. Therefore, the following observation can be made.

Observation 1. For any p-string x, the prev encoding $prev(x')$ of any substring x' of x contains at most π positions that are 0's, where π is the number of distinct symbols of Π in x.

3.1 *PSA* Construction

Based on this observation, we can see that the prev encoding of each suffix $x[i..n]$ can be partitioned into $z_i + 1 \leq \pi + 1$ *blocks*, where z_i is the number of 0's in $prev(x[i..n])$, and the jth block is the substring of $prev(x[i..n])$ that ends at the jth 0 in $prev(x[i..n])$ for $j = 1, \ldots, z_i$, and the (possibly empty) remaining suffix for $j = z_i + 1$. For technical reasons, we will append 0 to the last block as well. That is, we can write

$$prev(x[i..n])0 = B_{i,1} \cdots B_{i,z_i+1} \tag{1}$$

where, $B_{i,j}$ denotes the jth block of $prev(x[i..n])$. Also, for any $z_i + 1 < j \leq \pi + 1$, we let $B_{i,j} = 0$. Furthermore, for each j, let B_j denote the set of all jth blocks for all $i = 1, \ldots, n$, and let $C_{i,j}$ denote the lexicographic rank of $B_{i,j}$ in B_j. Note that if $B_{i,j} = B_{k,j}$ for some i, k, then $C_{i,j} = C_{k,j}$. Finally, let C_i denote the string over the alphabet $\{1, \ldots, n\}$ obtained by renaming each block $B_{i,j}$ of the string $prev(x[i..n])0$ with its lexicographic rank $C_{i,j}$. More formally,

$$B_j = \{B_{i,j} \mid i = 1, \ldots, n\}$$
$$C_{i,j} = |\{B_{i',j} \in B_j \mid B_{i',j} \prec B_{i,j}\}| + 1$$
$$C_i = C_{i,1} \cdots C_{i,z_i+1}.$$

Figure 2 shows an example.

i	$prev(x[i..\lvert r \rvert])$	$B_{i,1}$	$B_{i,2}$	$B_{i,3}$	C_i
1	0 0 2 1 A 4 3 1 A 2	0	0	2 1 A 4 3 1 A 2 0	1 1 4
2	0 0 1 A 4 3 1 A 2	0	0	1 A 4 3 1 A 2 0	1 1 3
3	0 1 A 0 3 1 A 2	0	1 A 0	3 1 A 2 0	1 2 5
4	0 A 0 3 1 A 2	0	A 0	3 1 A 2 0	1 4 5
5	A 0 0 1 A 2	A 0	0	1 A 2 0	2 1 2
6	0 0 1 A 2	0	0	1 A 2 0	1 1 2
7	0 1 A 2	0	1 A 2 0	0	1 3
8	0 A 2	0	A 2 0	0	1 5
9	A 0	A 0	0	0	2 1
10	0	0	0	0	1 1

Fig. 2. An example of C_i for a p-string $x = \mathtt{stssAtssAs}$, where $\Sigma = \{\mathtt{A}\}, \Pi = \{\mathtt{s}, \mathtt{t}\}$. $C_{4,2} = 4$, because $B_{4,2} = \mathtt{A0}$ is the lexicographically fourth smallest string in B_2, i.e., B_2 consists of 5 strings: $0 \prec \mathtt{1A0} \prec \mathtt{1A20} \prec \mathtt{A0} \prec \mathtt{A20}$.

Lemma 1. *For any* $1 \leq i_1, i_2 \leq n$,

$$prev(x[i_1..n]) \prec prev(x[i_2..n]) \iff C_{i_1} \prec C_{i_2}.$$

Proof. Notice that 0 is the smallest symbol in the two strings, so

$$prev(x[i_1..n]) \prec prev(x[i_2..n]) \Leftrightarrow prev(x[i_1..n])0 \prec prev(x[i_2..n])0$$
$$\Leftrightarrow B_{i_1,1} \cdots B_{i_1,z_{i_1}+1} \prec B_{i_2,1} \cdots B_{i_2,z_{i_2}+1}.$$

Also notice that since any block must end with a 0, if two blocks are not identical, it holds that one cannot be a prefix of the other. Thus, if $B_{i_1,1} \cdots B_{i_1,z_{i_1}+1} \prec B_{i_2,1} \cdots B_{i_2,z_{i_2}+1}$, this implies that there is some block k such that $B_{i_1,j} = B_{i_2,j}$, for all $1 \leq j < k$, and $B_{i_1,k} \prec B_{i_2,k}$, where $B_{i_1,k}$ is not a prefix of $B_{i_2,k}$. By definition, $B_{i_1,k} \preceq B_{i_2,k} \Leftrightarrow C_{i_1,k} \leq C_{i_2,k}$. Therefore, we have, $B_{i_1,1} \cdots B_{i_1,z_{i_1}+1} \prec B_{i_2,1} \cdots B_{i_2,z_{i_2}+1} \Leftrightarrow C_{i_1} \prec C_{i_2}$. $\qquad\square$

From Lemma 1, the problem of lexicographically sorting the set of strings $\{prev(x[1..n]), \ldots, prev(x[n..n])\}$ reduces to the problem of lexicographically sorting the set of strings $\{C_1, \ldots, C_n\}$. The latter can be done in $O(n\pi)$ time using radix sort, since the strings are over the alphabet $\{1, \ldots, n\}$ and the total length of the strings is at most $n\pi$.

What remains is to compute $C_{i,j}$ for all i, j in the same time bound. A problem is that the total length of all $B_{i,j}$ is $\Theta(n^2)$, so we cannot afford to naively process all of them.

Denote by $b_{i,j}$ and $e_{i,j}$ the beginning and end positions of $B_{i,j}$ with respect to their (global) position in x. Note that for any $1 \leq i \leq n$, we have $b_{i,1} = i$, and $b_{i,j} = e_{i,j-1} + 1$ for all $2 \leq j \leq z_i + 1$. Our algorithm depends on the following simple yet crucial lemma.

Lemma 2. *For any $1 < i \leq n$ and $1 \leq j \leq z_i + 1$, we have that either*

1. $b_{i,j} = e_{i-1,j} + 1$, *or,*
2. $b_{i,j} \geq b_{i-1,j}$, $e_{i,j} = e_{i-1,j}$, *and $B_{i,j}$ is a suffix of $B_{i-1,j}$*

holds.

Proof. If $x[i-1] \in \Sigma$, then, $prev(x[i..n])$ is a suffix of $prev(x[i-1..n])$, i.e., $prev(x[i..n]) = prev(x[i-1..n])[2..|n-i+2|]$ and $prev(x[i-1..n])[1] \neq 0$. Thus, $B_{i,1}$ is a suffix of $B_{i-1,1}$, and $B_{i,j} = B_{i-1,j}$ for all $2 \leq j \leq z_i$ and the second case of the claim holds.

If $x[i-1] \in \Pi$, the values in $prev(x[i..n])$ are equivalent to the corresponding values of $prev(x[i-1..n])[2..|n-i+2|]$, except possibly at some (global) position $k \geq i$ when there is a second occurrence of the symbol $x[i-1]$ at $x[k]$ which becomes the first occurrence in $x[i..n]$. (In other words, the value corresponding to $x[k]$ in $prev(x[i-1..n])$ is $k-i+1$.) Since there is no previous occurrence of $x[i-1]$ in $x[i-1..n]$, $prev(x[i-1..n])[1] = 0$. The situation is depicted in Fig. 2.

Let $B_{i-1,j'}$ be the block of $prev(x[i-1..n])$ that contains (global) position k. Because, as mentioned previously, $prev(x[i]..n)$ and $prev(x[i-1..n])[2..|n-i+2|]$ are equivalent except for the value corresponding to (global) position k, the block structure of $prev(x[i-1..n])$ is preserved in $prev(x[i..n])$, except that (1) the first block $B_{i-1,1}$ disappears, and (2) the block $B_{i-1,j'}$ is split into two blocks, corresponding to $B_{i,j'-1}$ and $B_{i,j'}$. Therefore, the first case of the claim is satisfied for $1 \leq j \leq j'$, since $b_{i,j} = b_{i-1,j+1} = e_{i-1,j} + 1$ for any $1 \leq j < j'$. Also, we can see that the second case of the claim is satisfied for $j' \leq j \leq z_i$, since $B_{i,j'}$ is a suffix of $B_{i-1,j'}$, and $B_{i,j} = B_{i-1,j}$ for $j' < j \leq z_i$ (Fig. 3).

Finally, the case when such k does not exist can be considered to be included above by simply assuming we are looking at a prefix of a longer string and

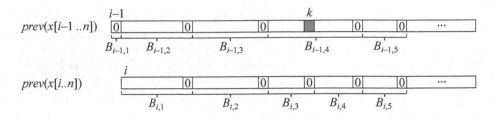

Fig. 3. A case in the proof of Lemma 2, where $x[i] \in \Pi$, and $x[k]$ is the first occurrence of $x[i]$ in $x[i + 1..n]$. The value corresponding to (global) position k in $prev(x[i..n])$, shown as a shaded box, is $k - i$, while it is 0 in $prev(x[i + 1..n])$. All other values in $prev(x[i..n])$ and $prev(x[i + 1..n])$ at the same (global) position are equivalent.

$k > |x|, j' > z_i$, since the prev encoding is preserved for prefixes, i.e., the prev encoding of a prefix of any p-string y is equivalent to the corresponding prefix of the prev encoding of y. Thus, the lemma holds. □

Lemma 2 implies that if we fix some j, we can represent $B_{i,j}$ for all i as suffixes (in the standard sense) of strings of total length $O(n)$.

Corollary 1. *For any j, there exists a set of strings S_j with total length $n + 1$ over the alphabet $\Sigma \cup \{0, \ldots, n - 1\}$ such that $B_{i,j}$ is a suffix of some string in S_j for all $i \in \{1, \ldots, n\}$.*

Proof. We include $B_{i,j}$ in S_j, if $i = 1$, or, if $i > 1$ and $B_{i,j}$ satisfies the first case of Lemma 2. Since the first case implies that the (global) positions $[b_{i-1,j}..e_{i-1,j}]$ and $[b_{i,j}...e_{i,j}]$ are disjoint, the total length of strings in S_j is at most $n + 1$ (including the 0 appended to B_{i,z_i+1}). On the other hand, if $B_{i,j}$ satisfies the second case is, it is a suffix of an already included string. □

Thus, computing $C_{i,j}$ for all i can be done by computing the generalized suffix array for the set S_j. This can be done in $O(n)$ time given S_j [15,17,18] and thus, for all j, the total is $O(n\pi)$ time.

Theorem 1. *The parameterized suffix array of a p-string of length n can be computed in $O(n\pi)$ time and $O(n)$ words of space.*

Proof. We compute a *forward encoding* of x, analogous to the prev encoding, defined as follows

$$fwd(x)[i] = \begin{cases} x[i] & \text{if } x[i] \in \Sigma, \\ \infty & \text{if } x[i] \in \Pi \text{ and } x[i] \neq x[j] \text{ for any } i < j \leq n, \\ j - i & \text{if } x[i] \in \Pi, x[i] = x[j] \text{ and } x[i] \neq x[k] \text{ for any } i < k < j. \end{cases}$$

This is done once, and can be computed in $O(n)$ time. Next, for any fixed j, we show how to compute the set S_j in linear time. This is done by using fwd and Lemma 2. We can first scan $prev(x)$ to obtain $B_{1,j}$. Suppose for some $i \geq 2$, we

know the beginning and end positions $b_{i-1,j}$, $e_{i-1,j}$ of $B_{i-1,j}$. Notice that when $x[i-1] \in \Pi$, k in the proof of Lemma 2 is $i+fwd(x)[i-1]-2$. Based on this value, we know that if $k < b_{i-1,j}$, then $B_{i,j} = B_{i-1,j}$ and if $b_{i-1,j} \leq k \leq e_{i-1,j}$ $B_{i,j}$ is a suffix of $B_{i-1,j}$, which corresponds to the second case of Lemma 2. When $k > e_{i-1,j}$, this corresponds to the first case of Lemma 2, so we scan $prev(x[i..n])$ starting from position corresponding to the global position $b_{i,j} = e_{i-1,j} + 1$ (i.e., $e_{i-1,j} - i$ in $prev(x[i..n])$) until we find the first 0, which gives us $B_{i,j}$ which we include in S_j. Since we only scan each position once, the total time for computing S_j is $O(n)$.

The time complexity follows from arguments for sorting C_j based on radix sort. Since, for a single step of the radix sort, we only require the values $C_{i,j}$ for a fixed j and all $1 \leq i \leq n$ and from Corollary 1, the space complexity is $O(n)$ words.   □

3.2 pLCP Construction

Given PSA, we describe below how to construct $pLCP$ in $O(n\pi)$ time and $O(n)$ words of space.

As a tool, we use longest common extension (LCE) queries. For a string x of length n, a *longest extension query*, given positions $1 \leq i, j \leq n$ asks for the longest common prefix between $x[i..n]$ and $x[j..n]$. It is known that the string can be pre-processed in $O(n)$ time so that the longest extension query can be answered in $O(1)$ time for any i, j (e.g. [10]).

We recompute S_j for $j = 1, \ldots, \pi$, and each time process it for LCE queries, so that the longest common prefix between $B_{i_1,j}$ and $B_{i_2,j}$ for some $1 \leq i_1, i_2 \leq n$ can be computed in constant time. This can be done in time linear in the total length of S_j, so in $O(n\pi)$ total time for all j. We compute the longest common prefix between each adjacent suffix in PSA block by block. Since each block takes constant time, and there are $O(\pi)$ blocks for each suffix, the total is $O(n\pi)$ time for all entries of the $pLCP$ array. The space complexity is $O(n)$ words, since, as for the case of PSA construction, we only process the jth block at each step.

A Note on a Previous Algorithm. A linear time algorithm for computing $pLCP$ given PSA, is claimed in [4, Theorem 20]. However, we note that their proof of correctness seems to be incomplete, and we could not verify the correctness of their algorithm.

Their algorithm follows the idea for computing the Longest Previous Factor (LPF) array by Crochemore and Ilie [6], where, for all $1 \leq i \leq n$,

$$LPF[i] = \max\{l \geq 0 \mid x[i..i+l-1] = x[j..j+l-1], 1 \leq j < i\}.$$

Their algorithm computes two arrays:

$$X = \max\{l \geq 0 \mid x[i..i+l-1] \approx x[j..j+l-1], 1 \leq j < i, PSA^{-1}[j] < PSA^{-1}[i]\},$$
$$Y = \max\{l \geq 0 \mid x[i..i+l-1] \approx x[j..j+l-1], i < j \leq n, PSA^{-1}[j] < PSA^{-1}[i]\},$$

where $PSA^{-1}[PSA[i]] = i$ for any $1 \leq i \leq n$. Then, $pLCP[i] = \max\{X[PSA^{-1}[i]],$ $Y[PSA^{-1}[i]]\}$. Here, X can also be denoted as $pLPF_<$, i.e., the longest prefix of the suffix $x[i..n]$ that occurs (in the parameterized sense) before i, and is a prefix of a lexicographically smaller suffix. Although the rationale for their algorithm is claimed to be ([4, Lemma 15])

$$pLPF[i] \geq pLPF[i-1] - 1,$$

which holds, it seems that the required property actually is

$$pLPF_<[i] \geq pLPF_<[i-1] - 1$$

which does not hold. For example, in the example stssAtssAs of Fig. 1, $pLPF_<[9] = 0$, while $pLPF_<[8] = 2$.

References

1. Baker, B.S.: A program for identifying duplicated code. Comput. Sci. Stat. **24**, 49–57 (1992)
2. Baker, B.S.: Parameterized pattern matching: algorithms and applications. J. Comput. Syst. Sci. **52**(1), 28–42 (1996). https://doi.org/10.1006/jcss.1996.0003. https://doi.org/10.1006/jcss.1996.0003
3. Baker, B.S.: Parameterized duplication in strings: algorithms and an application to software maintenance. SIAM J. Comput. **26**(5), 1343–1362 (1997). https://doi.org/10.1137/S0097539793246707
4. Beal, R., Adjeroh, D.: Variations of the parameterized longest previous factor. J. Discrete Algorithms **16**, 129–150 (2012). https://doi.org/10.1016/j.jda.2012.05.004. http://www.sciencedirect.com/science/article/pii/S1570866712001013, selected papers from the 22nd International Workshop on Combinatorial Algorithms (IWOCA 2011)
5. Beal, R., Adjeroh, D.A.: p-suffix sorting as arithmetic coding. J. Discrete Algorithms **16**, 151–169 (2012). https://doi.org/10.1016/j.jda.2012.05.001
6. Crochemore, M., Ilie, L.: Computing longest previous factor in linear time and applications. Inf. Process. Lett. **106**(2), 75–80 (2008). https://doi.org/10.1016/j.ipl.2007.10.006. http://www.sciencedirect.com/science/article/pii/S0020019007002979
7. Deguchi, S., Higashijima, F., Bannai, H., Inenaga, S., Takeda, M.: Parameterized suffix arrays for binary strings. In: Holub, J., Žďárek, J. (eds.) Proceedings of the Prague Stringology Conference 2008, Prague, Czech Republic, 1–3 September 2008. pp. 84–94. Prague Stringology Club, Department of Computer Science and Engineering, Faculty of Electrical Engineering, Czech Technical University in Prague (2008), http://www.stringology.org/event/2008/p08.html
8. Diptarama, Katsura, T., Otomo, Y., Narisawa, K., Shinohara, A.: Position heaps for parameterized strings. In: Kärkkäinen, J., Radoszewski, J., Rytter, W. (eds.) 28th Annual Symposium on Combinatorial Pattern Matching, CPM 2017, July 4–6, 2017, Warsaw, Poland. LIPIcs, vol. 78, pp. 8:1–8:13. Schloss Dagstuhl - Leibniz-Zentrum fuer Informatik (2017). https://doi.org/10.4230/LIPIcs.CPM.2017.8, https://doi.org/10.4230/LIPIcs.CPM.2017.8

9. Ehrenfeucht, A., McConnell, R.M., Osheim, N., Woo, S.W.: Position heaps: a simple and dynamic text indexing data structure. J. Discrete Algorithms **9**(1), 100–121 (2011). https://doi.org/10.1016/j.jda.2010.12. 001. http://www.sciencedirect.com/science/article/pii/S1570866710000535, 20th Anniversary Edition of the Annual Symposium on Combinatorial Pattern Matching (CPM 2009)

10. Fischer, J., Heun, V.: Theoretical and practical improvements on the RMQ-Problem, with applications to LCA and LCE. In: Lewenstein, M., Valiente, G. (eds.) CPM 2006. LNCS, vol. 4009, pp. 36–48. Springer, Heidelberg (2006). https://doi.org/10.1007/11780441_5

11. Fujisato, N., Nakashima, Y., Inenaga, S., Bannai, H., Takeda, M.: Right-to-left online construction of parameterized position heaps. CoRR abs/1808.01071 (2018). http://arxiv.org/abs/1808.01071

12. Ganguly, A., Hon, W., Huang, Y., Pissis, S.P., Shah, R., Thankachan, S.V.: Parameterized text indexing with one wildcard. In: 2019 Data Compression Conference (DCC), pp. 152–161. March 2019. https://doi.org/10.1109/DCC.2019.00023

13. Ganguly, A., Shah, R., Thankachan, S.V.: pbwt: achieving succinct data structures for parameterized pattern matching and related problems. In: Klein, P.N. (ed.) Proceedings of the Twenty-Eighth Annual ACM-SIAM Symposium on Discrete Algorithms, SODA 2017, Barcelona, Spain, Hotel Porta Fira, January 16–19, pp. 397–407. SIAM (2017). https://doi.org/10.1137/1.9781611974782.25, https://doi.org/10.1137/1.9781611974782.25

14. I, T., Deguchi, S., Bannai, H., Inenaga, S., Takeda, M.: Lightweight parameterized suffix array construction. In: Fiala, J., Kratochvíl, J., Miller, M. (eds.) IWOCA 2009. LNCS, vol. 5874, pp. 312–323. Springer, Heidelberg (2009). https://doi.org/10.1007/978-3-642-10217-2_31

15. Kärkkäinen, J., Sanders, P., Burkhardt, S.: Linear work suffix array construction. J. ACM **53**(6), 918–936 (2006). https://doi.org/10.1145/1217856.1217858

16. Kasai, T., Lee, G., Arimura, H., Arikawa, S., Park, K.: Linear-time longest-common-prefix computation in suffix arrays and its applications. In: Amir, A. (ed.) CPM 2001. LNCS, vol. 2089, pp. 181–192. Springer, Heidelberg (2001). https://doi.org/10.1007/3-540-48194-X_17

17. Kim, D.K., Sim, J.S., Park, H., Park, K.: Constructing suffix arrays in linear time. J. Discrete Algorithms **3**(2–4), 126–142 (2005). https://doi.org/10.1016/j.jda.2004.08.019

18. Ko, P., Aluru, S.: Space efficient linear time construction of suffix arrays. J. Discrete Algorithms **3**(2–4), 143–156 (2005). https://doi.org/10.1016/j.jda.2004.08.002

19. Kosaraju, S.R.: Faster algorithms for the construction of parameterized suffix trees (preliminary version). In: 36th Annual Symposium on Foundations of Computer Science, Milwaukee, Wisconsin, USA, 23–25 October 1995, pp. 631–637. IEEE Computer Society (1995). https://doi.org/10.1109/SFCS.1995.492664

20. Shibuya, T.: Generalization of a suffix tree for RNA structural pattern matching. Algorithmica **39**(1), 1–19 (2004). https://doi.org/10.1007/s00453-003-1067-9

21. Weiner, P.: Linear pattern matching algorithms. In: 14th Annual Symposium on Switching and Automata Theory, Iowa City, Iowa, USA, October 15–17, 1973, pp. 1–11. IEEE Computer Society (1973). https://doi.org/10.1109/SWAT.1973.13

Parallel External Memory Wavelet Tree and Wavelet Matrix Construction

Jonas Ellert and Florian Kurpicz$^{(\boxtimes)}$

Department of Computer Science,
Technische Universität Dortmund, Dortmund, Germany
{jonas.ellert,florian.kurpicz}@tu-dortmund.de

Abstract. We present the first parallel external memory wavelet tree and matrix construction algorithm. The algorithm's throughput is nearly the same as the hard disk drives' throughput, using six cores. We also present the fastest (parallel) semi-external construction algorithms for both wavelet trees and matrices.

Keywords: External memory · Parallel algorithm · Wavelet tree

1 Introduction

The *wavelet tree* [9] is a compact data structure that can answer access, rank, and select queries for a text over an alphabet $[0, \sigma)$ in $\mathcal{O}(\lg \sigma)$ time, while requiring only $n \lceil \lg \sigma \rceil (1 + o(1))$ bits of space. The *wavelet matrix* [3] is an alternative representation with the same space and time bounds for construction and answering queries. Both are used in many applications, e. g., text indexing [9], compression [10,15], and as an alternative to fractional cascading [13].

Our Contributions. First, we develop semi-external memory wavelet tree construction algorithms, where semi-external means we keep data that requires random access in main memory and all other data in external memory. Our implementations outperform the only previously available implementations by a factor of up to 1.43 regarding their running time, while using up to 16.6 times less memory. We then describe parallel fully external wavelet tree construction algorithms, which almost achieve a throughput that is only 1.19 times slower than the maximum throughput achievable on the hard disk drive using six threads. In general, we can achieve a speedup of up to 3.51 using six threads, and are mostly limited by I/Os. Finally, all algorithms are able to compute the wavelet matrix with the same space and time requirements, making them the fastest and first (semi-)external wavelet matrix construction algorithms.

This work was supported by the German Research Foundation (DFG) SPP 1736 priority programme "Algorithms for Big Data".

N. R. Brisaboa and S. J. Puglisi (Eds.): SPIRE 2019, LNCS 11811, pp. 392–406, 2019.
https://doi.org/10.1007/978-3-030-32686-9_28

Related Work. To our best knowledge, there exist no other external memory wavelet tree construction algorithms. The succinct data structure library [8] contains algorithms that can construct wavelet trees in semi-external memory.

Still, engineering of efficient wavelet tree and matrix construction algorithms in other models of computations is an ongoing problem with very recent advances. First, Fischer et al. [5] introduced bottom-up wavelet construction algorithms that are very fast and memory efficient in practice, and result in the fastest sequential and shared memory parallel wavelet tree and matrix construction algorithms. Also, Kaneta [11] recently presented a practical implementation of the $\mathcal{O}(n\lceil \lg \sigma / \sqrt{\lg n}\rceil)$-construction time algorithm, which uses word packing techniques in word-RAM, and has been (independently) introduced by Babenko et al. [2] and Munro et al. [16].

2 Preliminaries

(In this paper, we use the notation introduced by Fischer et al. [5].) Let $T = T[0] \ldots T[n-1]$ be a text of length n over an alphabet $\Sigma = [0, \sigma)$. Each character $T[i]$ can be represented using $\lceil \lg \sigma \rceil$ bits. The leftmost bit is the *most significant bit* (MSB), hence the *least significant bit* (LSB) is the rightmost bit. We denote the binary representation of a character $\alpha \in \Sigma$ that uses $\lceil \lg \sigma \rceil$ bits as bits(α). Whenever we write a binary representation of a value, we indicate it by a subscript two. The k-th bit (from MSB to LSB) of a character α is denoted by bit(k, α) for all $0 \leq k < \lceil \lg \sigma \rceil$. The *bit prefix* of size k of $\alpha \in \Sigma$ are the k MSBs, i.e., bit_prefix$(k, \alpha) = (\text{bit}(0, \alpha) \ldots \text{bit}(k-1, \alpha))_2$. We interpret sequences of bits as integer values. Let BV be a bit vector of size n. The operation $\text{rank}_0(\text{BV}, i)$ returns the number of 0's in $\text{BV}[0, i)$, whereas $\text{select}_0(\text{BV}, i)$ returns the position of the i-th 0 in BV. We define $\text{rank}_1(\text{BV}, i)$ and $\text{select}_1(\text{BV}, i)$ analogously. Both, rank and select queries on a bit vector of size n can be answered in $\mathcal{O}(1)$ time using succinct dictionary data structures that require only $o(n)$ bits space [17].

2.1 The Wavelet Tree

Let T be a text of length n over an alphabet $[0, \sigma)$. The *wavelet tree* (WT) [9] of T is a complete and balanced binary tree. Each node of the WT represents characters in $[\ell, r) \subseteq [0, \sigma)$. The root of the WT represents characters in $[0, \sigma)$, i.e., all characters. The left (or right) child of a node representing characters in $[\ell, r)$ represents the characters in $[\ell, (\ell + r)/2)$ (or $[(\ell + r)/2, r)$, respectively). A node is a leaf if $l + 2 \geq r$. The characters in $[\ell, r)$ at the corresponding node v are represented using a bit vector BV_v such that the i-th bit in BV_v is bit$(d(v), T_{[\ell,r)}[i])$, where $d(v)$ is the depth of v in the WT, i.e., the number of edges on the path from the root to v, and $T_{[\ell,r)}$ denotes the array containing the characters of T (in the same order) that are in $[\ell, r)$.

Wavelet trees can be used to generalize access, rank, and select queries from bit vectors to alphabets of size σ. Answering these queries then requires $\mathcal{O}(\lg \sigma)$ time. To do so, the bit vectors of the WT are augmented by binary rank and select data structures. For further information on queries, we point to [17].

	0	1	3	7	1	5	4	2	6	3
BV_0	0	0	0	1	0	1	1	0	1	0

	0	1	3	1	2	3	7	5	4	6
BV_1	0	0	1	0	1	1	1	1	0	0	1

	0	1	1	3	2	3	5	4	7	6
BV_2	0	1	1	1	0	1	1	0	1	0

(a) Level-wise wavelet tree.

	0	1	3	7	1	5	4	2	6	3	
BV_0	0	0	0	1	0	1	1	0	1	0	$Z[0] = 6$

	0	1	3	1	2	3	7	5	4	6	
BV_1	0	0	1	0	1	1	1	1	0	0	1	$Z[1] = 5$

	0	1	1	5	4	3	2	3	7	6	
BV_2	0	1	1	1	0	1	0	1	1	0	$Z[2] = 4$

(b) Wavelet matrix.

Fig. 1. The level-wise (a) WT and the WM (b) of $T = [0, 1, 3, 7, 1, 5, 4, 2, 6, 3]$. The light gray ( ) arrays contain the characters represented at the corresponding position in the bit vector and are not a part of the WT and WM. In (a), thick lines represent the borders of the intervals, which are not stored explicitly. In (b), thick lines represent the number of zeros, which are stored in the Z-array (Color figure online).

Level-Wise Wavelet Tree. In this paper, we consider *level-wise* wavelet trees. Here, we concatenate the bit vectors of all nodes at the same depth. Since we lose the tree topology, the resulting bit vectors correspond to a *level* that is equal to the depth of the concatenated nodes and the concatenated bit vectors correspond to *intervals* in the level. We store only a single bit vector BV_ℓ for each level $\ell \in [0, \lceil \lg \sigma \rceil)$, see Fig. 1a. This retains the functionality, but reduces the redundancy for the succinct dictionaries needed to answer rank and select queries on the bit vectors in constant time [13,14]. We can also easily identify the interval in which a character is represented at any level:

Observation 1. (Fuentes-Sepúlveda et al. [6]). *Given a character $T[i]$ for $i \in [0, n)$ and a level $\ell \in [1, \lceil \lg \sigma \rceil)$ of the WT, the interval pertinent to $T[i]$ in BV_ℓ can be computed by* bit_prefix$(\ell, T[i])$.

Wavelet Matrix. A variant of the wavelet tree, the *wavelet matrix* (WM), was introduced in 2011 by Claude [3]. It requires the same space as a WT and has the same asymptotic running times for access, rank, and select; but in practice it is often faster than a WT for rank and select queries [3], as it needs less calls to binary rank/select data structures. For the definition of the WM, we need additional notations: *Reversing* the significance of the bits is denoted by reverse, e. g., reverse$((001)_2) = (100)_2$. The *bit-reversal* permutation of order k (denoted by ρ_k) is a permutation of $[0, 2^k)$ with $\rho_k(i) = (\text{reverse}(\text{bits}(i)))_2$. For example, $\rho_2 = (0, 2, 1, 3) = ((00)_2, (10)_2, (01)_2, (11)_2)$. ρ_k and ρ_{k+1} can be computed from another, as $\rho_{k+1} = (2\rho_k(0), \ldots, 2\rho_k(2^k - 1), 2\rho_k(0) + 1, \ldots, 2\rho_k(2^k - 1) + 1)$.

In a WM the tree structure is discarded completely and we use the array $Z[0, \lceil \lg \sigma \rceil)$ to store the number of zeros at each level ℓ in $Z[\ell]$. BV_0 contains the MSBs of each character in T in text order (this is the same as the first level of a WT). For $\ell \geq 1$, BV_ℓ is defined as follows. Assume that a character α is represented at position i in $BV_{\ell-1}$. Then the position of its ℓ-th MSB in BV_ℓ depends on $BV_{\ell-1}[i]$ in the following way: if $BV_{\ell-1}[i] = 0$, bit(ℓ, α) is stored at position rank$_0(BV_{\ell-1}, i)$; otherwise $(BV_{\ell-1}[i] = 1)$, it is stored at position

$Z[\ell - 1] + \text{rank}_1(\text{BV}_{\ell-1}, i)$. For an example, see Fig. 1b. Similar to the intervals in the bit vectors of the WT, characters of T form intervals in BV_ℓ of the WM. Again, the intervals at level ℓ correspond to bit prefixes of size ℓ, but due to the construction of the WM, we have to consider the reversed bit prefixes:

Observation 2. *Given a character $T[i]$ for $i \in [0, n)$ and a level $\ell \in [1, \lceil \lg \sigma \rceil)$ of the WM, reverse(bit_prefix($\ell, T[i]$)) indicates the interval pertinent to $T[i]$ in BV_ℓ. Namely, $\text{BV}_\ell[i] = \text{bit}[\ell, S[i]]$, where S is T stably sorted using the reversed bit prefixes of length ℓ of the characters as key.*

2.2 The External Memory Model

The *external memory model* [1] measures the transfer of data between the main memory of size M (also called *internal memory*) and a secondary memory (also called *external memory*) that is assumed to be of unlimited size and slower in terms of memory access than the main memory. Also, data can only be transferred in *blocks* of size B between main and secondary memory. Transfers of blocks are called *I/O operations* (*I/Os* for short) and are the main cost measure of the external memory model.

For *semi-external* algorithms, we assume that we have random access on either the input or output—but not both. This relaxation allows for algorithms that cannot be efficiently be expressed in the external memory model. The model is used in practice, e. g., the succinct data structure library (SDSL) [8] provides semi-external WT construction algorithms (among others).

Computing the Effective Alphabet. We construct WTs and WMs using an *effective* alphabets, i. e., every character of the effective alphabet occurs in the text. Therefore, we can store $\lfloor w/\lg \sigma \rfloor$ characters in one w-bit computer word. To obtain the effective alphabet, we have to scan the text twice. First, we compute the histogram of all characters of the text, second we compute a transformation from the alphabet to the effective alphabet, and finally we scan the original text but store the text in the effective alphabet. We denote the number of blocks that must be transferred to scan the text by $\text{scan}(n\lceil \lg \sigma \rceil) = \lceil \lceil n\lceil \lg \sigma \rceil/w \rceil /B \rceil$.

In main memory, most implementations assume that the text is available in main memory over the effective alphabet. Our (semi-)external WT and WM construction algorithms mimic the behavior by assuming that the text is available in secondary memory over the effective alphabet. If that is not the case, all our algorithms require additional $2\lceil n/B \rceil + \text{scan}(n\lceil \lg \sigma \rceil)$ I/Os and $\mathcal{O}(n)$ time to compute the histogram and store the text over an effective alphabet.

3 Construction in Semi-External Memory

In this section, we briefly discuss how to adapt the fast WT and WM construction algorithms presented by Fischer et al. [5] to the semi-external memory model. To this end, we first discuss the *bottom-up* construction, the new approach to

compute the WT and WM [5]. Here, we construct the histogram of the text, and then use that histogram to compute histograms for all other bit prefixes without another text access. Generally speaking, if we have the histogram of length-ℓ bit prefixes, we can simply compute the histogram of the bit prefixes of length $\ell - 1$ by ignoring the last bit of the current prefix, e. g., the number of characters with bit prefix $(01)_2$ is the total number of characters with bit prefixes $(010)_2$ and $(011)_2$, since both share the bit prefix $(01)_2$. Using these histograms, we can compute the interval starting positions for all levels of the WT and WM.

Now, we have a look at the *space requirements* of this technique. The histogram of all characters requires $\sigma \lceil \lg n \rceil$ bits of space. We can always reuse that space for any histograms at a previous levels ℓ, which require $\lceil \sigma \lg n \rceil / 2^{\lceil \lg \sigma \rceil - \ell}$ bits of space. Storing the borders requires the same space as storing the histogram. Note that we do not require a histogram for the first level, and we also do not require the starting positions resulting from the histogram of all characters. Since we require at most $\sigma \lceil \lg n \rceil / 2$ bits of space for the histogram (of the last level) and the starting positions, and we can reuse the space when computing both for the following level, we require $\sigma \lceil \lg n \rceil$ bits of space for both the histogram and the starting positions in total. If we require access to all histograms and cannot reuse the space, we need $2 \lceil \sigma \lg n \rceil$ bits of space.

Random Access on the Output. Our first semi-external WT construction algorithm is the semi-external variant of the (single scan) prefix counting WT construction algorithm [5]. Here, we first compute the histogram for all characters in T and compute all histograms and interval borders without another scan of the text in $\mathcal{O}(n)$ time, $\mathrm{scan}(n \lceil \lg \sigma \rceil)$ I/Os, and $\sigma \lceil \lg n \rceil$ bits space, as described above. Next, we scan the text once again and fill all the bit vectors accordingly using the precomputed borders, i. e., for each symbol, we look at the border for each of the symbol's bit prefixes and set the corresponding bit in each bit vector accordingly (one bit per level) and then we update the borders. This requires $\mathcal{O}(n \lg \sigma)$ time in total for all levels. Setting the bits in the bit vectors still requires random access. Hence, we only read the text from the secondary memory. The number of I/Os is $2 \, \mathrm{scan}(n \lceil \lg \sigma \rceil)$. In terms of main memory, we need $n \lceil \lg \sigma \rceil$ bits for the bit vectors of the WT and $\sigma \lceil \log n \rceil$ bits for histograms that are later used for the starting positions of the intervals. We call this semi-external algorithm *se.pc*.

This algorithm can also be parallelized by parallelizing the computation of the initial histogram and writing the bit vectors for each level in parallel, which scales up to $\lceil \lg \sigma \rceil$ threads. We denote this algorithm by *se.par.pc*.

Random Access on the Input. Next, we consider a modified and semi-external version of the prefix sorting WT construction algorithm [5]. Here, each level of the WT is written in sequential order, which lets us efficiently stream the bit vectors of the WT. Again, we precompute all borders of the intervals. Then, for each level ℓ, we use counting sort with the length-ℓ bit prefixes as keys to sort the text, such that we can fill the bit vector from left to right. Counting sort requires $\mathcal{O}(n)$ time, given the borders array, hence the running time does not differ from se.pc. Since we require a stable sort, we cannot sort the text in

place [18] and thus need additional $n\lceil \lg \sigma \rceil$ bits of space. We write the output to disk exactly once and each level is written sequentially, therefore the number of I/Os is $\text{scan}(n\lceil \lg \sigma \rceil)$. We call this algorithm *se.ps*. To overcome the space requirements by sorting, we use a new in-place algorithm that rearranges the text as required by the WT in $\mathcal{O}(n)$ time. We decompose the text into $\Theta(\sqrt{n})$ blocks of size $\Theta(\sqrt{n})$ and use two buffers of the same size. Then, we separate the text using one buffer for symbols corresponding to a one bit and the other for the other bits. Whenever a buffer is full, we can write it to a part of the text, because the part is already written to the buffers. In the end, we have to rearrange the blocks (and shift some of them). We denote this variant by *se.ps.ip*. It requires less space, but is one of the slowest algorithms (see §5).

Lemma 1. *The semi-external algorithms se.pc, se.ps, and se.ps.ip compute the WT of a text of length n over an alphabet of size σ in $\mathcal{O}(n \lg \sigma)$ time using $\mathcal{O}(\text{scan}(n\lceil \lg \sigma \rceil))$ I/Os, and $n\lceil \lg \sigma \rceil + \sigma\lceil \lg n \rceil$ (se.pc) and $2n\lceil \lg \sigma \rceil + \sigma\lceil \lg n \rceil$ (se.ps) bits of main memory including input and output, respectively.*

Adaption to the Wavelet Matrix. Our semi-external memory WT construction algorithms can easily be extended to compute the WM instead. To this end, we only have to compute the borders in bit reversal permutation order and thus change the order of the intervals within the bit vectors of each level [5]. Also, this change does not affect the running time or the memory requirement; it only affects the content of the border array and subsequently the resulting bit vectors.

4   Wavelet Tree Construction in External Memory

If we replace the sorting in se.ps with any external memory sorting algorithm we obtain an external memory version of se.ps. However, sorting in external memory is (in practice) expensive. Now, we present dedicated external memory WT and WM construction algorithms. For the sequential algorithm we first explain how to build the WM, and then show how to adapt the algorithm to produce the WT.

4.1   Sequential Construction in External Memory

Each level ℓ of the WM can be interpreted as a reordered version T_ℓ of the original input text T, where the first level represents $T_0 = T$, and each text T_ℓ with $\ell > 0$ can be obtained by stable sorting the text $T_{\ell-1}$ of the previous level by the $(\ell-1)$-th bit. This property of the WM has been originally described as *all zeros of the level go left, and all the ones go right* [3]. If we know T_ℓ, then we can easily build BV_ℓ by taking the ℓ-th bit of each symbol of T_ℓ in left-to-right order. Thus, we can construct the entire WM by simply repeatedly sorting the text and extracting the bit vector of one level after each sort. Conveniently, the sorting key in each iteration is only a single bit. Therefore, we only have to create a binary partition of the text, where L_ℓ contains all the zeros of T_ℓ, and R_ℓ contains all the ones (retaining their order). Clearly, we have $T_{\ell+1} = L_\ell \cdot R_\ell$.

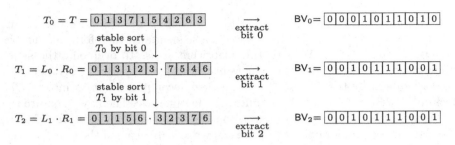

Fig. 2. Constructing the WM for our running example by partitioning the text and extraction of bits. The resulting WM can also be seen in Fig. 1b.

In the external memory setting we can realize the partitioning by performing a single scan over T_ℓ and appending all characters α with $\mathrm{bit}(\ell, \alpha) = 0$ to L_ℓ and all other characters to R_ℓ. Also, we can simultaneously write the bit vector BV_ℓ by appending $\mathrm{bit}(\ell, \alpha)$ to BV_ℓ. Note that after the scan no additional copying is needed to get $T_{\ell+1}$ from L_ℓ and R_ℓ, as we can simply scan directly over L_ℓ and R_ℓ in the next iteration, see Fig. 2. The number $\mathsf{Z}[\ell]$ of zeros in each level is $|L_\ell|$.

Adaptation to the Wavelet Tree. Our external WM construction algorithm can easily be adapted to construct the WT instead. As described in §2, the bit vector belonging to any node of the WT always occurs in the WM, too. Only the order of these intervals is different. Our L_ℓ and R_ℓ buffers therefore already contain all the correct nodes, but in wrong order. It is easy to see that L_ℓ contains exactly all of the left children, whereas R_ℓ contains the right children. Clearly, instead of defining $T_{\ell+1} = L_\ell \cdot R_\ell$ at the end of each scan, we can define $T_{\ell+1}$ by interleaving L_ℓ and R_ℓ such that left children and right children alternate. This way we will continue with the correct WT order in the next scan. To this end, we only need to know the size of each node, allowing us to always read the appropriate number of characters from L_ℓ or R_ℓ. Hence, we simply determine the last level's histogram during the initial scan. After the scan we can compute all histograms (see §2). We simply keep the histograms of all levels in main memory.

Analysis. We will first look at the I/O complexity of the WT/WM construction algorithm. The $\mathsf{Z}[\ell]$ values have size $\lceil \lg n \rceil \lceil \lg \sigma \rceil$ bits and are insignificant in terms of I/O complexity. Each reordered text T_ℓ has size $n\lceil \lg \sigma \rceil$ bits, which can be stored using $\mathrm{scan}(n\lceil \lg \sigma \rceil)$ blocks in external memory. We read each of these texts exactly once, and write all texts except for the initial text exactly once as well, resulting in $(2\lceil \lg \sigma \rceil - 1) \cdot \mathrm{scan}(n\lceil \lg \sigma \rceil)$ I/Os. The resulting WM or WT is written exactly once using another $\mathrm{scan}(n\lceil \lg \sigma \rceil)$ I/Os. All used structures in external memory are both written and read exclusively sequentially.

Now, we determine the time complexity and main memory bounds of our algorithm. Clearly, each of the $\lceil \lg \sigma \rceil$ scans takes $\mathcal{O}(n)$ time. Thus the overall time for the WM construction is $\mathcal{O}(n \lg \sigma)$. The WT construction needs additional $\mathcal{O}(\sigma)$ time to compute the histograms of all levels. In terms of space, the WM

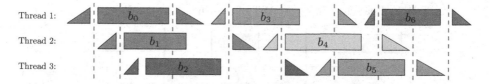

Thread 1:

Thread 2:

Thread 3:

Fig. 3. Domain decomposition for a text $T = \boxed{b_0|b_1|b_2|b_3|b_4|b_5|b_6}$ split into seven segments. Here, ◿ means loading b_i from external memory, ☐ means computing the local tree for b_i, and ◺ means writing the local tree of b_i to external memory. Only one of the three threads is allowed to read/write at a time, as indicated by the dashed synchronization barriers. (Best viewed in color.)

construction is fully external and only needs $\mathcal{O}(1)$ bits of main memory, since all data structures are kept in external memory. For the WT we need $2\lceil \sigma \lg n \rceil$ additional bits to store the histograms.

Lemma 2. *The fully external algorithm ext.ps computes the WT of a text of length n over an alphabet of size σ in $\mathcal{O}(n \lg \sigma + \sigma)$ time using a total of $2\lceil \lg \sigma \rceil \cdot scan(n \lceil \lg \sigma \rceil)$ I/Os and $2\lceil \sigma \lg n \rceil$ bits of main memory including input and output. For the WM the time is $\mathcal{O}(n \lg \sigma)$ and only $\mathcal{O}(1)$ bits of main memory are needed.*

4.2 Parallel Construction in External Memory

For a more generic approach, we present a meta-algorithm based on the internal memory domain decomposition, see, e. g., [5,7,12]. Let p be the number of available threads, then in the internal memory setting we split the text into p segments, and compute the WT of each segment on a different thread, using a sequential construction algorithm of our choice. After that, the so called *local trees* can be merged into one *global tree*. In the external memory setting the length of the segments depends on the amount M of main memory. Assume that the sequential construction algorithm needs $s(n, \sigma)$ bits of memory for a text of length n over the alphabet $[0, \sigma)$. Then, the length k of each segment must satisfy $s(k, \sigma) \leq M/p$. This way all threads can work simultaneously.

Each thread runs a simple loop: load the next text segment from external memory into internal memory, compute the WT of the segment, and write it back to external memory. Only one thread is allowed to read/write at a time (see Fig. 3). In terms of external memory layout, we store the local trees in text order, and each local tree as the concatenation of its levels (see LT in Fig. 4). When merging the local trees into the global tree, we simply perform a single scan over the local trees and *zip* the corresponding intervals together. Since the length of each interval must be known in order to copy the right amount of bits, we need the histograms of all text parts during the merge phase. However, many of the fastest sequential WT construction algorithms either build the histograms or can easily be modified to do so [5]. We are not using parallelism during the

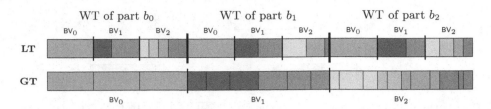

Fig. 4. External memory layout of local (LT) and global (GT) WTs for $T = b_0 b_1 b_2$. Best viewed in color, as colors indicate parts of local trees that are zipped together.

merge phase, since we are only copying bit vectors. In practice we are limited by the speed of the external memory, even when using only a single thread. Clearly, if we use a WM algorithm as a subroutine, our algorithm produces the WM instead.

Analysis. We will first look at the I/O complexity of our meta-algorithm. The input text, the concatenation of all local trees as well as the global tree are of size $n\lceil \lg \sigma \rceil$ bits each, which can be stored using $\mathrm{scan}(n\lceil \lg \sigma \rceil)$ blocks in external memory. We read the input text and write the local trees once, taking $2\,\mathrm{scan}(n\lceil \lg \sigma \rceil)$ sequential I/Os. Reading all local trees sequentially during the merge phase causes another $\mathrm{scan}(n\lceil \lg \sigma \rceil)$ I/Os. When writing the global tree we jump to a different external memory address for each interval of a local tree. Therefore, we need up to $\sigma\lceil n/k \rceil$ random I/Os in addition to the $\mathrm{scan}(n\lceil \lg \sigma \rceil)$ I/Os that are generally needed to write the local tree. Thus, the total number of I/Os is bound by $4\,\mathrm{scan}(n\lceil \lg \sigma \rceil) + \sigma\lceil n/k \rceil$. In practice we use the entire internal memory as a write buffer while merging the local trees. This way we maximize the length of sequential writes and keep random I/Os at a minimum.

Now we determine the time complexity of our algorithm as well as the internal memory bounds. Let $t(n,\sigma)$ and $s(n,\sigma)$ be the time and the bits of memory used by the sequential construction algorithm that we deploy as a subroutine. We know that at any given point in time there is either exactly one processor performing I/Os, or all threads are computing local trees. The total I/O time (including the merge phase) is bound by $\mathcal{O}(n + \sigma\lceil n/k \rceil)$. The time during which all threads are computing local trees is bound by $\lceil n/pk \rceil \cdot t(k,\sigma)$. In terms of main memory we use $p \cdot s(k,\sigma)$ bits for up to p simultaneous executions of our internal memory construction algorithm over text segments of size k. Additional $\mathcal{O}(\lceil n/k \rceil \sigma \lg n)$ bits are needed to store all histograms.

Lemma 3. *Let $t(n,\sigma)$ and $s(n,\sigma)$ be the time and space used by an internal memory WT construction algorithm, and let $p, k \in \mathbb{N}^+$. The external memory algorithm ext.dd computes the WT of a text of length n over an alphabet of size σ using $4\,scan(n\lceil \lg \sigma \rceil) + \sigma\lceil n/k \rceil$ I/Os. If p threads are available, it takes $\mathcal{O}(n + \sigma\lceil n/k \rceil) + \lceil n/pk \rceil \cdot t(k,\sigma)$ time and $\mathcal{O}(\lceil n/k \rceil \sigma \lg n) + p \cdot s(k,\sigma)$ bits of internal memory including input and output.*

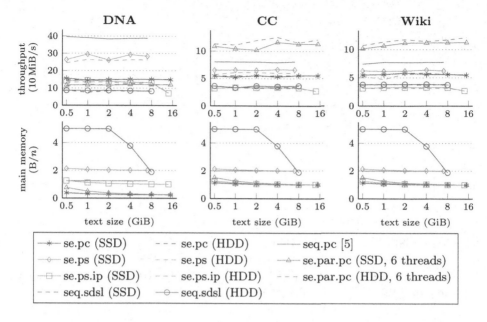

Fig. 5. Throughput and main memory peak of semi-external WT construction.

5   Experiments

For our experiments, we used a machine equipped with 16 GiB RAM, eight Hitachi HUA72302 HDDs each with a capacity of 1.8 TiB and two Samsung SSD 850 EVO SSDs each with a capacity of 465.8 GiB, and an Intel Xeon CPU i7-6800K (6 cores with frequency up to 3.4 GHz and cache sizes: 32 kB L1D and L1I, 256kB L2, and 15360 kB L3). The operating system is Ubuntu 16.04 (64-bit, Linux kernel 4.4). Our external memory algorithms use the STXXL [4] development snapshot (26-09-2017). We compiled all source code using g++ 7.4 with flags -O3 and -march=native, and express parallelism using OpenMP 4.5. We test our algorithms using both (a) four HDDs and (b) two SSDs. Before starting the timer, we compute the text over the effective alphabet and store it on disk, which is the input. Running times are the median of three executions. The implementations used for the evaluation are available from https://github.com/kurpicz/pwm.

We compare the following algorithms: **se.pc**, **se.par.pc**, **se.ps**, and **se.ps.ip** are the semi-external memory WT and WM algorithms described in §3, **seq.sdsl** is the semi-external memory algorithm contained in the SDSL, and **seq.pc** the fastest *main memory* WT algorithm [5], which we use as baseline. Our external (and parallel) WT and WM algorithms are the only external construction algorithms. Hence, we cannot compare **ext.ps** and **ext.dd** (see §4) with other algorithms in the same model. The construction times for WTs and WMs are nearly identical in both models.

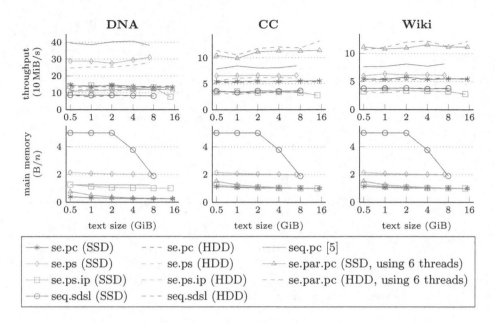

Fig. 6. Throughput and main memory peak of semi-external WM construction. Same experiment for the WM as reported in Fig. 5 for the WT.

We use the following real world inputs of sizes up to 128 GiB. When needing a smaller size, we consider a prefix of that size.

DNA ($\sigma = 4$) is a collection of DNA data from the 1000 Genomes Project (http://internationalgenome.org/data),

CC ($\sigma = 242$) contains websites (without HTML tags) that have been crawled by the Common Crawl corpus (http://commoncrawl.org), and

Wiki ($\sigma = 213$) are recent Wikipedia dumps containing XML files that are available from (https://dumps.wikimedia.org).

Semi-external Memory Construction Algorithms. An overview of the throughput and the required main memory of our semi-external WT construction algorithms can be found in Fig. 5. The results of the semi-external WM construction algorithms can be found in Fig. 6. Not plotted data means that the algorithm could not process the input size with the given main memory. The main memory algorithm seq.pc is used as base line and—as expected—always fastest sequential algorithm. The fastest semi-external memory algorithm on all inputs is se.ps. However, se.ps requires the most main memory—even more than seq.pc. The second fastest algorithm is se.pc. In addition, it is also the most memory efficient one, requiring less than all other tested algorithms. On DNA, se.ps.ip achieves a similar throughput to se.pc and is faster than seq.sdsl. On all other inputs se.ps.ip and seq.sdsl are always the slowest. Still, on all instances except for DNA, seq.sdsl requires five times more memory than se.pc. On DNA it even

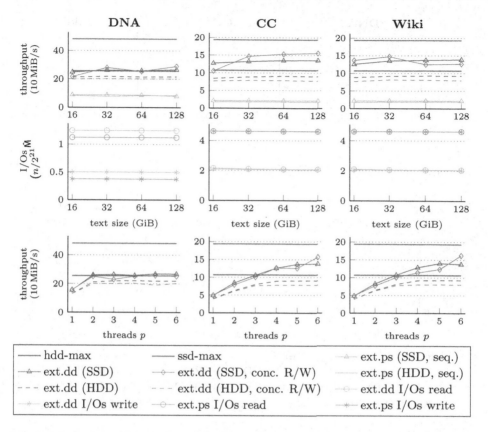

Fig. 7. Throughput (first row) and I/Os (second row) of external WT algorithms. Here, parallel algorithms uses all six threads. Throughput of ext.dd using 20 GiB input per thread (last row).

requires 16.6 times as much. The memory requirements of our semi-external algorithms per byte input is decreasing with larger inputs, as we use fixed-size buffers for our algorithms. When given inputs of size 4 GiB or more, seq.sdsl has to move the system swap, which explains the decrease in required memory. Therefore, se.pc is the fastest and most memory efficient semi-external memory algorithm. The throughput on SSDs is slightly better than on HDDs, except for se.par.pc, which has higher throughput on HDDs, which we cannot explain. It is also roughly twice as fast as se.pc, using slightly more memory, except on DNA (as expected).

External Memory Construction Algorithms. In Fig. 7 we show the throughput (first row) and I/Os (second row) of our external memory algorithms computing the WT. We show the same experiments for the WM computation in Fig. 8. We also give the maximum throughput (hdd-max and ssd-max) we achieved for reading the text and the WT once and writing the WT twice, which are exactly

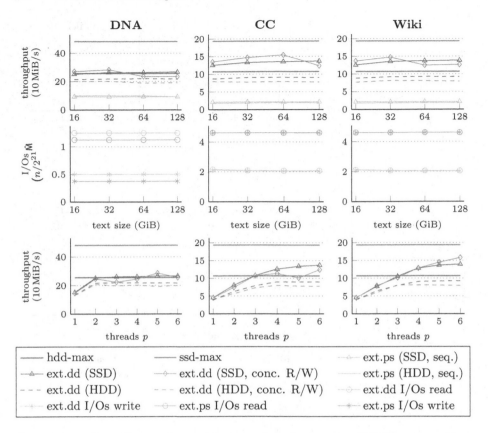

Fig. 8. Throughput (first row) and I/Os (second row) of external WT algorithms. Here, parallel algorithms uses all six threads. Throughput of ext.dd using 20 GiB input per thread (last row). Same experiment for the WM as reported in Fig. 7 for the WT.

the external memory operations conducted by ext.dd. All algorithms have a nearly constant throughput, which is independent of the input size. The same is true for I/Os (both read and write). We also allow ext.dd to read and write concurrently (conc. R/W), which increases the throughput for SSDs on CC for inputs larger than 16 GiB, inputs of size up to 32 GiB on Wiki, and in general on DNA. For HDDs, it reduces the throughput on all text sizes by 4.55 % (DNA) to 11.12 % (Wiki). In the last row of Fig. 7 we show a weak scaling experiment of ext.dd. Using one thread, ext.dd is faster than ext.ps by a factor between 1.64 (DNA) to 2.14 (CC). Hence it is the fastest external memory WT construction algorithm. It also scales reasonably well, achieving a speedup of up to 3.51 (Wiki with conc. R/W). Here, concurrent read and write only increases throughput on Wiki. On DNA, ext.dd does only scale for up to two threads, which is as expected as the number of threads that can efficiently be used is limited by the size of the alphabet. Also, we see using six threads ext.dd's throughput on HDDs is between 17.3 MiB/s (CC) and 38.9 MiB/s (DNA) less than the

Table 1. Characteristics of squential algorithms proposed in this paper.

Name	Time	I/Os	Memory in bits
se.pc	$\mathcal{O}(n \lg \sigma)$	$\mathcal{O}(\mathrm{scan}(n \lceil \lg \sigma \rceil))$	$n \lceil \lg \sigma \rceil + \sigma \lceil \lg n \rceil$
se.ps	$\mathcal{O}(n \lg \sigma)$	$\mathcal{O}(\mathrm{scan}(n \lceil \lg \sigma \rceil))$	$2n \lceil \lg \sigma \rceil + \sigma \lceil \lg n \rceil$
ext.ps (computing WT)	$\mathcal{O}(n \lg \sigma + \sigma)$	$2 \lceil \lg \sigma \rceil \cdot \mathrm{scan}(n \lceil \lg \sigma \rceil)$	$2 \sigma \lceil \lg n \rceil$
ext.ps (computing WM)	$\mathcal{O}(n \lg \sigma)$	$2 \lceil \lg \sigma \rceil \cdot \mathrm{scan}(n \lceil \lg \sigma \rceil)$	$\mathcal{O}(1)$

maximum throughput. Using SSDs, its throughput is between 157.3 MiB/s (Wiki) and 268.0 MiB/s (DNA).

On Shared Memory Wavelet Tree Construction. We have not included the throughput of the currently fastest parallel shared *internal* memory wavelet tree construction algorithm *dd.pc* [5] in any of the plots, due to the huge difference in speed (compared with our semi-external and external memory construction algorithms). For completeness, we now list the throughput of this algorithm on the same hardware and running on six threads. Note that we could only run dd.pc for inputs up to size 4 GiB, as a result of the memory usage of the algorithm.

On DNA, the maximum throughput is 1209.32 MiB/s, on CC it is 431.70 MiB/s, and on Wiki it is 416.26 MiB/s. All these throughputs are more than the theoretical best result an external memory algorithm can achieve on this machine.

6 Conclusion

We presented the fastest semi-external memory WT and WM construction algorithm and the first parallel semi-external memory WT and WM construction algorithm based on the main memory algorithms by Fischer et al. [5]. Then, we showed the first external memory WT and WM construction algorithm. A summary of the characteristics of these sequential algorithms is given in Table 1. In addition, we also parallelized the external memory algorithm. On HDDs, the parallel version of our external memory WT and WM construction achieves nearly perfect throughput, compared to the throughput that we obtain when we read and write the same amount that is read and written during the algorithm. It remains an open problem if there is a parallel algorithm that can obtain the same relative throughput on SSDs.

References

1. Aggarwal, A., Vitter, J.S.: The input/output complexity of sorting and related problems. Commun. ACM **31**(9), 1116–1127 (1988)
2. Babenko, M.A., Gawrychowski, P., Kociumaka, T., Starikovskaya, T.A.: Wavelet trees meet suffix trees. In: 26th Annual ACM-SIAM Symposium on Discrete Algorithms (SODA). pp. 572–591. SIAM (2015)

3. Claude, F., Navarro, G., Pereira, A.O.: The wavelet matrix: an efficient wavelet tree for large alphabets. Inf. Syst. **47**, 15–32 (2015)
4. Dementiev, R., Kettner, L., Sanders, P.: STXXL: standard template library for XXL data sets. Softw. Pract. Exper. **38**(6), 589–637 (2008)
5. Fischer, J., Kurpicz, F., Löbel, M.: Simple, fast and lightweight parallel wavelet tree construction. In: 20th Workshop on Algorithm Engineering and Experiments (ALENEX), pp. 9–20. SIAM (2018)
6. Fuentes-Sepúlveda, J., Elejalde, E., Ferres, L., Seco, D.: Efficient wavelet tree construction and querying for multicore architectures. In: Gudmundsson, J., Katajainen, J. (eds.) SEA 2014. LNCS, vol. 8504, pp. 150–161. Springer, Cham (2014). https://doi.org/10.1007/978-3-319-07959-2_13
7. Fuentes-Sepúlveda, J., Elejalde, E., Ferres, L., Seco, D.: Parallel construction of wavelet trees on multicore architectures. Knowl. Inf. Syst. **51**(3), 1043–1066 (2017)
8. Gog, S., Beller, T., Moffat, A., Petri, M.: From theory to practice: plug and play with succinct data structures. In: Gudmundsson, J., Katajainen, J. (eds.) SEA 2014. LNCS, vol. 8504, pp. 326–337. Springer, Cham (2014). https://doi.org/10.1007/978-3-319-07959-2_28
9. Grossi, R., Gupta, A., Vitter, J.S.: High-order entropy-compressed text indexes. In: 14th Annual ACM-SIAM Symposium on Discrete Algorithms (SODA), pp. 841–850. SIAM (2003)
10. Grossi, R., Vitter, J.S., Xu, B.: Wavelet trees: from theory to practice. In: International Conference on Data Compression, Communications and Processing (CCP), pp. 210–221. IEEE (2011)
11. Kaneta, Y.: Fast wavelet tree construction in practice. In: Gagie, T., Moffat, A., Navarro, G., Cuadros-Vargas, E. (eds.) SPIRE 2018. LNCS, vol. 11147, pp. 218–232. Springer, Cham (2018). https://doi.org/10.1007/978-3-030-00479-8_18
12. Labeit, J., Shun, J., Blelloch, G.E.: Parallel lightweight wavelet tree, suffix array and fm-index construction. J. Discrete Algorithms **43**, 2–17 (2017)
13. Mäkinen, V., Navarro, G.: Position-restricted substring searching. In: Correa, J.R., Hevia, A., Kiwi, M. (eds.) LATIN 2006. LNCS, vol. 3887, pp. 703–714. Springer, Heidelberg (2006). https://doi.org/10.1007/11682462_64
14. Mäkinen, V., Navarro, G.: Rank and select revisited and extended. Theor. Comput. Sci. **387**(3), 332–347 (2007)
15. Makris, C.: Wavelet trees: a survey. Comput. Sci. Inf. Syst. **9**(2), 585–625 (2012)
16. Munro, J.I., Nekrich, Y., Vitter, J.S.: Fast construction of wavelet trees. Theor. Comput. Sci. **638**, 91–97 (2016)
17. Navarro, G.: Compact Data Structures - A Practical Approach. Cambridge University Press, Cambridge (2016)
18. Sedgewick, R.: Algorithms in C - Parts 1–4: Fundamentals, Data Structures, Sorting, Searching, 3rd edn. Addison-Wesley-Longman, Boston (1998)

SACABench: Benchmarking Suffix Array Construction

Johannes Bahne, Nico Bertram, Marvin Böcker, Jonas Bode, Johannes Fischer, Hermann Foot, Florian Grieskamp, Florian Kurpicz[✉], Marvin Löbel, Oliver Magiera, Rosa Pink, David Piper, and Christopher Poeplau

Department of Computer Science, Technische Universität Dortmund, Dortmund, Germany
{johannes.bahne,nico.bertram,marvin.boecker,jonas.bode,johannes.fischer, hermann.foot,florian.grieskamp,florian.kurpicz,marvin.loebel, oliver.magiera,rosa.pink,david.piper,christopher.poeplau}@tu-dortmund.de, johannes.fischer@cs.tu-dortmund.de

Abstract. We present a practical comparison of suffix array construction algorithms on modern hardware. The benchmark is conducted using our new benchmark framework *SACABench*, which allows for an easy deployment of publicly available implementations, simple plotting of the results, and straight forward support to include new construction algorithms. We use the framework to develop a construction algorithm running on the GPU that is competitive with the fastest parallel algorithm in our test environment.

Keywords: Suffix array · Practical survey · Text indexing

1 Introduction

The suffix array (SA) [28] is one of the most versatile and well-researched full-text indices. Given a text T of length n, the SA is the permutation of $[1, n]$, such that $T[\mathsf{SA}[i]..n] < T[\mathsf{SA}[i+1]..n]$ for all $i \in [1, n-1]$, i.e., the starting positions of all suffixes of the text in lexicographical order.

There exist extensive surveys on SA construction algorithms (SACAs), starting with the one by Puglisi et al. [42] and ending currently with the one by Bingmann [4, p. 163–192]. However, none of these surveys address any practical results for SACAs in main memory. There are 24 main memory SACAs that we are aware of. However, not all SACAs have been implemented. It is generally accepted that the *Divsufsort* [12,33] is the fastest SACA—despite it having a superlinear running time. Different models of computation have also been considered for this problem: external memory, e.g., [5,9,18–20,38], shared memory, e.g., [20,25], distributed memory, e.g., [1,6,13,14,20,32,36], and GPGPU, e.g., [10,41,46,47].

In this paper, we first present a practical comparison of SACAs that have a publicly available implementation. This comparison has been conducted using

© Springer Nature Switzerland AG 2019
N. R. Brisaboa and S. J. Puglisi (Eds.): SPIRE 2019, LNCS 11811, pp. 407–416, 2019.
https://doi.org/10.1007/978-3-030-32686-9_29

our new SACA benchmark framework called *SACABench*, which allows for (a) an easy comparison of SACAs including the output of the results (running time and memory peak) in form of raw data in JSON format, as PDF, or LaTeX file, (b) a simple way to include new SACAs, such that the features mentioned before can be used, and (c) fast development of new SACAs due to a variety of building blocks needed for SACAs (such as prefix sorting, renaming techniques, etc.). The framework is available from https://github.com/sacabench/sacabench. It is coded in C++17 and contains 13 SACA implementations, which are to our best knowledge *all* SACAs having a publicly available implementation. See Fig. 1 for a list (and also the historical development) of the SACAs that are included in the framework. We then use the building blocks of SACABench to implement a new GPU-based SACA, which is competitive with the fastest parallel (shared memory) SACA par_DivSufSort [25]. Here, our GPU SACA achieves a speedup between 0.93 and 1.69 compared to par_DivSufSort for inputs fitting into the GPU's memory.

2   SACABench: A Suffix Array Construction Benchmark

In Fig. 1 we give an overview of different SACAs in main memory. There are four general types of SACAs: *Prefix Doubling* algorithms sort the length-2^i prefixes of all suffixes by using the length-2^{i-1} prefixes as keys, and stopping when all considered prefixes are unique. If carefully implemented, this results in a running time of $\mathcal{O}(n \lg n)$. *Induced Copying* algorithms first sample certain suffixes and only sort those suffixes. Based on the sorted sample, the lexicographical order of all other suffixes can be computed in a second phase, which usually has linear running time. Depending on which algorithms are used to sort the sample, induced copying algorithms have either linear or slightly superlinear running time. *Recursive* algorithms reduce the problem size during each recursive step until the problem is trivially solvable (e.g., when all suffixes start with unique characters). They can achieve linear running time and are sometimes used in induced copying algorithms in the first phase (to achieve linear running time). *Grouping* is a new approach somehow similar to induced copying. Here, all suffixes are first grouped together by presorting them according to some prefix (in the only algorithm using grouping [3], Lyndon words determine this prefix). Those groups are then refined using already sorted suffixes, similar to induced copying algorithms.

2.1   Experimental Setup

We conducted our experiments on a computer with two Intel E5-2640v4 (10 physical cores, Hyper-Threading is disabled (per default on the cluster that can not be changed by users), with frequencies up to 3.4 GHz, and cache sizes of 320 KiB (L1I and L1D), 2.5 MiB (L2) and 25 MiB (L3)), one NVidia Tesla K40 graphics card (2880 stream processors with frequencies up to 875 MHz and 12 GB GDDR5 SDRAM) and 64 GB of RAM. We compiled the code using g++ 8.3.0 and compiler flags -O3 and -march=native. Note that *Cilk* support was removed from g++ 8.0.0. Hence, we use *OpenMP* to express parallelism.

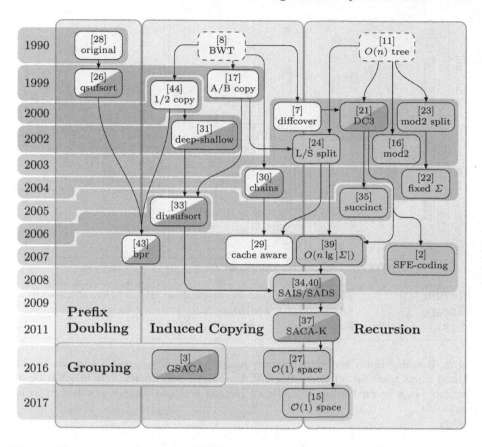

Fig. 1. Historical development of SACAs in main memory (enhanced and updated, based on [4,42]). For each algorithm, we cite its most recent publication, and the years on the left hand side show the year of its first publication. In some cases these years may not match, e.g., due to a later journal publication. SACAs are marked with a grey background (⬤), if they have linear running time, and a partly brown background (⬤), if an implementation is publicly available. All of the latter are also part of SACABench.

2.2 Evaluation of Sequential Suffix Array Construction Algorithms

For the evaluation of the sequential SACAs we use 1600 MiB prefixes of three texts. Note that we encode each symbol of the text using one byte, as this is required by most implementations. *1000G* ($\sigma = 4$, avg_lcp = 24, max_lcp = 353), which is a concatenation of DNA sequences provided by the 1000 Genomes Project (https://internationalgenome.org). We removed every character but A, C, G, and T. *CommonCrawl* ($\sigma = 242$, avg_lcp = 3,995, max_lcp = 605,632), which is a crawl of the web done by the CommonCrawl Corpus (http://commoncrawl.org) without any HTML tags. Here, we also removed all annotations added. Last, *Wiki* ($\sigma = 209$, avg_lcp = 32, max_lcp = 25,063), which is a concatenation of recent Wikipedia dumps in XML format (https://dumps.wikimedia.org).

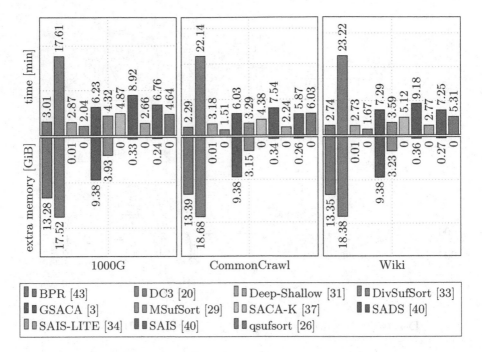

Fig. 2. Running times and extra memory usage (memory required in addition to the SA and input text) for all sequential SACAson real world inputs. The LATEX code of the plot was generated using SACABench (legend and size slightly modified to fit in this layout).

Here, max_lcp denotes the maximum size of a common prefix of two suffixes that are consecutive in the SA, and avg_lcp is the average of if all these sizes (rounded down).

We used this set of texts, as more popular corpora (e.g., Pizza & Chili http://pizzachili.dcc.uchile.cl or the Lightweight corpus http://people.unipmn. it/manzini/lightweight) do only contain one files larger than 1600 MiB and we want to test on larger inputs.

In addition, we also tested the algorithms on highly repetitive texts that are available from the Pizza & Chili corpus, as some suffix array construction algorithms behave differently on this kind of input. To be precise, we use *Cere* ($\sigma = 5, n = 461,286,644, \text{avg\_lcp} = 7,066, \text{max\_lcp} = 303,204$), *Einstein.en.txt* ($\sigma = 139, n = 467,626,544, \text{avg\_lcp} = 59,074, \text{max\_lcp} = 935,920$), and *Para* ($\sigma = 5, n = 429,265,758, \text{avg\_lcp} = 3,273, \text{max\_lcp} = 104,177$).

Running time and memory usage are automatically measured by the framework for each included algorithm. To this end, we use the timing functionality of C++ and have overwritten the malloc, realloc, and free functions to track the memory usage of all components and also already coded algorithms.

The running times and the additional memory required are shown in Fig. 2. It is easy to see that DivSufSort is the fastest sequential SACA running in main

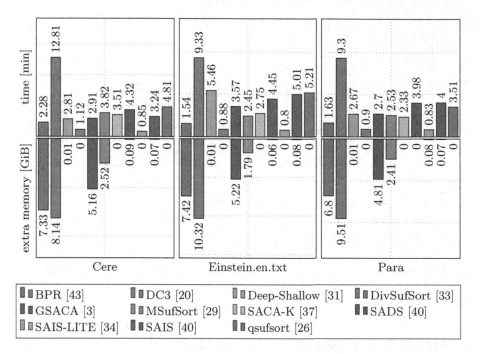

Fig. 3. Running times and extra memory usage (memory required in addition to the SA and input text) for all sequential SACAs on highly repetitive inputs. The LATEX code of the plot was generated using SACABench (legend and size slightly modified to fit in this layout).

memory on all input texts. Also, it is among the SACAs that require nearly no memory in addition to the space for the SA and the input text. Overall, DivSufSort is 1.3, 1.61, and 1.63 times faster than the second fastest SACA on DNA, CommonCrawl, and Wiki. SAIS-LITE, which also does not require additional memory, is the second fastest SACA on DNA and CommonCrawl. It is noteworthy that both DivSufSort and SAIS-LITE have been coded by Yuta Mori. On Wiki, Deep Shallow is the second fastest SACA, but it is just 0.01 s faster than BPR and 0.04 s faster than SAIS-LITE. Those two SACAs (BPR and Deep Shallow) are also the third and fourth fastest algorithm on CommonCrawl and the fourth and third fastest on DNA. BPR is the only algorithm among the fast ones that requires an extensive amount of additional memory. More than 13 GiB for an input of size 1600 MiB. BPR, DC3, GSACA, and MSufSort require more additional memory than the size of the input.

For the highly repetitive texts, we have similar results regarding the running time and memory peaks. We show the results of our experiments in Fig. 3. Surprisingly, SAIS-LITE is faster than DivSufSort on this kind of inputs. On Cere it is 24.11% faster, on Einstein.en.txt it is 9.1% faster, and on Para it is 8.78% faster. All this while requiring the same memory as DivSufSort.

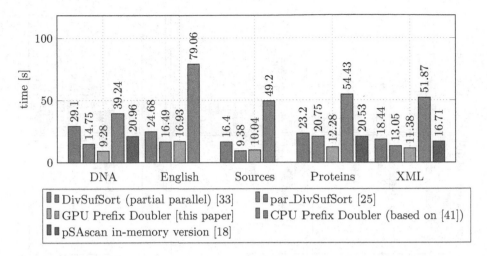

Fig. 4. Running time of the parallel (shared memory on 20 cores and GPU) SACAs on 200 MiB texts. The LATEX code of the plot was generated using SACABench (legend and size slightly modified to fit in this layout).

3   Suffix Array Construction on the GPU

Next to the well tuned SACAs compared above, SACABench also contains many experimental SACA implementations. The best performing one is a parallel prefix doubling algorithm that runs on GPUs and is based on Osipov's GPU SACA [41, p. 44–51]. The main idea is similar to the general *prefix doubling* approach as used by Manber/Myers [28] and Larsson/Sadakane [26]. In iteration i, we consider the length-2^i prefixes of all suffixes and group equal (using the prefix as key) suffixes together into *buckets*. To refine the new buckets in the next iteration, we use the bucket numbers of the suffixes starting 2^{i-1} text positions to the right. This allows us to compute the new buckets without additional access to the text. A bucket is *sorted* if it contains only a single suffix. Larsson and Sadakane [26] added a clever mechanism to ignore already sorted buckets, which is a practical improvement. However, this can lead to load imbalance when parallelizing the algorithm. Our implementation combines techniques from both approaches such that sorted buckets can be ignored, but load imbalance is avoided by marking sorted groups and making heavy use of parallel prefix sums to compute the number of smaller groups for each group.

The prefix doubling technique has proven to be effective in other models of computation, e.g., distributed memory [6,13,14] and external memory [9].

3.1   Evaluation of Parallel Suffix Array Construction Algorithms

We compare our GPU SACA with three shared memory parallel SACAs. We could not compare our algorithm with the most recent GPU-algorithm by Wang et al. [47], we could only run it successfully for inputs smaller than 100 KiB for

our text collection. (To test their code, they use somehow meaningless *random* input texts, which we could get to work in our test environment for sizes up to 170 MiB. However, even if we reduced the alphabet size of our real world texts to match the alphabet size of the random texts, we could not get this algorithm to work with inputs larger than 100 KiB.) Likewise, we could not compare against Osipov's CPU-SACA [41], as it does not have publicly available code and the author seems to be have left research and did not reply to our code requests. We are also aware of *parallelKS, parallelRange* that are available from the *Problem Based Benchmark Suite* [45], however we were not able to make them compute the correct suffix array on short notice for the final version of this paper.

As inputs we use the *Pizza & Chili* corpus, as it offers a variety of smaller text that have size at least 200 MiB: DNA ($\sigma = 16, \text{max\_lcp} = 14,836$), English ($\sigma = 225, \text{max\_lcp} = 109,394$), Sources ($\sigma = 230, \text{max\_lcp} = 71,651$), Proteins ($\sigma = 25, \text{max\_lcp} = 45,704$), and (dblp.)XML ($\sigma = 96, \text{max\_lcp} = 1,084$). More characteristics of the texts are available from http://pizzachili.dcc.uchile. cl. Again, max_lcp denotes the maximum size of a common prefix of two suffixes that are consecutive in the SA. We only use inputs of 200 MiB due to the memory requirements of our algorithm. On the given hardware it cannot compute the suffix array for larger inputs.

The results of our experiments are shown in Fig. 4, where *par_DivSufSort* denotes the *fully* parallel version of DivSufSort by Labeit et al. [25]. The *partially parallel* DivSufSort is Mori's [33] implementation of DivSufSort, where only the first phase is be parallelized. The *GPU Prefix Doubler* is the algorithm presented in this paper. The *CPU Prefix Doubler* is the same as the GPU one but it only uses the CPU, which we included as sanity check to see the speedup of the GPU. The running time of prefix doubling SACAs is $\mathcal{O}(n \lg \text{max\_lcp})$.

Our new algorithm is the fastest on DNA, Proteins, and XML, where max_lcp is comparatively small. Here, we are 1.58 (DNA), 1.69 (Proteins), and 1.15 (XML) times faster than par_DivSufSort. On English and Sources, par_DivSufSort is 1.03 (English) and 1.07 (Sources) times faster than our GPU Prefix Doubler on inputs with large max_lcp. Hence, it is only slightly faster.

We also included the in-memory version of the external memory suffix array construction algorithm *pSAscan* [18] in our framework. The available implementation could not handle all inputs by design, as it cannot handle text that contain the character 255, which occurs in English and Sources. It is 2.25 times, 1.67 times, and 1.46 times slower than our GPU Prefix Doubler.

Although a fair comparison against [41] is difficult due to the problems mentioned above, we hypothesize the following: Osipov [41] used an NVidia Fermi GTX 480 graphics cards with 480 and 1.5 GB RAM and an Intel i7 920 CPU with 4 cores and frequencies up to 2.93 GHz, where they achieved a speedup of at most 5.8 against partially sequential DivSufSort in the best case, but often a speedup of only around 2.5. Our speedup against the partial parallelized Div-SufSort. varies between 1.45 (English) and 3.14 (DNA). Given that the ratio between the GPU and CPU cores is nearly the same in both setups (120:1 in their experiment and 144:1 in ours), but that our CPU cores have a higher frequency, we speculate that our implementation is of similar speed as Osipov's original one.

4   Conclusion

We presented a framework for SACAs that allows for an easy comparison of SACAs regarding time and memory consumption during construction. The result of this comparison is an empirical proof that *DivSufSort* is still the fastest SACA. It also has (in practice) optimal space requirements, as the additional memory only depends on the size of the alphabet. In addition, new algorithms can effortless be included in the framework allowing all features of the framework to be used. We also presented a GPU SACA that is the fastest parallel SACA, but is limited by the memory size of the graphics card, and part of the framework.

Recently, linear time SACAs that require only a constant number of computer words in addition to SA and the input text have been presented [15,27], which is optimal. Now, the only open question regarding SACAs in main memory is: is there a SACA faster than DivSufSort, which is the fastest since 2006? And if there is a faster algorithm than SAIS-LITE for highly repetitive texts, as it is even faster than DivSufSort on those.

Acknowledgment. We would like to thank the anonymous reviewer who pointed us to additional parallel suffix array construction algorithms that we had not previously included in the framework.

References

1. Abdelhadi, A., Kandil, A., Abouelhoda, M.: Cloud-based parallel suffix array construction based on MPI. In: Middle East Conference on Biomedical Engineering (MECBME), pp. 334–337. IEEE (2014)
2. Adjeroh, D.A., Nan, F.: Suffix sorting via Shannon-Fano-Elias codes. In: Data Compression Conference (DCC), p. 502. IEEE (2008)
3. Baier, U.: Linear-time suffix sorting - a new approach for suffix array construction. In: 27th Annual Symposium on Combinatorial Pattern Matching (CPM). LIPIcs, vol. 54, pp. 23:1–23:12. Schloss Dagstuhl – Leibniz Center for Informatics (2016)
4. Bingmann, T.: Scalable string and suffix sorting: algorithms, techniques, and tools. Ph.D. thesis, Karlsruhe Institute of Technology, Germany (2018). https://doi.org/10.5445/IR/1000085031
5. Bingmann, T., Fischer, J., Osipov, V.: Inducing suffix and LCP arrays in external memory. ACM J. Exp. Algorithmics **21**(1), 2.3:1–2.3:27 (2016)
6. Bingmann, T., Gog, S., Kurpicz, F.: Scalable construction of text indexes with thrill. In: IEEE International Conference on Big Data, pp. 634–643. IEEE (2018)
7. Burkhardt, S., Kärkkäinen, J.: Fast lightweight suffix array construction and checking. In: Baeza-Yates, R., Chávez, E., Crochemore, M. (eds.) CPM 2003. LNCS, vol. 2676, pp. 55–69. Springer, Heidelberg (2003). https://doi.org/10.1007/3-540-44888-8_5
8. Burrows, M., Wheeler, D.J.: A block-sorting lossless data compression algorithm. Technical report, Digital Equipment Corporation (1994)
9. Dementiev, R., Kärkkäinen, J., Mehnert, J., Sanders, P.: Better external memory suffix array construction. ACM J. Exp. Algorithmics **12**, 3.4:1–3.4:24 (2008)

10. Deo, M., Keely, S.: Parallel suffix array and least common prefix for the GPU. In: ACM SIGPLAN Symposium on Principles and Practice of Parallel Programming (PPoPP), pp. 197–206. ACM (2013)

11. Farach, M.: Optimal suffix tree construction with large alphabets. In: 38th IEEE Annual Symposium on Foundations of Computer Science (FOCS), pp. 137–143. IEEE (1997)

12. Fischer, J., Kurpicz, F.: Dismantling DivSufSort. In: Prague Stringology Conference (PSC), pp. 62–76. Department of Theoretical Computer Science, Faculty of Information Technology, Czech Technical University in Prague (2017)

13. Fischer, J., Kurpicz, F.: Lightweight distributed suffix array construction. In: 21st Workshop on Algorithm Engineering and Experiments (ALENEX), pp. 27–38. SIAM (2019)

14. Flick, P., Aluru, S.: Parallel distributed memory construction of suffix and longest common prefix arrays. In: International Conference for High Performance Computing, Networking, Storage and Analysis (SC), pp. 16:1–16:10. ACM (2015)

15. Goto, K.: Optimal time and space construction of suffix arrays and LCP arrays for integer alphabets. CoRR arXiv:1703.01009 (2017)

16. Hon, W., Sadakane, K., Sung, W.: Breaking a time-and-space barrier in constructing full-text indices. SIAM J. Comput. **38**(6), 2162–2178 (2009)

17. Itoh, H., Tanaka, H.: An efficient method for in memory construction of suffix arrays. In: 6th International Symposium on String Processing and Information Retrieval (SPIRE), pp. 81–88. IEEE (1999)

18. Kärkkäinen, J., Kempa, D., Puglisi, S.J.: Parallel external memory suffix sorting. In: Cicalese, F., Porat, E., Vaccaro, U. (eds.) CPM 2015. LNCS, vol. 9133, pp. 329–342. Springer, Cham (2015). https://doi.org/10.1007/978-3-319-19929-0_28

19. Kärkkäinen, J., Kempa, D., Puglisi, S.J., Zhukova, B.: Engineering external memory induced suffix sorting. In: 19th Workshop on Algorithm Engineering and Experiments (ALENEX), pp. 98–108. SIAM (2017)

20. Kärkkäinen, J., Sanders, P.: Simple linear work suffix array construction. In: Baeten, J.C.M., Lenstra, J.K., Parrow, J., Woeginger, G.J. (eds.) ICALP 2003. LNCS, vol. 2719, pp. 943–955. Springer, Heidelberg (2003). https://doi.org/10.1007/3-540-45061-0_73

21. Kärkkäinen, J., Sanders, P., Burkhardt, S.: Linear work suffix array construction. J. ACM **53**(6), 918–936 (2006)

22. Kim, D.K., Jo, J., Park, H.: A fast algorithm for constructing suffix arrays for fixed-size alphabets. In: Ribeiro, C.C., Martins, S.L. (eds.) WEA 2004. LNCS, vol. 3059, pp. 301–314. Springer, Heidelberg (2004). https://doi.org/10.1007/978-3-540-24838-5_23

23. Kim, D.K., Sim, J.S., Park, H., Park, K.: Constructing suffix arrays in linear time. J. Discrete Algorithms **3**(2–4), 126–142 (2005)

24. Ko, P., Aluru, S.: Space efficient linear time construction of suffix arrays. J. Discrete Algorithms **3**(2–4), 143–156 (2005)

25. Labeit, J., Shun, J., Blelloch, G.E.: Parallel lightweight wavelet tree, suffix array and FM-index construction. J. Discrete Algorithms **43**, 2–17 (2017)

26. Larsson, N.J., Sadakane, K.: Faster suffix sorting. Theor. Comput. Sci. **387**(3), 258–272 (2007)

27. Li, Z., Li, J., Huo, H.: Optimal in-place suffix sorting. In: Data Compression Conference (DCC), p. 422. IEEE (2018)

28. Manber, U., Myers, E.W.: Suffix arrays: a new method for on-line string searches. SIAM J. Comput. **22**(5), 935–948 (1993)

29. Maniscalco, M.A., Puglisi, S.J.: An efficient, versatile approach to suffix sorting. ACM J. Exp. Algorithmics **12**, 1.2:1–1.2:23 (2007)
30. Manzini, G.: Two space saving tricks for linear time LCP array computation. In: Hagerup, T., Katajainen, J. (eds.) SWAT 2004. LNCS, vol. 3111, pp. 372–383. Springer, Heidelberg (2004). https://doi.org/10.1007/978-3-540-27810-8_32
31. Manzini, G., Ferragina, P.: Engineering a lightweight suffix array construction algorithm. Algorithmica **40**(1), 33–50 (2004)
32. Metwally, A.A., Kandil, A.H., Abouelhoda, M.: Distributed suffix array construction algorithms: comparison of two algorithms. In: Cairo International Biomedical Engineering Conference (CIBEC), pp. 27–30. IEEE (2016)
33. Mori, Y.: DivSufSort (2006). https://github.com/y-256/libdivsufsort
34. Mori, Y.: SAIS (2008). https://sites.google.com/site/yuta256/sais
35. Na, J.C.: Linear-time construction of compressed suffix arrays using $o(n \log n)$-bit working space for large alphabets. In: Apostolico, A., Crochemore, M., Park, K. (eds.) CPM 2005. LNCS, vol. 3537, pp. 57–67. Springer, Heidelberg (2005). https://doi.org/10.1007/11496656_6
36. Navarro, G., Kitajima, J.P., Ribeiro-Neto, B.A., Ziviani, N.: Distributed generation of suffix arrays. In: Apostolico, A., Hein, J. (eds.) CPM 1997. LNCS, vol. 1264, pp. 102–115. Springer, Heidelberg (1997). https://doi.org/10.1007/3-540-63220-4_54
37. Nong, G.: Practical linear-time O(1)-workspace suffix sorting for constant alphabets. ACM Trans. Inf. Syst. **31**(3), 15 (2013)
38. Nong, G., Chan, W.H., Hu, S.Q., Wu, Y.: Induced sorting suffixes in external memory. ACM Trans. Inf. Syst. **33**(3), 12:1–12:15 (2015)
39. Nong, G., Zhang, S.: Optimal lightweight construction of suffix arrays for constant alphabets. In: Dehne, F., Sack, J.-R., Zeh, N. (eds.) WADS 2007. LNCS, vol. 4619, pp. 613–624. Springer, Heidelberg (2007). https://doi.org/10.1007/978-3-540-73951-7_53
40. Nong, G., Zhang, S., Chan, W.H.: Two efficient algorithms for linear time suffix array construction. IEEE Trans. Comput **60**(10), 1471–1484 (2011)
41. Osipov, V.: Parallel suffix array construction for shared memory architectures. In: Calderón-Benavides, L., González-Caro, C., Chávez, E., Ziviani, N. (eds.) SPIRE 2012. LNCS, vol. 7608, pp. 379–384. Springer, Heidelberg (2012). https://doi.org/10.1007/978-3-642-34109-0_40
42. Puglisi, S.J., Smyth, W.F., Turpin, A.H.: A taxonomy of suffix array construction algorithms. ACM Comput. Surv. **39**(2) (2007). Article No. 4
43. Schürmann, K., Stoye, J.: An incomplex algorithm for fast suffix array construction. Softw. Pract. Exp. **37**(3), 309–329 (2007)
44. Seward, J.: On the performance of BWT sorting algorithms. In: Data Compression Conference (DCC), pp. 173–182. IEEE (2000)
45. Shun, J., et al.: Brief announcement: The problem based benchmark suite. In: 24th ACM Symposium on Parallelism in Algorithms and Architectures (SPAA), pp. 68–70. ACM (2012)
46. Sun, W., Ma, Z.: Parallel lexicographic names construction with CUDA. In: 15th IEEE International Conference on Parallel and Distributed Systems (ICPADS), pp. 913–918. IEEE (2009)
47. Wang, L., Baxter, S., Owens, J.D.: Fast parallel skew and prefix-doubling suffix array construction on the GPU. Concurr. Comput. Pract. Exp. **28**(12), 3466–3484 (2016)

Compressed Data Structures

Faster Dynamic Compressed d-ary Relations

Diego Arroyuelo[1,2], Guillermo de Bernardo[3(✉)], Travis Gagie[4],
and Gonzalo Navarro[1,5]

[1] Millennium Institute for Foundational Research on Data (IMFD), Santiago, Chile
[2] Department of Informatics, Universidad Técnica Federico Santa María,
Santiago, Chile
darroyue@inf.utfsm.cl
[3] Universidade da Coruña, Centro de investigación CITIC, A Coruña, Spain
gdebernardo@udc.es
[4] Faculty of Computer Science, Dalhousie University, Halifax, Canada
travis.gagie@gmail.com
[5] Department of Computer Science, University of Chile, Santiago, Chile
gnavarro@dcc.uchile.cl

Abstract. The k^2-tree is a successful compact representation of binary
relations that exhibit sparseness and/or clustering properties. It can be
extended to d dimensions, where it is called a k^d-tree. The representation
boils down to a long bitvector. We show that interpreting the k^d-tree as a
dynamic trie on the Morton codes of the points, instead of as a dynamic
representation of the bitvector as done in previous work, yields operation
times that are below the lower bound of dynamic bitvectors and offers
improved time performance in practice.

1 Introduction

The k^2-tree [14] is a compact data structure conceived to represent the adjacency
matrix of Web graphs, but its functionality was later extended to represent other
kinds of d-ary relations such as ternary relations [1], point grids [12], raster data
[9], RDF stores [2], temporal graphs [15], graph databases [3], etc.

The k^2-tree compactly represents an extension of a variant of the Quadtree
data structure [20], more precisely of the MX-Quadtree [25, Section 1.4.2.1]. The
MX-Quadtree splits the $n \times n$ grid into four submatrices of $n/2 \times n/2$ cells.
The root indicates which of the submatrices are nonempty of points, and a child
of the root recursively represents each nonempty submatrix. In the k^2-tree, the

Funded by the Millennium Institute for Foundational Research on Data (IMFD),
Chile and by European Union's Horizon 2020 research and innovation programme
under the Marie Sklodowska-Curie grant agreement No 690941 (project BIRDS). GdB
funded by Xunta de Galicia/FEDER-UE GEMA: IN852A 2018/14, CSI:ED431G/01
and GRC:ED431C 2017/58; by MINECO-AEI/FEDER-UE TIN2016-77158-C4-3-R
and TIN2015-69951-R; and by MICINN RTC-2017-5908-7.

© Springer Nature Switzerland AG 2019
N. R. Brisaboa and S. J. Puglisi (Eds.): SPIRE 2019, LNCS 11811, pp. 419–433, 2019.
https://doi.org/10.1007/978-3-030-32686-9_30

matrix is instead split into k^2 submatrices of $n/k \times n/k$ cells. In d dimensions, the structure becomes a k^d-tree, where the grid is divided into k^d submatrices of $n/k \times \cdots \times n/k$ cells. The height of the k^d-tree is then $\log_{k^d}(n^d) = \log_k n$.

Instead of using pointers to represent the tree topology, the k^d-tree uses a long bitvector $B[1..N]$, where each node stores only k^d bits indicating which of its submatrices are nonempty, and all the node bitvectors are concatenated level-wise into B. Bitvector B supports navigation towards children and parents in $O(1)$ time [14] by means of rank/select operations [16,21] on bitvector B. Query operations like retrieving all the neighbors or the reverse neighbors of a node (when representing graphs) or retrieving all the points in a range (when representing grids) then translate into traversals on the k^d-tree [14].

In various applications one would like the relations to be *dynamic*, that is, elements (graph edges, grid points) can be inserted and deleted from the relation. Each such update requires flipping bits or inserting/deleting chunks of k^d bits at each of the $\log_k n$ levels in B. Such operations can be supported using a dynamic bitvector representation [13]. There exists, however, an $\Omega(\log N/\log\log N)$ lower bound to support updates and rank/select operations on a bitvector of length N [17], and such slowdown factor multiplies every single operation carried out on the bitvector, both for traversals and for updates.

In this paper we take a different view of the k^d-tree representation. We regard the k^d-ary tree as a trie on the Morton codes [20] of the elements stored in the grid. The Morton code (in two dimensions, but the extension is immediate) is the concatenation of the $\log_k n$ identifiers of the consecutive subgrids chosen by a point until it is inserted at the last level. We then handle a trie of strings of length $\log_k n$ over an alphabet of size k^d. While such a view yields no advantage in the static case, it provides more efficient implementations in the dynamic scenario. For example, a succinct dynamic trie [4] on the Morton codes requires space similar to our bitvector representation, but it is much faster in supporting the operations: $o(d\log k)$ time, and constant for practical values of d and k.

In this paper we implement this idea and show that it is not only theoretically appealing but also competitive in practice with the preceding dynamic-bitvector-based representation [13]. In our way, we define a new depth-first deployment for tries that, unlike the level-wise one [14], cannot be traversed in constant time per edge. Yet, we show it turns out to be convenient in a dynamic scenario because we have to scan only small parts of the representation.

2   The k^2-Tree and Its Representation as a Trie

Let us focus on the case $k = 2$ and $d = 2$ for simplicity; $d = 2$ encompasses all the applications where we represent graphs, and the small value of k is the most practical in many cases. Given p points in an $n \times n$ matrix M, the k^2-tree is a k^2-ary (i.e., 4-ary) tree where each node represents a submatrix. Assume n is a power of k (i.e., of 2) for simplicity. The root then represents the whole matrix $M[0..n-1, 0..n-1]$. Given a node representing a submatrix $M[r_1..r_2, c_1..c_2]$, its 4 children represent the submatrices $M[r_1..r_m, c_1..c_m]$ (top-left), $M[r_1..r_m, c_m +$

$1..c_2]$ (top-right), $M[r_m+1..r_2, c_1..c_m]$ (bottom-left), and $M[r_m+1..r_2, c_m+1..c_2]$ (bottom-right), in that order, where $r_m = (r_1+r_2-1)/2$ and $c_m = (c_1+c_2-1)/2$. Each of the 4 submatrices of a node may be empty of points, in which case the node does not have the corresponding child. The node stores 4 bits indicating with a 1 that the corresponding matrix is nonempty, or with a 0 that it is empty. The k^2-tree is of height $\log_k n = \log_2 n$. The p matrix points correspond to the leaves marked with 1 at depth $\log_k n$. See Fig. 1.

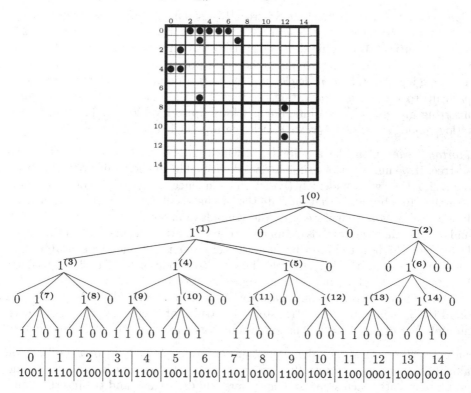

Fig. 1. Binary relation for the set $\{(0,2), (0,3), (0,4), (0,5), (0,6), (1,3), (1,7), (2,1), (4,0), (4,1), (7,3), (8,12), (11,12)\}$ (on top). The corresponding k^2-tree (in the middle) with level-order numbers shown in parentheses above each node, and its levelwise representation (on the bottom), again with level-order numbers above each node.

Succinct Representation. A simplified description of the compact k^2-tree representation [14] consists of a bitvector B where the tree is traversed levelwise, left to right, and the $k^2 = 4$ bits of all the nodes are concatenated. This is similar to a LOUDS tree representation [8,18], which is also computed via a BFS traversal. Then, if the tree has v nodes, the bitvector B is of length $k^2v = 4v$, $B[1..4v]$. Note that the nodes of depth $\log_k n = \log_2 n$ correspond to 4 cells, and therefore it is sufficient to store their 4 bits; their children are not represented. Given p points, the number of nodes of the k^2-tree is $v \leq p\log_4(n^2/p) + O(p)$ [22, Sec. 9.2].

Each k^2-tree node is identified by the position of the first of the 4 bits that describes its empty/nonempty children. To move from a node i to its t-th child $(0 \leq t \leq 3)$, the formula is simply $4 \cdot rank_1(B, i + t) + 1$, where $rank_1(B, i)$ counts the number of 1s in $B[1..i]$ and can be computed in $O(1)$ time using $o(v)$ space on top of B [16]. For example, we determine in $O(\log_k n)$ time whether a certain point exists in the grid. Other operations require traversal of selected subtrees [14].

Dynamic k^2-Trees. A dynamic k^2-tree [13] is obtained by representing B as a dynamic bitvector. Now operation $rank$ takes time $O(\log v / \log \log v)$ [23], which is optimal [17]. This slows down the structure with respect to the static variant. For example, determining whether a point exists takes time $O(\log_k n \cdot \log v / \log \log v) \subseteq O(\log^2 n / \log \log n)$. To insert a point (r, c), we must create its path up to the leaves, converting the first 0 in the path to a 1 and thereafter inserting groups of $k^2 = 4$ bits, one per level up to level $\log_k n$. This takes time $O(\log_k n \cdot \log v / \log \log v)$ as well. Deleting a point is analogous.

Morton Codes. Consider a point (r, c), which induces a root-to-leaf path in the k^2-tree. If we number the 4 submatrices described in the beginning of this section as $0,1,2,3$, then we can identify (r, c) with a sequence of $\log_4(n^2) = \log_2 n$ symbols over the alphabet $[0..3]$ that indicate the submatrix chosen by (r, c) at each level. In particular, note that if we write the symbols in binary, $0 = 00, 1 = 01, 2 = 10$, and $3 = 11$, then the row r is obtained by concatenating the first bits of the $\log_2 n$ levels, from highest to lowest bit, and the column c is obtained by concatenating the second bits of the $\log_2 n$ levels. The Morton code of (r, c) is then obtained by interlacing the bits of the binary representations of r and c.

As a consequence, we can regard the k^2-tree as the trie of the Morton codes of all the p points, that is, a trie storing p strings of length $\log_k n = \log_2 n$ over an alphabet of size $k^2 = 4$. The extension to general values of k^d is immediate.

Succinct Tries. A recent dynamic representation [4] of tries of v nodes over alphabet $[0..\sigma - 1]$ requires $v(2 + \log_2 \sigma) + o(v \log \sigma)$ bits. If σ is polylogarithmic in v, it simulates each step of a trie traversal in $O(1)$ time, and the insertion and deletion of each trie node in $O(1)$ amortized time. Used on our Morton codes, with alphabet size $\sigma = k^2 = 4$, the tries use $v(2 + 2\log_2 k) + o(v) = 4v + o(v)$ bits, exactly as the representation using the bitvector B. Instead, they support queries like whether a given point exists in time $O(\log_k n)$, and inserting or deleting a point in amortized time $O(\log_k n)$, way faster than on the dynamic bitvector B.

The General Case. With larger values of k and d, B requires $k^d v$ bits, and it may become sparse. By using sparse bitvector representations [24], the space becomes $O(p \log(n^d/p) + pd \log k)$ bits [22, Sec. 9.2], but the time of operation $rank$ becomes $O(d \log k)$, and this time penalty factor multiplies all the other operations. A dynamic representation of the compressed bitvector [23] uses the same space and requires $O(\log v / \log \log v)$ time for each operation. The space usage of the trie [4] on a general alphabet of size $\sigma = k^d$ is of the same order, $O(p \log(n^d/p) + pd \log k)$ bits, but the operations are supported in less time,

$O(\log \sigma / \log \log \sigma) = O(\log(k^d) / \log \log(k^d)) = O(d \log k / \log(d \log k))$ (amortized for updates). The insertion or deletion of a point, which affects $\log_k n$ tree edges, then requires $O(d \log n / \log(d \log k))$ amortized time. We state this simple result as a theorem.

Theorem 1. *A dynamic k^d-tree can represent p points on an n^d-size grid within $O(p \log(n^d/p) + pd \log k)$ bits, while supporting the traversal, insertion, or deletion of each tree edge in time $O(d \log k / \log(d \log k))$ (amortized for updates). If $k^d = O(\mathrm{polylog}\, p)$, then the times are $O(1)$ (also amortized for updates).*

3   Implementation of the Dynamic Trie

We now define a practical implementation of succinct dynamic tries, for the particular case of k^2-trees with $k = 2$. The whole trie is divided into blocks, each being a connected component of the trie. A block can have child blocks, so we can say that the trie is represented as a tree of blocks. Blocks will be of variable size. Let us define block sizes $N_1 < N_2 < \cdots < N_{max}$, such that $N_i = N_{i-1}/\alpha$, for $i = 2, \ldots, max$, for a given parameter $0 < \alpha < 1$, and $N_{max} = 4 \cdot N_1$. At any given time, a block B of size N_i is able to store at most N_i nodes. If new nodes are added to B such that the number of nodes exceeds N_i, then B is enlarged to have size N_j, for $j > i$, such that the new nodes can be stored. By defining the block sizes N_i as we do, we ensure that the fill ratio of each block is at least $1 - \alpha$ [5]; for example, if $\alpha = 0.05$, then every block is at least 95% full, which means that the space wasted is at most 5%.

Each block B stores the following components:

- T_B: the tree topology of the connected component represented by the block. Every node in the trie is either an internal node, a leaf node, or a *frontier* node in some T_B. The latter are seen as leaves in T_B, but they correspond to trie nodes whose subtree is stored in a descendant block. We mark such nodes in B and store a pointer to the corresponding child block, see next.
- F_B: a sorted array storing the preorder numbers of the frontier nodes.
- P_B: an array with the pointers to children blocks, in the same order of F_B.
- d_B: the depth (in the trie) of the root of T_B.

Unlike the classical k^2-tree representation [13,14], which deploys the nodes levelwise, we represent the tree topology T_B in depth-first order. This order is compatible with our block layout and speeds up the insertion of points, since the bits of all the edges to insert or remove are contiguous.

Representation. In T_B, each node is encoded using 4 bits, indicating which of its children are present. For instance, '0110' encodes a node that has two children, labeled by symbols 1 and 2. Therefore, the total number of bits used to encode the trees T_B is *exactly the same* as in the classical representations [13,14].

We store T_B using a simple array able to hold N_i nodes. A node is identified by its index within this array. Figure 2 shows an example top block for the k^2-tree of Fig. 1 and our array-based depth-first representation. Depth-first numbers are

shown along each node; these are also their indexes in the array storing T_B. In the example, nodes with depth-first number 2 and 3 are frontier nodes; they are underlined in the array representation.

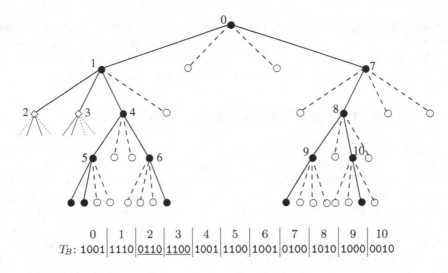

0	1	2	3	4	5	6	7	8	9	10

T_B: 1001 1110 <u>0110</u> <u>1100</u> 1001 1100 1001 0100 1010 1000 0010

Fig. 2. Example block of a k^2-tree and its depth-first representation. Depth-first numbers are shown along with each node, and they correspond with the index in the array representation. Nodes with numbers 2 and 3 (underlined in T_B) are frontier nodes.

Apart from T_B, each block B then requires 3 words to store d_B and its corresponding entries in the arrays $F_{B'}$ and $P_{B'}$ in its parent block B'. This implies a maximum overhead of $O(\log(n)/N_1)$ bits per node, assuming pointers of $\Theta(\log n)$ bits as in the transdichotomous RAM model of computation. Thus we have to choose $N_1 = \omega(\log N)$ for this overhead to be $o(n)$.

The depth-first order we use, however, corresponds more to the DFUDS representation [8], whereas the classical levelwise deployment is analogous to a LOUDS representation [18]. An important difference is that, whereas the fixed-arity variant of LOUDS is easy to traverse in constant time per edge, the DFUDS representation requires more space [8,22]: apart from the 4 bits, each node with c children uses $c + 1$ bits to mark its number of children.

As a consequence, our actual storage format cannot be traversed in constant time per edge. Rather, we will traverse the blocks sequentially and carry out all the edge traversals or updates on the block in a single left-to-right pass. This is not only cache-friendly, but convenient because we do not need to store nor recompute any sublinear-space data structure to speed up traversals [16].

A complication related to our format is that, when traversing the tree, we must maintain the current trie depth in order to identify the leaves (these are always at depth $\log_2 n$). Besides, as we traverse the block we must be aware of which are the frontier nodes, so as to skip them in the current block or switch to another block, depending on whether or not we want to enter into them.

Operation child. This is the main operation needed for traversing the tree. Let child(x, i) yield the child of node x by symbol $0 \le i \le 3$ (if it exists). Assume node x belongs to block B. Recall that x is actually the position of the node within the array that represents T_B. For computing child(x, i), we first check whether node x is in the frontier of B or not. To support this checking efficiently, we keep a finger i_f on array F_B, such that i_f is the smallest value for which $F_B[i_f]$ is greater or equal than the preorder of the current node in the traversal. Since we traverse in preorder, and F_B is sorted, increasing i_f as we traverse T_B is enough to keep i_f up to date. When the preorder of the current node exceeds $F_B[i_f]$, we increase i_f. If $F_B[i_f] = x$, then node x is in the frontier, hence we go down to block $P_B[i_f]$, start from the root node (which is x itself stored in the child block), and set $i_f \leftarrow 0$. Otherwise, x is not a frontier node, and we stay in B. In summary, operation child(x, i) is supported by an Euler tour on the part of the subtree of node x that belongs to block B. During the tour, we skip pointers associated to the frontier nodes we find before getting to the ith child of x.

Determining whether the i-th child of a node x exists requires a simple bit inspection. If it does, we must determine how many children of x (and their subtrees) must be skipped to get to child(x, i). We store a precomputed table that, for every 4-bit pattern and each $i = 0, \ldots, 3$, indicates how many subtrees must be skipped to get the desired child. For instance, if x is '1011' and $i = 2$, this table tells that one child of x must be skipped to get to the node labeled 2.

In our sequential traversal of B, corresponding to a depth-first traversal of T_B, we keep a stack S (initially empty) such that for every node in the path from node x to the current node, stores the number of children not yet traversed. We start looking for the desired child by moving to position $x + 1$, corresponding to the first child of x in preorder. At this point, we push the number of children of this node into S. The traversal is carried out by increasing an index on the array that stores T_B. The key for the traversal is to know the current depth at each step. As said before, we keep track of the current depth d, to know when we arrive at a leaf node. When traversing, we update d as follows. Every time we move to the next node (in preorder), we increase d only if (1) d is not the maximum depth (minus 1, recall that the last level is not represented), (2) the current node is not a frontier node, or (3) the current node is the last child of its parent. We use S to check the latter condition. Every time we reach a new node, we push in S its number of children if the node is not of maximum depth (minus 1), and it is not a frontier node. Otherwise, we instead decrease the value at the top of the stack, since the subtree of the corresponding node has been completely traversed. When the top value becomes 0, it means that a whole subtree has been traversed. In such a case we pop S, decrease the current depth d, and decrease the new value at the top (if this also becomes 0, we keep repeating the process, decreasing d and the top value).

Once the stack S becomes empty again, we have traversed the subtree of the first child. We repeat the same process from the current node, skipping as many children of x as needed.

Operation insert. To insert a point (c, r), we use the corresponding Morton code $M = yz$, for strings $y \in \{0, \dots, 3\}^*$ and $z \in \{0, \dots, 3\}^+$ to navigate the trie, until we cannot descend anymore. Assume that we have been able to get down to a node x (stored in block B) that represents string y, and at this node we have failed to descend using the first symbol of z. Then, we must insert string z in the subtree of node x. If the block has enough space for the $|z|$ new nodes, we simply find the insertion point from x (skipping subtrees as explained above), make room for the new nodes, and write them sequentially using a precomputed table that translates a given symbol of z to the 4-bit pattern corresponding to the unary node for that symbol. We also store a precomputed table that, given the encoding of x and the first symbol of string z, yields the new encoding for x.

If, on the other hand, the array used to store T_B has no room for the new nodes, we proceed as follows. If the array is currently able to store up to $N_i < N_{max}$ nodes, we reallocate it to make it of size N_j, for the smallest N_j such that $N_i + |z| \leq N_j$ holds. If, otherwise, $N_i = N_{max}$, or $N_i + |z| > N_{max}$, we must first *split* B to make room.

To minimize space usage, the splitting process should traverse T_B to choose the node w such that splitting T_B at w generates two trees whose size difference is minimum. We combine this criterion, however, with another one that optimizes traversal time. As explained, an advantage of our method is that we can traverse several edges in a single left-to-right scan of the block. Such scan, however, ends when we have to follow a pointer to another block. We try, therefore, to have those pointers as early as possible in the block so as to increase the probability that the left siblings of the ith child of node x are frontier nodes, so we avoid traversing their subtrees when computing child(x, i). Our splitting criterion, then, tries first to separate the leftmost node in the block whose subtree size is 25%–75% of the total block size.

After choosing node w, we carry out the split by generating two blocks, adding the corresponding pointer to the new child block, and adding w as a frontier node (storing its preorder in F_B and its pointer in P_B).

Increasing the Size of Deeper Blocks. A way to reduce the cost of traversing the blocks sequentially is to define a small maximum block size N_{max}. The cost is that this increases the space usage, because more blocks will be needed (thus increasing the number of pointers, and hence the space, of the data structure). We have the fortunate situation, however, that the most frequently traversed blocks are closer to the root, and these are relatively few. To exploit this fact, we define different maximum block sizes according to the depth of the corresponding block, with smaller maximum block sizes for smaller depths. We define parameters $0 \leq d_1 < d_2$ such that a block whose root has depth at most d_1 have maximum block size N''_{max}, a block whose root has depth at most d_2 have maximum block size N'_{max}, and the remaining blocks have maximum size N_{max}, for $N''_{max} < N'_{max} < N_{max}$. In this way, we aim to reduce the traversal cost, while using little space at deeper blocks (recall that, regardless the block size N_i, a block has at least $(1 - \alpha) \cdot N_i$ nodes, ensuring a good space utilization).

Pushing this idea to the extreme, we may set $N''_{max} = 1$, equivalent to allowing the top part of the tree to be represented with explicit pointers.

Analysis Again. Theorem 1 builds on a highly theoretical result [4]. The engineered structure defined in this section, on the other hand, obtains higher time complexities. In our implementation, each operation costs $O(N_{max})$ time, which we set close to $\log^2 N$ to obtain the same space redundancies of dynamic bitvectors. In turn, the implementation of dynamic bitvectors [13] takes $\Theta(\log^2 N)$ time per basic operation (edge traversal or update). An advantage of our implementation is that, during the $\Theta(\log^2 N)$-time traversal of a single block, we may be able to descend several levels in the root-to-leaves path of the k^2-tree, but this is not guaranteed. As a result, we can expect that our implementation will be about as fast as the dynamic bitvectors or significantly faster, depending on the tree topology. Our experiments in the next section confirm these expectations.

4 Experiments

4.1 Experimental Setup

We experimentally evaluate our proposal comparing it with the dynamic k^2-tree implementation based on dynamic bit vectors [13], to demonstrate the comparative performance of our technique. Other dynamic trie implementations exist [6,7,19] that are designed for storing general string dictionaries, and could store the points using their Morton codes. However, these techniques usually require space comparable to that of the original collection of strings (in our case, the Morton codes), which would be excessive in datasets like Web graphs that can be easily compressed using a few bits per edge. Moreover, since generic dictionaries are mainly optimized for word queries, or at most prefix queries, it would be harder to provide an efficient implementation of row/column queries in these structures, whereas those algorithms are very efficient in k^2-trees.

We use four different datasets in our experiments. Their basic information is described in Table 1. The graphs indochina and uk are Web graphs from the Laboratory for Web Algorithmics[1] [10,11], known to be very sparse and compressible. The datasets triples-med and triples-dense are selected predicates of the DBPedia 3.5.1[2], transformed through vertical partitioning as in previous work [2]; they are also sparse matrices but much less regular, and more difficult to compress.

For our structure we use $k = 2$ and the following configuration parameters: $N''_{max} = 1$ (i.e., we use explicit pointers in the first few levels of the trie), $N'_{max} = 96$, and show results for different configurations varying N_{max}, from 256 to 1024. We also show the tradeoff using values of d_1 8 and 12, and values of d_2 from 10 to 16 depending on d_1.

For the approach based on dynamic bitvectors (dyn-bitmap), we show results of the practical implementation with the default setup (base block size 512 for

---

[1] http://law.di.unimi.it/datasets.php.
[2] https://wiki.dbpedia.org/services-resources/datasets/data-set-35/data-set-351.

Table 1. Datasets used in our experiments.

Type	Dataset	Rows/cols (millions)	Points (millions)
Web graph	indochina-2004	7.4	194.1
	uk-2002	18.5	298.1
RDF	triples-med	67.0	7.9
	triples-dense	67.0	98.7

the nodes that compose the dynamic bitmap; $k = 4$ in the first 3 levels of the tree and $k = 2$ in the remaining levels). Using a higher value of k in a few upper levels of the conceptual tree has negligible effect on compression, since only a few nodes at the beginning of the bitmap use it, but it has a positive effect in query times since it reduces the height of the tree. When relevant, we also display results for another configuration with smaller block size 128 and $k = 4$ in the first 5 levels, keeping $k = 2$ in the remaining levels of the tree.

We also performed tests with dynamic path-decomposed tries [19] (dyn-PDT) and HAT tries [6] (HAT-Trie), applied to the Morton codes. For dyn-PDT we used the default configuration provided by the authors, but set λ to 4 as suggested by the authors for DNA, since Morton codes also use a very small alphabet (results were not significantly different with $\lambda = 16$). We tested the variants with bitmap management, varying ℓ from 8 to 64. For HAT-Trie we also used the default configuration. As explained previously, these are generic implementations not designed for this specific problem, and therefore their results are not competitive with ours. Nevertheless, we will outline the results obtained with these structures for completeness.

We run our experiments in a machine with 4 Intel i7-6500@2.5 GHz cores and 8 GB RAM, running Ubuntu 16.04.6. Our code is implemented in C++ and compiled with g++ 5.5.0 using the -O9 optimization flag. Our implementation is publicly available at https://github.com/darroyue/k2-dyn-tries.

4.2   Results

In order to test the compression and performance of our techniques, we start by building the representations from the original datasets. To do this, we shuffle the points in the dataset into a random order, and insert them in the structures one by one. Then, we measure the average insertion time during construction of the complete dataset, as well as the space used by the structure after construction.

Figure 3 displays insertion times during construction and final space for all the datasets and tested configurations. The results show that in Web graphs (indochina and uk) our representations can be created significantly faster than the dynamic bitvectors while requiring negligible additional space, for example 20–25% faster using 3% more space. Moreover, our representations provide a wide space-time tradeoff that the technique based on dynamic bitvectors does

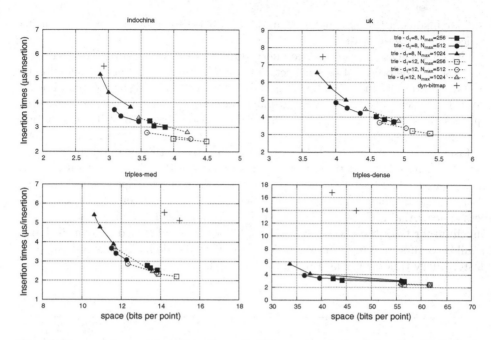

Fig. 3. Compression and insertion times (in bits per inserted point and μs/insertion)

not match (in Web graphs we only show results for the default configuration of dyn-bitmap, because the configuration with smaller blocks is both larger and slower). The configuration to achieve this tradeoff is also quite intuitive: larger(smaller) blocks in the lower levels lead to slower(faster), but more(less) compact structures.

In the RDF datasets (triples-med and triples-dense), our structures are even more competitive, using far less space and time than the dynamic bitvectors. In triples-med, our structures are 2.5 times faster when using similar space, or use 25% less space for the same speed. In triples-dense we are about 5 times faster when using the same space, and still 3 times faster than dynamic bitvectors when using 20% less space. Notice that the main difference between RDF and Web graph datasets is the regularity and clusterization of the points in the matrix, which is much higher in Web graphs than in RDF datasets. This also explains the worse space results achieved in these datasets compared to Web graphs. A similar difference in regularity exists between triples-med and triples-dense, where the latter is much more difficult to compress.

Dyn-PDT and HAT-Trie are not displayed in the figure, since they require much more space than any of the displayed variants and us, whereas their insertion times are similar to ours. Particularly, Dyn-PDT uses 2–4 times our space in triples-dense, 5–10 times our space in triples-med, and 12–20 times our space in Web graphs. It obtains insertion times similar to ours, ranging from 2.5 to 4.5 μs/insertion depending on the dataset and configuration, and query times very similar to insertion times and also similar to ours. HAT-Trie also obtains

similar insertion times, ranging from 3.5 to 4.5 μs/insertion, but requires up to twice the space of dyn-PDT.

Next, we measure the average query times to retrieve a point. To do this, we again select the points of each collection in random order, limiting our selection to 100 million points in the larger datasets, and measure the average query time to search for each of them. Figure 4 displays the query times for these cell retrieval queries. Results are analogous to those of insertion times. In Web graphs, our tries obtain even better performance compared to dynamic bitvectors. In RDF datasets the times are slightly closer but our tries still outperform dynamic bitvectors in space and time: In triples-med tries are 70% faster when using the same space, or 20% smaller when taking the same time. In triples-dense tries are 4 times faster when using the same space, and 3 times faster when using 20% less space. Again, Dyn-PDT and HAT-Trie are not displayed, because they are much larger than the displayed variants. Query times obtained by dyn-PDT are similar to ours, ranging from 2.5 to 5 μs/query. HAT-Trie has similar but slightly faster query times, ranging from 2.2 to 3.4 μs/query.

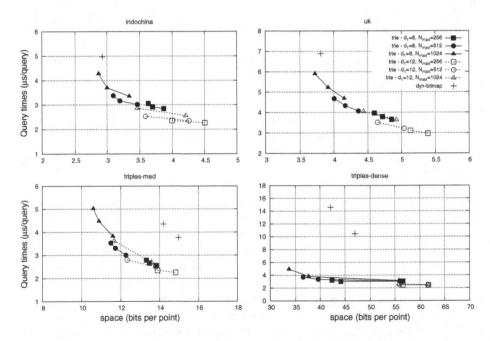

Fig. 4. Query times to retrieve cells (in μs/query)

We also perform tests querying for 100 million randomly selected cells. In practice, most of these cells will not belong to the collection, and they will probably be relatively far from existing points, hence allowing the structures to stop the traversal in the upper levels of the tree. These kind of queries are much faster and almost identical for all the trie configurations tested in each dataset. Figure 5

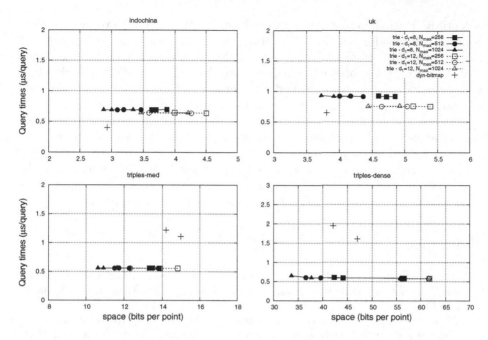

Fig. 5. Query times to retrieve random cells (in μs/query)

displays the query times for these queries. In Web graphs, the dynamic bitvectors obtain better query times for these queries (0.4–0.6 μs/query in indochina and uk, while our tries take around 0.6–0.7 and 0.75–0.95 μs/query, respectively). In RDF datasets, our tries are still significantly faster (around 0.55–0.6 μs/query in both datasets, whereas dynamic bitvectors take 1.1–1.2 μs/query in triples-med and 1.5–1.9 μs/query in triples-dense). This points to the depth of the tree search as a relevant factor in query complexity: our tries seem to have more stable query times, and are faster in queries that involve traversal of the full tree depth. In Web graphs, where points are usually clustered, non-existing points are detected in upper levels of the tree, and query times are usually better. In the RDF datasets, where points are more randomly distributed, the depth of the search is expected to be higher on average even if the dataset is still very sparse.

5 Conclusions

Regarding the k^2-tree as a trie on the Morton codes of the points it represents yields a new view that differs from the classical one based on bitvectors [14]. We have shown that this makes an important difference in the dynamic scenario, because dynamic tries can break lower bounds on maintaining dynamic bitvectors. Apart from the theoretical result, we have implemented a dynamic trie specialized in representing k^2-trees, where the trie is cut into a tree of blocks, each block representing a connected component of the trie. The dynamic trie uses

a depth-first search deployment of the trie, unlike the classical level-wise deployment. While this format cannot be traversed in constant time per trie edge, it is convenient for a dynamic trie representation because it is consistent with the tree of blocks, update operations require local changes, a single left-to-right block scan processes several downward edge traversals, and such scan is cache-friendly and does not require rebuilding any speed-up data structure.

Our experimental results show that our representation significantly outperforms the one based on dynamic bitvectors [13] on some datasets, in space, time, or both, depending on the nature of the dataset.

In the final version we will include experiments on other operations like extracting all the neighbors of a node. A future goal is to explore applications of our dynamic k^2-tree representation, in particular for graph databases [3].

References

1. Álvarez-Garcia, S., de Bernardo, G., Brisaboa, N., Navarro, G.: A succinct data structure for self-indexing ternary relations. J. Discrete Algorithms **43**, 38–53 (2017)
2. Álvarez-García, S., Brisaboa, N., Fernández, J., Martínez-Prieto, M., Navarro, G.: Compressed vertical partitioning for efficient RDF management. Knowl. Inf. Syst. **44**(2), 439–474 (2015)
3. Álvarez-García, S., Freire, B., Ladra, S., Pedreira, O.: Compact and efficient representation of general graph databases. CoRR abs/1812.10977 (2018, to appear). in Knowledge and Information Systems
4. Arroyuelo, D., Davoodi, P., Satti, S.R.: Succinct dynamic cardinal trees. Algorithmica **74**(2), 742–777 (2016)
5. Arroyuelo, D., Navarro, G.: Space-efficient construction of Lempel-Ziv compressed text indexes. Inf. Comput. **209**(7), 1070–1102 (2011)
6. Askitis, N., Sinha, R.: Engineering scalable, cache and space efficient tries for strings. VLDB J. **19**(5), 633–660 (2010)
7. Belazzougui, D., Boldi, P., Vigna, S.: Dynamic z-fast tries. In: Chavez, E., Lonardi, S. (eds.) SPIRE 2010. LNCS, vol. 6393, pp. 159–172. Springer, Heidelberg (2010). https://doi.org/10.1007/978-3-642-16321-0_15
8. Benoit, D., Demaine, E., Munro, J.I., Raman, R., Raman, V., Rao, S.S.: Representing trees of higher degree. Algorithmica **43**(4), 275–292 (2005)
9. de Bernardo, G., Álvarez-García, S., Brisaboa, N.R., Navarro, G., Pedreira, O.: Compact querieable representations of raster data. In: Kurland, O., Lewenstein, M., Porat, E. (eds.) SPIRE 2013. LNCS, vol. 8214, pp. 96–108. Springer, Cham (2013). https://doi.org/10.1007/978-3-319-02432-5_14
10. Boldi, P., Rosa, M., Santini, M., Vigna, S.: Layered label propagation: a multiresolution coordinate-free ordering for compressing social networks. In: Srinivasan, S., Ramamritham, K., Kumar, A., Ravindra, M.P., Bertino, E., Kumar, R. (eds.) Proceedings of the 20th International Conference on World Wide Web, pp. 587–596. ACM Press (2011)
11. Boldi, P., Vigna, S.: The WebGraph framework I: Compression techniques. In: Proceedings of the Thirteenth International World Wide Web Conference (WWW 2004), pp. 595–601. ACM Press, Manhattan (2004)

12. Brisaboa, N., de Bernardo, G., Konow, R., Navarro, G., Seco, D.: Aggregated 2D range queries on clustered points. Inf. Syst. **60**, 34–49 (2016)
13. Brisaboa, N., Cerdeira-Pena, A., de Bernardo, G., Navarro, G.: Compressed representation of dynamic binary relations with applications. Inf. Syst. **69**, 106–123 (2017)
14. Brisaboa, N.R., Ladra, S., Navarro, G.: Compact representation of Web graphs with extended functionality. Inf. Syst. **39**(1), 152–174 (2014)
15. Cerdeira-Pena, A., de Bernardo, G., Fariña, A., Paramá, J.R., Silva-Coira, F.: Towards a compact representation of temporal rasters. In: Gagie, T., Moffat, A., Navarro, G., Cuadros-Vargas, E. (eds.) SPIRE 2018. LNCS, vol. 11147, pp. 117–130. Springer, Cham (2018). https://doi.org/10.1007/978-3-030-00479-8_10
16. Clark, D.R.: Compact PAT Trees. Ph.D. thesis, University of Waterloo, Canada (1996)
17. Fredman, M., Saks, M.: The cell probe complexity of dynamic data structures. In: Proceedings of the 21st Annual ACM Symposium on Theory of Computing (STOC), pp. 345–354 (1989)
18. Jacobson, G.: Space-efficient static trees and graphs. In: Proceedings of the 30th IEEE Symposium on Foundations of Computer Science (FOCS), pp. 549–554 (1989)
19. Kanda, S., Morita, K., Fuketa, M.: Practical implementation of space-efficient dynamic keyword dictionaries. In: Fici, G., Sciortino, M., Venturini, R. (eds.) SPIRE 2017. LNCS, vol. 10508, pp. 221–233. Springer, Cham (2017). https://doi.org/10.1007/978-3-319-67428-5_19
20. Morton, G.M.: A computer oriented geodetic data base; and a new technique in file sequencing. Technical report, IBM Ltd. (1966)
21. Munro, J.I.: Tables. In: Proceedings of the 16th Conference on Foundations of Software Technology and Theoretical Computer Science (FSTTCS), pp. 37–42 (1996)
22. Navarro, G.: Compact Data Structures - A Practical Approach. Cambridge University Press, Cambridge (2016)
23. Navarro, G., Sadakane, K.: Fully-functional static and dynamic succinct trees. ACM Trans. Algorithms 10(3), article 16 (2014)
24. Okanohara, D., Sadakane, K.: Practical entropy-compressed rank/select dictionary. In: Proceedings of the 9th Workshop on Algorithm Engineering and Experiments (ALENEX), pp. 60–70 (2007)
25. Samet, H.: Foundations of Multidimensional and Metric Data Structures. Morgan Kaufmann, San Francisco (2006)

Faster Repetition-Aware Compressed Suffix Trees Based on Block Trees

Manuel Cáceres$^{(\boxtimes)}$ and Gonzalo Navarro

CeBiB — Center for Biotechnology and Bioengineering,
Department of Computer Science, University of Chile, Santiago, Chile
{mcaceres,gnavarro}@dcc.uchile.cl

Abstract. Suffix trees are a fundamental data structure in stringology, but their space usage, though linear, is an important problem in applications. We design and implement a new compressed suffix tree targeted to highly repetitive texts, such as large genomic collections of the same species. Our suffix tree builds on Block Trees, a recent Lempel-Ziv bounded data structure that captures the repetitiveness of its input. We use Block Trees to compress the topology of the suffix tree, and augment the Block Tree nodes with data that speeds up suffix tree navigation.

Our compressed suffix tree is slightly larger than previous repetition-aware suffix trees based on grammars, but outperforms them in time, often by orders of magnitude. The component that represents the tree topology achieves a speed comparable to that of general-purpose compressed trees, while using 2–10 times less space, and might be of independent interest.

1 Introduction

Suffix trees [22,36,37] are one of the most appreciated data structures in Stringology [3] and in application areas like Bioinformatics [13], enabling efficient solutions to complex problems such as (approximate) pattern matching, pattern discovery, finding repeated substrings, computing matching statistics, computing maximal matches, and many others. In other collections, like natural language and software repositories, suffix trees are useful for plagiarism detection [23], authorship attribution [38], document retrieval [14], and others.

While their linear space complexity is regarded as acceptable in classical terms, their actual space usage brings serious problems in application areas. From an Information Theory standpoint, on a text of length n over alphabet $[1, \sigma]$, classical suffix tree representations use $\Theta(n \lg n)$ bits, whereas the information contained in the text is, in the worst case, just $n \lg \sigma$ bits. From a practical point of view, even carefully engineered implementations [17] require at least 10 bytes per symbol, which forces many applications to run the suffix tree on (orders of magnitude slower) secondary memory.

Funded by Fondecyt Grant 1-170048 and by Basal Funds FB0001, Conicyt, Chile.

N. R. Brisaboa and S. J. Puglisi (Eds.): SPIRE 2019, LNCS 11811, pp. 434–451, 2019.
https://doi.org/10.1007/978-3-030-32686-9_31

Consider for example Bioinformatics, where various complex analyses require the use of sophisticated data structures, suffix trees being among the most important ones. DNA sequences range over $\sigma = 4$ different nucleotides represented with $\lg 4 = 2$ bits each, whereas the suffix tree uses at least 10 bytes = 80 bits per base, that is, 4000% of the text size. A human genome fits in approximately 715 MB, whereas its suffix tree requires about 30 GB. The space problem becomes daunting when we consider the DNA analysis of large groups of individuals; consider for example the 100,000-human-genomes project (www.genomicsengland. co.uk).

One solution to the problem is to build suffix trees on secondary memory [7,9]. Most suffix tree algorithms, however, require traversing them across arbitrary access paths, which makes secondary memory solutions many orders of magnitude slower than in main memory. Another approach replaces the suffix trees with suffix arrays [21], which decreases space usage to 4 bytes (32 bits) per character but loses some functionality like the suffix links, which are essential to solve various complex problems. This functionality can be recovered [2] by raising the space to about 6 bytes (48 bits) per character.

A promising line of research is the construction of compact representations of suffix trees, named *Compressed Suffix Trees (CSTs)*, which simulate all the suffix tree functionality within space bounded not only by $O(n \lg \sigma)$ bits, but by the information content (or text entropy) of the sequence. An important theoretical achievement was a CST using $O(n)$ bits on top of the text entropy that supports all the operations within an $O(\text{polylog } n)$ time penalty factor [34]. A recent implementation [28] uses, on DNA, about 10 bits per base and supports the operations in a few microseconds. While even smaller CSTs have been proposed, reaching as little as 5 bits per base [32], their operation times raise to milliseconds, thus becoming nearly as slow as a secondary-memory deployment.

Still, further space reductions are desirable when facing large genome repositories. Fortunately many of the largest text collections are highly repetitive; for example DNA sequences of two humans differ by less than 0.5% [35]. This repetitiveness is not well captured by statistical based compression methods [16], on which most of the CSTs are based. Lempel-Ziv [19] and grammar [15] based compression techniques, among others, do better in this scenario [24], but only recently we have seen CSTs building on them, both in theory [5,11] and in practice [1,26]. The most successful CSTs in practice on repetitive collections are the grammar-compressed suffix trees (GCSTs), which on DNA use about 2 bits per base and support the operations in tens to hundreds of microseconds.

GCSTs use grammar compression on the parentheses sequence that represents the suffix tree topology [31], which inherits the repetitiveness of the text collection. While Lempel-Ziv compression is stronger, it does not support easy access to the sequence. In this paper we explore an alternative to grammar compression called Block Trees [6,29], which offer similar approximation ratios to Lempel-Ziv compression, but promise faster access.

Our main contribution is the BT-CT, a Block-Tree-based representation of tree topologies, which enriches Block Trees to support the required navigation

Table 1. List of typical operations implemented by suffix trees; str(v) represents the concatenation of the strings in the root-to-v path.

Operation	Description
root()	The root of the suffix tree
is-leaf(v)	True if v is a leaf node
first-child(v)	The first child of v in lexicographical order
tree-depth(v)	The number of edges from root() to v
next-sibling(v)	The next sibling of v in lexicographical order
previous-sibling(v)	The previous sibling of v in lexicographical order
parent(v)	The parent of v
is-ancestor(v, u)	True if v is ancestor of u
level-ancestor(v, d)	The ancestor of v at tree depth d
lca(v, u)	The lowest common ancestor between v and u
letter(v, i)	str(v)[i]
string-depth(v)	|str(v)|
suffix-link(v)	The node u s.t. str(u) = str(v)[2,string-depth(v)]
string-ancestor(v, d)	The highest ancestor u of v s.t. string-depth(u) $\geq d$
child(v, c)	The child u of v s.t. str(u)[string-depth(v)+1] = c

operations. Although we are unable to prove useful upper bounds on the operation times, the BT-CT performs very well in practice: while using 0.3–1.5 bits per node in our repetitive suffix trees, it implements the navigation operations in a few microseconds, becoming very close to the performance of plain 2.8-bit-per-node representations that are blind to repetitiveness [27]. We use the BT-CT to represent suffix tree topologies in this paper, but it might also be useful in other scenarios, such as representing the topology of repetitive XML collections [4].

As said, our new suffix tree, BT-CST, uses the BT-CT to represent the suffix tree topology. Although larger than the GCST, it still requires about 3 bits per base in highly repetitive DNA collections. In exchange, it is faster than the GCST, often by an order of magnitude. This owes to the BT-CT directly, but also indirectly: Its faster navigation enables the binary search for the "child by letter" operation in suffix trees, which is by far the slowest one. While with the GCST a linear traversal of the children is advisable [26], a binary search pays off in the BT-CST, making it faster especially on large alphabets.

2   Preliminaries and Related Work

A text $T[1, n] = T[1] \ldots T[n]$ is a sequence of symbols over an alphabet $\Sigma = [1, \sigma]$, terminated by a special symbol \$ that is lexicographically smaller than any symbol of Σ. A substring of T is denoted $T[i, j] = T[i] \ldots T[j]$. A substring $T[i, j]$ is a prefix if $i = 1$ and a suffix if $j = n$.

The *suffix tree* [22,36,37] of a text T is a trie of its suffixes in which unary paths are collapsed into a single edge. The tree then has less than $2n$ nodes. The suffix tree supports a set of operations (see Table 1) that suffices to solve a large number of problems in Stringology [3] and Bioinformatics [13].

The *suffix array* [21] $A[1, n]$ of a text $T[1, n]$ is a permutation of $[1, n]$ such that $A[i]$ is the starting position of the ith suffix in increasing lexicographical order. The leaves descending from a suffix tree node span a range of suffixes in A.

The function $lcp(X, Y)$ is the length of the longest common prefix (lcp) of strings X and Y. The *LCP array* [21], $LCP[1, n]$, is defined as $LCP[1] = 0$ and $LCP[i] = lcp(T[A[i-1], n], T[A[i], n])$ for all $i > 1$, that is, it stores the lengths of the lcps between lexicographically consecutive suffixes of $T[1, n]$.

2.1 Succinct Tree Representations

A balanced parentheses (BP) representation (there are others [31]) of the topology of an ordinal tree $\mathcal{T}$ of t nodes is a binary sequence (or bitvector) $P[1, 2t]$ built as follows: we traverse $\mathcal{T}$ in preorder, writing an opening parenthesis (a bit 1) when we first arrive at a node, and a closing one (a bit 0) when we leave its subtree. For example, a leaf looks like "10". The following primitives can be defined on P:

- $access(i) = P[i]$
- $rank_{0|1}(i) = |\{1 \le j \le i; P[j] = 0|1\}|$
- $excess(i) = rank_1(i) - rank_0(i)$
- $select_{0|1}(i) = \min(\{j; rank_{0|1}(j) = i\} \cup \{\infty\})$
- $leaf\text{-}rank(i) = rank_{10}(i) = |\{1 \le j \le i-1; P[j] = 1 \wedge P[j+1] = 0\}|$
- $leaf\text{-}select(i) = select_{10}(i) = \min(\{j; leaf\text{-}rank(j+1) = i\} \cup \{\infty\})$
- $fwd\text{-}search(i, d) = \min(\{j > i; excess(j) = excess(i) + d\} \cup \{\infty\})$
- $bwd\text{-}search(i, d) = \max(\{j < i; excess(j) = excess(i) + d\} \cup \{-\infty\})$
- $min\text{-}excess(i, j) = \min(\{excess(k) - excess(i-1); i \le k \le j\} \cup \{\infty\})$

These primitives suffice to implement a large number of tree navigation operations, and can all be supported in constant time using $o(t)$ bits on top of P [27]. These include the operations needed by suffix trees. For example, interpreting nodes as the position of their opening parenthesis in P, it holds that $parent(v) = bwd\text{-}search(i, -2)+1$, $next\text{-}sibling(v) = fwd\text{-}search(v, -1)+1$ and the lowest common ancestor of two nodes $v \le u$ is $lca(v, u) = parent(fwd\text{-}search(v - 1, min\text{-}excess(v, u)) + 1)$.

2.2 Compressed Suffix Arrays

A milestone in the area was the emergence of Compressed Suffix Arrays (CSAs) [25], which using space proportional to that of the compressed sequence managed to answer access queries to the original suffix array and its inverse (i.e., return any $A[i]$ and $A^{-1}[j]$), to the indexed sequence (i.e., return any $T[i..j]$), and access to a novel array, $\Psi[i] = A^{-1}[(A[i] \bmod n) + 1]$, which lets us move from a text suffix $T[j, n]$ to the next one, $T[j+1, n]$, yet indexing the suffixes by their lexicographic rank, $A^{-1}[j]$. This function plays a key role in the design of CSTs, as seen next.

2.3   Compressed Suffix Trees

Sadakane [34] designed the first CST, on top of a CSA, using $|CSA| + O(n)$ bits and solving all the suffix tree operations in time $O(\text{polylog } n)$. He makes up a CST from three components: a CSA, for which he uses his own proposal [33]; a BP representation of the suffix tree topology, using at most $4n + o(n)$ bits; and a compressed representation of LCP, which is a bitvector $H[1, 2n]$ encoding the array $PLCP[i] = LCP[A^{-1}[i]]$ (i.e., the LCP array in text order). A recent implementation [28] of this index requires about 10 bits per character and takes a few microseconds per operation.

Russo et al. [32] managed to use just $o(n)$ bits on top of the CSA, by storing only a sample of the suffix tree nodes. An implementation of this index [32] uses as little as 5 bits per character, but the operations take milliseconds, as slow as running in secondary storage.

Yet another approach [10] also obtains $o(n)$ on top of a CSA by getting rid of the tree topology and expressing the tree operations on the corresponding suffix array intervals. The operations now use primitives on the LCP array: find the previous/next smaller value (psv/nsv) and find minima in ranges (rmq). They also noted that bitvector H contains $2r$ runs, where r is the number of runs of consecutive increasing values in Ψ, and used this fact to run-length compress H. Abeliuk et al. [1] designed a practical version of this idea, obtaining about 8 bits per character and getting a time performance of hundreds of microseconds per operation, an interesting tradeoff between the other two options.

Engineered adaptations of these three ideas were implemented in the SDSL library [12], and are named `cst_sada`, `cst_fully`, and `cst_sct3`, respectively. We will use and adapt them in our experimental comparison.

2.4   Repetition-Aware Compressed Suffix Trees

Abeliuk et. al [1] also presented the first CST for repetitive collections. They built on the third approach above [10], so they do not represent the tree topology. They use the RLCSA [20], a repetition-aware CSA with size proportional to r, which is very low on repetitive texts. They use grammar compression on the *differential* LCP array, $DLCP[i] = LCP[i] - LCP[i-1]$. The nodes of the parsing tree (obtained with Re-Pair [18]) are enriched with further data to support the operations psv/nsv and rmq. To speed up simple LCP accesses, the bitvector H is also stored, whose size is also proportional to r. Their index uses 1–2 bits per character on repetitive collections. It is rather slow, however, operating within (many) milliseconds.

Navarro and Ordóñez [26] include again the tree topology. Since text repetitiveness induces isomorphic subtrees in the suffix tree, they grammar-compressed the BP representation. The nonterminals are enriched to support the tree navigation operations enumerated in Sect. 2.1. Since they do not need psv/nsv/rmq operations on LCP, they just use the bitvector H, which has a few runs and thus is very small. Their index uses slightly more space, closer to 2 bits per character, but it is up to 3 orders of magnitude faster than that of Abeliuk et al. [1]: their

structure operates in tens to hundreds of microseconds per operation, getting closer to the times of general-purpose CSTs.

Less related or theoretical work [5,8,11] is not discussed for lack of space.

3   Block Trees

A Block Tree [6] is a full r-ary tree that represents a (repetitive) sequence $P[1,p]$ in compressed space while offering access and other operations in logarithmic time. The nodes at depth d (the root being depth 0) represent blocks of P of length $b = |P|/r^d$, where we pad P to ensure these numbers are integers. Such a node v, representing some block $v.blk = P[i, i + b - 1]$, can be of three types:

LeafBlock: If $b \leq mll$, where mll is a parameter, then v is a leaf of the Block Tree, and it stores the string $v.blk$ explicitly.

BackBlock: Otherwise, if $P[i - b, i + b - 1]$ and $P[i, i + 2b - 1]$ are not their leftmost occurrences in P, then the block is replaced by its leftmost occurrence in P: node v stores a pointer $v.ptr = u$ to the node u such that the first occurrence of $v.blk$ starts inside $u.blk = P[j, j + b - 1]$, more precisely it occurs in $P[j + o, j + o + b - 1]$. This offset inside $u.blk$ is stored at $v.off = o$. Node v is not considered at deeper levels.

InternalBlock: Otherwise, the block is split into r equal parts, handled in the next level by the children of v. The node v then stores a pointer to its children.

The Block Tree can return any $P[i]$ in logarithmic time, by starting at position i in the root block. Recursively, the position i is translated in constant time into an offset inside a child node (for InternalBlocks), or inside a leftward node in the same level (for BackBlocks, at most once per level). At leaves, the symbol is stored explicitly.

If we augment the nodes of the Block Tree with rank information for the σ symbols of the alphabet, the Block Tree answers rank and select queries on P in logarithmic time as well. Specifically, for every $c \in [1, \sigma]$, we store in every node v the number $v.c$ of cs in $v.blk$. Further, every BackBlock node v pointing to u stores the number of cs in $u.blk[1, v.off - 1]$.

Our new repetition-aware CST will represent the BP topology with a Block Tree. The basic structure supports operations $access(i)$, $rank_{0|1}(i)$, $excess(i)$ and $select_{0|1}(i)$. In the next section we show how to solve the remaining operations.

4   Our Repetition-Aware Compressed Suffix Tree

Following the scheme of Sadakane [34] we propose a three-component structure to implement a new CST tailored to highly repetitive inputs. We use the RLCSA [20] as our CSA. For the LCP, we use the compressed version of the bitvector H [10]. For the topology, we use BP and represent the sequence with a Block Tree, adding new fields to the Block Tree nodes to efficiently answer all the queries we need (Sect. 2.1). We call this representation Block Tree CST (BT-CST). Section 4.1 describes BT-CT, our extension to Block Trees, and Sect. 4.2 our improved operation $child(v, a)$ for the BT-CST.

4.1   Block Tree Compressed Topology (BT-CT)

We describe our main data structure, *Block Tree Compressed Topology (BT-CT)*, which compresses a parentheses sequence and supports navigation on it.

Stored Fields. We augment the nodes of the Block Tree with the following fields:

- For every node v that represents the block $v.blk = P[i, i + b - 1]$:
  - $rank_1$, the number of 1s in $v.blk$, i.e., $rank_1(i + b - 1) - rank_1(i - 1)$ in P.
  - *lrank* (leaf rank), the number of 10s (i.e., leaves in BP) that finish inside $v.blk$, i.e., $leaf\text{-}rank(i + b - 1) - leaf\text{-}rank(i - 1)$ in P.
  - *lbreaker* (leaf breaker), a bit telling whether the first symbol of $v.blk$ is a 0 and the preceding symbol in P is a 1, i.e., whether $P[i - 1, i] = 10$.
  - *mexcess*, the minimum excess in $v.blk$, i.e., $min\text{-}excess(i, i + b - 1)$ in P.
- For every BackBlock node v that represents $v.blk = P[i, i + b - 1]$ and points to its first occurrence $O = P[j + o, j + o + b - 1]$ inside $u.blk = P[j, j + b - 1]$ with offset $v.off = o$:
  - $fb\text{-}rank_1$, the number of 1s in the prefix of O contained in $u.blk$ ($O \cap u.blk$, the 1st block spanned by O), i.e., $rank_1(j + b - 1) - rank_1(j + o - 1)$ in P.
  - *fb-lrank*, the number of 10s that finish in $O \cap u.blk$, i.e., $leaf\text{-}rank(j + b - 1) - leaf\text{-}rank(j + o - 1)$ in P.
  - *fb-lbreaker*, a bit telling whether the first symbol of O is a 0 and the preceding symbol is a 1, i.e., whether $P[j + o - 1, j + o] = 10$.
  - *fb-mexcess*, the minimum excess reached in $O \cap u.blk$, i.e., $min\text{-}excess(j + o, j + b - 1)$.
  - *m-fb*, a bit telling whether the minimum excess of $u.blk$ is reached in $O \cap u.blk$, i.e., whether $min\text{-}excess(i, i+b-1) = min\text{-}excess(j+o, j+b-1)$.

Fields Computed on the Fly. In the description of the operations we will use other fields that are computed in constant time from those we already store:

- For every node v that represents $v.blk = P[i, i + b - 1]$
  - $rank_0$, the number of 0s in $v.blk$, i.e., $b - v.rank_1$.
  - *excess*, the excess of 1s over 0s in $v.blk$, i.e., $v.rank_1 - v.rank_0 = 2 \cdot v.rank_1 - b$.
- For every BackBlock node v that represents $v.blk = P[i, i + b - 1]$ and points to its first occurrence $O = P[j + o, j + o + b - 1]$ inside $u.blk = P[j, j + b - 1]$ with offset $v.off = o$:
  - $fb\text{-}rank_0$, the number of 0s in $O \cap v.blk$, i.e., $(b - o) - v.fb\text{-}rank_1$.
  - $pfb\text{-}rank_{0|1}$, the number of 0s|1s in the prefix of $u.blk$ that precedes O ($u.blk - O$), i.e., $u.rank_{0|1} - v.fb\text{-}rank_{0|1}$.
  - *fb-excess*, the excess in $O \cap u.blk$, i.e., $v.fb\text{-}rank_1 - v.fb\text{-}rank_0$.
  - *sb-excess*, the excess in $O - u.blk$ (2nd block spanned by O), i.e., $v.excess - v.fb\text{-}excess$.
  - *pfb-lrank*, the number of 10s that finish in $u.blk - O$, i.e., $u.lrank - v.fb\text{-}lrank$.

- *sb-mexcess*, the minimum excess in $O - u.blk$, i.e., $min\text{-}excess(j + b, j + b + o - 1)$ in P. We store either $v.fb\text{-}mexcess$ or $v.sb\text{-}mexcess$, the one that differs from $v.mexcess$. To deduce the non-stored field we use *mexcess*, *fb-excess* and *m-fb*.

Complex Operations. Apart from the basic operations solved in the original Block Tree we need, as described in Sect. 2.1, more sophisticated ones to support navigation in the parentheses sequence.

leaf-rank(i) **and** *leaf-select*(i)**.** The implementations of these operations are analogous to those for $rank_c(i)$ and $select_c(i)$ respectively, in the base Block Tree. The only two differences are that in LeafBlocks we consider the *lbreaker* field to check whether the block starts with a leaf, and in BackBlocks we consider fields *lbreaker* and *fb-lbreaker* to check whether we have to add or remove one leaf when moving to a leftward node. Like $rank_c(i)$ and $select_c(i)$, our operations work $O(1)$ per level, and then have their same time complexity, given in Sect. 3.

fwd-search(i, d) **and** *bwd-search*(i, d)**.** We only show how to solve *fwd-search* (i, d) with $d < 0$; the other cases are similar (some combinations not needed for our CST require further fields). Thus we aim to find the smallest position $j > i$ where the excess of $P[i + 1..j]$ is d.

We describe our solution as a recursive procedure *fwd-search*(i, j) with two global variables: d from the input, and e. Variables i and j are the limits of the search for the currently processed node, and e is the accumulated excess of the part of the range that has already been processed. The procedure is initially called at the Block Tree root with *fwd-search*(i, n) and with $e = 0$. If at some point e reaches d, we have found the answer to the search. The general idea is to traverse the range of the current node v left to right, using the fields $v.mexcess$, $v.fb\text{-}mexcess$ and $v.sb\text{-}mexcess$ to speed up the procedure:

- If the search range spans the entire block $v.blk$ (i.e., $j - i = b$) and the answer is not reached inside v (i.e., $e + v.mexcess > d$), then we increase e by $v.excess$ and return ∞.
- If v is a LeafBlock we scan $v.blk$ bitwise, increasing/decreasing e for each 1/0. If e reaches d at some index k, we return k; otherwise we return ∞.
- If v is an InternalBlock, we identify the k-th child of v, which contains position $i + 1$, and the m-th, which contains position j (it could be that $k = m$). We then call *fwd-search* recursively on the k-th to the m-th children, intersecting the query range with the extent of each child (the search range will completely cover the children after the k-th and before the m-th). As soon as any of these calls returns a non-∞ value, we adjust (i.e., shift) and return it. If all of them return ∞, we also return ∞.
- If v is a BackBlock we must translate the query to the original block O, which starts at offset $v.off$ in $u.blk$, where $u = v.ptr$. We first check whether the query covers the prefix of $v.blk$ contained in $u.blk$, $O \cap u.blk$ (i.e., if $i = 0$ and $j \geq b - v.off$). If so, we check whether we can skip $O \cap u.blk$, namely if $e + v.fb\text{-}mexcess > d$. If we can skip it, we just update e to $e + v.fb\text{-}excess$,

otherwise we call *fwd-search* recursively on the intersection of $u.blk$ and the translated query range. If the answer is not ∞, we adjust and return it. Otherwise, we turn our attention to the node u' next to u. Again, we check whether the query covers the suffix of $v.blk$ contained in $u'.blk$, $O - u.blk$ (i.e., $j = b$ and $i \leq b - v.off$). If so, we check whether we can skip $O - u.blk$, namely if $e + v.sb\text{-}mexcess > d$. If we can skip it, we just update e to $e + v.sb\text{-}excess$, otherwise we call *fwd-search* recursively on the intersection of $u'.blk$ and the translated query range. If the answer is not ∞, we adjust and return it. Otherwise, we return ∞.

min-excess(i, j). We will also start at the root with the global variable e set to zero. A local variable m will keep track of the minimum excess seen in the current node, and will be initialized at $m = 1$ in each recursive call. The idea is the same as for *fwd-search*: traverse the node left to right and use the fields $v.mexcess$, $v.fb\text{-}mexcess$ and $v.sb\text{-}mexcess$ to speed up the traversal.

- If the query covers the entire block $v.blk$ (i.e., $j - i + 1 = b$), we increase e by $v.excess$ and return $v.mexcess$.
- If v is a LeafBlock we record the initial excess in $e' = e$ and scan $v.blk$ bitwise, updating e for each bit read as in operation *fwd-search*. Every time we have $e - e' < m$, we update $m = e - e'$. At the end of the scan we return m.
- If v is an InternalBlock, we identify the k-th child of v, which contains position i, and the m-th, which contains position j (it could be that $k = m$). We then call *min-excess* recursively on the k-th to the m-th children, intersecting the query range with the extent of each child (the search range will completely cover the children after the k-th and before the m-th, so these will take constant time). We return the minimum between all their answers (composed with their correspondent prefix excesses).
- If v is a BackBlock we translate the query to the original block O, which starts at offset $v.off$ in $u.blk$, where $u = v.ptr$. We first check whether the query covers the prefix of $v.blk$ contained in $u.blk$, $O \cap u.blk$ (i.e., if $i = 1$ and $j \geq b - v.off - 1$). If so, we simply set $m = v.fb\text{-}mexcess$ and update e to $e + v.fb\text{-}excess$. Otherwise we call *min-excess* recursively on the intersection of $u.blk$ and the translated query range, and record its answer in m. We now consider the block u' next to u and again check whether the query covers the suffix of $v.blk$ contained in $u'.blk$, $O - u.blk$ (i.e., if $j = b$ and $i \leq b - v.off + 1$). If so, we just set $m = \min(m, v.fb\text{-}excess + v.sb\text{-}mexcess)$ and update e to $e + v.sb\text{-}excess$. Otherwise, we call *min-excess* on the intersection of $u'.blk$ and the translated query range, record its answer in m', and set $m = \min(m, v.fb\text{-}excess + m')$. Finally, we return m.

Note that, although we look for various opportunities to use the precomputed data to skip parts of the query range, the operations *fwd-search*, *bwd-search*, and *min-excess* are not guaranteed to work proportionally to the height of the Block Tree. The instances we built that break this time complexity, however, are unlikely to occur. Our experiments will show that the algorithms perform well in practice.

4.2 Operation Child

The fast operations enabled by our BT-CT structure give space for an improved algorithm to solve operation child(v, a). Most previous CSTs first compute $d =$ string-depth(v) and then linearly traverse the children of v from $u =$ first-child(v) with operation next-sibling, checking for each child u whether letter$(u, d+1) = a$, and stopping as soon as we find or exceed a. Since computing letter is significantly more expensive than our next-sibling, we consider the variant of first identifying all the children u of v, and then binary searching them for a, using letter. We then perform $O(\sigma)$ next-sibling operations, but only $O(\lg \sigma)$ letter operations.

5 Experiments and Results

We measured the time/space performance of our new BT-CST and compared it with the state of the art. Our code and testbed is available at https://github.com/elarielcl/BT-CST.

5.1 Experimental Setup

Compared CSTs. We compare the following CST implementations.

BT-CST. Our new Compressed Suffix Tree with the described components. For the BT-CT component we vary $r \in \{2, 4, 8\}$ and $mll \in \{4, 8, 16, 32, 64, 128, 256\}$.

GCST. The Grammar-based Compressed Suffix Tree [26]. We vary parameters *rule-sampling* and *C-sampling* as they suggest.

CST_SADA, CST_SCT3, CST_FULLY. Adaptation and improvements from the SDSL library[1] on the indexes of Sadakane [34], Fischer et al. [10] and Russo et al. [32], respectively. CST_SADA maximizes speed using Sadakane's CSA [33] and a non-compressed version of bitvector H. CST_SCT3 uses instead a Huffman-shaped wavelet tree of the BWT as the suffix array, and a compressed representation [30] for bitvector H and those of the wavelet tree. This bitvector representation exploits the runs and makes the space sensitive to repetitiveness, but it is slower. CST_FULLY uses the same BWT representation. For all these suffix arrays we set *sa-sampling* = 32 and *isa-sampling* = 64.

CST_SADA_RLCSA, CST_SCT3_RLCSA. Same as the preceding implementations but (further) adapted to repetitive collections: We replace the suffix array by the RLCSA [20] and use a run-length-compressed representation of bitvector H [10].

For the CSTs using the RLCSA, we fix their parameters to 32 for the sampling of Ψ and 128 for the text sampling. We only show the Pareto-optimal results of each structure. Note that we do not include the CST of Abeliuk et al. [1] in the comparison because it was already outperformed by several orders of magnitude by GCST [26].

---

[1] Succinct data structures library (SDSL), https://github.com/simongog/sdsl-lite.

Text Collection and Queries. Our input sequences come from the Repetitive Corpus of *Pizza & Chili* (http://pizzachili.dcc.uchile.cl/repcorpus). We selected `einstein`, containing all the versions (up to January 12, 2010) of the German Wikipedia Article of *Albert Einstein* (89 MB, compressible by p7zip to 0.11%); `influenza`, a collection of 78,041 H. influenzae genomes (148 MB, compressible by p7zip to 1.69%); and `kernel`, a set of 36 versions of the Linux Kernel (247 MB, compressible by p7zip to 2.56%).

Data points are the average of 100,000 random queries, similar to the scheme used in previous work on Compressed Suffix Trees [1, 26] to choose the nodes on which the operations are called: For *next-sibling* and *parent* we collect the nodes in leaf-to-root paths starting from random leaves. For *lca* we choose random leaf pairs. For *suffix-link* we collect the nodes on traversals starting from random leaves, and taking suffix-links until reaching the root. For *child* we choose random leaves and collect the nodes in the traversals to the root, discarding the nodes with less than 3 children, and we choose the initial letter of a random child of the node.

Computer. The experiments ran on an isolated Intel(R) Xeon(R) CPU E5-2407 @ 2.40 GHz with 256 GB of RAM and 10 MB of L3 cache. The operating system is GNU/Linux, Debian 2, with kernel 4.9.0-8-amd64. The implementations use a single thread and all of them are coded in C++. The compiler is gcc version 4.6.3, with -O9 optimization flag set (except CST_SADA, CST_SCT3 and CST_FULLY, which use their own set of optimization flags).

Operations. We implemented all the suffix tree operations of Table 1. From those, for lack of space, we present the performance comparison with other CSTs on five important operations: next-sibling, parent, child, suffix-link, and lca. To test our suffix tree in more complex scenarios we implemented the suffix-tree-based algorithm to solve the "maximal substrings" problem [26] on all of the above implementations except for CST_FULLY (because of its poor time performance). We use their same setup [26], that is, `influenza` from *Pizza & Chili* as our larger sequence and a substring of size m ($m = 3000$ and $m = 2\,MB$) of another `influenza` sequence taken from https://ftp.ncbi.nih.gov/genomes/INFLUENZA. BT-CST uses $r = 2$ and $mll = 128$ and GCST uses $rule\text{-}sampling = 1$ and $C\text{-}sampling = 2^{10}$. The tradeoffs refer to $sa\text{-}sampling \in \{64, 128, 256\}$ for the RLCSAs.

5.2 Results and Discussion

Figures 1, 2 and 3 show the space and time for all the indexes and all the operations. The smallest structure is GCST, which takes as little as 0.5–2 bits per symbol (bps). The next smallest indexes are BT-CST, using 1–3 bps, and CST_FULLY, using 2.0–2.5 bps. The compressed indexes not designed for repetitive collections use 4–7 bps if combined with a RLCSA, and 6–10.5 bps in their original versions (though we also adapted the bitvectors of CST_SCT3).

From the BT-CST space, component H takes just 2%–9%, the RLCSA takes 23%–47%, and the rest is the BT-CT (using a sweetpoint configuration).

This component takes 0.30 bits per node (bpn) on `einstein`, 1.06 bpn on `influenza`, and 1.50 bpn on `kernel`. The grammar-compressed topology of GCST takes, respectively, 0.05, 0.81, and 0.39 bpn.

In operations next-sibling and parent, which rely most heavily on the suffix tree topology, our BT-CT component building on Block Trees makes BT-CST excel in time: The operations take nearly one microsecond (μsec), at least 10 times less than the grammar-based topology representation of GCST. CST_FULLY is three orders of magnitude slower on this operation, taking over a millisecond (msec). Interestingly, the larger representations, including those where the tree topology is represented using 2.79 bits per node (CST_SADA[_RLCSA]), are only marginally faster than BT-CST, whereas the indexes CST_SCT3[_RLCSA] are a bit slower than CST_SADA[_RLCSA] because they do not store an explicit tree topology. Note that these operations, in BT-CT, make use of the operations *fwd-search* and *bwd-search*, thereby showing that they are fast although we cannot prove worst-case upper bounds on their time.

Operation lca, which on BT-CST involves essentially the primitive *min-excess*, is costlier, taking around $10\,\mu$s in almost all the indexes, including ours. This includes again those where the tree topology is represented using 2.79 bits per node (CST_SADA[_RLCSA]). Thus, although we cannot prove upper bounds on the time of *min-excess*, it is in practice as fast as on perfectly balanced structures, where it can be proved to be logarithmic-time. The variants CST_SCT3[_RLCSA] also require an operation very similar to *min-excess*, so they perform almost like CST_SADA[_RLCSA]. For this operation, CST_FULLY is equally fast, owing to the fact that operation lca is a basic primitive in this representation. Only GCST is several times slower than BT-CST, taking several tens of μsec.

Operation suffix-link involves *min-excess* and several other operations on the topology, but also the operation Ψ on the corresponding CSA. Since the latter is relatively fast, BT-CST also takes nearly $10\,\mu$s, whereas the additional operations on the topology drive GCST over $100\,\mu$s, and CST_FULLY over the msec. This time the topology representations that are blind to repetitivess are several times faster than BT-CST, taking a few μsec, possibly because they take more advantage of the smaller ranges for *min-excess* involved when choosing random nodes (most nodes have small ranges). The CST_SCT3[_RLCSA] variants also solve this operation with a fast and simple formula.

Finally, operation child is the most expensive one, requiring one application of string-depth and several of next-sibling and letter, thereby heavily relying on the CSA. BT-CST-bin and CST_SCT3[_RLCSA] binary search the children; the others scan them linearly. The indexes using a CSA that adapts to repetitiveness require nearly 1 ms on large alphabets, whereas those using a larger and faster CSA are up to 10 (CST_SCT3) and 100 (CST_SADA) times faster. Our BT-CST-bin variant is faster than the base BT-CST by 15% on `einstein` and 18% on `kernel`, and outperforms the RLCSA-based indexes. On DNA, instead, most of the indexes take nearly $100\,\mu$s, except for CST_SADA, which is several times faster; GCSA, which is a few times slower; and CST_FULLY, which stays in the msec.

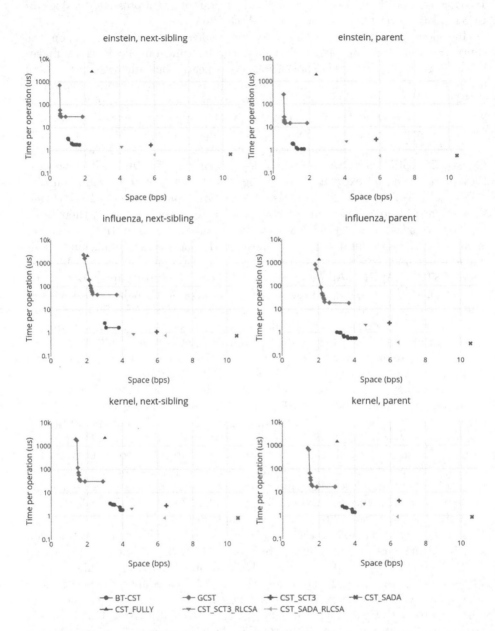

Fig. 1. Performance of CSTs for operations next-sibling and parent. The y-axis is in log-scale.

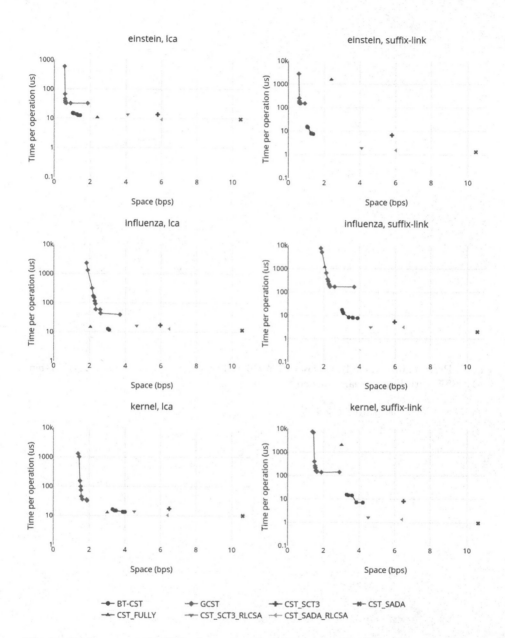

Fig. 2. Performance of CSTs for operations lca and suffix-link. The y-axis is in log-scale.

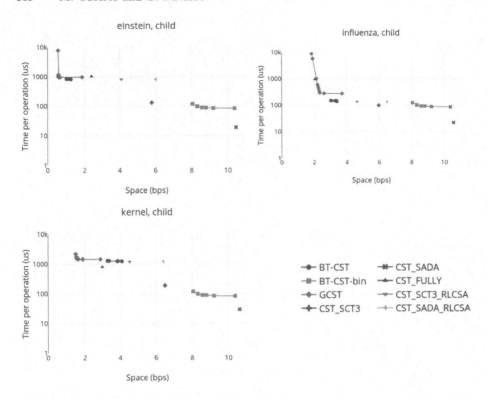

Fig. 3. Performance of CSTs for operation child. The y-axis is in log-scale. BT-CST-bin is BT-CST with binary search for child.

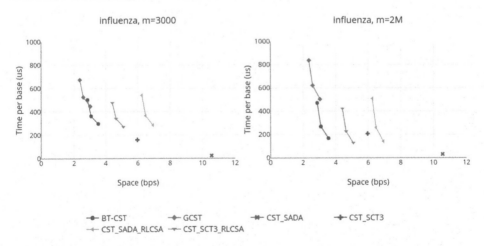

Fig. 4. Performance of CSTs when solving the maximal substrings problem. The y-axis is time in microseconds per base in the smaller sequence (of length m).

Figure 4 shows the results for the maximal substrings problem. BT-CST sharply dominates an important part of the Pareto-curve, including the sweet point at 3.5 bps and 200–300 μs per symbol. The other structures for repetitive collections take either much more time and slightly less space (GCSA, 1.5–2.5 times slower), or significantly more space and slightly less time (CST_SCT3, 45% more space and around 200 μs). CST_SADA is around 10 times faster, the same as its CSA when solving the dominant operation, child.

6   Conclusions and Future Work

We have introduced the Block-Tree Compressed Suffix Tree (BT-CST), a new compressed suffix tree aimed at indexing highly repetitive text collections. Its main feature is the BT-CT component, which uses Block Trees to represent the parentheses-based topology of the suffix tree and exploit the repetitiveness it inherits from the text collection. Block Trees [6] represent a sequence in space close to its Lempel-Ziv complexity (with a logarithmic-factor penalty), in a way that logarithmic-time access to any element is supported. The BT-CT enhances Block Trees with the more complex operations needed to simulate tree navigation on the parentheses sequence, as needed by the suffix tree operations.

Our experimental results show that the BT-CST requires 1–3 bits per symbol in highly repetitive text collections, which is slightly larger than the best previous alternatives [26], but also significantly faster (often by an order of magnitude). In particular, the BT-CT component uses 0.3–1.5 bits per node on these suffix trees and it takes a few microseconds to simulate the tree navigation operations, which is close to the time obtained by the classical 2.8-bit-per-node representation that is blind to repetitiveness [27]. This structure may be interesting to represent other repetitive trees beyond compressed suffix tree topologies, for example those arising in XML datasets, JSON repositories, and many others.

Although we have shown that in practice they perform as well as their classical counterpart [27], an interesting open problem is whether the operations *fwd-search*, *bwd-search*, and *min-excess* can be supported in polylogarithmic time on Block Trees. This was possible on perfectly balanced trees [27] and even on balanced-grammar parse trees [26], but the ability of Block Trees to refer to a prefix or a suffix of a block makes this more challenging. We note that the algorithm described by Belazzougui et al. [6] claiming logarithmic time for *min-excess* does not work (as checked with coauthor T. Gagie).

References

1. Abeliuk, A., Cánovas, R., Navarro, G.: Practical compressed suffix trees. Algorithms **6**(2), 319–351 (2013)
2. Abouelhoda, M.I., Kurtz, S., Ohlebusch, E.: Replacing suffix trees with enhanced suffix arrays. J. Discrete Algorithms **2**(1), 53–86 (2004)
3. Apostolico, A.: The myriad virtues of subword trees. In: Apostolico, A., Galil, Z. (eds.) Combinatorial Algorithms on Words, pp. 85–96. Springer, Heidelberg (1985). https://doi.org/10.1007/978-3-642-82456-2_6

4. Arroyuelo, D., et al.: Fast in-memory XPath search using compressed indexes. Softw. Pract. Exp. **45**(3), 399–434 (2015)
5. Belazzougui, D., Cunial, F.: Representing the suffix tree with the CDAWG. In: Proceedings of 28th Annual Symposium on Combinatorial Pattern Matching (CPM), pp. 7:1–7:13 (2017)
6. Belazzougui, D., et al.: Queries on LZ-bounded encodings. In: Proceedings of Data Compression Conference (DCC), pp. 83–92 (2015)
7. Clark, D.R., Ian Munro, J.: Efficient suffix trees on secondary storage. In: Proceedings of 17th Annual ACM-SIAM Symposium on Discrete Algorithms (SODA), pp. 383–391 (1996)
8. Farruggia, A., Gagie, T., Navarro, G., Puglisi, S.J., Sirén, J.: Relative suffix trees. Comput. J. **61**(5), 773–788 (2018)
9. Ferragina, P., Grossi, R.: The string B-tree: a new data structure for string search in external memory and its applications. J. ACM **46**(2), 236–280 (1999)
10. Fischer, J., Mäkinen, V., Navarro, G.: Faster entropy-bounded compressed suffix trees. Theor. Comput. Sci. **410**(51), 5354–5364 (2009)
11. Gagie, T., Navarro, G., Prezza, N.: Optimal-time text indexing in BWT-runs bounded space. CoRR, 1705.10382 (2017). arxiv.org/abs/1705.10382
12. Gog, S.: Compressed suffix trees: design, construction, and applications. Ph.D. thesis, University of Ulm, Germany (2011)
13. Gusfield, D.: Algorithms on Strings, Trees, and Sequences: Computer Science and Computational Biology. Cambridge University Press, Cambridge (1997)
14. Hon, W.-K., Shah, R., Thankachan, S.V., Vitter, J.S.: Space-efficient frameworks for top-k string retrieval. J. ACM **61**(2), 9:1–9:36 (2014)
15. Kieffer, J.C., Yang, E.-H.: Grammar-based codes: a new class of universal lossless source codes. IEEE Trans. Inf. Theory **46**(3), 737–754 (2000)
16. Kreft, S., Navarro, G.: On compressing and indexing repetitive sequences. Theor. Comput. Sci. **483**, 115–133 (2013)
17. Kurtz, S.: Reducing the space requirement of suffix trees. Softw. Pract. Exp. **29**(13), 1149–1171 (1999)
18. Larsson, J., Moffat, A.: Off-line dictionary-based compression. Proc. IEEE **88**(11), 1722–1732 (2000)
19. Lempel, A., Ziv, J.: On the complexity of finite sequences. IEEE Trans. Inf. Theory **22**(1), 75–81 (1976)
20. Mäkinen, V., Navarro, G., Sirén, J., Välimäki, N.: Storage and retrieval of highly repetitive sequence collections. J. Comput. Biol. **17**(3), 281–308 (2010)
21. Manber, U., Myers, G.: Suffix arrays: a new method for on-line string searches. SIAM J. Comput. **22**(5), 935–948 (1993)
22. McCreight, E.M.: A space-economical suffix tree construction algorithm. J. ACM **23**(2), 262–272 (1976)
23. Mozgovoy, M., Fredriksson, K., White, D., Joy, M., Sutinen, E.: Fast plagiarism detection system. In: Proceedings of 12th International Symposium on String Processing and Information Retrieval (SPIRE), pp. 267–270 (2005)
24. Navarro, G.: Indexing highly repetitive collections. In: Proceedings of 23rd International Workshop on Combinatorial Algorithms (IWOCA), pp. 274–279 (2012)
25. Navarro, G., Mäkinen, V.: Compressed full-text indexes. ACM Comput. Surv. **39**, 1 (2007)
26. Navarro, G., Ordóñez, A.: Faster compressed suffix trees for repetitive collections. J. Exp. Algorithmics **21**(1), 1–8 (2016)
27. Navarro, G., Sadakane, K.: Fully functional static and dynamic succinct trees. ACM Trans. Algorithms **10**(3), 16 (2014)

28. Ohlebusch, E., Fischer, J., Gog, S.: CST++. In: Proceedings of 17th International Conference on String Processing and Information Retrieval (SPIRE), pp. 322–333 (2010)
29. Ordóñez, A.: Statistical and repetition-based compressed data structures. Ph.D. thesis, Universidade da Coruña (2016)
30. Raman, R., Raman, V., Satti, S.R.: Succinct indexable dictionaries with applications to encoding k-ary trees, prefix sums and multisets. ACM Trans. Algorithms **3**(4), 43 (2007)
31. Raman, R., Rao, S.S.: Succinct representations of ordinal trees. In: Brodnik, A., López-Ortiz, A., Raman, V., Viola, A. (eds.) Space-Efficient Data Structures, Streams, and Algorithms. LNCS, vol. 8066, pp. 319–332. Springer, Heidelberg (2013). https://doi.org/10.1007/978-3-642-40273-9_20
32. Russo, L.M.S., Navarro, G., Oliveira, A.L.: Fully compressed suffix trees. ACM Trans. Algorithms **7**(4), 53:1–53:34 (2011)
33. Sadakane, K.: New text indexing functionalities of the compressed suffix arrays. J. Algorithms **48**(2), 294–313 (2003)
34. Sadakane, K.: Compressed suffix trees with full functionality. Theory Comput. Syst. **41**(4), 589–607 (2007)
35. Tishkoff, S.A., Kidd, K.K.: Implications of biogeography of human populations for 'race' and medicine. Nat. Genet. **36**, S21–S27 (2004)
36. Ukkonen, E.: On-line construction of suffix trees. Algorithmica **14**(3), 249–260 (1995)
37. Weiner, P.: Linear pattern matching algorithms. In: Proceedings of 14th Annual Symposium on Switching and Automata Theory (FOCS), pp. 1–11 (1973)
38. Zhang, D., Lee, W.S.: Extracting key-substring-group features for text classification. In: Proceedings of 12th Annual International Conference on Knowledge Discovery and Data Mining (SIGKDD), pp. 474–483 (2006)

A Practical Alphabet-Partitioning Rank/Select Data Structure

Diego Arroyuelo[1,2(✉)] and Erick Sepúlveda[2]

[1] Millennium Institute for Foundational Research on Data (IMFD), Santiago, Chile
[2] Department of Informatics, Universidad Técnica Federico Santa María,
Vicuña Mackenna 3939, Santiago, Chile
darroyue@inf.utfsm.cl, erick.sepulvedav@gmail.com

Abstract. This paper proposes a practical implementation of an *alphabet-partitioning* compressed data structure, which represents a string within compressed space and supports the fundamental operations rank and select efficiently. We show experimental results that indicate that our implementation outperforms the current realizations of the alphabet-partitioning approach (which is one of the most efficient approaches in practice). In particular, the time for operation select can be reduced by about 80%, using only 11% more space than current alphabet-partitioning schemes. We also show the impact of our data structure on several applications, like the intersection of inverted lists (where improvements of up to 60% are achieved, using only 2% of extra space), and the distributed-computation processing of rank and select operations. As far as we know, this is the first study about the support of rank/select operations on a distributed-computing environment.

1 Introduction

Given a string $s[1..n]$, over an alphabet $\Sigma = \{0, \ldots, \sigma - 1\}$, operation $s.\mathsf{rank}_c(i)$ computes the number of occurrences of symbol $c \in \Sigma$ in $s[1..i]$. Operation $s.\mathsf{select}_c(j)$, on the other hand, yields the position of the j-th occurrence of symbol c in s. Finally, operation $s.\mathsf{access}(i)$ yields symbol $s[i]$.

These operations are fundamental for many applications [17], such as snippet extraction in text databases [2], query processing in information retrieval [1,3], cardinal trees, text search, and graph representation [5], among others.

Since the amount of data managed by these applications is usually large, *space-efficient* data structures to support these operations are vital [17]. *Succinct data structures* use space close to the information theory minimum, while supporting operations efficiently. *Compressed data structures*, on the other hand, take advantage of certain regularities in the data to further reduce the space usage. These are the focus of this paper. In particular, we propose a surprisingly simple and practical implementation of the *alphabet-partition* approach [5] for compressing a string while supporting operations rank, select, and access.

Funded by the Millennium Institute for Foundational Research on Data (IMFD).

N. R. Brisaboa and S. J. Puglisi (Eds.): SPIRE 2019, LNCS 11811, pp. 452–466, 2019.
https://doi.org/10.1007/978-3-030-32686-9_32

We show that our data structure introduces interesting trade-offs for supporting these operations. Also, we show how our data structure impacts several important applications, and show proof-of-concept experiments on the distributed support of rank and select. This is the first such study we are aware of.

2   Related Work

2.1   Succinct Data Structures for Bit Vectors

In this paper we will need to use a succinct data structure to represent bit vectors $B[1..n]$ with m 1 bits, and support operations rank and select. In particular, we are interested in the SDarray data structure from Okanohara and Sadakane [18], which uses $m \lg \frac{n}{m} + 2m + o(m)$ bits of space, and supports select in $O(1)$ time (provided we replace the rank/select bit vector data structure on which SDarray is based by a constant-time select data structure [15]). Operations rank and access are supported in $O\left(\lg \frac{n}{m}\right)$ time.

2.2   Compressed Data Structures

A *compressed data structure* uses space proportional to some compression measure of the data, e.g., the 0-th order empirical entropy of a string $s[1..n]$ over an alphabet of size σ, which is denoted by $H_0(s)$ and defined as:

$$H_0(s) = \sum_{c \in \Sigma} \frac{n_c}{n} \lg \frac{n}{n_c}, \tag{1}$$

where n_c is the number of occurrences of symbol c in s. The sum includes only those symbols c that do occur in s, so that $n_c > 0$. The value $H_0(s) \le \lg \sigma$ is the average number of bits needed to encode each string symbol, provided we encode them using $\lg \frac{n}{n_c}$ bits.

2.3   Rank/Select Data Structures on Strings

Wavelet Trees. A *wavelet tree* [12] (WT for short) is a succinct data structure that supports rank and select operations, among many virtues [16]. The space requirement is $n \lg \sigma + o(n \lg \sigma)$ bits [12], while operations rank, select, and access take $O(\lg \sigma)$ time. To achieve compressed space, the WT can be given the shape of the Huffman tree, obtaining $nH_0(s) + o(nH_0(s)) + O(n)$ bits of space. Operations take $O(\lg n)$ worst-case, or $O(1 + H_0(s))$ on average [17]. Alternatively, one can use compressed bit vectors [20] to represent each WT node. The space usage is $nH_0(s) + o(n \lg \sigma)$ bits, and operations take $O(\lg \sigma)$ time.

The approach by Ferragina et al. [7], which is based on multiary WTs, supports the operations in $O\left(1 + \frac{\lg \sigma}{\lg \lg n}\right)$ worst-case time, and the space usage is $nH_0(S) + o(n \lg \sigma)$ bits. Later, Golynski et al. [11] improved the (lower-order term of the) space usage to $nH_0(s) + o(n)$ bits. Notice that if $\sigma = O(\text{polylog}(n))$, these solutions allow one to compute the operations in $O(1)$ time.

Reducing to Permutations. Golynski et al. [10] introduce an approach that is more effective than the previous ones for larger alphabets. Their solution requires $n(\lg \sigma + o(\lg \sigma))$ bits of space, supporting operation rank in $O(\lg \lg \sigma)$ time, whereas operations select and access are supported in $O(1)$ time (among other trade-offs, see the original paper for details). Later, Grossi et al. [13] achieve higher-order compression, that is $nH_k(s) + o(n \lg \sigma)$ bits. Operations rank and select are supported in $O(\lg \lg \sigma)$, whereas access is supported in $O(1)$ time.

Alphabet Partitioning. We are particularly interested in this paper in the alphabet-partitioning approach [5]. Given an alphabet $\Sigma = \{0, \ldots, \sigma - 1\}$, the aim of *alphabet partitioning* is to divide Σ into p subalphabets $\Sigma_0, \Sigma_1, \ldots, \Sigma_{p-1}$, such that $\bigcup_{i=0}^{p-1} \Sigma_i = \Sigma$, and $\Sigma_i \cap \Sigma_j = \emptyset$ for all $i \neq j$.

The Mapping from Alphabet to Subalphabet. The data structure [5] consists of an alphabet mapping $m[1..\sigma]$ such that $m[i] = j$ iff i has been mapped to subalphabet Σ_j. Within Σ_j, symbols are re-enumerated from 0 to $|\Sigma_j| - 1$ as follows: if there are k symbols smaller than i that have been mapped to Σ_j, then i is encoded as k in Σ_j. Formally, $k = m.\mathsf{rank}_j(i)$. Let $n_j = |\{i, \ m[s[i]] = j\}|$ be the number of symbols of string s that have been mapped to subalphabet Σ_j. A way of defining the partitioning (which is called sparse [5]) is:

$$m[\alpha] = \left\lceil \lg \left(\frac{n}{n_\alpha} \right) \lg n \right\rceil, \tag{2}$$

where symbol $\alpha \in \Sigma$ occurs n_α times in s. Notice that $m[\alpha] \leq \lceil \lg^2 n \rceil$.

The Subalphabet Strings. For each subalphabet Σ_ℓ, we store the subsequence $s_\ell[1..n_\ell]$, with the symbols of the original string s that have been mapped to subalphabet Σ_ℓ.

The Mapping from String Symbols to Subalphabets. In order to retain the original string, we store a sequence $t[1..n]$, which maps every symbol $s[i]$ into the corresponding subalphabet. That is, $t[i] = m[s[i]]$. If $\ell = t[i]$, then the corresponding symbol $s[i]$ has been mapped to subalphabet Σ_ℓ, and has been stored at position $t.\mathsf{rank}_\ell(i)$ in s_ℓ. Also, symbol $s[i]$ in Σ corresponds to symbol $m.\mathsf{rank}_\ell(s[i])$ in Σ_ℓ. Thus, we have $s_\ell[t.\mathsf{rank}_\ell(i)] = m.\mathsf{rank}_\ell(s[i])$.

Notice that t has alphabet of size p. Also, there are n_0 occurrences of symbol 0 in t, n_1 occurrences of symbol 1, and so on. Hence, we define:

$$H_0(t) = \sum_{i=0}^{p-1} \frac{n_i}{n} \lg \frac{n}{n_i}. \tag{3}$$

Computing the Operations. One can compute the desired operations as follows, assuming that m, t, and the sequences s_ℓ have been represented using

appropriate rank/select data structures (details about this later). For $\alpha \in \Sigma$, let $\ell = m.\text{access}(\alpha)$ and $c = m.\text{rank}_\ell(\alpha)$. Hence,

$$s.\text{rank}_\alpha(i) \equiv s_\ell.\text{rank}_c(t.\text{rank}_\ell(i)),$$

and

$$s.\text{select}_\alpha(j) \equiv t.\text{select}_\ell(s_\ell.\text{select}_c(j)).$$

If we now define $\ell = t[i]$, then we have

$$s.\text{access}(i) \equiv m.\text{select}_\ell(s_\ell.\text{access}(t.\text{rank}_\ell(i))).$$

Space Usage and Operation Times. Barbay et al. [5] have shown that $nH_0(t) + \sum_{i=0}^{p-1} n_\ell \lg \sigma_\ell \leq nH_0(s) + o(n)$. This means that if we use a zero-order compressed rank/select data structure for t, and then represent every s_ℓ even in uncompressed form, we obtain zero-order compression for the input string s. Recall that $p \leq \lceil \lg^2 n \rceil$, hence the alphabets of t and m are poly-logarithmic. Thus, a multi-ary wavelet tree [11] is used for t and m, obtaining $O(1)$ time for rank, select, and access. The space usage is $nH_0(t) + o(n)$ bits for t, and $O\left(\frac{n \lg \lg n}{\lg n}\right) H_0(s) = o(n)H_0(s)$ bits for m. For s_ℓ, if we use Golynski et al. data structure [10] we obtain a space usage of $n_\ell \lg \sigma_\ell + O\left(\frac{n_\ell \lg \sigma_\ell}{\lg \lg \lg n}\right)$ bits per partition, and support operation select in $O(1)$ time for s_ℓ, whereas rank and access are supported in $O(\lg \lg \sigma)$ time for s_ℓ.

Overall, the space is $nH_0(s) + o(n)(H_0(s) + 1)$ bits, operation select is supported in $O(1)$ time, whereas operations rank and access on the input string s are supported in $O(\lg \lg \sigma)$ time (see [5] for details and further trade-offs).

Practical Considerations. In practice, the sparse partitioning defined in Eq. (2) is replaced by an scheme such that for any $\alpha \in \Sigma$, $m[\alpha] = \lfloor \lg r(\alpha) \rfloor$. Here $r(\alpha)$ denotes the ranking of symbol α according to its frequency (that is, the most-frequent symbol has ranking 1, and the least-frequent one has ranking σ). Thus, the first partition contains only one symbol (the most-frequent one), the second partition contains two symbols, the third contains four symbols, and so on. Hence, there are $p = \lfloor \lg \sigma \rfloor$ partitions. This approach is called dense [5]. Another practical consideration is to have a parameter ℓ_{min} for dense, such that the top-$2^{\ell_{min}}$ symbols in the ranking are represented directly in t. That is, they are not represented in any partition. Notice that the original dense partitioning can be achieved by setting $\ell_{min} = 1$.

3   A Practical Alphabet-Partitioning Rank/Select Data Structure

The alphabet-partitioning approach was originally devised to speed-up decompression [21]. Barbay et al. [5] showed that alphabet partitioning is also effective for supporting operations rank and select on strings, being also one of the most

competitive approaches in practice. Next we introduce an alternative implementation of alphabet partitioning, which is able to trade operation-access efficiency for rank/select efficiency.

The main idea is as follows: in the original proposal, mapping t (introduced in Sect. 2.3) is represented with a multiary wavelet tree [11], supporting rank, select, and access in $O(1)$ time, since t has alphabet of size $O(\text{polylog}(n))$. However, as far as we know, there is no efficient implementation of multiary wavelet trees in practice. Indeed, Barbay et al. [5] use a WT in their experiments, whereas the sdsl library [9] uses a Huffman-shaped WT by default for t. We propose an implementation of the alphabet-partitioning approach that is faster in practice: rather than having a global mapping t, we distribute the workload among partitions. This shall allow us to use a simpler and faster approach (for instance, a single bit vector per partition) that replaces t.

3.1   Data Structure Definition

Our scheme consists of the mapping m and the subalphabets subsequences s_ℓ for each partition ℓ, just as originally defined in Sect. 2.3. In our case, however, we disregard mapping t, and replace it by a bit vector $B_\ell[1..n]$ per partition ℓ of the original alphabet. We set $B_\ell[i] = 1$ iff $s[i] \in \Sigma_\ell$ (or, equivalently, it holds that $m.\text{access}(s[i]) = \ell$). Notice that B_j has n_j 1s.

Given a symbol $\alpha \in \Sigma$ mapped to subalphabet $\ell = m.\text{access}(\alpha)$, let $c = m.\text{rank}_\ell(\alpha)$ be its representation in Σ_ℓ. Hence, we define:

$$s.\text{rank}_\alpha(i) \equiv s_\ell.\text{rank}_c(B_\ell.\text{rank}_1(i)).$$

Also,

$$s.\text{select}_\alpha(j) \equiv B_\ell.\text{select}_1(s_\ell.\text{select}_c(j)).$$

Unfortunately, operation $s.\text{access}(i)$ cannot be supported efficiently by our approach: since we do not know symbol $s[i]$, we do not know the partition j such that $B_j[i] = 1$. The alternative is to check every partition, until for a given ℓ it holds that $B_\ell[i] = 1$. Once this partition ℓ has been determined, we compute

$$s.\text{access}(i) \equiv m.\text{select}_\ell(s_\ell.\text{access}(B_\ell.\text{rank}_1(i))).$$

Although in general our implementation does not support access efficiently, there are still relevant applications where this operation is not needed, such as computing the intersection of inverted lists [3,5,17], or computing the term positions for phrase searching and positional ranking functions [2].

Besides, many applications need operation access to obtain not just a single symbol, but a snippet $s[i..i + L - 1]$—e.g., snippet-generation tasks [2,22]. In this case, one needs operation access to obtain not just a single symbol, but a snippet $s[i..i+L-1]$ of L consecutive symbols in s. Let us call $s.\text{snippet}(i, L)$ the corresponding operation. Rather than using operation access L times to obtain the desired symbols, we define Algorithm 1. The idea is that for each partition $j = 0, \ldots, p-1$, we obtain the symbols contained in $s[i..i+L-1]$ that correspond

to this partition. Line 4 of the algorithm computes the number of symbols to be extracted from this partition. Operation select on B_j is used to determine the position of each symbol within the snippet, as it can be seen in Line 7.

Algorithm 1. snippet(i, L)

1: Let $S[1..l]$ be an array of symbols in Σ.
2: **for** $j = 0$ **to** $p - 1$ **do**
3:     $cur \leftarrow B_j.\mathsf{rank}_1(i - 1)$
4:     $count \leftarrow B_j.\mathsf{rank}_1(i + L - 1) - cur$
5:     **for** $k = 1$ **to** $count$ **do**
6:         $cur \leftarrow cur + 1$
7:         $S[B_j.\mathsf{select}(cur) - i + 1] \leftarrow m.\mathsf{select}_j(s_j.\mathsf{access}(cur))$
8:     **end for**
9: **end for**
10: **return** S

3.2   Analysis of Space Usage and Query Time

If we use the SDarray representation of Okanohara and Sadakane [18] to represent the bit vectors B_ℓ, their total space usage is $\sum_{i=0}^{p-1} (n_i \lg \frac{n}{n_i} + 2n_i + o(n_i))$. Notice that $\sum_{i=0}^{p-1} n_i \lg \frac{n}{n_i} = nH_0(t)$, according to Eqs. (1) and (3). Also, we have that $2 \sum_{i=0}^{p-1} n_i = 2n$. Finally, for $\sum_{i=0}^{p-1} o(n_i)$ we have that each term in the sum is actually $O(n_i / \lg n_i)$ [18]. In the worst case, we have that every partition has $n_i = n/p$ symbols. Hence, $n_i / \lg n_i = n/(p \lg \frac{n}{p})$, which for p partitions yields a total space of $O(n/ \lg \frac{n}{p})$ bits. This is $o(n)$ since $\lg \frac{n}{p} \in w(1)$. In our case, $p \leq \lg^2 n$, hence $\sum_{i=0}^{p-1} o(n_i) \in o(n)$.

Summarizing, bit vectors B_ℓ require $n(H_0(t) + 2 + o(1))$ bits of space. This is 2 extra bits per symbol when compared to mapping t from Barbay et al.'s original approach [5]. The whole data structure uses $nH_0(s) + 2n + o(n)(H_0(s) + 1)$ bits.

Regarding operation times, $s.\mathsf{select}$ can be supported in $O(1)$ time (by using the SDarray from Sect. 2.1). Operation $s.\mathsf{rank}$ can be supported in $O(\lg n)$ worst-case time: if $n_i = O(\sqrt{n})$, operation $B_i.\mathsf{rank}$ takes $O(\lg \frac{n}{n_i}) = O(\lg n)$ time. Algorithm snippet takes $O(\sum_{i=0}^{p-1} \lg \frac{n}{n_i} + L \lg \lg \sigma)$ time. The sum $\sum_{i=0}^{p-1} \lg \frac{n}{n_i}$ is maximized when $n_i = n/p = n/ \lg^2 n$. Hence, $\sum_{i=0}^{p-1} \lg \frac{n}{n_i} = O(\lg^2 n \cdot \lg \lg n)$, thus the total time for snippet is $O(\lg^2 n \cdot \lg \lg n + L \lg \lg \sigma) = O((L + \lg^2 n) \lg \lg n)$. As a comparison, using operation access to extract the snippet would yield time $O(pL) = O(L \lg^2 n)$. When $L = \Theta(\lg \lg n)$, both approaches are asymptotically similar. However, as soon as $L = w(\lg \lg n)$, the time for algorithm snippet is $O((w(\lg \lg n) + \lg^2 n) \lg \lg n)$, versus $O(w(\lg \lg n) \cdot \lg^2 n)$ of operation access. Thus, for sufficiently long snippets, operation snippet is faster than using access.

Regarding construction time, bit vectors B_i can be constructed in linear time: we traverse string s from left to right; for each symbol $s[j]$, determine

its partition i and push-back the corresponding symbol in s_i, and position j into an extendible array [6]. Afterwards, the SDarray for B_i is constructed from this array. Extendible arrays can be implemented to obtain good performance in practice [14], so this imposes no restrictions to our approach.

4 Experimental Results and Applications

4.1 Experimental Setup

We implemented our data structure following the sdsl library [9]. Our source codes were compiled using g++ with flags -std=c++11 and -O3. Our source code can be downloaded from https://github.com/ericksepulveda/asap. We run our experiments on an HP Proliant server running an Intel(R) Xeon(R) CPU E5-2630 at 2.30 GHz, with 6 cores, 15 MB of cache, and 48 GB of RAM.

We used a 3.0 GB prefix of the Wikipedia (dump from August 2016). We removed the XML tags, leaving just the text. The text has 8,468,328 distinct words. We represent every word using a 32-bit unsigned integer, resulting in 1.9 GB of space. The zero-order empirical entropy of this string is 12.45 bits.

We tested sparse and dense partitioning, the latter with parameter $\ell_{min} = 1$ (i.e., the original dense partitioning), and $\ell_{min} = \lg \lg \sigma = \lg 23$ (which corresponds to the partitioning scheme currently implemented in sdsl). The number of partitions generated is 476 for sparse, 24 for dense $\ell_{min} = 1$, and 46 for dense $\ell_{min} = \lg 23$.

4.2 Experimental Results for Basic Operations

For operations rank and select, we tested two alternatives for choosing the symbols on which to base the queries:

Random Symbols: 30,000 alphabet symbols generated uniformly at random.
Query-log Symbols: we use the query log from the TREC 2007 Million Query Track[1]. We removed stopwords, and used only the words that exist in the alphabet. Overall we obtained 29,711 words (not necessarily unique).

For rank operation, we generate uniformly at random the positions where the query is issued. For select, we search for the j-th occurrence of a given symbol, with j generated at random (we are sure that there are at least j occurrences of the queried symbol). For operation access, we generate text positions at random.

Figures 1 and 2 show the experimental results for operations rank and select, comparing with the most efficient approaches from the sdsl. We name ASAP our approach, after agile and succinct alphabet partitioning. We show several combinations for mapping m and sequences s_ℓ, as well as several ways to carry out the alphabet partitioning. For instance, the label ASAP gmr-wm (D 23) corresponds to the scheme using Golynski et al. data structure [10] (gmr_wt<> in

---

[1] https://trec.nist.gov/data/million.query07.html.

sdsl) for s_ℓ, a wavelet matrix for mapping m, and dense $\ell_{min} = \lg 23$ partitioning. Label ASAP gmr-wm (D) is the same approach as before, this time using the original dense partitioning. The sparse partitioning is indicated with "(S)" in the labels. Bit vectors B_ℓ are implemented using sd_vector<> from sdsl, which corresponds to the SDArray data structure [18]. We show only the most competitive combinations. The original alphabet partitioning scheme is labeled AP in the plots. We used the default scheme from sdsl, which implements mappings t and m using Huffman-shaped WTs, and the sequences s_ℓ using wavelet matrices. The alphabet partitioning used is dense $\ell_{min} = \lg 23$. This was the most competitive combination for AP in our tests.

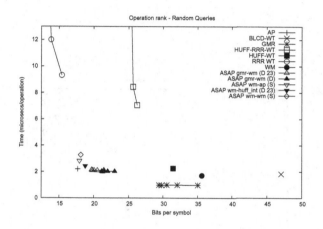

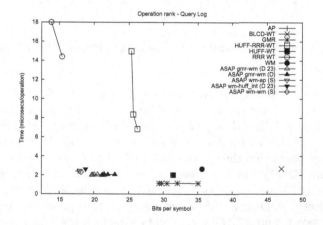

Fig. 1. Experimental results for rank. The x axis starts at $H_0(s) = 12.45$ bits.

As it can be seen, ASAP yields interesting trade-offs. In particular, for select on random symbols, alternative ASAP gmr-wm (D 23) uses 1.11 times the space

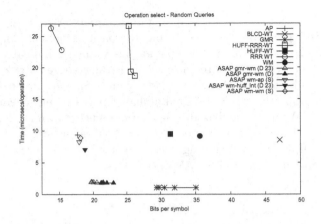

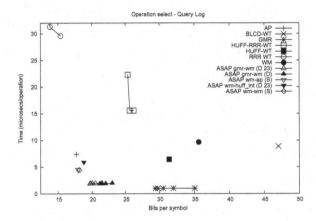

Fig. 2. Experimental results for select. The x axis starts at $H_0(s) = 12.45$ bits.

of AP, and reduces the average time per select by 79.50% (from 9.37 to 1.92 μs per select). For query-log symbols, we obtain similar results. However, in this case there is another interesting alternative: ASAP wm-ap (S) uses only 1.01 times the space of AP, and reduces the average select time by 38.80%. For rank queries we improve query time between 4.78% (ASAP wm-wm (S)) to 17.34% (ASAP gmr-wm (D 23)). In this case the improvements are smaller compared to select. This is because operation rank on bit vectors sd_vector<> is not as efficient as select [18]. Overall, ASAP gmr-wm (D 23) improves ASAP by 79.50% for operation select, and 17.34% for operation rank, using 1.11 times the space of ASAP.

Figure 3 shows experimental results for operation access. As expected, we cannot compete with the original AP scheme. However, we are still faster than RRR WT, and competitive with GMR [10] (yet using much less space).

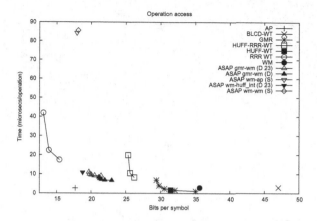

Fig. 3. Experimental results for access. The x axis starts at $H_0(s) = 12.45$ bits.

4.3 Application 1: Snippet Extraction

We study next the snippet extraction task, common in text search engines [2,22]. In our experiments we tested with $L = 100$ (see Fig. 4). As it can be seen, we are able to reduce the time per symbol considerably (approximately by 75%) when compared with operation access, making our approach more competitive for snippet extraction. It is important to note that operation B_j.select in line 7 of Algorithm 1 is implemented using the select operation provided by the sd_vector<> implementation. A more efficient approach in practice would be to have an iterator that allows one to obtain the desired 1 bits in segment $[i..i + L - 1]$ of the bit vectors, without repeatedly using operation select. This iterator is still not provided by the sdsl library and would not be effective for sd_vector<>, it could be a good idea for another kind of bit vectors.

4.4 Application 2: Intersection of Inverted Lists

Another relevant application of rank/select data structures is that of intersecting inverted lists. A previous work [3] has shown that one can carry out the intersection of inverted lists by representing the document collection (seen as a single string) with a rank/select data structure. Figure 5 shows experimental results for intersecting inverted lists. We implemented the variant of intersection algorithm tested by Barbay et al. [5]. As it can be seen, ASAP yields important improvements in this application: using only 2% extra space, ASAP wm-wm (S) is able to reduce the intersection time of AP by 60.67%. This is a promising result: our query time (of around 16 ms per query) is competitive with that of inverted indexes for BM25 query processing in IR (around 12 ms per query is the usual time reported in the literature [2,4]). There are, also, faster and smaller compression approaches for inverted indexes for the case of re-enumerated document collections [4], like the highly efficient Partitioned Elias-Fano (PEF) [19]. In this particular case, 16 ms per query is not competitive with PEF. Neither is

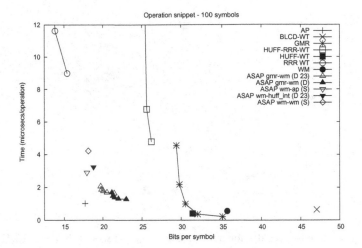

Fig. 4. Experimental results for extracting snippets of length $L = 100$. The x axis starts at $H_0(s) = 12.45$ bits.

the space usage. However, it is worth to remind that within the space of the H_0-compressed text, we are also implicitly storing the inverted index. This can be used not only to extract snippets, but also to look for positional information, substituting positional inverted indexes (or full inversions) [2], all within the same space. This would make our data structure more competitive.

4.5   Application 3: Distributed Computation of rank and select

The partitions generated by the alphabet partitioning approach are amenable for the distributed computation of batches of rank and select operations. In a distributed query processing system, a specialized node is in charge of receiving the query requests (this is called the *broker*), and then distributes the computation among the computation nodes (or simply *processors*). We study how to support the fast computation of batches of rank and select queries using the original AP approach and our proposal (ASAP).

A simple approach for operation rank would be to partition the input string into equal-size chunks, and let each processor deal with one chunk. To support the distributed computation of rank, the processor storing the ith chunk, from the left, also stores the rank of each symbol up to chunk $i - 1$. The space needed to store these ranks is $O(\sigma \lg n)$ bits per processor, which is impractical for big alphabets (as the ones we are testing in this paper). Also, this works only for operation rank, yet not for select. Next, we consider more efficient approaches in general.

A Distributed Query-Processing System Based on AP. The subalphabet sequences s_ℓ are distributed among the computation nodes, hence we have p processors in the system. We also have a specialized broker, storing mappings

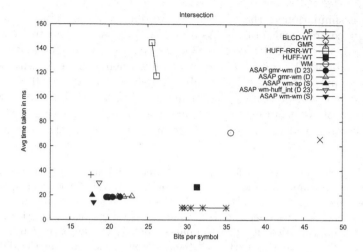

Fig. 5. Experimental results for inverted list intersection. Times are in milliseconds. The x axis starts at $H_0(s) = 12.45$ bits.

m and t. This is a drawback of this approach, as these mappings become a bottleneck for the distributed computation.

A Distributed Query-Processing System Based on ASAP. In this case, the sub-alphabet sequences s_ℓ and the bit vectors B_ℓ are distributed among the computation nodes. Unlike AP, now each computation node acts as a broker: we only need to replicate mapping m on them. The overall space usage is $O(p\sigma \lg \lg p)$ if we use an uncompressed WT for m. This is only $O(\sigma \lg \lg p) = o(n)H_0(s)$ bits per processor [5]. In this simple way, we avoid having a specialized broker, but distribute the broker task among the computation nodes. This avoids bottlenecks at the broker, and can make the system more fault tolerant.

Queries arrive at a computation node, which uses mapping m to distribute it to the corresponding computation node. For operation $s.\mathsf{access}(i)$, we carry out a broadcast operation, in order to determine for which processor ℓ, $B_\ell[i] = 1$; this is the one that must answer the query. For extracting snippets, on the other hand, we also broadcast the operation to all processors, which collaborate to construct the desired snippet using the symbols stored at each partition.

Comparison. The main drawback of the scheme based on AP is that it needs a specialized broker for m and t. Thus, the computation on these mappings is not distributed, lowering the performance of the system. The scheme based on ASAP, on the other hand, allows a better distribution: we only need to replicate mapping m in each processor, with a small space penalty in practice. To achieve a similar distribution with AP, we would need to replicate m and t in each processor, increasing the space usage considerably (mainly because of t). Thus, given a fixed amount of main memory for the system, the scheme based on ASAP would be likely able to represent a bigger string than AP. Table 1 shows experimental

results on a simulation of these schemes. We only consider computation time, disregarding communication time. As it can be seen, ASAP uses the distributed system in a better way. The average time per operation for rank and select are reduced by about 71% and 76%, respectively, when compared with AP. For extracting snippets, the time per symbol is reduced by about 50%. Although the speedup for 46 nodes might seem not too impressive (around 7–8), it is important to note that our experiments are just a proof of concept. For instance, the symbols could be distributed in such a way that the load balance is improved.

Table 1. Experimental results for the distributed computation of operations on a string. Times are in microseconds per operation, on average (for extracting snippets, it is microseconds per symbol). For rank and select, the symbols used are from our query log. Scheme ASAP implements the sequences s_ℓ using wavelet matrices, whereas mapping m is implemented using a Huffman-shaped WT. The partitioning is dense $\ell_{min} = \lg 23$. The number of partitions (i.e., computation nodes in the distributed system) is 46.

Operation	ASAP		AP		AP/ASAP
	Time	Speedup	Time	Speedup	
rank	0.373	8.03	1.310	1.91	3.51
select	0.706	8.41	2.970	2.55	4.21
access	1.390	8.11	2.130	1.45	1.53
snippet	0.466	6.96	0.939	1.25	2.02

5   Conclusions

Our alphabet-partitioning rank/select data structure offers interesting trade-offs in practice. Using slightly more space than the original alphabet-partitioning data structure from [5], we are able to reduce the time for operation select by about 80%. The performance for rank can be improved between 4% and 17%. For the inverted-list intersection problem, we showed improvements of about 60% for query processing time, using only 2% extra space when compared to the original alphabet-partitioning data structure. This makes this kind of data structures more attractive for this relevant application in information retrieval tasks. We also studied how the alphabet-partitioning data structures can be used for the distributed computation of batches of rank, select, access, and snippet operations. As far as we know, this is the first study about the support of these operation on a distributed-computing environment. In our experiments, we obtained speedups from 6.96 to 8.41, for 46 processors. This compared to 1.25–2.55 for the original alphabet-partitioning data structure. Our results were obtained simulating the distributed computation, hence considering only computation time (and disregarding communication time). The good performance observed in the experiments allows us to think about a real distributed-computing implementation.

This is left for future work, as well as a more in-depth study that includes aspects like load balance and total communication time, among others. As another interesting future work, it would be interesting to study how our data structure behaves with different alphabet sizes, as well as how it compares with approaches like the one used by Gog et al. [8].

References

1. Arroyuelo, D., Gil-Costa, V., González, S., Marin, M., Oyarzún, M.: Distributed search based on self-indexed compressed text. Inf. Process. Manag. **48**(5), 819–827 (2012)
2. Arroyuelo, D., González, S., Marín, M., Oyarzún, M., Suel, T.: To index or not to index: time-space trade-offs in search engines with positional ranking functions. In: Proceedings of 35th International ACM SIGIR Conference on Research and Development in Information Retrieval, pp. 255–264 (2012)
3. Arroyuelo, D., González, S., Oyarzún, M.: Compressed self-indices supporting conjunctive queries on document collections. In: Chavez, E., Lonardi, S. (eds.) SPIRE 2010. LNCS, vol. 6393, pp. 43–54. Springer, Heidelberg (2010). https://doi.org/10.1007/978-3-642-16321-0_5
4. Arroyuelo, D., Oyarzún, M., González, S., Sepulveda, V.: Hybrid compression of inverted lists for reordered document collections. Inf. Process. Manag. **54**(6), 1308–1324 (2018)
5. Barbay, J., Claude, F., Gagie, T., Navarro, G., Nekrich, Y.: Efficient fully-compressed sequence representations. Algorithmica **69**(1), 232–268 (2014)
6. Brodnik, A., Carlsson, S., Demaine, E.D., Ian Ian Munro, J., Sedgewick, R.: Resizable arrays in optimal time and space. In: Dehne, F., Sack, J.-R., Gupta, A., Tamassia, R. (eds.) WADS 1999. LNCS, vol. 1663, pp. 37–48. Springer, Heidelberg (1999). https://doi.org/10.1007/3-540-48447-7_4
7. Ferragina, P., Manzini, G., Mäkinen, V., Navarro, G.: Compressed representations of sequences and full-text indexes. ACM Trans. Algorithms **3**(2), 20 (2007)
8. Gog, S., Moffat, A., Petri, M.: CSA++: fast pattern search for large alphabets. In: Proceedings of 19th Workshop on Algorithm Engineering and Experiments (ALENEX), pp. 73–82 (2017)
9. Gog, S., Petri, M.: Optimized succinct data structures for massive data. Softw. Pract. Exper. **44**(11), 1287–1314 (2014)
10. Golynski, A., Munro, J.I., Srinivasa Rao, S.: Rank/select operations on large alphabets: a tool for text indexing. In: Proceedings of 17th Annual ACM-SIAM Symposium on Discrete Algorithms (SODA), pp. 368–373 (2006)
11. Golynski, A., Raman, R., Srinivasa Rao, S.: On the redundancy of succinct data structures. In: Proceedings of 11th Scandinavian Workshop on Algorithm Theory (SWAT), pp. 148–159 (2008)
12. Grossi, R., Gupta, A., Vitter, J.S.: High-order entropy-compressed text indexes. In: Proceedings of 14th Annual ACM-SIAM Symposium on Discrete Algorithms (SODA), pp. 841–850 (2003)
13. Grossi, R., Orlandi, A., Raman, R.: Optimal trade-offs for succinct string indexes. In: Abramsky, S., Gavoille, C., Kirchner, C., Meyer auf der Heide, F., Spirakis, P.G. (eds.) ICALP 2010. LNCS, vol. 6198, pp. 678–689. Springer, Heidelberg (2010). https://doi.org/10.1007/978-3-642-14165-2_57

14. Joannou, S., Raman, R.: An empirical evaluation of extendible arrays. In: Pardalos, P.M., Rebennack, S. (eds.) SEA 2011. LNCS, vol. 6630, pp. 447–458. Springer, Heidelberg (2011). https://doi.org/10.1007/978-3-642-20662-7_38
15. Munro, J.I.: Tables. In: Chandru, V., Vinay, V. (eds.) FSTTCS 1996. LNCS, vol. 1180, pp. 37–42. Springer, Heidelberg (1996). https://doi.org/10.1007/3-540-62034-6_35
16. Navarro, G.: Wavelet trees for all. J. Discrete Algorithms **25**, 2–20 (2014)
17. Navarro, G.: Compact Data Structures - A Practical Approach. Cambridge University Press, Cambridge (2016)
18. Okanohara, D., Sadakane, K.: Practical entropy-compressed rank/select dictionary. In: Proceedings of 9th Workshop on Algorithm Engineering and Experiments (ALENEX), pp. 60–70 (2007)
19. Ottaviano, G., Venturini, R.: Partitioned Elias-Fano indexes. In: Proceedings of 37th International ACM SIGIR Conference on Research and Development in Information Retrieval, pp. 273–282 (2014)
20. Raman, R., Raman, V., Rao Satti, S.: Succinct indexable dictionaries with applications to encoding k-ary trees, prefix sums and multisets. ACM Trans. Algorithms **3**(4), 43 (2007)
21. Said, A.: Efficient alphabet partitioning algorithms for low-complexity entropy coding. In: Proceedings of 15th Data Compression Conference (DCC), pp. 183–192 (2005)
22. Turpin, A., Tsegay, Y., Hawking, D., Williams, H.: Fast generation of result snippets in web search. In: Proceedings of 30th Annual International ACM SIGIR Conference on Research and Development in Information Retrieval, pp. 127–134 (2007)

Adaptive Succinctness

Diego Arroyuelo[1,2]([✉]) and Rajeev Raman[3]

[1] Millennium Institute for Foundational Research on Data (IMFD), Santiago, Chile
[2] Department of Informatics, Universidad Técnica Federico Santa María,
Santiago, Chile
darroyue@inf.utfsm.cl
[3] Department of Informatics, University of Leicester,
University Road, Leicester LE1 7RH, UK
r.raman@leicester.ac.uk

Abstract. Representing a static set of integers S, $|S| = n$ from a finite universe $U = [1..u]$ is a fundamental task in computer science. Our concern is to represent S in small space while supporting the operations of rank and select on S; if S is viewed as its characteristi c vector, the problem becomes that of representing a bit-vector, which is arguably the most fundamental building block of succinct data structures.

Although there is an information-theoretic lower bound of $\mathcal{B}(n, u) = \lg \binom{u}{n}$ bits on the space needed to represent S, this applies to worst-case (random) sets S, and sets found in practical applications are compressible. We focus on the case where elements of S contain non-trivial *runs* of consecutive elements, one that occurs in many practical situations.

Let $\mathcal{C}_n$ denote the class of $\binom{u}{n}$ distinct sets of n elements over the universe $[1..u]$. Let also $\mathcal{C}_g^n \subset \mathcal{C}_n$ contain the sets whose n elements are arranged in $g \leq n$ runs of $\ell_i \geq 1$ consecutive elements from U for $i = 1, \ldots, g$, and let $\mathcal{C}_{g,r}^n \subset \mathcal{C}_g^n$ contain all sets that consist of g runs, such that $r \leq g$ of them have at least 2 elements.

- We introduce new compressibility measures for sets, including:
  - $\mathcal{L}_1 = \lg |\mathcal{C}_g^n| = \lg \binom{u-n+1}{g} + \lg \binom{n-1}{g-1}$ and
  - $\mathcal{L}_2 = \lg |\mathcal{C}_{g,r}^n| = \lg \binom{u-n+1}{g} + \lg \binom{n-g-1}{r-1} + \lg \binom{g}{r}$

  We show that $\mathcal{L}_2 \leq \mathcal{L}_1 \leq \mathcal{B}(n, u)$.
- We give data structures that use space close to bounds $\mathcal{L}_1$ and $\mathcal{L}_2$ and support rank and select in $O(1)$ time.
- We provide additional measures involving entropy-coding run lengths and gaps between items, data structures to support these measures, and show experimentally that these approaches are promising for real-world datasets.

1 Introduction

Given a static sorted set $S = \{x_1, \ldots, x_n\}$ of n elements from a finite universe $U = [1..u] \subset \mathbb{N}$, such that $1 \leq x_1 < \cdots < x_n \leq u$, we want to support the following fundamental operations:

Funded by the Millennium Institute for Foundational Research on Data (IMFD).

N. R. Brisaboa and S. J. Puglisi (Eds.): SPIRE 2019, LNCS 11811, pp. 467–481, 2019.
https://doi.org/10.1007/978-3-030-32686-9_33

- rank(S, x), which for $x \in U$, yields $|\{x_i \in S, \ x_i \leq x\}|$, and
- select(S, k), which for $k \in \mathbb{N}$, yields $x \in S$ such that rank$(S, x) = k$.

If S is viewed as its characteristic bit vector (cbv for short) $C_S[1..u]$, such that $C_S[i] = 1$ iff $i \in S$, the problem becomes that of representing a bit vector with operations rank$_1(C_S, x)$, which yields the number of bits 1 in $C_S[1..x]$, and select$_1(C_S, k)$, that finds the position of the kth 1 bit in C_S. These are arguably the most fundamental building block of succinct data structures [28]. Also, they allow one to compute the fundamental operation predecessor$(S, x) \equiv$ select$(S, \text{rank}(S, x - 1))$ (among others). We assume the transdichotomous word RAM model with word size $w = O(\lg u) = \Omega(\lg n)$. Arithmetic, logic, and bitwise operations, as well as accesses to w-bit memory cells, take constant time.

Succinct data structures use space close to the corresponding information-theoretic lower bound while supporting operations efficiently. For instance, there are $\binom{u}{n}$ different subsets of U of size n. Hence, the information-theoretic lower bound on the number of bits needed to represent any such sets is $\mathcal{B}(n, u) = \lceil \lg \binom{u}{n} \rceil$ bits. If $n \ll u$, $\mathcal{B}(n, u) \approx n \lg e + n \lg \frac{u}{n} - O(\lg u)$ bits. Compressed data structures, on the other hand, exploit regularities in specific instances of data to go below the information-theoretic lower bound. The space usage of compressed data structures on a specific instance is evaluated relative to some measure of compressibility of that instance.

Starting with the seminal work of Jacobson [25], much effort has gone into representing sets succinctly while supporting rank and select in $O(1)$ time. Clark and Munro [9,10] were the first to achieve $O(1)$ time rank and select, using $u + O(u/\lg \lg u)$ bits of space. Raman et al. [36] achieved succinct space, $\mathcal{B}(n, u) + O(u \lg \lg u / \lg u)$ bits, with $O(1)$-time operations. In order to capture the compressibility of sets better, researchers have considered the *empirical higher-order entropy* of C_S, denoted by $H_k(C_S)$, which achieves good compression (beyond $\mathcal{B}(n, u)$). Several researchers, including Sadakane and Grossi [37] showed how to achieve constant-time operations while using $uH_k(S) + O(u((k+1) + \lg \lg u)/\lg u)$ bits. An important drawback is that for big universes (i.e., $n \ll u$), $H_k(C_S)$ decreases slowly as k grows. In addition, using small k it is not possible to capture longer-range dependencies (e.g. for any (long) string x, $H_k(x) \approx H_k(xx)$).

Another well-studied measure is GAP(S). If $S = \{x_1, \ldots, x_n\}$, with $x_1 < \ldots < x_n$, then define $g_1 = x_1 - 1$ and, for $i > 1$, $g_i = x_i - x_{i-1} - 1$, and

$$\text{GAP}(S) = \sum_{i=1}^{n} \{\lfloor \lg g_i \rfloor + 1\}.$$

Although GAP(S) is not an *achievable* measure[1], GAP(S) exploits variation in the gaps between elements. It can be seen that GAP$(S) < \mathcal{B}(n, u)$, and GAP(S) approaches $\mathcal{B}(n, u)$ only when $g_i = \frac{u}{n}$ (for $i = 1, \ldots, n$). Gupta et al. [23] showed

---

[1] For example, if we choose every element in U to be in S with probability 0.5, then GAP$(S) \sim 0.81u$, less than the Shannon lower bound for S.

how to represent S in $\text{GAP}(S) + \text{O}(n \lg \frac{u}{n}/\text{poly} \lg (n)) \leq \mathcal{B}(n, u)(1 + o(1))$ bits while supporting rank and select quickly (albeit not in constant time).

In this paper we focus on applications where set elements are clustered, forming runs of successive elements. Some applications are interval intersection in computational biology [34], web-graph compression [3,4], IR and query processing in reordered databases [2,26], valid-time joins in temporal databases [5,13,16,39], ancestor checking in trees [6,7], data structures for set intersection [8], and bit vectors of wavelet trees [14,22] of the Burrows-Wheeler transform of highly-repetitive texts [15,24,27]. Although $\text{GAP}(S)$ addresses this kind of non-uniform distribution, in the presence of runs, run-length encoding (RLE) [18] is more appropriate. Here, a set S with cbv $C_S[1..u] = \mathbf{0}^{z_1}\mathbf{1}^{l_1}\mathbf{0}^{z_2}\mathbf{1}^{l_2}\cdots\mathbf{0}^{z_g}\mathbf{1}^{l_g}$ is represented through the sequences $\langle z_1, \ldots, z_g \rangle$ and $\langle l_1, \ldots, l_g \rangle$ of $2g$ lengths of the alternating 0/1-runs in C_S (assume wlog that C_S begins with $\mathbf{0}$ and ends with $\mathbf{1}$). Then:

$$\text{RLE}(S) = \sum_{i=1}^{g} \{\lfloor \lg (z_i - 1) \rfloor + 1\} + \sum_{i=1}^{g} \{\lfloor \lg (l_i - 1) \rfloor + 1\}.$$

It holds that if $n < u/2$, $\text{RLE}(S) < \mathcal{B}(n, u) + n + \text{O}(1)$ [14]. Note that $\text{RLE}(S)$ is also not an achievable measure, but handles sets S that contain runs better than $\text{GAP}(S)$—a set S with cbv $\mathbf{0}^{u-n}\mathbf{1}^n$ has $\text{GAP}(S) = \Theta(n + \lg u)$ but $\text{RLE}(S) = \Theta(\lg n + \lg u)$.

In practice, $\text{GAP}(S)$, $\text{RLE}(S)$ and H_k each perform well on specific data sets S. In the important case when S is a posting list in an inverted index, a recent breakthrough [30] showed that so-called *partitioned* Elias-Fano (PEF) indices are very effective in compressing sets and can support select in $\text{O}(1)$ time. However, we are not aware of any compressibility measure associated with these indices, and it appears rank cannot be supported in constant time.

Since we wish not only to compress sets, but also to support operations on them, the *overall* space usage (including any space needed to support operations) is important. Firstly, since predecessor queries can be answered using rank and select, any lower bound for the former applies also to the joint use of the latter operations. Pătraşcu and Thorup [33] showed that if we use $\Theta(s \lg u)$ bits of space, the time to answer predecessor queries is given by:

$$\text{PT}(u, n, a) = \Theta(\min \{\tfrac{\lg n}{\lg \lg u}, \lg \tfrac{\lg (u/n)}{a}, \lg \tfrac{\lg u}{a} / \lg \tfrac{a \lg \frac{\lg u}{a}}{\lg n}, \lg \tfrac{\lg u}{a} / \lg \tfrac{\lg \frac{\lg u}{a}}{a} \}),$$

where $a = \lg \frac{s \lg u}{n}$. It follows from this that even if we are allowed to use $\text{O}(\text{poly } n)$ words of space, constant-time operations are possible only for relatively small universes, i.e. $u = \text{O}(n \cdot \text{poly} \lg(n))$, or for very small sets, i.e. $n = (\lg u)^{\text{O}(1)}$. Fortunately, the first case, which is our main focus, is also very commonly seen in applications.

A more refined analysis looks at the *redundancy* of a data structure, which is the space used by a data structure over and above the corresponding space bound for representing the set itself. Pătraşcu [32] improved on earlier work [19,21] and showed the following:

Theorem 1 ([32]). *For any $c > 0$, a set S can be represented using $\mathcal{B}(n, u) + O(u/\lg^c (u/c)) + O(u^{3/4} \text{poly} \lg(u))$ bits and support* rank *and* select *in $O(c)$ time.*

We will use Theorem 1 only when $c = O(1)$. For the parameter values of interest, namely $u = O(n \cdot \text{poly} \lg(n))$ and $c = O(1)$, the redundancy of Theorem 1 was shown to be optimal by Pǎtraşcu and Viola [31]. In the so-called *systematic* model, [20] gave matching upper and lower bounds on redundancy.

Contributions. Our contributions are as follows. Firstly, in Sects. 2 and 3, we give a surprisingly simple adaptive approach that stores S in potentially better than $\mathcal{B}(n, u)$ space, while still supporting constant-time rank/select. This is based on the intuition that within the class $\mathcal{C}_n$ of $\binom{u}{n}$ distinct sets of n elements from U, there are sets that can be represented more succinctly than others. For instance, in a extreme case where the elements form a single interval $[i..i + n - 1]$ of size n, why would one use $\lg \binom{u}{n}$ bits to describe this set? The smallest set element i and the set size n are enough to represent such a set.

Let class $\mathcal{C}_g^n \subset \mathcal{C}_n$ contain the sets whose elements are arranged in $g \le n$ runs of $l_i \ge 1$ successive elements from U, for $i = 1, \ldots, g$. Also, let $\mathcal{C}_{g,r}^n \subset \mathcal{C}_g^n$ be a further refinement of class $\mathcal{C}_g^n$, which contains all sets that consist of g runs, such that $r \le g$ of them have at least 2 elements.

- We introduce new compressibility measures for sets:
  - $\mathcal{L}_1 = \lg |\mathcal{C}_g^n| = \lg \binom{u-n+1}{g} + \lg \binom{n-1}{g-1}$ and
  - $\mathcal{L}_2 = \lg |\mathcal{C}_{g,r}^n| = \lg \binom{u-n+1}{g} + \lg \binom{n-g-1}{r-1} + \lg \binom{g}{r}$

  We show that $\mathcal{L}_2 \le \mathcal{L}_1 \le \mathcal{B}(n, u)$.
- We give data structures that use space close to bounds $\mathcal{L}_1$ and $\mathcal{L}_2$, namely $\lg \binom{u}{g} + \lg \binom{n}{g} + o(u) \approx \mathcal{L}_1 + o(u)$ bits of space and $\lg \binom{u}{g} + \lg \binom{n}{r} + \lg \binom{g}{r} + o(u) \approx \mathcal{L}_2 + o(u)$ bits of space, and support rank and select in $O(1)$ time.

Next, in Sect. 4, we revisit GAP(S) and RLE(S) measures in the following sense. The GAP(S) and RLE(S) measures encode a gap/run of length x using $1 + \lfloor \lg x \rfloor$ bits respectively. By Shannon's theorem, coding x using $1 + \lfloor \lg x \rfloor$ bits is tailoring the code length to a particular, and fixed, distribution of gap/run lengths[2]. We therefore propose two new measures of compressibility: we encode gap sizes and run lengths using their empirical zeroth-order entropy. That is, we treat the sequence of runs as a string of length $2g$ from the alphabet $[1..n]$, and encode each run using the Shannon optimal number of bits based upon the number of times this run length is seen (and analogously for gaps). On any set S, such approaches should outperform RLE(S) and GAP(S). For example, given a set S with $C_S = 0^4 1^{20} 0^4 1^{20} \cdots 0^4 1^{20}$, RLE($S$) is far inferior to H_0-coding the runs, which would use only one bit for encoding each run. We introduce two new measures of compressibility, $H_0^{\text{run}}(S)$ and $H_0^{\text{gap}}(S)$, to address this. We give data structures that support rank/select in $O(1)$ time using space close to $H_0^{\text{gap}}(S)$, and select on both S and its complement in $O(1)$ time using space close to $H_0^{\text{run}}(S)$. In this section, we also give additional compressibility measures.

---

[2] Since GAP(S) and RLE(S) are not achievable, this statement is imprecise.

Finally, in Sect. 5, we show experimentally that these approaches are promising for real-world datasets.

2   Adaptive Succinctness

We prove adaptive lower bounds for space needed to represent a set S.

Given a set $S = \{x_1, \ldots, x_n\} \subseteq U$ of n elements $1 \leq x_1 < \cdots < x_n \leq u$, a *maximal run of successive elements* $G \subseteq S$ contains $|G| \geq 1$ elements $x_i, x_i + 1, \ldots, x_i + |G| - 1$, such that $x_i - 1 \notin S$ and $x_i + |G| \notin S$. Let $G_1, \ldots, G_g$ be the partition of $S = \{x_1, \ldots, x_n\}$ into maximal runs of successive elements, such that $\forall x \in G_i, \forall y \in G_j, x < y \Leftrightarrow i < j$. Let us assume that $r \leq g$ of these G_i are of size $|G_i| \geq 2$.

Let $\mathcal{C}_n$ denote the class of sets of n elements from U. Notice that $|\mathcal{C}_n| = \binom{u}{n}$. We define class $\mathcal{C}_g^n \subset \mathcal{C}_n$, a refinement of $\mathcal{C}_n$ containing sets whose n elements are arranged in $g \leq n$ maximal runs of successive elements. Notice that $\bigcup_{g=1}^{n} \mathcal{C}_g^n = \mathcal{C}_n$. Let class $\mathcal{C}_{g,r}^n \subset \mathcal{C}_g^n$ be a further refinement consisting of all sets such that $r \leq g$ out of the g maximal runs have size ≥ 2. It holds that $\bigcup_{r=1}^{g} \mathcal{C}_{g,r}^n = \mathcal{C}_g^n$.

The cbv C_S of set $S \in \mathcal{C}_{g,r}^n$ consists of g 0-runs of lengths $z_1, \ldots, z_g$ such that $z_1 = \min\{G_1\} - 1$, and $z_i = \min\{G_i\} - \max\{G_{i-1}\} - 1$, and g 1-runs of length $l_i = |G_i|$, for $i = 1, \ldots, g$. We call *solitary* the elements in a run of size 1. Let $o_1 = l_{j_1} - 1, \ldots, o_r = l_{j_r} - 1$, for $j_1 < \cdots < j_r$, denote the sorted sequence of lengths (-1) of the r runs of length $l_{j_i} \geq 2$.

We show that within $\mathcal{C}_n$, there are sets that can be represented more succinctly than others, depending on which subclass they belong to. Less bit are needed to represent S if $S \in \mathcal{C}_g^n$ or, moreover, $S \in \mathcal{C}_{g,r}^n$. This is because the number of different such sets is smaller than $\binom{u}{n}$, so distinguishing them is easier. For instance, for $u = 6$ and $n = 3$ there are $\binom{6}{3} = 20$ different sets, yet if $g = 3$, the number of sets is only 4 ($\{1, 3, 5\}, \{1, 3, 6\}, \{1, 4, 6\}, \{2, 4, 6\}$). We formalize this fact next:

Theorem 2. *There are* $|\mathcal{C}_g^n| = \binom{n-1}{g-1}\binom{u-n+1}{g}$ *different sets whose n elements from U can be partitioned into g maximal runs of successive elements.*

Hence, we have:

Corollary 1. $\mathcal{L}_1 = \lg|\mathcal{C}_g^n| = \lg\binom{u-n+1}{g} + \lg\binom{n-1}{g-1}$ *bits are necessary to represent any set* $S \in \mathcal{C}_g^n$.

Next, we determine $|\mathcal{C}_{g,r}^n|$.

Theorem 3. *There are* $|\mathcal{C}_{g,r}^n| = \binom{n-g-1}{r-1}\binom{g}{r}\binom{u-n+1}{g}$ *different sets whose n elements from U can be partitioned into g maximal runs of successive elements, such that $r \leq g$ of these groups have at least 2 elements.*

Corollary 2. $\mathcal{L}_2 = \lg|\mathcal{C}_{g,r}^n| = \lg\binom{u-n+1}{g} + \lg\binom{n-g-1}{r-1} + \lg\binom{g}{r}$ *bits are necessary to represent any set* $S \in \mathcal{C}_{g,r}^n$.

3   Adaptive Succinct **rank/select** Data Structures

Given set $S \in \mathcal{C}_{g,r}^n$, let us define its *essential* sets:

1. $\hat{P} = \{p_1, \ldots, p_g\} \subseteq S$ such that $p_i = \min\{G_i\}$, for $i = 1, \ldots, g$. This set has universe u. We call each element p_i a *pioneer*;
2. $\hat{L} = \{l_1, \ldots, l_g\}$, such that $l_j = \sum_{i=1}^{j} |G_i|$. The corresponding characteristic bit vector is $C_{\hat{L}} = \mathbf{0}^{|G_1|-1}\mathbf{1}\mathbf{0}^{|G_2|-1}\mathbf{1}\cdots\mathbf{0}^{|G_g|-1}\mathbf{1}$ of length (universe of $\hat{L}$) n and g 1s. These are the unary encodings of $|G_i|$s.
3. $\hat{G} = \{y_1, \ldots, y_g\}$ such that $y_j = \sum_{i=1}^{j} z_i$, for $j = 1, \ldots, g$. The corresponding characteristic bit vector is $C_{\hat{G}} = \mathbf{0}^{z_1-1}\mathbf{1}\mathbf{0}^{z_2-1}\mathbf{1}\cdots\mathbf{0}^{z_g-1}\mathbf{1}$, of length (universe of $\hat{G}$) $\sum_{i=1}^{g} z_i = u - n$ (i.e., the number of 0s in C_S), and it has g 1s. These are the unary encodings of values z_i.
4. $\hat{R} = \{q_1, \ldots, q_r\}$ such that $q_j = \sum_{i=1}^{j} o_i$, for $j = 1, \ldots, r$. The corresponding characteristic bit vector is $C_{\hat{R}} = \mathbf{0}^{o_1-1}\mathbf{1}\mathbf{0}^{o_2-1}\mathbf{1}\cdots\mathbf{0}^{o_r-1}\mathbf{1}$, of length (universe of $\hat{R}$) $n - g$, and r 1s. These are the unary encodings of values o_i.
5. $\hat{V} = \{v_1, \ldots, v_r\}$ such that $|G_{v_i}| > 1$, for $i = 1, \ldots, r$. The corresponding characteristic bit vector $C_{\hat{V}}[1..g]$ has length g and r 1s. It holds that $C_{\hat{V}}[i] = 1$ iff the run G_i of the ith pioneer has $|G_i| > 1$.

A set S can be unambiguously described with the following combinations of essential sets: (Scheme 1) Sets $\hat{P}$ and $\hat{L}$; (Scheme 2) Sets $\hat{P}$, $\hat{R}$, and $\hat{V}$; (Scheme 3) Sets $\hat{G}$, $\hat{R}$, and $\hat{V}$; and (Scheme 4) Sets $\hat{G}$ and $\hat{L}$. Notice that, for instance, the Elias-Fano encoding of sets $\hat{G}$ and $\hat{L}$ (Scheme 4) yields space close to $\mathcal{L}_1$. The Elias-Fano encoding of Scheme 3, alternatively, yields space close to $\mathcal{L}_2$. In what follows, we build on above schemes to obtain adaptive succinct data structures.

3.1   Using Space Close to $\mathcal{L}_1$

Given a set $S \in \mathcal{C}_g^n$, consider the interval $[p_i..p_{i+1})$, for $1 \leq i < g$, between two consecutive pioneers. This is the *locus* of pioneer p_i [12]: all $\mathsf{rank}(S, x)$ queries within this interval (i.e., $p_i \leq x < p_{i+1}$) have similar answer, obtainable from p_i and $|G_i|$ (similarly for **select**). We use Scheme 1 above, which builds on sets $\hat{P}$ and $\hat{L}$. Building on Scheme 4 would use space closer to $\mathcal{L}_1$, however it does not allow (seemingly) for constant-time **rank/select**.

Operation $\mathsf{rank}(S, x)$. Notice that $\forall x$ such that $p_i \leq x < p_i + |G_i|$, $\mathsf{rank}(S, x) \equiv \mathsf{rank}(S, p_i) + x - p_i$; otherwise, if $p_i + |G_i| \leq x < p_{i+1}$, $\mathsf{rank}(S, x) \equiv \mathsf{rank}(S, p_i) + |G_i| - 1$. So, we show how to compute p_i, $\mathsf{rank}(S, p_i)$, and $|G_i|$ from sets $\hat{P}$ and $\hat{L}$. Let $i = \mathsf{rank}(\hat{P}, x)$ be the number of pioneers that are smaller (or equal) than x, then $p_i = \mathsf{select}(\hat{P}, i)$. Hence, $\mathsf{rank}(S, p_i) \equiv \mathsf{select}(\hat{L}, i - 1) + 1$, since $\mathsf{select}(\hat{L}, i - 1) = \sum_{j=1}^{i-1} |G_j|$. Finally, $|G_i| = \mathsf{select}(\hat{L}, i) - \mathsf{select}(\hat{L}, i - 1)$.

Operation $\mathsf{select}(S, k)$. Assume that for the element x_k we are looking for, it holds that $p_i \leq x_k < p_{i+1}$. Then, $\mathsf{select}(S, k) \equiv p_i + k - \mathsf{rank}(S, p_i)$. This time, $i = \mathsf{rank}(\hat{L}, k) + [k \notin \hat{L}]$[3], and $p_i = \mathsf{select}(\hat{p}, i)$. Finally, as explained above for operation rank, $\mathsf{rank}(S, p_i) \equiv \mathsf{select}(\hat{L}, i - 1) + 1$.

---

[3] $[k \notin \hat{L}]$ is Iverson brackets notation, which equals 1 iff $k \notin \hat{L}$ is true, 0 otherwise.

We represent $\hat{P}$ and $\hat{L}$ using Theorem 1. This uses $\lg \binom{u}{g} + \lg \binom{n}{g} + O(u/\lg^c u)$ bits, for any constant $c \geq 1$, and supports rank and select in $O(1)$ time.

Theorem 4. *There exists a data structure that represents any set $S \in C_g^n$ of n elements from universe U, using $\lg \binom{u}{g} + \lg \binom{n}{g} + O(u/\lg^c u)$ bits, for any constant $c \geq 1$, while supporting operations* rank *and* select *in $O(1)$ time.*

3.2   Using Space Close to $\mathcal{L}_2$

We use sets $\hat{P}$, $\hat{V}$, and the following variant of set $\hat{R}$: $\hat{R}' = \{q_1', \ldots, q_r'\}$ such that $q_j' = \sum_{i=1}^{j} (o_i + \text{select}(\hat{V}, i) - \text{select}(\hat{V}, i - 1))$. This set has universe $[1..n]$, and has r elements. To understand how this set works, let us see at its characteristic bit vector $C_{\hat{R}'}[1..n]$. It has r **1**s, each corresponding to a run G_i. Each such **1** is preceded by $|G_i| - 1$ **0**s. Consider runs G_i and G_{i+l}, for $i, l \geq 1$, such that runs $G_{i+1}, \ldots, G_{i+l-1}$ are each of size 1 (i.e., they correspond to solitary elements in S). Then, in $C_{\hat{R}'}$ there are $l - 1 + |G_i| - 1$ **0**s between the **1**s corresponding to G_i and G_{i+l}.

Operation rank(S, x). Let us assume that $p_i \leq x < p_{i+1}$, for $p_i \in \hat{P}$. First, consider the case where $|G_i| = 1$. That is, p_i is a solitary pioneer. Notice that rank$(S, x) \equiv$ rank$(S, p_{i-l}) + |G_{i-l}| - 1 + l$, for $l \geq 1$, such that p_{i-l} is the greatest pioneer smaller than p_i such that $|G_{i-l}| > 1$ (assume $p_{i-l} = 0$ if there is none). Let $i = \text{rank}(\hat{P}, x)$, and $p_i = \text{select}(\hat{P}, i)$. Then, $l = i - \text{select}(\hat{V}, \text{rank}(\hat{V}, i))$. Hence, rank$(S, p_{i-l}) + |G_{i-l}| - 1 \equiv \text{select}(\hat{R}', \text{rank}(\hat{V}, i))$, and we are done. Otherwise, $|G_i| > 1$, so we must distinguish two cases: (1) $p_i \leq x < p_i + |G_i|$, in whose case rank$(S, x) \equiv$ rank$(S, p_i) + x - p_i$; or (2) $p_i + |G_i| \leq x$, hence rank$(S, x) \equiv$ rank$(S, p_i) + |G_i| - 1$. Notice that rank$(S, p_i) \equiv$ rank$(S, p_{i-l}) + |G_{i-l}| - 1 + l$, which has been already computed. Finally, $|G_i| \equiv \text{select}(\hat{R}', \text{rank}(\hat{V}, i)) - \text{select}(\hat{R}', \text{rank}(\hat{V}, i) - 1) - l$.

Operation select(S, k). We must determine whether x_k is a solitary element or lies within a run of successive elements of length > 1. Let us regard runs G_v and G_i, both of size > 1, such that there is no other run of size > 1 between them, and $p_v + |G_v| - 1 < x_k \leq p_i + |G_i| - 1$. Here, $j = \text{rank}(\hat{R}', k) - [k \in \hat{R}']$ and $v = \text{select}(\hat{V}, j)$. The number of solitary pioneers between runs G_v and G_i is $l = \text{select}(\hat{V}, j + 1) - \text{select}(\hat{V}, j) - 1$. Let $s = \text{select}(\hat{R}', j)$ be the rank up to position $p_v + |G_v| - 1$ (i.e., up to the last element in G_v). Notice that if $k - s \leq l$, x_k is the $(k - s)$th pioneer after G_v Otherwise, if $k - s > l$, x_k lies within G_i.

We represent sets $\hat{P}$, $\hat{V}$, and $\hat{R}'$ using Theorem 1 and obtain:

Theorem 5. *There exists a data structure that represents any set $S \in C_{g,r}^n$ of n elements from universe U, using $\lg \binom{u}{g} + \lg \binom{n}{r} + \lg \binom{g}{r} + O(u/\lg^c u)$ bits, for any constant $c \geq 1$, while supporting operations* rank *and* select *in $O(1)$ time.*

4   Further Squeezing rank/select Data Structures

In this section, we study the extent to which a static rank/select data structure can be squeezed, while still supporting operations efficiently. Let t_r denote the

time complexity of operation rank, and t_s that of select. Since predecessor(S, x) can be reduced to select$(S, \text{rank}(S, x-1))$, Pătraşcu and Thorup's [33] predecessor lower bound is also a lower bound for $t_r + t_s$. It is natural, then, to compare with this lower bound to see how the time deteriorates as we squeeze.

4.1   GAP(S)

Gupta et al. [23] introduce a data structure using $\text{GAP}(S) + \text{O}(n \lg \frac{u}{n} / \lg n) + \text{O}(n \lg \lg \frac{u}{n})$ bits of space. Originally, Andersson and Thorup's predecessor data structure [1] is used as building block (using $\text{O}(n \lg \frac{u}{n} / \lg n)$ bits of space), yet it can be easily modified to use Pătraşcu and Thorup's data structure [33]; to get the same space usage, we set $a = \lg (\lg u / \lg^2 n)$. Operation select is supported in $\text{O}(\lg \lg n)$ time, and rank in $\text{PT}(u, n, \lg (\lg u / \lg^2 n)) + \text{O}(\lg \lg n)$ time.

4.2   RLE(S)

We now consider RLE(S), and begin by noting that $\text{RLE}(S) = \text{GAP}(\hat{G}) + \text{GAP}(\hat{L})$.

Property 1. For any set $S \in \mathcal{C}_g^n$ it holds that $\text{RLE}(S) \leq \lg \binom{u-n+1}{g} + \lg \binom{n}{g}$.

Since $\text{RLE}(S) = \text{GAP}(\hat{G}) + \text{GAP}(\hat{L})$, the proof is immediate. Set $\hat{G}$ has g elements over universe of size $u - n$, it holds that $\text{GAP}(\hat{G}) \leq \lg \binom{u-n}{g} \leq \lg \binom{u-n+1}{g}$. Similarly, $\text{GAP}(\hat{L}) \leq \lg \binom{n}{g}$.

Theorem 6. *There exists a data structure that represents any set $S \in \mathcal{C}_g^n$ over the universe U, using $\text{RLE}(S) + \text{O}(g \lg \lg \frac{u}{g} / \lg g) + \text{O}(g \lg \lg \frac{u}{g})$ bits of space, and supporting operation select in $\text{PT}(u - n, g, \lg \frac{\lg u - n}{\lg^2 g})$ time, whereas operation rank is supported in $\text{PT}(u - n, g, \lg \frac{\lg u - n}{\lg^2 g}) + \text{O}((\lg \lg g)^2)$ time.*

This is achieved by using the data structure from Sect. 4.1 on sets $\hat{G}$ and $\hat{L}$.

4.3 H_0 Coding of Gaps and Runs

In this section, we first describe our new measures of compressing sets based on H_0 coding the gaps or runs (called $H_0^{\text{gap}}(S)$ and $H_0^{\text{run}}(S)$). We then describe the main result of this section: a data structure that supports rank and select on a set S in constant time while using close to $H_0^{\text{gap}}(S)$ space. We then obtain as a corollary a representation of S using close to $H_0^{\text{run}}(S)$ space, but supporting only select on S and $U \setminus S$. Finally, we relate these measures to Theorem 3.

Definition 1. *Let $S = \{x_1, \ldots, x_n\}$, with $1 < x_1 < \ldots < x_n = u$. Define $g_1 = x_1 - 1$ and, for $i > 1$, $g_i = x_i - x_{i-1}$. Let $\mathcal{G}(S) = \{h_1^{m(h_1)}, h_2^{m(h_2)}, \ldots, h_t^{m(h_t)}\}$ be the multiset of values in the sequence $\langle g_1, \ldots, g_n \rangle$, where $m(h_i)$ is the multiplicity of h_i in the sequence of gaps. Then:*

$$nH_0^{\text{gap}}(S) = \lg \binom{n}{m(h_1), m(h_2), \ldots, m(h_t)}.$$

Letting $\hat{L}$ and $\hat{G}$ be as defined at the start of Sect. 3, we define:

$$H_0^{\mathrm{run}}(S) = H_0^{\mathrm{gap}}(\hat{L}) + H_0^{\mathrm{gap}}(\hat{G}).$$

Remark. Note that $H_0^{\mathrm{gap}}(S)$ is (almost) an achievable measure: we can apply arithmetic coding to the sequence of gaps, which would take $H_0^{\mathrm{gap}}(S)$ bits. However, we would also need to output the model of the arithmetic coder, which can be $\mathcal{G}(S)$ itself, stored in $t(\lg u + \lg n)$ bits, specifying which runs are present and their multiplicities. Since $t = O(\sqrt{u})$, this is not excessive for many applications. Similar remarks apply to $H_0^{\mathrm{run}}(S)$. Observe that $|\hat{L}| = |\hat{G}| = g$, so we would aim to compress S to $gH_0^{\mathrm{run}}(S)$ bits.

$\boldsymbol{H_0^{\mathrm{gap}}(S)}$. In this section we show the following:

Theorem 7. *Given a set $S \subseteq [1..u]$, $|S| = n$, we can represent it to support* rank *and* select *in $O(1)$ time using $nH_0^{\mathrm{gap}}(S) + O(n) + o(u)$ bits.*

Proof. Let the elements of S be $\{x_i\}_{i=1}^n$, sequence of gaps be $\{g_i\}_{i=1}^n$, and the multiset of gaps be $\mathcal{G}(S) = \{h_i^{m(h_i)}\}_{i=1}^t$. We first note that the result is achieved trivially if $n = O(u/\lg u)$: we represent S as the bit-string $0^{g_1-1}1 \ldots 0^{g_n-1}1$, which is of length u and has n 1s. If stored using Theorem 1, this bit-string will use $\frac{u}{B}\lg\frac{u}{nB} + O(\frac{u}{B}) + \frac{u}{(\lg u)^{O(1)}} = O(u\lg\lg u/\lg u) = o(u)$ bits and rank and select operations can be supported in $O(1)$ time, which proves the theorem. We therefore henceforth assume that $n \geq cu/\lg u$ for some sufficiently large $c \geq 1$.

We begin by converting S to a new set S' with $n' \leq n + u/B = O(n)$ numbers from U, for some integer parameter $B = \Theta((\lg n/\lg\lg n)^2)$, by setting $S' = S \cup \{iB | 1 \leq i \leq u/B\}$. Let the elements, the sequence of gaps, and the multiset of gaps of S' be $\{x_i'\}_{i=1}^{n'}$, $\{g_i'\}_{i=1}^{n'}$, and $\mathcal{G}(S') = \{h_i'^{m'(h_i')}\}_{i=1}^{t'}$. We now show that $H_0^{\mathrm{gap}}(S')$ is close to $H_0^{\mathrm{gap}}(S)$, using the following [11, Theorem 17.3.3]:

Theorem 8. *Let p and q be two probability mass functions on a set T such that $\|p - q\|_1 = \sum_{x \in T} |p(x) - q(x)| \leq \frac{1}{2}$. Then $|H(p) - H(q)| \leq \|p - q\|_1 \lg \frac{|T|}{\|p-q\|_1}$.*

Let T be the underlying set of the multiset $\mathcal{G}(S) \cup \mathcal{G}(S')$. For any integer $x \in T$ let $p(x) = m(x)/n$ and $q(x) = m'(x)/n'$; $m(x) = 0$ if $x \notin \mathcal{G}(S)$, and and similarly $m'(x)$. It is easy to see that $\|p - q\|_1 = O(u/(nB))$, since at most u/B gaps in S are changed during the conversion, and $n = \Theta(n')$. Since we assume $n \geq cu/\lg u$ for any constant c, we can ensure that $\|p - q\|_1 \leq 1/2$. Noting that $\|p - q\|_1 = \Omega(1/n)$ (unless $p = q$, in which case the RHS of Theorem 8 equals 0 as well), we see that $\lg\frac{|T|}{\|p-q\|_1} = O(\lg n|T|) = O(\lg u)$. It follows that $|H_0^{\mathrm{gap}}(S) - H_0^{\mathrm{gap}}(S')| = O(\frac{u\lg u}{nB})$ and hence that $|nH_0^{\mathrm{gap}}(S) - n'H_0^{\mathrm{gap}}(S')| = O(\frac{u\lg u}{B}) = O(u\frac{(\lg\lg u)^2}{\lg u}) = o(u)$.

The data structure comprises two parts: first, we divide the bit-string representing S into blocks of B consecutive bits (the i-th block corresponds to the interval $[(i-1)B + 1..iB]$). Next, we create a bit-string O which encodes, for each block, the count of the number of elements of S that lie in each block,

written in unary. O will have u/B **0**s and n **1**s. If stored using Theorem 1, O will use $\frac{u}{B} \lg \frac{u}{nB} + O(\frac{u}{B}) + \frac{u}{(\lg u)^{O(1)}} = O(u \lg \lg u / \lg u) = o(u)$ bits and support rank and select operations on both **0** and **1** in $O(1)$ time. It is easy to see (details omitted) how to reduce rank/select operations on S to rank/select operations on individual blocks using O.

We now describe the representation of an individual block. Each block is encoded independently; by Jensen's inequality, if the gaps in each block are encoded using H_0 bits, the total space usage of all block encodings is $H_0^{\text{gap}}(S')$. We also need to store the arithmetic coding model for each block, which requires $O(\sigma \lg B)$ bits, where $\sigma = O(\sqrt{B})$ is the number of distinct integers in a block. The overhead of the models in each block is therefore $O(\frac{u\sqrt{B}\lg B}{B}) = O(u\frac{(\lg \lg u)^2}{\lg u})$ bits. We now fix $B = (c \lg u / \lg \lg u)^2$ for sufficiently small constant c, and make the following observations that allow all operations in a block to be performed in $O(1)$ time using table lookup (details omitted): (i) all integers in a block are $O(\lg \lg u)$ bits long (ii) if we group the integers in a block with into sub-blocks of $c \lg u / \lg \lg u$ bits, we can ensure that the H_0 code of a sub-block is no more than $c' \lg u$ bits long for some $c' < 1$ (iii) the encoding of the arithmetic coding model for each block also fits into $c' \lg n$ bits (this is important since to decode the encoding of a sub-block, we must also use the arithmetic coding model as an argument to the table lookup). □

$H_0^{\textbf{run}}(\boldsymbol{S})$. Given a set $S \subseteq U$, let $\hat{L}$ and $\hat{G}$ be defined as in Sect. 3. Applying Theorem 7 to $\hat{L}$ and $\hat{G}$ and using [35, Theorem 1(c)], we obtain:

Corollary 3. *Given a set $S \subseteq U = [1..u]$ such that $|S| = n$ and $|\hat{L}| = g$, S can be represented in $gH_0^{\text{run}}(S) + O(g) + o(u)$ bits and support* select *on S and on $U \setminus S$ in $O(1)$ time.*

Discussion. Theorem 7 and Corollary 3 refine the results from Sect. 4.1 and Theorem 6 in terms of the space bound. In Sect. 3.2, we described Scheme 4, which comprises the sets $\hat{P}$, $\hat{V}$, and $\hat{R}'$. An alternative view of Scheme 1 from Sect. 3.1, and how it leads to Scheme 4 and then towards $H_0^{\text{run}}(S)$ is as follows. Scheme 1 identifies the start positions of the runs of 1s using $\hat{P}$, then encodes their lengths using $\hat{L}$; each run is encoded using $\lg(n/g) + O(1)$ bits, i.e., each run of 1s is encoded using a number of bits equal to the log of the average run length. This is clearly non-optimal if the distribution of the lengths of the runs is non-uniform, which can happen in many situations (for example, in a random bit-string, run-lengths are geometrically distributed). Scheme 4 improves upon Scheme 1 by encoding runs of length 1 using the Shannon optimal number of bits, based upon the number of times run length 1 length is seen. Choosing to focus on runs of length 1 can be non-optimal: e.g. in a set where the runs were of length $1, 2, 2, \ldots, 2$, Scheme 4 would offer no improvement over Scheme 1. The next step, that we consider experimentally, is to modify Scheme 4 to encode $\hat{L}$ adaptively using Theorem 7. Such a modification should give superior compression to Scheme 4, while supporting $O(1)$-time *rank/select*. Finally, $H_0^{\text{run}}(S)$

is the logical conclusion, replacing the "non-adaptive" encoding of $\hat{P}$ with an adaptive encoding of $\hat{G}$.

4.4 HYB(S)

Next, we study the following compression measure.

Definition 2. *Given set* $S \in \mathcal{C}_{g,r}^n$, *we define the entropy measure:*

$$\text{HYB}(S) = \sum_{i=1}^{g} \{\lfloor \lg(z_i - 1) \rfloor + 1\} + \sum_{i=1}^{r} \{\lfloor \lg(o_i - 1) \rfloor + 1\}$$
$$= \text{GAP}(\hat{G}) + \text{GAP}(\hat{R}).$$

Similar to Property 1, we have:

Property 2. For any set $S \in \mathcal{C}_{g,r}^n$ it holds that $\text{HYB}(S) \leq \lg \binom{u-n+1}{g} + \lg \binom{n-g}{r}$.

Property 3. Given set $S \in \mathcal{C}_{g,r}^n$, $\text{HYB}(S) \leq \min\{\text{GAP}(S), \text{RLE}(S)\}$.

Proof omited for lack of space.

Theorem 9. *There exists a data structure that represents any set* $S \in \mathcal{C}_{g,r}^n$ *over the universe* U, *using* $\text{HYB}(S)(1 + o(1)) + \text{O}(g \lg \frac{u-n}{g} / \lg g) + \text{O}(r \lg \frac{n-g}{r} / \lg r) + \lg \binom{g}{r}$ *bits. Operations* rank *and* select *are supported in* $\text{PT}(u - n, \frac{g}{\lg^2 g}, \lg \frac{\lg u - n}{\lg^2 g}) + \text{O}((\lg \lg g)^2)$ *worst-case time.*

4.5 $\widetilde{\text{HYB}}$(S)

Next, we use gap compression on sets $\hat{P}$ and $\hat{L}$ to obtain space smaller than $\text{GAP}(S)$ in some cases, with query time that equals that of Sect. 4.1 (and hence improving Theorems 6 and 9).

Definition 3. *Given a set* $S \in \mathcal{C}_{g,r}^n$ *with pioneers* $\hat{P} = \{p_1, \ldots, p_g\} \subseteq S$ *and 1-run lengths* $\hat{L} = \{l_1, \ldots, l_g\}$, *we define the following compression measure:*

$$\widetilde{\text{HYB}}(S) = \sum_{i=1}^{g} \{\lfloor \lg(p_i - p_{i-1} - 1) \rfloor + 1\} + \sum_{i=1}^{r} \{\lfloor \lg(l_i - 1) \rfloor + 1\}$$
$$= \text{GAP}(\hat{P}) + \text{GAP}(\hat{L}).$$

We can prove:

Lemma 1. *Given a set* $S \in \mathcal{C}_{g,r}^n$, *it holds that* $\widetilde{\text{HYB}}(S) \leq \lg \binom{u}{g} + \lg \binom{n}{g}$.

Theorem 10. *There exists a data structure that represents any set* $S \in \mathcal{C}_{g,r}^n$ *over the universe* U, *using* $\widetilde{\text{HYB}}(S)(1 + o(1)) + o(g \lg \frac{u}{g})$ *bits of space, and supports operation* select *in* $\text{O}(\lg \lg g)$ *time, and* rank *in* $\text{PT}(u, g, \lg \frac{\lg u}{\lg^2 g}) + \text{O}(\lg \lg g)$ *time.*

Proof. Use the data structure from Sect. 4.1 to represent sets $\hat{P}$ and $\hat{L}$, and use the O(1)-time support for rank and select described in Sect. 3.1.   □

5   Experimental Results

We show preliminary experiments on the space usage of the compressed approaches proposed in this paper. We use the URL-sorted GOV2 inverted index [38] as input, as this document order tends to generate runs in the posting lists. The universe size is 25,138,630 (this is the number of documents in the collection). We consider posting lists of size at least 100,000. We average the total space of each approach over 5,055,078,461 total postings. Table 1 shows the average number of bits per element for different state-of-the-art rank/select data structures (upper table) and different compression measures (bottom).

Table 1. Average number of bits per element for different rank/select data structures (on top) and compression measures (bottom), for URL-sorted GOV2 posting lists.

sd_vector	rrr<127>	rrr<63>	rrr<31>	rrr<15>	PEF	$\mathcal{L}_1$ d.s.	$\mathcal{L}_2$ d.s.
7.29	4.53	6.28	9.4	14.78	2.8	4.62	4.14

$\mathcal{L}_1$	$\mathcal{L}_2$	GAP(S)	RLE(S)	HYB(S)	$\widetilde{\text{HYB}}(S)$	$nH_0^{\text{gap}}(S)$	$gH_0^{\text{run}}(S)$
3.65	3.43	3.14	2.77	2.81	3.29	2.77	2.46

We considered the most efficient data structures from the sdsl library [17]: Elias-Fano sd_vector [29], Raman et al. [36] rrr data structure (using blocks of size 15, 31, 63, and 127). We also compared with the (very space-efficient) partitioned Elias-Fano approach [30] (PEF in the table). We used sd_vector to represent the sets that comprise the $\mathcal{L}_1$ and $\mathcal{L}_2$ data structures. The space for these data structures is reported at the top of Table 1.

In the bottom of Table 1, we report on compression measures applied to the above dataset (without making any allowance for space needed to support rank or select). Note that, as defined, the compression measures GAP(S), RLE(S), HYB(S), and $\widetilde{\text{HYB}}(S)$ are not realizable. To make a fair comparison with a realizable compression measure, we assume that gaps/runs are encoded using Elias-δ coding, i.e. our reported GAP(S) equals GAP(S) $= \sum_{i=1}^{n} \lfloor \lg g_i \rfloor + 2\lfloor \lg(\lfloor \lg g_i \rfloor + 1) \rfloor + 1$, and similarly for RLE($S$). Also, HYB($S$) includes space for the Elias-Fano representation of $\hat{V}$ (which uses space slightly more than $\lg \binom{g}{r}$ bits). As it can be seen, the results are promising in practice.

6   Conclusion and Open Problems

We have presented new measures of the compressibility of sets that are suitable when the elements of the sets are clustered in runs. In addition, when the sets are relatively dense (i.e. $n = u/(\lg u)^{O(1)}$) we present data structures whose space usage is close to these measures, but which support rank and select operations

in $O(1)$ time. Our preliminary experimental results show that our approaches yield space-efficient set representations.

There are a number of open directions that could be pursued. For example, we believe that an analogue of Theorem 7 for RLE(S) is well within reach. Other interesting directions would be to close the gap between the space bounds $\mathcal{L}_1$ and $\mathcal{L}_2$ and their corresponding data structures. Finally, the data structure of Theorem 7 is unlikely to be practical; finding a practical variant with small redundancy is another interesting question.

References

1. Andersson, A., Thorup, M.: Dynamic ordered sets with exponential search trees. J. ACM **54**(3), 13 (2007)
2. Arroyuelo, D., Oyarzún, M., González, S., Sepulveda, V.: Hybrid compression of inverted lists for reordered document collections. Inf. Process. Manag. **54**(6), 1308–1324 (2018)
3. Boldi, P., Vigna, S.: The webgraph framework I: compression techniques. In: Proceedings of the 13th International Conference on World Wide Web (WWW), pp. 595–602 (2004)
4. Boldi, P., Vigna, S.: The webgraph framework II: codes for the world-wide web. In: Proceedings of the Data Compression Conference (DCC), p. 528 (2004)
5. Cafagna, F., Böhlen, M.H.: Disjoint interval partitioning. VLDB J. **26**(3), 447–466 (2017)
6. Chen, Y., Chen, Y.: An efficient algorithm for answering graph reachability queries. In: Proceedings of the 24th International Conference on Data Engineering (ICDE), pp. 893–902 (2008)
7. Chen, Y., Chen, Y.: Decomposing DAGs into spanning trees: a new way to compress transitive closures. In: Proceedings of the 27th International Conference on Data Engineering (ICDE), pp. 1007–1018 (2011)
8. Chen, Y., Shen, W.: An efficient method to evaluate intersections on big data sets. Theor. Comput. Sci. **647**, 1–21 (2016)
9. Clark, D.: Compact PAT trees. Ph.D. thesis, University of Waterloo (1997)
10. Clark, D.R., Munro, J.I.: Efficient suffix trees on secondary storage (extended abstract). In: Proceedings of the 7th Annual ACM-SIAM Symposium on Discrete Algorithms (SODA), pp. 383–391 (1996)
11. Cover, T.M., Thomas, J.A.: Elements of Information Theory. Wiley Interscience, Hoboken (2006)
12. de Berg, M., Cheong, O., van Kreveld, M.J., Overmars, M.H.: Computational Geometry: Algorithms and Applications, 3rd edn. Springer, Heidelberg (2008). https://doi.org/10.1007/978-3-540-77974-2
13. Dignös, A., Böhlen, M.H., Gamper, J.: Overlap interval partition join. In: Proceedings of the 2014 International Conference on Management of Data (SIGMOD), pp. 1459–1470 (2014)
14. Foschini, L., Grossi, R., Gupta, A., Vitter, J.S.: When indexing equals compression: experiments with compressing suffix arrays and applications. ACM Trans. Algorithms **2**(4), 611–639 (2006)
15. Gagie, T., Navarro, G., Prezza, N.: Optimal-time text indexing in BWT-runs bounded space. In: Proceedings of the 29h Annual ACM-SIAM Symposium on Discrete Algorithms (SODA), pp. 1459–1477 (2018)

16. Gao, D., Jensen, C.S., Snodgrass, R.T., Soo, M.D.: Join operations in temporal databases. VLDB J. **14**(1), 2–29 (2005)
17. Gog, S., Petri, M.: Optimized succinct data structures for massive data. Softw. Practrice Exp. **44**(11), 1287–1314 (2014)
18. Golomb, S.: Run-length encodings (corresp.). IEEE Trans. Inf. Theory **12**(3), 399–401 (1966)
19. Golynski, A., Grossi, R., Gupta, A., Raman, R., Rao, S.S.: On the size of succinct indices. In: Arge, L., Hoffmann, M., Welzl, E. (eds.) ESA 2007. LNCS, vol. 4698, pp. 371–382. Springer, Heidelberg (2007). https://doi.org/10.1007/978-3-540-75520-3_34
20. Golynski, A., Orlandi, A., Raman, R., Srinivasa Rao, S.: Optimal indexes for sparse bit vectors. Algorithmica **69**(4), 906–924 (2014)
21. Golynski, A., Raman, R., Rao, S.S.: On the redundancy of succinct data structures. In: Gudmundsson, J. (ed.) SWAT 2008. LNCS, vol. 5124, pp. 148–159. Springer, Heidelberg (2008). https://doi.org/10.1007/978-3-540-69903-3_15
22. Grossi, R., Gupta, A., Vitter, J.S.: High-order entropy-compressed text indexes. In: Proceedings of the 14th Annual ACM-SIAM Symposium on Discrete Algorithms (SODA), pp. 841–850 (2003)
23. Gupta, A., Hon, W.-K., Shah, R., Vitter, J.S.: Compressed data structures: dictionaries and data-aware measures. Theor. Comput. Sci. **387**(3), 313–331 (2007)
24. Huo, H., Chen, L., Zhao, H., Vitter, J.S., Nekrich, Y., Yu, Q.: A data-aware fm-index. In: Proceedings of the 17th Workshop on Algorithm Engineering and Experiments (ALENEX), pp. 10–23 (2015)
25. Jacobson, G.: Space-efficient static trees and graphs. In: Proceedings of the 30th Annual Symposium on Foundations of Computer Science (FOCS), pp. 549–554 (1989)
26. Johnson, D.S., Krishnan, S., Chhugani, J., Kumar, S., Venkatasubramanian, S.: Compressing large boolean matrices using reordering techniques. In: Proceedings of the 30th International Conference on Very Large Data Bases (VLDB), pp. 13–23 (2004)
27. Mäkinen, V., Navarro, G.: Succinct suffix arrays based on run-length encoding. Nord. J. Comput. **12**(1), 40–66 (2005)
28. Navarro, G.: Compact Data Structures - A Practical Approach. Cambridge University Press, Cambridge (2016)
29. Okanohara, D., Sadakane, K.: Practical entropy-compressed rank/select dictionary. In: Proceedings of the 9th Workshop on Algorithm Engineering and Experiments (ALENEX), pp. 60–70 (2007)
30. Ottaviano, G., Venturini, R.: Partitioned Elias-fano indexes. In: Proceedings of the 37th International ACM SIGIR Conference on Research and Development in Information Retrieval, pp. 273–282 (2014)
31. Pătraşcu, M., Viola, E.: Cell-probe lower bounds for succinct partial sums. In: Proceedings of the 21st Annual ACM-SIAM Symposium on Discrete Algorithms (SODA), pp. 117–122 (2010)
32. Pătraşcu, M.: Succincter. In: Proceedings of the 49th Annual IEEE Symposium on Foundations of Computer Science (FOCS), pp. 305–313 (2008)
33. Pătraşcu, M., Thorup, M.: Time-space trade-offs for predecessor search. In: Proceedings of the 38th Annual ACM Symposium on Theory of Computing (STOC), pp. 232–240 (2006)
34. Quinlan, A.R., Robins, G., Hall, I.M., Skadron, K., Layer, R.M.: Binary Interval Search: a scalable algorithm for counting interval intersections. Bioinformatics **29**(1), 1–7 (2012)

35. Rahman, N., Raman, R.: Rank and select operations on binary strings. In: Kao, M.Y. (ed.) Encyclopedia of Algorithms. Springer, Boston (2008). https://doi.org/10.1007/978-0-387-30162-4

36. Raman, R., Raman, V., Rao Satti, S.: Succinct indexable dictionaries with applications to encoding k-ary trees, prefix sums and multisets. ACM Trans. Algorithms **3**(4), 43 (2007)

37. Sadakane,K., Grossi, R.: Squeezing succinct data structures into entropy bounds. In: Proceedings of the 17th Annual ACM-SIAM Symposium on Discrete Algorithms (SODA), pp. 1230–1239 (2006)

38. Silvestri, F.: Sorting out the document identifier assignment problem. In: Amati, G., Carpineto, C., Romano, G. (eds.) ECIR 2007. LNCS, vol. 4425, pp. 101–112. Springer, Heidelberg (2007). https://doi.org/10.1007/978-3-540-71496-5_12

39. Soo, M.D., Snodgrass, R.T., Jensen, C.S.: Efficient evaluation of the valid-time natural join. In: Proceedings of the 10th International Conference on Data Engineering (ICDE), pp. 282–292 (1994)

Fast, Small, and Simple Document Listing on Repetitive Text Collections

Dustin Cobas$^{(\boxtimes)}$ and Gonzalo Navarro

CeBiB — Center for Biotechnology and Bioengineering,
Department of Computer Science, University of Chile, Santiago, Chile
{dcobas,gnavarro}@dcc.uchile.cl

Abstract. Document listing on string collections is the task of finding all documents where a pattern appears. It is regarded as the most fundamental document retrieval problem, and is useful in various applications. Many of the fastest-growing string collections are composed of very similar documents, such as versioned code and document collections, genome repositories, etc. Plain pattern-matching indexes designed for repetitive text collections achieve orders-of-magnitude reductions in space. Instead, there are not many analogous indexes for document retrieval. In this paper we present a simple document listing index for repetitive string collections of total length n that lists the $ndoc$ distinct documents where a pattern of length m appears in time $\mathcal{O}(m + ndoc \cdot \lg n)$. We exploit the repetitiveness of the *document array* (i.e., the suffix array coarsened to document identifiers) to grammar-compress it while precomputing the answers to nonterminals, and store them in grammar-compressed form as well. Our experimental results show that our index sharply outperforms existing alternatives in the space/time tradeoff map.

1 Introduction

Document retrieval is a family of problems aimed at retrieving *documents* from a set that are relevant to a query *pattern*. In a general setting, both documents and patterns are arbitrary strings. This encompasses the well-known application of natural language and Web searching, but also many others of interest in bioinformatics, software development, multimedia retrieval, etc. [22].

The most fundamental document retrieval problem, on top of which more sophisticated ones are built, is *document listing*. This problem aims at simply returning the list of documents where the pattern appears. An obvious solution to document listing resorts to *pattern matching*: find all the occ positions where the pattern appears, and then return the $ndoc$ different documents where those positions lie. This solution requires time $\Omega(occ)$ and the output is of size $\mathcal{O}(ndoc)$, so the approach is very inefficient if $ndoc \ll occ$ (i.e., if the pattern appears many times in the same documents). A better solution, which however applies only in natural language settings, resorts to *inverted indexes* [1]. These restrict

Funded with basal funds FB0001 and by Fondecyt Grant 1-170048, Conicyt, Chile.

N. R. Brisaboa and S. J. Puglisi (Eds.): SPIRE 2019, LNCS 11811, pp. 482–498, 2019.
https://doi.org/10.1007/978-3-030-32686-9_34

the possible patterns to sequences of words and store the list of the documents where each word appears, thereby solving document listing via intersections of the lists of the pattern words.

Muthukrishnan [20] designed the first linear-space and optimal-time index for general string collections. Given a collection of total length n, he builds an index of $\mathcal{O}(n)$ words that lists the $ndoc$ documents where a pattern of length m appears in time $\mathcal{O}(m + ndoc)$. While linear space is deemed as sufficiently small in classic scenarios, the solution is impractical for very large text collections unless one resorts to disk, which is orders of magnitude slower. Sadakane [26] showed how to reduce the space of Muthukrishnan's index to that of the statistically-compressed text plus $\mathcal{O}(n)$ *bits*, while raising the time complexity to only $\mathcal{O}(m + ndoc \cdot \lg n)$ if the appropriate underlying pattern-matching index is used [2].

The sharp growth of text collections is a concern in many recent applications, outperforming Moore's Law in some cases [27]. Fortunately, many of the fastest-growing text collections are *highly repetitive*: each document can be obtained from a few large blocks of other documents. These collections arise in different areas, such as repositories of genomes of the same species (which differ from each other by a small percentage only) like the 100K-genome project[1], software repositories that store all the versions of the code arranged in a tree or acyclic graph like GitHub[2], versioned document repositories where each document has a timeline of versions like Wikipedia[3], etc. On such text collections, statistical compression is ineffective [14] and even $\mathcal{O}(n)$ bits of extra space can be unaffordable.

Repetitiveness is the key to tackle the fast growth of these collections: their amount of new material grows much slower than their size. For example, version control systems compress those collections by storing the list of edits with respect to some reference document that is stored in plain form, and reconstruct it by applying the edits to the reference version. Much more challenging, however, is to *index* those collections in small space so as to support fast pattern matching or document retrieval tasks. To date, there exist several pattern matching indexes for repetitive text collections (see a couple of studies [10,21] and references therein). However, there are not many document retrieval indexes for repetitive text collections [5,8,23]. Most of these indexes [8,26] rely on a pattern-matching index needs $\Omega(n)$ bits in order to offer $\mathcal{O}(\lg n)$ time per retrieved document.

In this paper we introduce new simple and efficient document listing indexes aimed at highly repetitive text collections. Like various preceding indexes, we achieve $\mathcal{O}(m + ndoc \cdot \lg n)$ search time, yet our indexes are way faster and/or smaller than previous ones on various repetitive datasets, because they escape from the space/time tradeoff of the pattern-matching index. Our main idea is as follows: we use the *document array* $DA[1..n]$ [20], which projects the entries

---

[1] www.genomicsengland.co.uk/about-genomics-england/the-100000-genomes-project.

[2] github.com/search?q=is:public.

[3] en.wikipedia.org/wiki/Wikipedia:Size_of_Wikipedia.

of the *suffix array* [19] to the document where each position belongs. Document listing boils down to listing the distinct integers in a range $DA[sp..ep]$, where sp and ep are found in time $\mathcal{O}(m)$. Array DA must be grammar-compressible since the differential suffix array is known to be so on repetitive texts [10,11]. We then build a *balanced* binary context-free grammar that generates (only) DA. This allows us retrieve any individual cell of DA in time $\mathcal{O}(\lg n)$ and any range $DA[sp..ep]$ in time $\mathcal{O}(ep - sp + \lg n)$. We can then implement existing indexes [8,26] within much less space and without affecting their time complexities. Further, we propose a new simple index based on the grammar-compressed array DA. Our compression guarantees that any range $DA[sp..ep]$ is covered by $\mathcal{O}(\lg n)$ nonterminals. For each nonterminal of the grammar, we store the list of the distinct documents appearing in it. The set of all the lists is grammar-compressed as well, as done in previous work [5,8]. We then merge the lists of the $\mathcal{O}(\lg n)$ nonterminals that cover $DA[sp..ep]$, in time $\mathcal{O}(ndoc \cdot \lg n)$.

2   Preliminaries

A *document* T is a sequence of symbols over an alphabet $\Sigma = [1..\sigma]$, terminated by a special symbol \$ that is lexicographically smaller than any symbol of Σ.

A *collection* $\mathcal{D}$ is a set of d documents $\mathcal{D} = \{T_1, \ldots, T_d\}$. $\mathcal{D}$ is commonly represented as the concatenation $\mathcal{T} = T_1 T_2 \ldots T_d$, of length $|\mathcal{T}| = n$.

A *pattern* P is a string over the same alphabet Σ with length $|P| = m$. It occurs *occ* times in $\mathcal{T}$, and appears in *ndoc* documents.

Text Indexes. The *suffix tree* [28] of a string $\mathcal{T}$ is a compressed digital tree storing all the suffixes $\mathcal{T}[i..n]$, for all $1 \leq i \leq n$. The suffix tree node reached by following the symbols of a pattern P is called the *locus* of P and is the ancestor of all the *occ* leaves corresponding to the positions of P in $\mathcal{T}$. The suffix tree uses $\mathcal{O}(n \lg n)$ bits and lists all the occurrences of P in time $\mathcal{O}(m + occ)$.

The *suffix array* [19] $SA[1..n]$ of a string $\mathcal{T}[1..n]$ is a permutation of the starting positions of all the suffixes of $\mathcal{T}$ in lexicographic order, $\mathcal{T}[SA[i], n] < \mathcal{T}[SA[i + 1], n]$ for all $1 \leq i < n$. SA can be binary searched to obtain the range $SA[sp..ep]$ of all the suffixes prefixed by P (note $occ = ep - sp + 1$). Thus the occurrences of P can be listed in time $\mathcal{O}(m \lg n + occ)$. The suffix array takes $n \lg n$ bits.

Compressed suffix arrays (CSAs) [24] are space-efficient representations of the suffix array. They find the interval $[sp..ep]$ corresponding to $P[1..m]$ in time $t_{search}(m)$, and access any cell $SA[i]$ in time $t_{lookup}(n)$. Their size in bits, $|\,CSA\,|$, is usually bounded by $\mathcal{O}(n \lg \sigma)$.

Grammar Compression. Grammar compression of a string $S[1..n]$ replaces it by a context-free grammar (CFG) $\mathcal{G}$ that uniquely generates S. This CFG $\mathcal{G}$ may require less space than the original sequence S, especially when S is repetitive.

Finding the smallest CFG $\mathcal{G}^*$ generating the input S is NP-hard [16], but various $\mathcal{O}(\lg(n/|\mathcal{G}^*|))$-approximations exist. In particular, we are interested in

approximations that are *binary* (i.e., the maximum arity of the parse tree is 2) and *balanced* (i.e., any substring is covered by $\mathcal{O}(\lg n)$ maximal nodes of the parse tree) [3,13,25].

3 Related Work

Muthukrishnan [20] proposed the first optimal-time linear-space solution to the document listing problem. He defines the *document array* $DA[1..n]$ of $\mathcal{T}$, where $DA[i]$ stores the identifier of the document to which $\mathcal{T}[SA[i]]$ belongs. The document listing problem is then translated into computing the *ndoc* distinct identifiers in the interval $DA[sp..ep]$ corresponding to the pattern P. He uses a suffix tree to find sp and ep in time $\mathcal{O}(m)$, and then an algorithm that finds the *ndoc* distinct numbers in the range in time $\mathcal{O}(ndoc)$.

Sadakane [26] adapts the method to use much less space. He replaces the suffix tree by a CSA, and mimics the algorithm to find the distinct numbers in $DA[sp..ep]$ using only $\mathcal{O}(n)$ bits of space. Within $|\,CSA\,|+\mathcal{O}(n)$ bits, he performs document listing in time $\mathcal{O}(t_{\text{search}}(m)+ndoc\cdot t_{\text{lookup}}(n))$. Using a particular CSA [2] the space is $n\lg\sigma+o(n\lg\sigma)+\mathcal{O}(n)$ bits and the time is $\mathcal{O}(m+ndoc\cdot\lg n)$.

There are many other classical and compact indexes for document listing. We refer the reader to a survey [22] and focus on those aimed at repetitive texts.

Gagie et al. [8] proposed a technique adapting Sadakane's solution to highly repetitive collections. They show that the technique to find the distinct elements of $DA[sp..ep]$ can be applied almost verbatim on an array they call *interleaved longest-common-prefix array (ILCP)*. On repetitive collections, this array can be decomposed into a small number ρ of equal values, which allows them represent it in little space. The *ILCP* index requires $|\,CSA\,|+\mathcal{O}((\rho+d)\lg n)$ bits of space and solves document listing in time $\mathcal{O}(t_{\text{search}}(m)+ndoc\cdot t_{\text{lookup}}(n))$.

Gagie et al. [8] proposed another radically different approach, called *Precomputed Document Lists (PDL)*. The idea is to store the list of the documents where (the corresponding substring of) each suffix tree node appears. Then the search consists of finding the locus of P and returning its list. To reduce space, however, only some sampled nodes store their lists, and so document listing requires *merging* the lists of the maximal sampled nodes descending from the locus node. To further save space, the lists are grammar-compressed.

To bound the query time, the deepest sampled nodes cover at most b leaves, and a factor β restricts the work done per merged document in the unions of the lists. The index then uses $|\,CSA\,|+\mathcal{O}((n/b)\lg n)$ bits, and the document listing time is $\mathcal{O}(t_{\text{search}}(m)+ndoc\cdot\beta\cdot h+b\cdot t_{\text{lookup}}(n))$, h being the suffix tree height.

A problem in all revisited CSA-based solutions are the $\Theta((n\lg n)/t_{\text{lookup}}(n))$ extra bits that must be included in $|\,CSA\,|$ in order to get $\Theta(t_{\text{lookup}}(n))$ time per document. This space does not decrease with repetitiveness, forcing all these indexes to use $\Omega(n)$ bits to obtain time $\mathcal{O}(t_{\text{search}}(m)+ndoc\cdot\lg n)$, for example.

Claude and Munro [5] propose the first index for document listing based on grammar compression, which escapes from the problem above. They extend a grammar-based pattern-matching index [6] by storing the list of the documents

where each nonterminal appears. Those lists are grammar-compressed as well. The index searches for the minimal nonterminals that contain P and merges their lists. While it does not offer relevant space or query time guarantees, the index performs well in practice. Navarro [23] extends this index in order to obtain space guarantees and $\mathcal{O}(m^2 + m\lg^2 n)$ time, but the scheme is difficult to implement.

4   Our Document Listing Index

Like most of the previous work, we solve the document listing problem by computing the $ndoc$ distinct documents in the interval $DA[sp..ep]$ corresponding to the pattern P, found with a CSA in time $\mathcal{O}(t_{search}(m))$. Instead of also using the CSA to compute the values of DA (and thus facing the problem of using $\Theta((n\lg n)/t_{lookup}(n))$ bits to compute a cell in time $\Theta(t_{lookup}(n))$, as it happens in previous work [8,26]), we store the array DA directly, yet in grammar-compressed form. This is promising because the suffix array of repetitive collections is known to have large areas $SA[i..i+\ell]$ that appear shifted by 1 elsewhere, $SA[j..j+\ell]$, that is, $SA[i+k] = SA[j+k]+1$ for all $0 \leq k \leq \ell$ [10,18]. Except for the d entries of SA that point to the ends of the documents, it also holds that $DA[i+k] = DA[j+k]$. Grammar compression is then expected to exploit those large repeated areas in DA.

To answer the queries efficiently, we use an idea similar to the one introduced in PDL [8] and the Grammar-index [5]: precomputing and storing the answers of document listing queries, and grammar-compressing those lists as well. An important difference with them is that PDL stores lists for suffix tree nodes and the Grammar-index stores lists for nonterminals of the grammar of $\mathcal{T}$. Our index, instead, stores lists for the nonterminals of the grammar of DA. This is much simpler because we do not store a suffix tree topology (like PDL) nor a complex grammar-based pattern-matching index (like the Grammar-index): we simply find the interval $DA[sp..ep]$ using the CSA, fetch the nonterminals covering it, and merge their lists. By using a binary balanced grammar on DA, we ensure that any document is obtained in the merging only $\mathcal{O}(\lg n)$ times, which leads to our worst-case bound of $\mathcal{O}(ndoc \cdot \lg n)$. PDL and the Grammar-index cannot offer such a logarithmic-time guarantee.

4.1   Structure

The first component of our index is a CSA suitable for repetitive collections, of which we are only interested in the functionality of finding the interval $SA[sp..ep]$ corresponding to a pattern $P[1..m]$. For example, we can use the Run-Length CSA (RLCSA) variant of Gagie et al. [10], which offers times $t_{search}(m) = \mathcal{O}(m\lg\lg_w \sigma)$ within $\mathcal{O}(r\lg n)$ bits, or $t_{search}(m) = \mathcal{O}(m)$ within $\mathcal{O}(r\lg(n/r)\lg n)$ bits, where r is the number of equal-letter runs in the Burrows-Wheeler Transform of $\mathcal{T}$. This also upper-bounds the number of areas $SA[i..i+\ell]$ into which SA can be divided such that each area appears elsewhere shifted by 1 [17].

The second component is the grammar $\mathcal{G}$ that generates $DA[1..n]$, which must be binary and balanced. Such grammars can be built so as to ensure that their total size is $\mathcal{O}(r \lg(n/r) \lg n)$ bits [9], of the same order of the first component.

The third component are the lists D_v of the distinct documents that appear in the expansion of each nonterminal v of $\mathcal{G}$. These lists are stored in ascending order to merge them easily. To reduce their size, the set of sequences $D_1, \ldots, D_g$ are grammar-compressed as a whole in a new grammar $\mathcal{G}'$, ensuring that no nonterminal of $\mathcal{G}'$ expands beyond a list D_v. Each list D_v can then be obtained in optimal time, $\mathcal{O}(|D_v|)$, from a nonterminal of $\mathcal{G}'$.

4.2 Document Listing

Given a pattern $P[1..m]$, we use the CSA to find the range $[sp..ep]$ where the occurrences of P lie in the suffix array, in time $\mathcal{O}(t_{\text{search}}(m))$. We then find the maximal nodes of the parse tree of DA that cover $DA[sp..ep]$. Finally, we decompress the lists of the nonterminals corresponding to those maximal nodes, and compute their union.

Since $\mathcal{G}$ is binary and balanced, there are $\mathcal{O}(\lg n)$ maximal nonterminals that cover $DA[sp..ep]$ in the parse tree. By storing the length to which each nonterminal of $\mathcal{G}$ expands, we can easily find those $\mathcal{O}(\lg n)$ maximal nonterminals in time $\mathcal{O}(\lg n)$, by (virtually) descending in the parse tree from the initial symbol of $\mathcal{G}$ towards the area $DA[sp..ep]$.

To merge the $\mathcal{O}(\lg n)$ lists of documents in ascending order, we use an atomic heap [7] (see practical considerations in the next section). This data structure performs **insert** and **extractmin** operations in constant amortized time, when storing $\mathcal{O}(\lg^2 n)$ elements. We then insert the heads of the $\mathcal{O}(\lg n)$ lists in the atomic heap, extract the minimum, and insert the next element of its list. If we extract the same document many times, we report only one copy. We then expand and merge the lists $D_{v_1}, \ldots, D_{v_k}$ in time $\mathcal{O}(|D_{v_1}| + \cdots + |D_{v_k}|)$.

Since each distinct document we report may appear in the $\mathcal{O}(\lg n)$ lists, our document listing solution takes time $\mathcal{O}(t_{\text{search}}(m) + ndoc \cdot \lg n)$. By using the RLCSA that occupies $\mathcal{O}(r \lg(n/r) \lg n)$ bits, the total time is $\mathcal{O}(m + ndoc \cdot \lg n)$.

4.3 Example

Figure 1 shows an example with 3 documents, T_1 = MINIMUM\$, T_2 = MINIMAL\$, and T_3 = MINIMIZES\$. The rightmost column shows $\mathcal{T}$. The preceding columns show the sorted suffixes, the suffix array SA, and the document array DA, a sequence over $\{1, 2, 3\}$. To the left of DA we show the syntax tree of the grammar we built, with nonterminal symbols 4 to 18. Associated with each nonterminal we write the list of distinct documents to which the nonterminal expands.

A search for the pattern $P = $ I identifies the suffix array interval $SA[6..12]$, thus we have to report all the distinct documents in $DA[6..12]$. These correspond to two nodes in the grammar, the nonterminals 5 and 6. Thus we merge their

lists, $\{1,2,3\}$ and $\{1,2,3\}$, to obtain the answer $\{1,2,3\}$. Note that each of the 3 documents we report was found twice in the lists that cover $DA[6..12]$.

4.4   Plugging-in Other Indexes

Our grammar-compressed DA, without the lists D_v, can be used to replace the CSA component that requires $\Theta((n \lg n)/t_{\text{lookup}}(n))$ bits to compute a cell in time $\Theta(t_{\text{lookup}}(n))$. These indexes actually access cells $SA[i]$ in order to obtain $DA[i]$. Our compressed DA offers $\mathcal{O}(\lg n)$ access time in $O(r \lg(n/r) \lg n)$ bits.

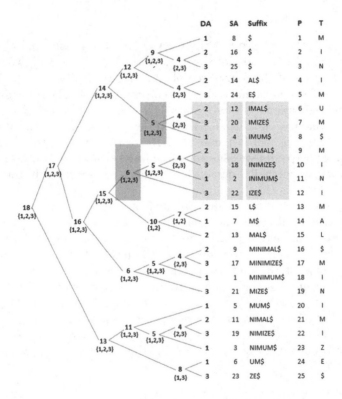

Fig. 1. An example of our document listing structure.

Thus, we can implement Sadakane's solution [26], as well as ILCP and PDL [8] all answering in time $\mathcal{O}(m + ndoc \cdot \lg n)$, and replacing the $\mathcal{O}((n \lg n)/t_{\text{lookup}}(n))$ part of their $|\text{CSA}|$ space by $O(r \lg(n/r) \lg n)$ bits (which also accounts for the RLCSA variant that finds $[sp..ep]$ in time $\mathcal{O}(m)$. We can also implement the brute-force solution in time $O(m + occ + \lg n)$ and $O(r \lg(n/r) \lg n)$ bits by extracting the whole $DA[sp..ep]$.

5   Practical Considerations

5.1   Compressed Suffix Array

We use a practical RLCSA [18, called RLFM+ in there] that uses $(r \lg \sigma + 2r \lg(n/r))(1+o(1))$ bits of space and offers search time $t_{\text{search}}(m)$ in $\mathcal{O}(m \lg r) \subseteq \mathcal{O}(m \lg n)$. Since we do not need to compute cells of SA with this structure, we do not need to spend the $\mathcal{O}((n \lg n)/t_{\text{lookup}}(n))$ bits, and as a result the contribution of the RLCSA to the total space is negligible.

5.2   Grammar Compressor

We choose Re-Pair [15] to obtain both $\mathcal{G}$ and $\mathcal{G}'$, since it performs very well in practice. Re-Pair repeatedly replaces the most frequent pair of adjacent symbols with a new nonterminal, until every pair is unique. Upon ties in frequency, we give priority to the pairs whose symbols have been generated earlier, which in practice yielded rather balanced grammars in all the cases we have tried.

Re-Pair yields a binary grammar, but the top-level is a long sequence. We then complete the grammar by artificially adding a parse tree on top of the final sequence left by Re-Pair. To minimize the height of the resulting grammar, we merge first the pairs of nonterminals with shorter parse trees.

We store the g grammar rules as an array G taking $2g \lg(g + d)$ bits, so that if A_i is the ith nonterminal of the grammar, it holds that $A_i \rightarrow A_{G[2i]} A_{G[2i+1]}$.

When building $\mathcal{G}'$, we concatenate all the lists D_v and separate them with unique numbers larger than d, to ensure that Re-Pair will not produce nonterminals that cross from one list to another. After running Re-Pair, we remove the separators but do not complete the grammars, as all we need is to decompress any D_v in optimal time. We represent all the reduced sets D'_v as a sequence D', marking the beginning of each set in a bitvector B. The beginning of D'_v is found with operation $select(B, v)$, which finds the vth 1 in B. This operation can be implemented in constant time using $o(|B|)$ further bits [4].

5.3   Sampling

The largest component of our index is the set of compressed lists D'_v. To reduce this space, we store those lists only for sampled nonterminals v of $\mathcal{G}$. The list of a nonsampled nonterminal v is then obtained by merging those of the highest sampled descendants of v in the parse tree, which yields a space/time tradeoff.

We use a strategy similar to PDL [8], based on parameters b and β. We define a *sampled tree* by sampling some nodes from the parse tree. First, no leaf v of the sampled tree can have an expansion larger than b, so that we spend time $\mathcal{O}(b \lg b)$ to obtain its sorted list directly from $\mathcal{G}$. To this aim, we sample all the nonterminals v of $\mathcal{G}$ with parent w such that $|D_v| \leq b < |D_w|$. Those are the leaves of the sampled tree, which form a partition of DA.

Second, for any nonsampled node v with $|D_v| > b$, we must be able to build D_v by merging other precomputed lists of total length $\leq \beta|D_v|$. This implies

Table 1. Statistics for document collections (small, medium, and large variants): *Collection* name; *Size* in megabytes; *RLCSA* bits per symbol (bps); *Docs*, number of documents; *Doc size*, average document length; number of *Patterns*; *Occs*, average number of occurrences; *Doc occs*, average number of document occurrences; *Occs/doc*, average ratio of occurrences to document occurrences. For the synthetic collections (second group), most of the statistics vary greatly among the variants that use 10 or 100 base documents with the different mutation probabilities.

Collection	Size (n)	RLCSA (bps)	Docs (d)	Doc size (n/d)	Patterns	Occs (occ)	Doc occs $(ndoc)$	Occs/doc $(\frac{occ}{ndoc})$
Page	110	0.18	60	1 919 382	7658	781	3	242.75
	641	0.11	190	3 534 921	14 286	2601	6	444.79
	1037	0.13	280	3 883 145	20 536	2889	7	429.04
Revision	110	0.18	8834	13 005	7658	776	371	2.09
	640	0.11	31 208	21 490	14 284	2592	1065	2.43
	1035	0.13	65 565	16 552	20 536	2876	1188	2.42
Influenza	137	0.32	100 000	1436	269	532 739	88 525	6.02
	321	0.26	227 356	1480	269	1 248 428	202 437	6.17
Concat	95		10	10^7	7538–10 832			
	95		100	10^6	10614–13 165			
Version	95		10 000	10 000	7537–13 165			

that generating D_v costs $\mathcal{O}(\beta \lg n)$ times more than having D'_v stored and just decompressing it.

We first assume the sampled tree contains all the ancestors of the sampled leaves and then proceed bottom-up in the sampled tree, removing some nodes from it. Any node v with parent w and children $u_1, ..., u_k$ is removed if $\sum_{i=1}^{k} |D_{u_i}| \leq \beta \cdot |D_v|$; the nodes u_i then become children of w in the sampled tree.

At query time, if a node v of interest is not sampled, we collect all the lists of its highest sampled descendants. Therefore, on a parse tree of height h we may end up merging many more than the original $\mathcal{O}(h)$ lists $D_1, ..., D_k$, but have the guarantee that the merged lists add up to size at most $\beta \cdot (|D_1| + \cdots + |D_k|)$. To merge the lists we use a classical binary heap instead of an atomic heap, so the cost per merged element is $\mathcal{O}(\lg n)$.

We may then spend $k \cdot b \lg b = \mathcal{O}(hb \lg b)$ time in extracting and sorting the lists D_v of size below b. The other lists D_v may lead to merging $\beta|D_v|$ elements. The total cost over the $k = \mathcal{O}(h)$ lists is then $\mathcal{O}(hb \lg b + \beta(|D_1| + \cdots + |D_k|)) \lg n) \subseteq \mathcal{O}(hb \lg b + ndoc \cdot \beta h \lg n)$. In terms of complexity, if we choose for example $b = \mathcal{O}(\lg n / \lg \lg n)$, $\beta = \mathcal{O}(1)$, and the grammar is balanced, $h = \mathcal{O}(\lg n)$, then the total cost of merging is $\mathcal{O}(ndoc \cdot \lg^2 n)$.

6   Experiments and Results

We evaluate different variants of our indexes and compare them with the state of the art. We use the experimental framework proposed by Gagie et al. [8].

6.1 Document Collections

To test various kinds of repetitiveness scenarios, we performed several experiments with real and synthetic datasets. We used the same document collections tested by Gagie et al. [8], available at jltsiren.kapsi.fi/rlcsa. Table 1 summarizes some statistics on the collections and the patterns used in the queries.

Real Collections. Page and Revision are collections formed by all the revisions of some selected pages from the Wikipedia in Finnish language. In Page, there is a document for each selected article, that also includes all of its revisions. In the case of Revision, each page revision becomes a separate document. Influenza is a repetitive collection composed of sequences of the H. influenzae virus genomes.

Synthetic Collections. We also used two types of synthetic collections to explore the effect of collection repetitiveness on document listing performance in more detail. Concat and Version are similar to Page and Revision, respectively. We use 10 and 100 base documents of length 1000 each, extracted at random from the English file of Pizza&Chili (pizzachili.dcc.uchile.cl). Besides, we include variants of each base document, generated using different mutation probabilities (0.001, 0.003, 0.01, and 0.03). A mutation is a replacement by a different random symbol. In collection Version, each variant becomes a separate document. In Concat, all variants of the same base document form a single document.

Queries. The query patterns for Page and Revision datasets are Finnish words of length ≥ 5 that occur in the collections. For Influenza, the queries are substrings of length 4 extracted from the dataset. In the case of Concat and Version, the patterns are terms selected from an MSN query log. See Gagie et al. [8] for a more detailed description.

6.2 Compared Indexes

Grammar-Compressed Document Array ($GCDA$). This is our main proposal. We use the balanced Re-Pair compressor implemented by Navarro (www.dcc.uchile.cl/gnavarro/software/repair.tgz). To sample the parse tree, we test several parameter configurations for the block size b and factor β.

Brute Force ($Brute$). This family of algorithms is the most basic solution to the document listing problem. They use a CSA to retrieve all the document identifiers in $DA[sp..ep]$, sort them, and report each of them once. Brute-L uses the CSA to extract the values $DA[i]$. Brute-D, instead, uses an explicit document array DA. Finally, Brute-C is our variant using the grammar-compressed DA. From the grammar tree of height h and storing the length of the expansion of each nonterminal, we extract the range $DA[sp..ep]$ in time $\mathcal{O}(h + ep - sp)$.

Sadakane (*Sada*). Sada-L is the original index of Sadakane [26]. Sada-D speeds up the query time by explicitly storing DA. Sada-C stores DA in grammar-compressed form, where each individual cell $DA[i]$ is extracted in time $\mathcal{O}(h)$.

Interleaved Longest Common Prefix (*ILCP*). ILCP-L implements the ILCP index of Gagie et al. [8] using a run-length encoded ILCP array. ILCP-D is a variant that uses the document array instead of the CSA functionality. ILCP-C uses, instead, our grammar-compressed DA, which accesses any cell in time $\mathcal{O}(h)$.

Precomputed Document Lists (*PDL*). PDL-BC and PDL-RP implement the PDL algorithm proposed by Gagie et al. [8]. PDL-BC uses a Web graph compressor [12] on the set of lists, whereas PDL-RP uses Re-Pair compression. Both use block size $b = 256$ and factor $\beta = 16$, as recommended by their authors.

Grammar-Based (*Grammar*). This is an implementation of the index by Claude and Munro [5]. It uses Re-Pair on the collection $\mathcal{T}$ and on the set of lists. This index is the only tested solution that does not use a CSA.

We implemented GCDA on C++, using several succinct data structures from the SDSL library (github.com/simongog/sdsl-lite). We used existing C++ implementations of the indexes Brute, Sada, ILCP and PDL, which were tested by Gagie et al. [8] (jltsiren.kapsi.fi/software/doclist.tgz), and modified the versions -C by using DA in grammar-compressed instead of in plain form.

All tested indexes except Grammar use a suffix array to compute the interval $[sp..ep]$ corresponding to pattern P. We used a RLCSA implementation (jltsiren.kapsi.fi/rlcsa) that is optimized for repetitive text collections. To compute entries $SA[i]$, the RLCSA uses a suffix array sampling, which requires significant space as explained. Our index does not use this operation, but it is required for the indexes Brute-L, Sada-L, ILCP-L, and both variants of PDL. We use 32 as the value for this sample rate, as it gave good results in previous tests [8]. The exception is Brute-L, which uses a RLCSA optimized to extract whole ranges $SA[sp..ep]$ [10] (github.com/nicolaprezza/r-index). The column $RLCSA$ of Table 1 gives the space used by the RLCSA without suffix array samples.

Our machine has two Intel(R) Xeon(R) CPU E5-2407 processors @ 2.40 GHz and 250 GiB RAM. The operating system was Debian Linux kernel 4.9.0-8-amd64. All indexes were compiled using g++ version 6.3.0 with flags -O3 -DNDEBUG.

6.3 Tuning Our Main Index

Figure 2 shows the tradeoff between time and space of GCDA on small real collections. We tested GCDA with 4 different sizes of block b: 128, 256, 512, and 1024. For each block size, we used 3 different factors β (4, 8, and 16), which are represented with increasing color darkness in the plots. The configuration

$b = 512$ and $\beta = 4$ shows to be a good general-purpose choice of parameter values, and we stick to it from now on.

The lower-right plot of Fig. 2 shows the space required by the main components of our index. As the number of documents in the collection grows and their size decreases, the weight of the grammar-compressed *DA*, and even more, of the grammar-compressed lists of documents, becomes dominant. Note also that `Influenza` is the least repetitive collection.

6.4   Comparison on Real Collections

Figures 3 and 4 show the tradeoff between time and space for all tested indexes on the real collections. Our main index, GCDA, and the -C variants of the other indexes we adapted, are clearly dominant in a large portion of the space/time map. Most of the previous indexes are way slower, way larger, or both, than ours. The best previous tradeoffs, PDL-BC and PDL-RP [8], are much closer, but still they are almost always slower and larger than GCDA.

For all versions of `Page`, where there are few large documents and our grammars compress very well, GCDA requires only 0.48–0.56 bits per symbol (bps)

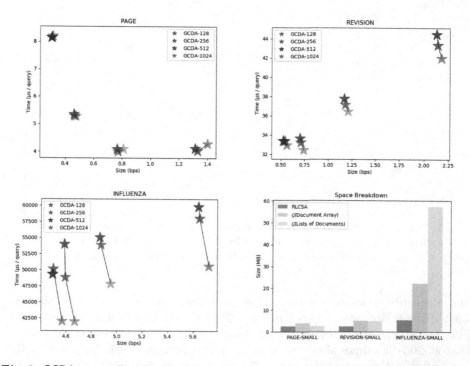

Fig. 2. GCDA on small real collections with different configurations. The x axis shows the total size of the index in bps. The y axis shows the average time per query in μsec. Beware that the plots do not start at zero. The lower-right plot shows the size of the main components of GCDA on the small collections; the y axis shows the size in megabytes.

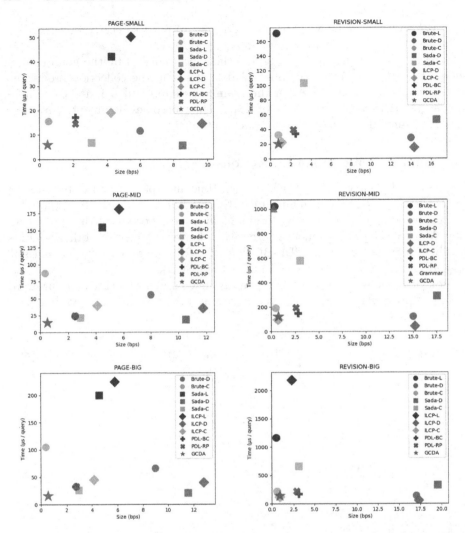

Fig. 3. Document listing indexes on real repetitive collections Page and Revision. The x axis shows the total size of the index in bps. The y axis shows the average time per query. Combinations with excessively high time are omitted in some plots.

and answers queries in less than 16 microseconds (μsec). The index using the least space is Grammar, which requires 0.21–0.35 bps. Grammar is way out of the plot, however, because it requires 1.2–3.4 milliseconds (msec) to solve the queries, that is, 205–235 times slower than GCDA (as in previous work [8], Grammar did not build on the largest dataset of Page). The next smallest index is our variant Brute-C, which uses 0.35–0.55 bps and is generally smaller than GCDA, but slower by a factor of 2.6–6.7. Brute-L, occupying 0.38–0.60 bps, is also smaller in some cases, but much slower (180–1080 μsec, out of the plot). GCDA sharply outperforms all the other indexes in space, and also in time (only Sada-D is 6%

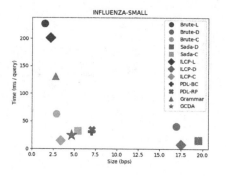 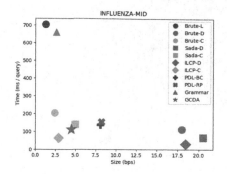

Fig. 4. Document listing indexes on real repetitive collection `Influenza`. The x axis shows the total size of the index in bps. The y axis shows the average time per query. Combinations with excessively high time are omitted in some plots.

faster in the small collection, yet using 18 times more space). The closest competitors, PDL-BC and PDL-RP, are 4.4–5.0 times larger and 2.8–5.0 times slower than GCDA.

In the case of `Revision`, where there are more and smaller documents, GCDA uses 0.73–0.88 bps and answers queries in less than 150 μsec. Again Grammar uses the least space, 0.26–0.42 bps, but once again at the price of being 8–30 times slower than GCDA. The case of Brute-L is analogous: 0.38–0.60 bps but over 8 times slower than GCDA. Instead, our variant Brute-C is a relevant competitor, using 0.45–0.76 bps and being less than 60% slower than GCDA. The other relevant index is our variant ILCP-C, using almost the same space and time of GCDA. The group GCDA/Brute-C/ILCP-C forms a clear sweetpoint in this collection. The closest competitors, again PDL-BC and PDL-RP, are 3.1–3.8 times larger and 1.2–1.9 times slower than GCDA.

`Influenza`, with many small documents, is the worst case for the indexes. GCDA uses 4.46–4.67 bps and answers queries within 115 msec. Many indexes are smaller than GCDA, but only our variants form a relevant space/time tradeoff: ILCP-C uses 2.88–3.37 bps, Brute-C uses 2.42–2.86 bps, and Sada-C uses 4.96–5.40 bps. All the -C variants obtain competitive times, and ILCP-C even dominates GCDA (it answers queries within 65 msec, taking less than 60% of the time of GCDA). The other indexes outperforming GCDA in time are -D variants, which are at least 3.7 times larger than GCDA and 5.2 times larger than ILCP-C.

6.5   Comparison on Synthetic Collections

Figure 5 compares the indexes on synthetic collections. These allow us study how the indexes evolve as the repetitiveness decreases, in a scenario of few large documents (`Concat`) and many smaller documents (`Version`). We combine in a single plot the results for different mutation rates of a given collection and number of base documents. The plots show the increasing mutation rates using variations of the same color, from lighter to darker. All the -L variants and Grammar are omitted because they were significantly slower.

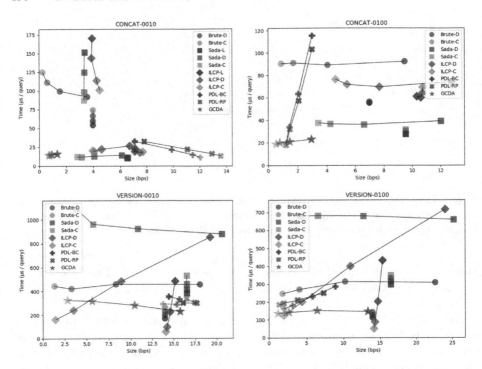

Fig. 5. Document listing on synthetic collections. The x axis shows the total size of the index in bps. The y axis shows the average time per query in μsec. Combinations with excessively high time are omitted in some plots.

On collection Concat, GCDA essentially outperforms all the other indexes. In the case of the version composed by 10 base documents, our index obtains the best space/time tradeoff by a wide margin. Only Brute-C is smaller than GCDA, but 8–9 times slower. On the other hand, various indexes are slightly faster than GCDA, but much larger (from Sada-D, which is up to 30% faster but 7 times larger, to Sada-C, which is 15% faster but at least 4 times larger). With the other variant of Concat (100 base documents), our index offers the best space and time for all mutation rates. Only PDL-RP is 6% faster in its best case, but 2.2 times larger. Further, GCDA retains its space/time performance as repetitiveness decreases, whereas the competing indexes worsen fast in one or both aspects.

On Version, composed by 10 000 documents of length 1000, GCDA is also a dominant solution, retaining its time performance as repetitiveness decreases and outperforming all the -D variants in space up to a mutation rate of 1%. Other competing indexes are our variants Brute-C and ILCP-C (the only one dominating GCDA in some cases), as well as PDL-BC and PDL-RP in the case of 100 base documents. The strange behavior of the PDL indexes in both collections with 10 base documents is briefly discussed in the original article [8].

7   Conclusions

We have presented simple and efficient indexes for document listing on repetitive string collections. They find the *ndoc* documents where a pattern of length m appears in a collection of size n in time $\mathcal{O}(m + ndoc \cdot \lg n)$. The indexes uses grammar-compression of the document array, and perform better as the collection is more repetitive.

Our experimental results show that our main index, GCDA, outperforms the best previous solutions by a fair margin in time and/or space on various repetitive collections. From the previous indexes, only PDL [8] gets close, but it is almost always dominated by GCDA in both space and time. GCDA performs well in space for mutation rates up to 1%, whereas its query time is mostly insensitive to the repetitiveness. Other previous solutions (especially ILCP [8] and brute force) that we adapted to run on our grammar-compressed document array also display unprecedented performance on repetitive texts, competing with GCDA.

For the final version of this paper, we plan to combine the PDL indexes with a grammar-compressed document array as well, which we omitted for lack of time. A line of future work is to further reduce the space of GCDA and our index variants that use the grammar-compressed document array, by using a more clever encoding of the grammars that may nearly halve their space at a modest increase in time [11]. Another line is to extend the index to support *top-k document retrieval*, that is, find the k documents where P appears most often.

References

1. Baeza-Yates, R., Ribeiro-Neto, B.: Modern Information Retrieval, 2nd edn. Addison-Wesley, New York (2011)
2. Belazzougui, D., Navarro, G.: Alphabet-independent compressed text indexing. ACM Trans. Algorithms **10**(4), article 23 (2014)
3. Charikar, M., et al.: The smallest grammar problem. IEEE Trans. Inf. Theory **51**(7), 2554–2576 (2005)
4. Clark, D.R.: Compact PAT Trees. Ph.D. thesis, University of Waterloo, Canada (1996)
5. Claude, F., Munro, J.I.: Document listing on versioned documents. In: Kurland, O., Lewenstein, M., Porat, E. (eds.) SPIRE 2013. LNCS, vol. 8214, pp. 72–83. Springer, Cham (2013). https://doi.org/10.1007/978-3-319-02432-5_12
6. Claude, F., Navarro, G.: Self-indexed grammar-based compression. Fundamenta Informaticae **111**(3), 313–337 (2010)
7. Fredman, M.L., Willard, D.E.: Trans-dichotomous algorithms for minimum spanning trees and shortest paths. J. Comput. Syst. Sci. **48**(3), 533–551 (1994)
8. Gagie, T., et al.: Document retrieval on repetitive collections. Inf. Retr. **20**, 253–291 (2017)
9. Gagie, T., Navarro, G., Prezza, N.: Fully-functional suffix trees and optimal text searching in BWT-runs bounded space. CoRR abs/1809.02792 (2018)
10. Gagie, T., Navarro, G., Prezza, N.: Optimal-time text indexing in BWT-runs bounded space. In: Proceedings of the 29th Annual ACM-SIAM Symposium on Discrete Algorithms (SODA), pp. 1459–1477 (2018)

11. González, R., Navarro, G., Ferrada, H.: Locally compressed suffix arrays. ACM J. Exp. Algorithmics **19**(1), article 1 (2014)
12. Hernández, C., Navarro, G.: Compressed representations for web and social graphs. Knowl. Inf. Syst. **40**(2), 279–313 (2014)
13. Jez, A.: A really simple approximation of smallest grammar. Theor. Comput. Sci. **616**, 141–150 (2016)
14. Kreft, S., Navarro, G.: On compressing and indexing repetitive sequences. Theor. Comput. Sci. **483**, 115–133 (2013)
15. Larsson, J., Moffat, A.: Off-line dictionary-based compression. Proc. IEEE **88**(11), 1722–1732 (2000)
16. Lehman, E., Shelat, A.: Approximation algorithms for grammar-based compression. In: Proceedings of the 13th Annual ACM-SIAM Symposium on Discrete Algorithms (SODA), pp. 205–212 (2002)
17. Mäkinen, V., Navarro, G.: Succinct suffix arrays based on run-length encoding. Nord. J. Comput. **12**(1), 40–66 (2005)
18. Mäkinen, V., Navarro, G., Sirén, J., Välimäki, N.: Storage and retrieval of highly repetitive sequence collections. J. Comput. Biol. **17**(3), 281–308 (2010)
19. Manber, U., Myers, G.: Suffix arrays: a new method for on-line string searches. SIAM J. Comput. **22**(5), 935–948 (1993)
20. Muthukrishnan, S.: Efficient algorithms for document retrieval problems. In: Proceedings of the 13th Annual ACM-SIAM Symposium on Discrete Algorithms (SODA), pp. 657–666 (2002)
21. Navarro, G.: Indexing highly repetitive collections. In: Arumugam, S., Smyth, W.F. (eds.) IWOCA 2012. LNCS, vol. 7643, pp. 274–279. Springer, Heidelberg (2012). https://doi.org/10.1007/978-3-642-35926-2_29
22. Navarro, G.: Spaces, trees and colors: the algorithmic landscape of document retrieval on sequences. ACM Comput. Surv. **46**(4), article 52 (2014)
23. Navarro, G.: Document listing on repetitive collections with guaranteed performance. In: Proceedings of the 28th Annual Symposium on Combinatorial Pattern Matching (CPM). LIPIcs , vol. 78, article 4 (2017)
24. Navarro, G., Mäkinen, V.: Compressed full-text indexes. ACM Comput. Surv. **39**(1), article 2 (2007)
25. Rytter, W.: Application of Lempel-Ziv factorization to the approximation of grammar-based compression. Theor. Comput. Sci. **302**(1–3), 211–222 (2003)
26. Sadakane, K.: Succinct data structures for flexible text retrieval systems. J. Discrete Algorithms **5**, 12–22 (2007)
27. Sthephens, Z.D., et al.: Big data: astronomical or genomical? PLoS Biol. **17**(7), e1002195 (2015)
28. Weiner, P.: Linear pattern matching algorithm. In: Proceedings of the 14th Annual IEEE Symposium on Switching and Automata Theory, pp. 1–11 (1973)

Implementing the Topological Model Succinctly

José Fuentes-Sepúlveda[1(✉)], Gonzalo Navarro[1,3], and Diego Seco[2,3]

[1] Department of Computer Science, University of Chile, Santiago, Chile
{jfuentes,gnavarro}@dcc.uchile.cl
[2] Department of Computer Science, Universidad de Concepción, Concepción, Chile
dseco@udec.cl
[3] IMFD — Millennium Institute for Foundational Research on Data, Santiago, Chile

Abstract. We show that the topological model, a semantically rich standard to represent GIS data, can be encoded succinctly while efficiently answering a number of topology-related queries. We build on recent succinct planar graph representations so as to encode a model with m edges within $4m + o(m)$ bits and answer various queries relating nodes, edges, and faces in $o(\log \log m)$ time, or any time in $\omega(\log m)$ for a few complex ones.

1 Introduction

Low-cost sensors are generating huge volumes of geographically referenced data, which are valuable in applications such as urban planning, smart-cities, self-driving cars, disaster response, and many others. Geographic Information Systems (GIS) that enable *capture, modeling, manipulation, retrieval, analysis and presentation* [13] of such data are thus gaining research attention. GIS models can be classified at different levels. For example, on the conceptual level, entity- and field-based approaches exist, whereas on the logical level, vector and raster are the most popular models. In this work we focus on the representation of the geometry of a collection of vector objects, such as points, lines, and polygons.

There are three common representations of collections of vector objects, called *spaghetti, network,* and *topological* model, which mainly differ in the expression of topological relationships among the objects [11]. In the spaghetti model, the geometry of each object is represented independently of the others and no explicit topological relations are stored. Despite its drawbacks, this is the most used model in practice because of its simplicity and the lack of efficient implementations of the other models. Those other two models are similar, and explicitly store topological relationships among objects. The network model is

Funded by the Millennium Institute for Foundational Research on Data (IMFD), Chile and by European Union's Horizon 2020 research and innovation programme under the Marie Sklodowska-Curie grant agreement No 690941 (project BIRDS). The first author received funding from Conicyt Fondecyt grant 3170534. The last author received funding from Conicyt Fondecyt regular grant 1170497.

N. R. Brisaboa and S. J. Puglisi (Eds.): SPIRE 2019, LNCS 11811, pp. 499–512, 2019.
https://doi.org/10.1007/978-3-030-32686-9_35

tailored to graph-based applications, such as transportation networks, whereas the topological model focuses on planar networks (e.g., all sorts of maps). This model is more efficient to answer topological queries, which are usually expensive, and thus it is gaining popularity in spatial databases such as Oracle Spatial.

In this work we focus on those topological queries where this model stands out, and show that they can be efficiently answered within very little space. We build on recent results on connected planar graphs [6] in order to provide a succinct-space representation of the topological model ($4m + o(m)$ bits, where m is the number of edges) that efficiently support a rich set of topological queries (most of them in $o(\log \log m)$ time), which include those defined in current standards and flagship implementations. Our main technical result is a new $O(\frac{\log \log m}{\log \log \log m})$ time algorithm to determine if two nodes are neighbors; then many other results are derived via analogous structures and exploiting duality. These results improve upon those of the planar graph representation on which we build [6] (see also that article for a wider coverage of previous work).

2   The Topological Model and Our Contribution

The topological model represents a planar subdivision into adjacent polygons. Hereinafter, we will refer to these polygons as *faces*. A face is represented as a sequence of *edges*, each of them being shared with an adjacent face, which may be the outer face. An edge connects two *nodes*, which are associated with a point in space, usually the Euclidean space. Edges also have a geometry, which represents the boundary shared between its two faces. This eliminates redundancy in the stored geometries and also reduces inconsistencies. In Fig. 1, faces are named with capital letters, A to H, A being the outer face. Face F is defined by the sequence of nodes $\langle 1, 8, 7, 6 \rangle$, and edge $(6, 7)$ is shared by faces D and F. Note, however, that a pair of nodes is insufficient in general to name an edge, because multiple edges may exist between two nodes.

Those topological concepts are related with geographic entities. The basic geographic entity is the point, defined by two coordinates. Each node in the topological model is associated with a point, and each edge is associated with a sequence of points describing a sequence of segments that form the boundary between the two faces that share such edge. Each face is related to the area limited by its edges (the external face is infinite).

The international standard ISO/IEC 13249-3:2016 [1] defines a basic set of primitive operations for the model, which are also implemented in flagship database systems[1]. Some of the queries relate the geometry with the topology, for example, find the face covering a point given its coordinates. Those queries require data structures that store coordinates, and are therefore bound to use considerable space. Instead, we focus on *pure topological* queries, which can be solved within much less space and can encompass many problems once mapped to topological space. We also restrict our work to a static version of the model, in which case our representation supports a much richer set of access operations.

---

[1] http://postgis.net/docs/Topology.html.

Table 1. The queries we consider on the topological model and the best results within succinct space. Our (sometimes partial) contributions are in boldface.

1. Relations between entities of the same type		
(1.a) Do edges e and e' share a node?	$O(1)$	[6] + Lemma 2
(1.b) Do edges e and e' border the same face?	$O(1)$	[6] + Lemma 2
(1.c) Do nodes u and v share an edge?	$O(\frac{\log\log m}{\log\log\log m})$	Lemma 3
(1.d) Do faces x and y share an edge?	$O(\frac{\log\log m}{\log\log\log m})$	Lemma 4
(1.e) Do nodes u and v border the same face?	any in $\omega(\sqrt{m}\log m)$	Lemma 7
(1.f) Do faces x and y share a node?	any in $\omega(\sqrt{m}\log m)$	Lemma 7
2. Relations between entities of different type		
(2.a) Is edge e incident on node u?	$O(1)$	[6] + Lemma 2
(2.b) Is edge e on the border of face x?	$O(1)$	[6] + Lemma 2
(2.c) Is face x incident on node u?	any in $\omega(\log m)$	Lemma 6
3. Listing related entities (time per element output)		
(3.a) Endpoints of edge e	$O(1)$	[6] + Lemma 2
(3.b) Faces divided by edge e	$O(1)$	[6] + Lemma 2
(3.c) Nodes/edges neighbors of node u	$O(1)$	[6]
(3.d) Faces bordering face x	$O(1)$	[6] and duality
(3.e) Faces incident on node u	$O(1)$	Lemma 5
(3.f) Nodes/edges bordering face x	$O(1)$	Lemma 5
4. Counting related entities (nodes/faces counted with duplicities)		
(4.a) Nodes/edges/faces neighbors of node u	any in $\omega(1)$	[6] extended
(4.b) Faces/edges/nodes bordering face x	any in $\omega(1)$	[6] and duality

Topological queries can be also solved using the geometries, but such approach is computationally very expensive. We propose instead an approach in which most of the work is done on an in-memory compact index on the topology, resorting to the geometric data only when necessary. Such an approach enables handling geometries that do not fit in main memory, but whose topologies do, and still solving queries on them with reasonable efficiency because secondary-memory accesses are limited. To illustrate this, consider the example of *given the coordinates of two query points, tell if they lie on adjacent faces, and if so, which edge separates them*. In our approach, this type of query can be solved with just two mappings from the geographical space to the topological space, and then using pure topological queries.

Table 1 lists a set of topological queries we consider on the topological model. They comprehensively consider querying about relations between two given entities of the same or different type, and listing or counting entities related to a given one. The set considerably extends the queries available in standards or flagship implementations, which comprise just `intersects` (1.d and 1.f), `GetNodeEdges` (3.c), and `ST_GetFaceEdges` (3.e).

A preliminary result essentially hinted in previous work [6], Lemma 2, sorts out a number of simple queries (all [123].[ab]) in constant time. Our main result is Lemma 3, which shows how to determine if two given nodes are connected by an edge (1.c) in time $O(\frac{\log \log m}{\log \log \log m})$, adding only $o(m)$ bits to the main structure. The same procedure on the dual graph, Lemma 4, determines in the same time if two given faces share an edge (1.d, a variant of the standard query intersects). Another consequence of Lemma 2 is Lemma 5, which extends previous work [6] listing the neighbors of a node (3.c, GetNodeEdges) in optimal time to list the faces incident on a node (3.e) and, by duality, list the faces or edges bordering a face (3.d, ST_GetFaceEdges) and the nodes bordering a face (3.f), all in optimal time. We also extend previous results [6] that count the edges incident on a node (4.a) in any time in $\omega(1)$ to count nodes, edges, or faces incident on a node or bordering a face (4.b).

Finally, our solution to determine if a given node is in the frontier of a given face (2.c) is costlier, in $\omega(\log m)$, and that to determine if two given nodes border the same face (1.e) or if two given faces share some node (1.f, a variant of query intersects) cost even more, in $\omega(\sqrt{m} \log m)$. The last two solutions build on Lemmas 3 and 4, and we conjecture that their times cannot be easily improved.

3   Succinct Data Structures

3.1   Sequences and Parentheses

Given a sequence $S[1..n]$ defined over an alphabet of size σ, the operation $rank_a(S, i)$ returns the number of occurrences of the symbol a in the prefix $S[1..i]$, and the operation $select_a(S, i)$ returns the position in S of the ith occurrence of the symbol a. For binary alphabets, $\sigma = 2$, S can be stored in $n + o(n)$ bits supporting $rank$ and $select$ in $O(1)$ time [3]. If S has m 1-bits, then it can be represented in $m \lg \frac{n}{m} + O(m) + o(n)$ bits, maintaining $O(1)$-time $rank$ and $select$ [10]. For $\sigma = O(\text{polylog } n)$, S can be represented in $n \log \sigma + o(n)$ bits, still supporting $O(1)$-time $rank$ and $select$ [5]. Binary sequences can be used to represent balanced parentheses sequences. Given a balanced parenthesis sequence S, $open(S, i)/close(S, i)$ returns the position in S of the closing/opening paren- . thesis matching the parenthesis $S[i]$, and $enclose(S, i)$ returns the rightmost position j such that $j \leq i \leq close(S, j)$. If S is used to represent an ordered tree, we find the parent of the node represented by the opening parenthesis $S[i]$ as $parent(S, i) = enclose(S, i)$. The sequence S can be represented in $n + o(n)$ bits, supporting $open$, $close$ and $enclose$ in $O(1)$ time [7]. Such representation can be extended to represent k superimposed balanced parenthesis sequences in the same space and time complexities, for any constant k [8, Sect. 7.3].

3.2   Planar Graphs

A planar graph is a graph that can be drawn in the plane without crossing edges. The topology of a specific drawing of a planar graph in the plane is

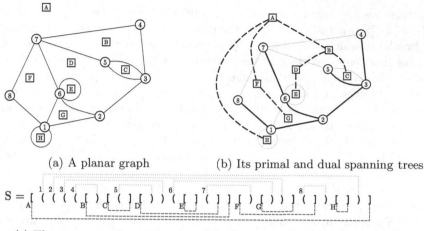

(a) A planar graph   (b) Its primal and dual spanning trees

$$S = {}_A[\ (\ (\ (\ (\ ({}_B[\ ]\ )\ {}_C[\ (\ )\ ]\ {}_D)\ )\ (\ {}_E[\ ]\ (\ )\ ]\ {}_F)\ {}_G[\ )\ )\ ]\ {}_H[\ ]\ )\ {}_H[\ ]\ )\ ]$$

(c) The sequence of parentheses and brackets encoding the planar graph

Fig. 1. Example of the succinct planar graph representation of Ferres *et al.* [6].

called a *planar embedding*. We use planar embeddings to represent topological models. In particular, we use Turán's representation [12], which can represent any planar embedding of m edges in $4m$ bits. Ferres *et al.* [6] extended Turán's representation with $o(m)$ extra bits in order to support fast navigation, providing the simple and efficient representation of planar embeddings we build on.

Given a planar embedding of a connected planar graph G, the computation of a spanning tree T of G induces a spanning tree T^* in the dual graph of G [2]. The edges of T^* correspond to the edges in the dual graph crossing edges in $G \setminus T$. Figure 1b shows a primal (thick continuous edges) and a dual (thick dashed edges) spanning trees for the planar graph of Fig. 1a. Lemma 1 states a key observation: a depth-first traversal of T induces a depth-first traversal in T^*.

Lemma 1. ([6]). *Consider any planar embedding of a planar graph G, any spanning tree T of G and the complementary spanning tree T^* of the dual of G. Suppose we perform a depth-first traversal of T starting from any node on the outer face of G and always process the edges incident to the node v we are visiting in counter-clockwise order. At the root, we arbitrarily choose an incidence of the outer face in the root and start from the last edge of the incidence in counterclockwise order; at any other node, we start from the edge immediately after the one to that node's parent. Then each edge not in T corresponds to the next edge we cross in a depth-first traversal of T^*.*

Here, an incidence of the outer face in the root means a place where the root and the outer face are in contact. For instance, in Fig. 1b, the traversal can start at edge $(1, 1)$, $(1, 2)$, or $(1, 8)$, taking node 1 as the root of the spanning tree.

The compact representation [6,12] is based on the traversal of Lemma 1. Starting at the root of any suitable spanning tree T, each time we visit for the first time an edge e, we write a "(" if e belongs to T, or a "[" otherwise.

Each time we visit an edge e for the second time, we write a ")" if e belongs to T or a "]" otherwise. We call S the resulting sequence of $2m$ parentheses and brackets, which are enclosed by an additional pair of parentheses and of brackets that represent the root and the outer face, respectively. Ranks of opening parentheses act as node identifiers, whereas ranks of opening brackets act as face identifiers. Further, positions in S act as edge identifiers: each edge is identified by an opening parenthesis or bracket, and also by its corresponding closing parenthesis or bracket. Fig. 1c shows the sequence S for the planar graph of Fig. 1b, starting the traversal at the edge $(1, 2)$. Observe that the parentheses of S encode the balanced-parentheses representation of T and the brackets the balanced-parentheses representation of the dual spanning tree T^*. In the succinct representation of Ferres *et al.* [6], the sequence S is stored in three bitvectors, $A[1..2(m+2)]$, $B[1..2n]$, and $B^*[1..2(m-n+2)]$. It holds that $A[i] = 1$ if the ith entry of S is a parenthesis, and $A[i] = 0$ if it is a bracket. Bitvector B stores the balanced sequence of parentheses of S, storing a 0 for each opening parenthesis and a 1 for each closing parenthesis. Bitvector B^* stores the balanced sequence of brackets of S in a similar way.

Adding support for *rank* and *select* operations on A, B and B^*, and for *open*, *close* and *enclose* (i.e., *parent*) operations on B and B^*, the succinct representation of Ferres *et al.* [6] supports constant-time operations to navigate the embedding. Precisely, the succinct representation supports $first(v)/last(v)$ (the position in S of the first/last visited edge of the node v), $mate(i)$ (the position in S of the other occurrence of the ith visited edge), $next(i)/prev(i)$ (the position of the next/previous edge after visiting the ith edge of a node v in counter-clockwise order), and $node(i)$ (the index of the source node when visiting the ith edge). Notice that the index v of the nodes corresponds to their order in the depth-first traversal of the spanning tree T, whereas the index i of a visited edge is just a position in S (i.e., each edge is visited twice). According to Lemma 1, the first visited edge of a node v is the edge immediately after the edge to the parent of v in T (except for the root of T), thus $first(v) = select_1(A, select_0(B, v)) + 1$. The implementation of $last(v)$ is similar. The operation $mate(i)$ is transformed to an *open* operation if $S[i]$ is a closing parenthesis or bracket (i.e., $\hat{B}[rank_{A[i]}(A, i)] = 1$): $mate(i) = select_{A[i]}(A, open(\hat{B}, rank_{A[i]}(A, i)))$, or to a *close* operation otherwise (i.e., $\hat{B}[rank_{A[i]}(A, i)] = 0$): $mate(i) = select_{A[i]}(A, close(\hat{B}, rank_{A[i]}(A, i)))$, where $\hat{B} = B$ if $A[i] = 1$ and $\hat{B} = B^*$ if $A[i] = 0$. The implementation of $next(i)$ depends on whether the ith visited edge belongs to T or not. Specifically, $next(i) = i + 1$ unless i is an opening parenthesis (i.e., $A[i] = 1$ and $B[rank_1(A, i)] = 0$), in which case it is instead $next(i) = mate(i) + 1$; $prev(i)$ is analogous. Operation $node(i)$ also depends on whether $S[i]$ is a parenthesis or a bracket. In the first case $(A[i] = 1)$, $node(i) = rank_0(B, enclose(B, rank_1(A, i)))$ if $B[rank_1(A, i)] = 0$ and $node(i) = rank_0(B, open(B, rank_1(A, i)))$ otherwise. On brackets $(A[i] = 0)$, $node(i) = rank_0(B, rank_1(A, i))$ if $B[rank_1(A, i)] = 0$, otherwise $node(i) = rank_0(B, enclose(B, open(B, rank_1(A, i))))$.

With the operations described above, we can implement more complex operations in optimal time, such as listing all the incident edges (and the corresponding neighbor nodes) of a node v in constant time per returned element, and listing all the edges or nodes bordering a face given an edge of the face, spending constant time per returned element. Other operations, such as the degree of a node and checking if two nodes are neighbors, are not supported in constant time. For the degree of a node v, $degree(v)$, the representation supports any time in $\omega(1)$, whereas for the adjacency test of two nodes u and v, $neighbor(u,v)$, they achieve any time in $\omega(\log m)$. In Sect. 4 we give an $O(\frac{\log\log m}{\log\log\log m})$-time solution for $neighbor(u,v)$, and introduce several other operations in Sect. 5. Theorem 1 summarizes the results of Ferres et $al.$

Theorem 1 ([6]). *An embedding of a connected planar graph with m edges can be represented in $4m + o(m)$ bits, supporting the listing in clockwise or counterclockwise order of the neighbors of a node and the nodes bordering a face in $O(1)$ time per returned node. Additionally, one can find the degree of a node in any time in $\omega(1)$, and check the adjacency of two nodes in any time in $\omega(\log m)$.*

3.3 Obtaining the Nodes and Faces of an Edge

Before presenting our main results, we show how to obtain the nodes connected by a given edge, and its dual, the faces separated by the edge. These results are somewhat implicit in the preceding work [6], but we prefer to present them clearly here. They trivially answer queries (1.a) and its dual (1.b), (2.a) and its dual (2.b), (3.a) and its dual (3.b), all in constant time.

Note that our edge representation, as positions in S, is valid for both G and G^* (the spanning tree edges of G, marked with parentheses in S, are exactly the non-spanning tree edges of G^*, and vice versa, the brackets in S are the spanning-tree edges of G^*). The two nodes corresponding to an edge i in G are obtaining analogously to operation $node(i)$: if i is a parenthesis ($A[i] = 1$), then $p \leftarrow rank_1(A, i)$ is its position in B. If it is closing ($B[p] = 1$), we set $p \leftarrow open(B, p)$. The two nodes are then $rank_0(B, p)$ and $rank_0(B, enclose(B, p))$. On brackets ($A[i] = 0$), we find two positions in B, $p_1 \leftarrow rank_1(A, i)$ and $p_2 \leftarrow rank_1(A, mate(i))$. If any is a closing parenthesis ($B[p_1] = 1$ or $B[p_2] = 1$), we take its parent, $p_1 \leftarrow enclose(B, open(B, p_1))$ and/or $p_2 \leftarrow enclose(B, open(B, p_2))$. Finally, the answers are the resulting nodes, $rank_0(B, p_1)$ and $rank_0(B, p_2)$. The identifiers of the two faces divided by the edge are obtained almost with the same formulas, replacing the meaning of 0 and 1 in A, and using B^* instead of B.

Lemma 2. *The representation of Theorem 1 can determine in time $O(1)$ the two nodes connected by an edge, and the two faces separated by an edge.*

4   Determining if Two Nodes Are Connected

Ferres et $al.$ [6] show how we can determine if two given nodes u and v are connected in any time $f(m) \in \omega(\log m)$. First, they check in constant time if they

are connected by an edge of the spanning tree T: one must be the parent of the other. Otherwise, the nodes can be connected by an edge not in T, represented by a pair of brackets. Their idea is to mark in a bitvector $D[1..n]$ the nodes having $f(m)$ neighbors or more. The subgraph G' induced by the marked nodes, where they also eliminate self-loops and multi-edges, has $n' \leq 2m/f(m)$ nodes, because at least $f(m)$ edges are incident on each marked node and each of the m edges can be incident on at most 2 nodes. Since G' is planar and simple, it can have only $m' < 3n' \leq 6m/f(m)$ edges.[2] They represent G' using adjacency lists, which use $o(m)$ bits as long as $f(m) \in \omega(\log m)$. Given two nodes u and v, if either of them is not marked in D, they simply enumerate its neighbors in time $O(f(m))$ to check for the other node. Otherwise, they map both to G' using $rank_1(D)$, and binary search the adjacency list of one of the nodes for the presence of the other, in time $O(\log m) = o(f(m))$. Bitvector D has $n' \leq 2m/f(m)$ bits set out of $n \leq m+1$ (this second inequality holds because G is connected), and therefore it can be represented using $(2m/f(m))\log(f(m)/2) + O(m/f(m)) + o(m) = o(m)$ bits while answering $rank$ queries in constant time [10].

In order to improve this time, we apply the idea for more than one level. This requires a more complex mapping, however, because only in the last level we can afford to represent the node identifiers in explicit form. The intermediate graphs, where we cannot afford to store a renumbering of nodes, will be represented using an extension of the idea of a sequence of parentheses and brackets, in order to maintain the order of the node identifiers.

Concretely, let us call $G_0 = G$ the original graph of $n_0 = n$ nodes and $m_0 = m$ edges, and $S_0[1..2(m_0 + 2)] = S[1..2(m + 2)]$ its representation using parentheses and brackets. A bitvector $D_0[1..n_0]$ marks which nodes of G_0 belong to $G_1 = G'$. When a certain node u is removed from G_0 to form G_1, we also remove all its edges, which are of two kinds:

- Not belonging to the spanning tree T. These are represented by a pair of brackets $[\cdots]$, opening and closing, which are simply removed from S_0 in order to form S_1.
- Belonging to the spanning tree T. These are *implicit* in the parent-child relation induced by the parentheses. By removing the parentheses of u we remove the node, but this implicitly makes the children of u to be interpreted as new children of v, the parent of u in T. To avoid this misinterpretation, we replace the two parentheses of u by angles: $(\cdots)$ becomes $\langle \cdots \rangle$.

In order to obtain the desired space/time performance, the angles must be reduced to the minimum necessary. In particular, we enforce the following rules:

1. Elements under consecutive angles are grouped inside a single one: $\langle X \rangle \langle Y \rangle$ becomes $\langle X Y \rangle$.
2. An angle containing only one angle is simplified: $\langle \langle X \rangle \rangle$ becomes $\langle X \rangle$.
3. Angles containing nothing disappear: $\langle \rangle$ is removed.

---

[2] In fact, they do not specify how to handle queries of the form (u, u) given that they remove self-loops. They could leave one self-loop around each node that has one or more, and the bound would be $m' \leq 4n' \leq 8m/f(m)$. We do this in our extension.

As seen, G_1 contains $n_1 \leq 2m_0/f(m)$ nodes and, since it contains no multiple edges, $m_1 < 8m_0/f(m)$ edges. Its representation, S_1, then contains $2n_1$ parentheses and $2(m_1 - n_1 + 2)$ brackets. It also contains angles, but by rules 2 and 3, each angle pair contains at least one distinct maximal pair of parentheses[3], and thus there are at most $2n_1$ angles. The length of S_1 is then $2(n_1 + m_1 + 2) < 20m_0/f(m) + 4$.

We represent S_1 using an array $A_1[1..2(n_1 + m_1 + 2)]$ over an alphabet of size 3 (to distinguish brackets $= 0$, parentheses $= 1$, and angles $= 2$), and the projected balanced sequences $B[1..2n_1]$ of parentheses, $B^*[1..2(m_1 - n_1 + 2)]$ of brackets, and $B^-[1..2n_1]$ of angles. We can then support constant-time $rank$ and $select$ on A_1 using $o(m_1)$ extra bits [5], and $open$, $close$, and $enclose$ on B, B^*, and B^- also using $o(m_1)$ extra bits. Thus we can support operations $mate(\cdot)$ and $node(\cdot)$ on S_1 in constant time, just as described in Sect. 3.2.

In order to determine if u_1 and v_1 are neighbors in G_1 we may visit the neighbors of u_1: We sequentially traverse the area between the parentheses of u_1, $S_1[p..p'] = (\cdots)$, starting from $p \leftarrow p + 1$, analogously as the neighbor traversal described in Sect. 3.2. If we see an opening parenthesis, $S[p] = $ "(", we skip it with $p \leftarrow mate(p) + 1$ because we are only checking for neighbors via brackets (we already know that the nodes are not neighbors via edges in T). If we see an opening angle, $S[p] = $ "⟨", we also skip it with $p \leftarrow mate(p) + 1$ because this encloses eliminated nodes and no top-level brackets of u_1 can be enclosed in those angles, as explained. If we see a bracket, $S[p] = $ "[" or $S[p] = $ "]", we find its mate, $j = mate(i)$, then the node containing it, $v = node(j)$, and check if $v = v_1$. Note that a bracket cannot lead us to an eliminated node, because brackets of eliminated nodes were effectively removed from S_1. This procedure takes time proportional to the number of neighbors of u_1 in G_1: although we may spend time in traversing angles, by rule 1 above, every angle we skip is followed by a non-angle or by the final closing parenthesis $S[p'] = $ ")".

Our construction does not end in G_1, however. We repeat the construction process in G_1, so that G_2 is the subgraph of G_1 induced by its nodes with $f(m)$ incident edges or more. We continue for $k(m)$ iterations, obtaining the sequences $S_0, \ldots, S_{k(m)-1}$ and the graph $G_{k(m)}$. In $G_{k(m)}$, we store the neighbors of each node in a perfect hash table. Figure 2 shows the resulting graphs G_1 and G_2 after applying two recursive calls over the planar graph of Fig. 1.

The algorithm to determine if $u_0 = u$ and $v_0 = v$ are neighbors, once we check that none is a child of the other in T, is then as follows. If $D_0[u_0] = 0$, we traverse the neighbors of u_0 as described (the top-level sequence, S_0, does not contain angles, though), to see if v_0 is mentioned. This takes time $O(f(m))$ because u_0 has less than $f(m)$ neighbors. Otherwise, if $D_0[v_0] = 0$, we proceed analogously with v_0, in time $O(f(m))$. Otherwise, both nodes are mapped to G_1, to $u_1 = rank_1(D_0, u_0)$ and $v_1 = rank_1(D_0, v_0)$, and we continue similarly with u_1 and v_1 in G_1. If, after $k(m)$ steps, we arrive at $G_{k(m)}$ without determining if

---

[3] Not brackets: a top-level bracket inside angles would correspond to an edge inciding on the removed node, and thus must have been removed when forming S_1.

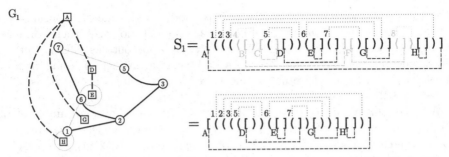

G_1

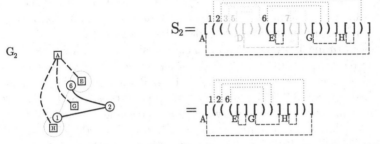

(a) Graph G_1 and its sequence S_1 after one recursive iteration with $f(m) = 3$

G_2

(b) Graph G_2 and its sequence S_2 after two recursive iterations with $f(m) = 3$

Fig. 2. Intermediate planar graphs and their sequences to support the operation $neighbor(\cdot, \cdot)$. Symbols in light-gray represent deleted elements.

they are neighbors, we look for $v_{k(m)}$ in the perfect hash table of the neighbors of $u_{k(m)}$, in constant time. Overall, the query time is $O(k(m) + f(m))$.

As for the space, G_i has $n_i \le 2m_{i-1}/f(m)$ nodes and $m_i < 4n_i \le 8m_{i-1}/f(m)$ edges (because G_i has no multiple edges for all $i > 0$), and thus $m_i < m \cdot (8/f(m))^i$ and $n_i \le (1/4)m \cdot (8/f(m))^i$. The length of S_i is then less than $2(n_i + m_i + 2) < (5/2)m \cdot (8/f(m))^i + 4$. The previous expression, summed over all $1 \le i < k(m)$, yields a total length for all $S_1, \ldots, S_{k(m)-1}$ below $20m/(f(m) - 8) + 4k(m) = O(m/f(m) + k(m))$, for any $f(m) \ge 9$. Since the S_i have constant-size alphabets, they can be represented within $O(m/f(m) + k(m))$ bits, with the constant-time support for *rank*, *select*, *open*, and *close*. On the other hand, the explicit representation of $G_{k(m)}$ requires $O(m_{k(m)} \log m) = O(m \log m \cdot (8/f(m))^{k(m)})$ bits. For all this space to be $o(m)$ we need that $k(m) = o(m)$, $f(m) = \omega(1)$, and $(f(m)/8)^{k(m)} = \omega(\log m)$.

The choice $f(m) = k(m) = \max(9, \frac{(1+\epsilon) \log \log m}{\log \log \log m})$, for any constant $\epsilon > 0$, yields a time complexity in $O(\frac{\log \log m}{\log \log \log m})$ and an extra space in $o(m)$ bits.

If we wish to retrieve the positions $S[b..b']$ of a pair of brackets that connect u and v, when the edge does not trivially belong to T, we enrich our structure with bitvectors $C_0, \ldots, C_{k(m)-1}$, where $C_i[1..m_i - n_i + 2]$ tells which face identifiers (i.e., ranks of opening brackets) survive in G_{i+1}. Once we find, in some G_i, that u_i and v_i are neighbors connected by the edge $S_i[p..p'] = [\cdots]$, we have that the

opening bracket number $b_i = rank_{\text{``}[\text{''}}(S_i, p) = rank_0(B^*, rank_0(A_i, p))$ connects them in G_i. We then identify the edge in G_{i-1} with $b_{i-1} = select_1(C_{i-1}, b_i)$, and continue upwards until finding the answers, $b = b_0$ and $b' = mate(b)$, all in $O(k(m))$ additional time.

The lengths of all bitvectors, for $i > 0$, is $|D_i| + |C_i| = m_i + 2$, so they add up to $o(m)$. For D_0 and C_0, note that they have n_1 and m_1, both in $O(m/f(m))$, 1s out of $n \le m + 1$ or $m - n + 2 \le m$, respectively. Therefore, they can be represented in $O(m \log(f(m))/f(m)) + o(m)$ bits [10]. We thus solve query (1.c).

Lemma 3. *The representation of Theorem 1 can be enriched with $o(m)$ bits so that we can determine whether two nodes are connected in time $O(\frac{\log\log m}{\log\log\log m})$.*

5   Other Results Exploiting Analogies and Duality

Determining Adjacency of Faces. By exchanging the interpretation of parentheses and brackets, the same sequence S represents the dual G^* of G, where the roles of nodes and faces are exchanged. We can then use the same solution of Lemma 3 to determine whether two faces are adjacent (1.d). We do not explicitly store the sequence S^* representing G^*, since we can operate it using S. Instead, we build a structure on S^* analogous to the one we built on S, creating sequences $S_1^*, \ldots, S_{k(m)-1}^*$, $D_0^*, \ldots, D_{k(m)-1}^*$, $C_0^*, \ldots, C_{k(m)-1}^*$, and the final explicit dual graph $G_{k(m)}^*$, so as to determine, within the same space and time complexities, whether two faces of G share an edge, and retrieve one of these edges. This time, the input to the query are the ranks of the opening brackets representing both faces (i.e., node identifiers in G^*).

Lemma 4. *The representation of Theorem 1 can be enriched with $o(m)$ bits so that we can determine whether two faces are adjacent in time $O(\frac{\log\log m}{\log\log\log m})$.*

Listing Related Nodes or Faces. Listing the faces bordering a given face (3.d) can be done as the dual of listing the neighbors of a node (3.c), by exchanging the roles of brackets and parentheses in Theorem 1. Listing the faces incident on a node (3.e) can also be done as a subproduct of Theorem 1. For each edge e incident on u, obtained in counter-clockwise order, we obtain the faces e divides using Lemma 2. This lists all the faces incident on u, in counter-clockwise order, with the only particularity that each face is listed twice, consecutively. Analogously, given a face identifier x, we can list the nodes found in the frontier of the face (3.f). This query is not exactly the same as in Theorem 1, because there we must start from an edge bordering the desired face.

Lemma 5. *The representation of Theorem 1 suffices to list, given a node u, the faces incident on u in counter-clockwise order from its parent in T, each in $O(1)$ time, or given a face x, the nodes in the frontier of x in clockwise order from its parent in T^*, each in $O(1)$ time.*

Determining Incidence of a Face in a Node. Given a node u and a face x, the problem is to determine whether x is incident on u (2.c). Since with Lemma 5 we can list each face incident on u in constant time, or each node bordering x in constant time, we can use a scheme combining those of Lemmas 3 and 4: If u has less than $f(m)$ neighbors, we traverse them looking for x. Otherwise, if x has less than $f(m)$ bordering nodes, we traverse them looking for u. Otherwise, we search for (u, x) in a perfect hash table where we store all the faces y (bounded by $f(m)$ or more nodes) incident on nodes v (having $f(m)$ or more neighbors).

To see that this hash table contains $O(m/f(m))$ elements, consider the bipartite planar graph $G^+(V^+, E^+)$ where $V^+ = V \cup F$ (F being the faces of our original graph $G(V, E)$) and $E^+ = \{(u, x), x \in F$ is incident on $u \in V$ in $G\}$. G^+ is planar because it can easily be drawn from an embedding of G, by placing the nodes $x \in F \subseteq V^+$ inside the face x of G and drawing its edges without having them cut. Note that the nodes $u \in V$ preserve their degree in G^+, whereas the degree of nodes $x \in F$ is the number of edges bordering their corresponding face in G. Therefore G^+ has $n^+ = |V| + |F| = m + 2$ nodes (as per Euler's formula $|F| = m - n + 2$) and $m^+ = 2m$ edges (one per edge limiting each face, so each edge of G contributes twice). If we remove from G^+ all the nodes (of either type) connected with less than $f(m)$ neighbors, and remove multiple edges, each surviving edge corresponds precisely with an entry (v, y) of our perfect hash table. By the same argument used in Sect. 4, at most $4m/f(m)$ nodes survive and, since the reduced graph has no multiple edges, at most $4 \cdot (4m/f(m)) = O(m/f(m))$ edges survive. Hence, we obtain extra space $o(m)$ by choosing any $f(m) \in \omega(\log m)$.

Lemma 6. *The representation of Theorem 1 can be enriched with $o(m)$ bits so that, given a node u and a face x, it answers in $O(f(m))$ time whether u is in the frontier of x, for any $f(m) \in \omega(\log m)$.*

Counting Neighbors. Ferres *et al.* [6] count the number of edges incident on a node u (4.a) in $O(f(m))$ time using $O(m \log f(m)/f(m))$ bits. For nodes with degree below $f(m)$, they traverse the neighbors; for the others, they store the degree explicitly. Neighboring nodes or faces can be counted similarly, except that we can reach several times the same node or face. Thus, we need time $O(f(m) \log f(m))$ on nodes with degree below $f(m)$ in order to remove repetitions; for higher-degree nodes we store the correct number explicitly. We then obtain $O(f(m) \log f(m))$ time using $O(m \log f(m)/f(m))$ bits, which still achieves any time in $\omega(1)$ in $o(m)$ bits. By building the structure on the dual of G, we count the number of edges, nodes, or faces in the frontier of a face x (4.b).

6   More Expensive Solutions

We left for the end other solutions that are likely impractical compared to using brute force, but that nevertheless have theoretical value. These more expensive solutions also encompass some more sophisticated queries not included in Table 1.

Determining if Two Nodes Border the Same Face. Given two nodes u and v, if either has less than $f(m)$ neighbors we can traverse its incident faces one by one and, for each face x, use Lemma 6 to determine if x is incident on the other node in time $\omega(\log m)$. For all the pairs of nodes (u, v) where both have $f(m)$ neighbors or more, we store a binary matrix telling whether or not they share a face. This requires $(2m/f(m))^2$ bits, which is $o(m)$ for any $f(m) = \omega(\sqrt{m})$. Thus we can solve query (1.e) and, by duality, query (1.f), in any time in $\omega(\sqrt{m}\log m)$.

Lemma 7. *The representation of Theorem 1 can be enriched with $o(m)$ bits so that, given two nodes or two faces, it answers in $O(f(m))$ time whether they share a face or a node, respectively, for any $f(m) \in \omega(\sqrt{m}\log m)$.*

If we want to know the identity of the shared face (or, respectively, node), this can be stored in the matrix, which now requires $O((m/f(m))^2\log m)$ bits. We can then reach any time in $\omega(\sqrt{m}\log^{3/2} m)$.

Determining if Two Nodes/Faces are Connected with the Same node/face. Given two nodes u and v, if either has less than $f(m)$ neighbors we can traverse its neighbors w and, using Lemma 3, determine if w is a neighbor of v. This takes $O(f(m) \cdot \frac{\log\log m}{\log\log\log m})$ time. For all the pairs of nodes (u, v) where both have $f(m)$ neighbors or more, we store a binary matrix telling whether or not they share a neighbor. By duality, we can tell if two faces share edges with the same face.

Lemma 8. *The representation of Theorem 1 can be enriched with $o(m)$ bits so that, given two nodes or two faces, it answers in $O(f(m))$ time whether they are connected with a node or a face, respectively, for any $f(m) \in \omega(\sqrt{m} \cdot \frac{\log\log m}{\log\log\log m})$.*

As before, to know the identity of the shared node or face, the time raises to $f(m) \in \omega(\sqrt{m} \cdot \frac{\sqrt{\log m}\log\log m}{\log\log\log m})$.

7   Conclusions

We built on a recent extension [6] of Turán's representation [12] for planar graphs to support queries on the topological model in succinct space. Starting with an improved solution to determine if two nodes are neighbors, we exploit analogies and duality to support a broad set of operations, most in time $O(\frac{\log\log m}{\log\log\log m})$.

One remaining challenge is the the support for the standard query `intersects` (whether two given faces touch each other). If this is interpreted as the faces sharing an edge, then this is query (1.d), which we solve in time $O(\frac{\log\log m}{\log\log\log m})$. If, instead, it suffices with the faces sharing a node, this is query (1.f), which we solve in any time in $\omega(\sqrt{m}\log m)$. We conjecture that this second interpretation is intersection-hard [4,9], and thus no significant improvement can be expected even if using non-compact space.

References

1. ISO/IEC 13249-3:2016. Information technology - Database languages - SQL multimedia and application packages - Part 3: Spatial. Technical report (2016)
2. Biggs, N.: Spanning trees of dual graphs. J. Comb. Theor. B **11**(2), 127–131 (1971)
3. Clark, D.R.: Compact PAT Trees. Ph.D. thesis, University of Waterloo, Canada (1996)
4. Cohen, H., Porat, E.: Fast set intersection and two-patterns matching. Theor. Comp. Sci. **411**(40–42), 3795–3800 (2010)
5. Ferragina, P., Manzini, G., Mäkinen, V., Navarro, G.: Compressed representations of sequences and full-text indexes. ACM Trans. Alg. **3**(2), 20 (2007)
6. Ferres, L., Fuentes-Sepúlveda, J., Gagie, T., He, M., Navarro, G.: Fast and compact planar embeddings. Comput. Geom. Theor. App. (2018), to appear. Available at https://arxiv.org/abs/1610.00130. Preliminary version in Proceedings WADS 2017
7. Munro, J.I., Raman, V.: Succinct representation of balanced parentheses and static trees. SIAM J. Comput. **31**(3), 762–776 (2001)
8. Navarro, G.: Compact Data Structures: A Practical Approach. CUP, Cambridge (2016)
9. Patrascu, M., Roditty, L.: Distance oracles beyond the Thorup-Zwick bound. SIAM J. Comput. **43**(1), 300–311 (2014)
10. Raman, R., Raman, V., Satti, S.: Succinct indexable dictionaries with applications to encoding k-ary trees, prefix sums and multisets. ACM Trans. Alg. **3**(4), 43 (2007)
11. Scholl, M.O.: Spatial Databases with Application to GIS. MK, Uganda (2002)
12. Turán, G.: On the succinct representation of graphs. Discr. Appl. Math. **8**(3), 289–294 (1984)
13. Worboys, M., Duckham, M.: GIS: A Computing Perspective, 2nd edn. CRC, Boca Raton (2004)

Space- and Time-Efficient Storage of LiDAR Point Clouds

Susana Ladra⬤, Miguel R. Luaces⬤, José R. Paramá⬤,
and Fernando Silva-Coira(✉)⬤

Universidade da Coruña, CITIC, A Coruña, Spain
{susana.ladra,luaces,jose.parama,fernando.silva}@udc.es

Abstract. LiDAR devices obtain a 3D representation of a space. Due to the large size of the resulting datasets, there already exist storage methods that use compression and present some properties that resemble those of compact data structures. Specifically, LAZ format allows accesses to a given datum or portion of the data without having to decompress the whole dataset and provides indexation of the stored data. However, LAZ format still has some drawbacks that need to be addressed. In this work, we propose a new compact data structure for the representation of a cloud of LiDAR points that supports efficient queries, providing indexing capabilities that are superior to those of the LAZ format.

Keywords: LiDAR point clouds · Compression · Indexing

1 Introduction

Light Detection and Ranging Technology (LiDAR) has been used during the last four decades in Geosciences as a geomatic method to obtain the 3D geometry of the surface of objects [6]. LiDAR uses a laser beam to compute the distance between a device and an object that reflects the beam. When the laser is used to scan the entire field of view at a high speed, the result is a dense cloud of points centered at the device, with each point having additional information such as the intensity of the laser beam reflection. If the device has additional sensors, each point in the cloud will have additional attributes (e.g., if the device includes a camera, each point will have a color value associated).

The decrease in cost of laser scanning devices has helped to drastically increase the application fields for point clouds. For instance, laser scanning has been used to classify and recognize objects in urban environments [23], in natural

This research has received funding from the European Union's Horizon 2020 research and innovation programme under the Marie Sklodowska-Curie [grant agreement No 690941]; from the Ministerio de Economía y Competitividad (PGE and ERDF) [grant numbers TIN2016-78011-C4-1-R; TIN2016-77158-C4-3-R]; and from Xunta de Galicia (co-founded with ERDF) [grant numbers ED431C 2017/58; ED431G/01].

© Springer Nature Switzerland AG 2019
N. R. Brisaboa and S. J. Puglisi (Eds.): SPIRE 2019, LNCS 11811, pp. 513–527, 2019.
https://doi.org/10.1007/978-3-030-32686-9_36

environments (e.g., landslides [10], or forests [8]), or even underwater environments [17]. Hence, huge datasets are being produced that require immense computing resources to be processed. To give two examples, the ISPRS benchmark on indoor modelling consists of five point clouds containing 98.1×10^6 points [11] and the mobile laser scanning test data "MLS 1 - TUM City Campus" contains more than 1.7×10^9 points collected in 15 min.

Due to the size of the obtained data, the use of compression is almost mandatory. However, the traditional approach of keeping the data compressed in disk and decompressing the whole dataset before any processing is not effective. Therefore, the classical method for storing LiDAR data (LAZ format [21]) is able to compress the data but, in addition, permits accessing a datum or portions of the data without the need of decompressing the whole dataset. In addition, it is equipped with an index to accelerate the queries. Observe that this setup is very similar to that of many modern compact data structures [15].

The use of compression for spatial data is not exclusive of LAZ format. In the case of raster data, Geo-Tiff and NetCDF are able to store the data in compressed form, and in the case of NetCDF, it also permits querying the data directly in that format. Therefore, the application of the knowledge acquired in compact data structures soon led to a new research line.

The wavelet tree [7] was the first compact data structure used in the scope of spatial data. The work in [3] proposes a new point access method based on it. In [16], it is used to represent a set of points in the two-dimensional space, each one with an associated value given by an integer function. Another family of compact data structures based on the quadtree also arose. In [2], a compact version of the region quadtree was adapted for storing and indexing compressed rasters. In the same line, Ladra et al. [13] presented an improvement on the previous works.

In this work we continue in this line, now tackling even more complex spatial data. We present a compact data structure to represent LiDAR point clouds, denoted k^3-*lidar*, which compresses and indexes the data. The improvements with respect to the LAZ format are in two aspects. The LAZ format relies on differential encoding plus an entropy encoder, which compresses/decompresses data by blocks, and thus, in order to retrieve a small region, one or more complete blocks must be decompressed. Another drawback of the LAZ format is that it uses a quadtree to index the points, and this only accelerates the queries by the x and y coordinates. Our new k^3-lidar is able to retrieve/decompress a given datum and indexes the three dimensions of the space.

2   Related Work: LAS and LAZ Format

The American Society for Photogrammetry and Remote Sensing[1] defined in 2003 the LASer (LAS) file format, an open data exchange format for LiDAR point data records. The format contains binary data consisting of a header block

---

[1] https://www.asprs.org/.

that describes general information of the point cloud (e.g., number of points, bounding box), variable-length records to describe additional information such as georeferencing information or other metatada, and point records. Each point record consists of a collection of fields describing the point (e.g., the point x, y and z Cartesian coordinates represented as 4-byte integers, the return intensity represented as a 2-byte integer, or the laser pulse return number). The standard defines a Point Data Record Format 0, and allows for additional formats to be defined with additional data (e.g., the Point Data Record Format 1 adds the GPS time at which the point was acquired as an 8-byte double).

The version 1.4 R14 of the LAS file format specification has been released in 2019 [22]. It defines 11 point data record formats (some of them to provide legacy support), a mechanism to customize the LAS file format to meet application-specific needs by adding point classes and attributes, extended variable-length records to carry larger payloads, and additional types of variable-length records (e.g., georeferencing using Well Known Text descriptions, textual description of the LAS file content, or extra bytes for each point record).

Even though the LAS file format avoids using unnecessary space by storing the coordinates as scaled and offset integers, a LAS file requires much storage space (e.g., a 13.2M point cloud requires 254 MB of disk space, see Table 2). The LAZ file format [9], defined by the LASzip lossless compressor for LiDAR, achieves high compression rates supporting streaming, and random access decompression. LASzip encodes the points in the cloud using chunks of 50,000 points. For each chunk, the first point is stored as raw bytes and it is used as the initial value for subsequent prediction schemes. Each additional point is compressed using an entropy coder (an adaptive, context-based arithmetic coding [19]). These techniques make the LAZ file format very efficient in terms of space. Isenburg [9] showed that the compression ratio is similar or better than general-purpose compression formats such as ZIP or RAR, while maintaining the possibility of processing the file as a stream of points or directly accessing a particular point without having to decompress the complete file.

The LAZ file format is also very efficient answering range queries on the x and y dimensions because LAZ files can be indexed using an adaptative quadtree over these coordinates. Each quadtree leaf contains a list of point indexes that can be used to determine the chunk that contains the point. To resolve a range query, the quadtree is first traversed to determine the candidate point indexes, then, the relevant chunks are retrieved, and finally the chunks have to be decompressed and sequentially scanned to determine the points in the result.

Considering additional types of queries, the LAZ file format is highly inefficient on three-dimensional queries or queries over attribute data because a sequential scan has to be performed over the points in a chunk. These queries are becoming more common because classification algorithms over point clouds quite often require to locate close points in the three-dimensional space, or close points in a two-dimensional space with a similar attribute value (e.g., having the same intensity). Both types of queries can be efficiently answered if a three-dimensional index is built over the data.

3   Background: k^2-Trees and k^3-Trees

The k^2-tree [5] is a time- and space-efficient version of a region quadtree [12,20]. Considering a binary matrix of size $n \times n$, it is divided into 2^2 *quadboxes* (submatrices) of size $n/2 \times n/2$. Each quadbox produces a child of the root node. The label of the node is 1-bit if the corresponding quadbox contains at least one 1-bit, and 0-bit otherwise. The quadboxes having at least one 1-bit are divided using the same procedure until reaching a quadbox full of 0-bits, or reaching the cells of the original matrix.

The k^2-tree, instead of using a classical pointer-based representation of the tree, represents the quadtree using only sequences of bits. More concretely, it uses two bitmaps, denoted as T and L, where T is formed by a breadth-first traversal of the internal nodes, whereas the L is formed by the leaves of the tree. From this basic version, several other improvements yield better space and time performance [5]. Among them, one is that instead of diving each quadbox into 2^2 quadboxes of size $n/2 \times n/2$, each division produces k^2 quadboxes of size $n/k \times n/k$, where k is a parameter that can be adapted for each level of the tree. This is usually used to obtain shorter and wider trees that, at the price of a slightly worse space consumption, are faster when querying.

As in the case of the quadtree, where the simple addition of a third dimension produces the octree [14], the k^3-tree [2] is simply a 3-dimensional k^2-tree. Figure 1 shows a 3-dimensional binary matrix, its corresponding octree, and the k^3-tree represented using bitmaps T and L.

The k^3-tree can be efficiently navigated using *rank* and *select* operations[2] over T and L (see [1, Section 6.2.1]).

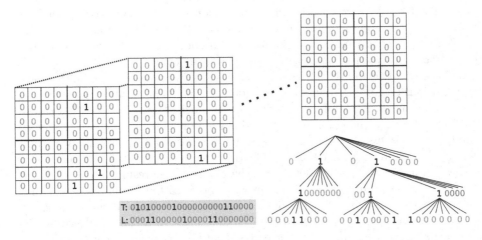

Fig. 1. k^3-tree.

---

[2] Given a bitmap B, $rank_b(B, i)$ is the number of occurrences of bit b in $B[1, i]$ and $select_b(B, j)$ is the j-th occurrence of bit b in B.

4   Our Proposal: k^3-Lidar

In this section, we present the k^3-lidar, a data structure to represent LiDAR point clouds in compact space, which allows us to perform efficient queries.

4.1   Conceptual Description

Consider a 3D matrix of size $n \times n \times n$ that stores a set of LiDAR points. In addition to its coordinates, a point contains several values that correspond to each of its attributes (e.g., intensity, scan angle, color, etc.).

Observe that the k^3-tree was designed to store and index a matrix with a bit value for all positions, that is, a 3D binary raster. However, when dealing with LiDAR datasets, we have to change that raster approach to a point-based data structure. In this scenario, many positions of a LiDAR matrix can be empty, that is, there are no points in those positions.

The k^3-lidar recursively divides the matrix into several equal-sized submatrices, by following the same strategy used by the k^3-tree. Again, this recursive subdivision is represented as a tree, where each node corresponds to a submatrix. As in the original k^3-tree, the division of the submatrices stops when the submatrix is empty (full of 0-bits, in the case of the k^3-tree) and when the process reaches the cells of the original matrix (submatrices of sixe $1 \times 1 \times 1$), placed at level $\lceil log_k n \rceil$. In addition to these cases, the subdivision of the k^3-lidar also stops when the number of points in the processed submatrix is less than or equal to a given threshold l.

Regarding the attributes of each point, their values are compactly stored in our structure and can be efficiently retrieved when obtaining a point. Our structure stores the attributes defined in Point Data Record Format 0 (LAS Specification 1.4 [22]) but it can be easily adapted to allow attributes defined in other formats.

4.2   Data Structures

We use several data structures to represent the conceptual tree previously described:

- **Tree structures (T and H):** We use two data structures to represent the topology of the tree. T is a bitmap, similar to that of the k^3-tree, but a 0 means that the submatrix is empty or the number of points does not exceed the threshold l. The L bitmap of k^3-tree is not used. This is due to: (i) in many cases, the k^3-lidar does not reach the last level of subdivision (level $\lceil log_k n \rceil$), as the division frequently stops before that level due to empty submatrices or submatrices containing l points or less; and (ii) in case of reaching the last level of the subdivision, leaf points require a more complex data structure than just a bitmap.
  In addition to T, we use another bitmap, H, which has one bit for each 0-bit in T, plus, if the last level is reached, as many bits as cells in that

last level. Each bit differentiates empty submatrices with those that contain some points. That is, for each i such that $T[i] \leftarrow 0$, we set $H[rank_0(T, i)] \leftarrow 0$ iff the submatrix is empty or $H[rank_0(T, i)] \leftarrow 1$ in other case. In the case of positions corresponding to the last level $\lceil log_k n \rceil$, a 1-bit means that the corresponding cell has points.

- **Number of points (N):** We use a bitmap N to store the number of points that each leaf contains. When $H[rank_0(T, i)] = 1$, we store in N the number of points, in unary.

- **Arrays of coordinates (X, Y, and Z):** Coordinates of a point $\langle x, y, z \rangle$ are stored in arrays X, Y, Z respectively. Points in level $\lceil log_k n \rceil$ are not stored, since their position can be calculated during the descent through the tree. These values are encoded as local coordinates with respect to the node to which they belong. This results in smaller values than the original ones, and thus they are represented with DACs [4]. DACs provide efficient random access to any position and a good compression ratio with small values.

- **Attributes:** Attributes, as intensity, return number, etc., are stored in separated arrays. They follow the same order as the coordinate arrays. The sequence of values are encoded with DACs or with a bitmap when the attribute can be represented with just one bit per point.

- **Scale factor and Offset:** We convert real coordinates into positive integer values. Therefore, we use one scale factor and one different offset for each dimension. The scale factor is a float value that allows us to transform float values into integer values. The offset is an integer value that allows us to translate points to a coordinate system that starts at $\langle 0, 0, 0 \rangle$. In addition to converting negative numbers into positives, the offset also allows us to obtain smaller numbers in the values of the coordinates. For example, if the minimum value in X is 1000, we can move all points 1000 positions to the left, that is, the coordinate 1000 would be the 0, the 1001 would be the 1, and so on. These parameters use the same strategy than LAS/LAZ format.

4.3   Construction

Figure 2 shows the recursive division of a cloud of LiDAR points (left), and the k^3-lidar representation (right-top). This examples uses $k = 2$ and the maximum number of points in a leaf (l) is 3. Circles represent LiDAR points and they are identified with a unique identifier. In this case, each point only contains an intensity value labeled I_{id}, where id is the identifier of that point. The algorithm starts by dividing the cube in $k^3 = 2^3 = 8$ submatrices of equal size. We add a child node of the root for each submatrix. The first child (top-left) has 4 points (*5*, *8*, *7* and *6*) and, since $4 > l$, we set $T[0] \leftarrow 1$. This node is then enqueued to be processed later. The second node is empty, thus $T[1] \leftarrow 0$ and $H[0] \leftarrow 0$. Note that the third submatrix (bottom-left) has 3 points (*2*, *3* and *1*), thus, since $3 <= l$, we set $T[6] \leftarrow 0$, $H[1] \leftarrow 1$, and add 3 in unary (001) to N. Moreover, local coordinates are stored into arrays X, Y, and Z following the space filling curve z−order. Point *2* having global coordinates $\langle 5, 0, 0 \rangle$ becomes

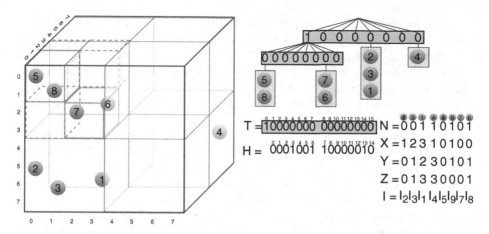

Fig. 2. Example of a cloud of LiDAR points (left) where each point (circle) has an intensity I_i. Conceptual tree representation (top-right) and compact structures involved (bottom-right) in the construction of the k^3-lidar. This example uses $k = 2$ and $l = 3$ (maximum number of points at a leaf). For this example, only the intensity attribute is stored. The scale factor and the offset are not represent for clarity.

$\langle 1, 0, 0 \rangle$. Hence, $X[0] \leftarrow 1$, $Y[0] \leftarrow 0$, and $Z[0] \leftarrow 0$. This point has an intensity value of I_2, so we set $I[0] \leftarrow I_2$. We repeat the same process with nodes 3 and 1, in that order. The algorithm continues with the rest of submatrices until reaching leaf nodes. In case of having more attributes, we would create a sequence A for each attribute in an analogous way to the sequence of intensities I.

At the bottom-right of Fig. 2 we include the final structures that represent the k^3-lidar of the example.

4.4 Query

In this section, we describe two queries designed for the k^3-lidar, which are of interest for LiDAR point clouds.

Obtaining All Points of a Region (*getRegion*): The k^3-lidar is able to obtain all points of a given region $\langle x_i, y_i, z_i \rangle \times \langle x_e, y_e, z_e \rangle$ by performing a top-down traversal of the tree from the root node. We follow the branches corresponding to submatrices that overlap with the region of interest. When a leaf is found, the coordinates of its points are checked and the attributes are only retrieved if the point is within the region. Due to the fact that points follow a z-order, some mechanisms are included to decrease the number of points checked. For instance, when we reach a point $\langle x', y', z' \rangle$ and $x' > x_e$, the process stops as we can assure that there are no more valid points in that node.

Algorithm 1 shows a pseudocode of the algorithm to solve this query. Let T, H, N, X, Y, Z, and $1s\_in\_T$ be global parameters and $n \times n \times n$ the size of the dataset. Parameter $1s\_in\_T$ stores the total number of 1s in bitmap T and was

calculated previously. The parameter *result* is the list of points returned by the query. Given a region $\langle x_i, y_i, z_i \rangle \times \langle x_e, y_e, z_e \rangle$, the first call of the algorithm is **getRegion**$(n, x_i, y_i, z_i, x_e, y_e, z_e, 0)$.

Lines 1–3 run through each child node that overlaps the defined region. *c_pos* (Line 4) is the position in T of the current child node. The condition in Line 5 determines if the node is at level $\lceil log_k n \rceil$ (which corresponds to submatrices of size $1 \times 1 \times 1$) or not.

Line 6 checks if the node is an internal node or a leaf node. In the first case, a recursive call is invoke (Lines 7–10). Function **getLocalCoordinates** converts the given region into local coordinates with respect to the current node. The position of its children is calculated as $rank(T, c\_pos) \cdot k^3$.

Lines 11–22 are executed when the algorithm reaches a leaf node. Line 12 counts the number of 0-bits in T until position *c_pos*, i.e the number of leaves until that position (*#leaves*). The condition in Line 13 checks if the node is not empty. Line 14 counts the number of leaf nodes containing points (*#ones*) until position *#leaves*.

Lines 15–17 obtain the positions of the first point and the last point of the current child node in arrays X, Y, and Z. Since the number of points has been inserted in unary code, each node corresponds to a 1-bit in the bitmap. With the *select* operation, we obtain the position of the 1-bit corresponding to the previous node and the 1-bit of the current node. Intermediate positions are points of the current node. Finally, in Lines 18–22, the algorithm gets the coordinates of each point and checks if it belongs to the region of interest. If affirmative, the corresponding attributes are added and the point is inserted into the final result.

When the algorithm reaches a node in the level $\lceil log_k n \rceil$, Lines 24–32 are executed. Line 24 calculates the position in H, recall that level $\lceil log_k n \rceil$ is not represented in T. Lines 26–29 are equal to lines 14–17. Finally, for each point in the node, the algorithm retrieves its information and adds the point to the list. Observe that it is not necessary to obtain the local coordinates of the vectors X, Y and Z.

Obtaining All Points of a Region Filtered by Attribute Value (*filterAttRegion*): Given region $\langle x_i, y_i, z_i \rangle \times \langle x_e, y_e, z_e \rangle$ and a range of values for an attribute $[A_i, A_e]$, this query obtains all points within the defined region with values for the attribute between A_i and A_e. Again, this query performs a top-down traversal of the tree. Unlike the query **getRegion**, when the algorithm reaches a leaf node, in addition to the coordinates, it also retrieves the attribute value of the point.

Algorithm 1. getRegion$(n, x_i, y_i, z_i, x_e, y_e, z_e, children\_pos)$ returns all LIDAR points from region $\langle x_i, y_i, z_i \rangle \times \langle x_e, y_e, z_e \rangle$

```
 1  for x' ← ⌊xi/(n/k)⌋ ... ⌊xe/(n/k)⌋ do
 2    for y' ← ⌊yi/(n/k)⌋ ... ⌊ye/(n/k)⌋ do
 3      for z' ← ⌊zi/(n/k)⌋ ... ⌊ze/(n/k)⌋ do
 4        c_pos ← children_pos + k · k · x' + k · y' + z'
 5        if (n/k) ≠ 1 then /* not at level ⌈logk n⌉ */
 6          if T[c_pos] = 1 then /* internal node */
 7            ⟨x'i, x'e⟩ ← getLocalCoordinates(x', xi, xe, (n/k))
 8            ⟨y'i, y'e⟩ ← getLocalCoordinates(y', yi, ye, (n/k))
 9            ⟨z'i, z'e⟩ ← getLocalCoordinates(z', zi, ze, (n/k))
10            getRegion(n/k, x'i, y'i, z'i, x'e, y'e, z'e, rank(T, c_pos) · k³)
11          else /* leaf node */
12            #leaves ← rank0(T, c_pos)
13            if H[#leaves] = 1 then /* Node with points */
14              #ones ← rank(H, #leaves)
15              if #ones = 0 then  p_init ← 0 ;
16              else p_init ← select(N, #ones) + 1;
17              p_end ← select(N, #ones + 1) + 1
18              for p' ← p_init ... p_end do
19                ⟨px, py, pz⟩ ← ⟨X[p], Y[p], Z[p]⟩
20                if ⟨px, py, pz⟩ within ⟨xi, yi, zi⟩ × ⟨xe, ye, ze⟩ then
21                  point ← retrieve_attributes(p, ⟨px, py, pz⟩)
22                  ADD point to result

23        else /* last level */
24          #leaves ← c_pos − 1s_in_T
25          if H[#leaves] = 1 then /* Node with points */
26            #ones ← rank(H, #leaves)
27            if #ones = 0 then  p_init ← 0 ;
28            else p_init ← select(N, #ones) + 1;
29            p_end ← select(N, #ones + 1) + 1
30            for p' ← p_init ... p_end do
31              point ← retrieve_attributes(p, ⟨x', y', z'⟩)
32              ADD point to result
```

---

5   Experimental Evaluation

We ran some experiments as a proof of concept of the good properties of our proposed data format. More concretely, we compare the space and time results obtained by k^3-lidar, to those obtained when using LAS/LAZ formats.

Table 1 shows the description of the LiDAR point clouds used in the experimental evaluation. The last three rows show the values of the coordinates after converting to integers with the procedure explained in Sect. 4.2. We use five different datasets coming from two different sources, an airborne LiDAR and a mobile laser scanning:

- Three datasets were created from the union of different files of the *Plan Nacional de Observación del Territorio*[3] (PNOA). Each tile (file) represents an area of Spanish territory of size 2×2 km with a minimum density of

---

[3] http://pnoa.ign.es/productos_lidar.

0.5 points/m². `PNOA-small` is composed of 4 tiles and represents an area of 16 km², `PNOA-medium` is composed of 9 tiles and represents an area of 36 km², and `PNOA-large` is composed of 23 tiles and represents an area of 92 km². The number of points can vary from one tile to another. These datasets contain the following attributes: intensity (with values between 0 and 255), return number (with values between 1 and 4), number of returns (with values between 1 and 4), edge of flight line (with values between 0 and 1), scan direction flag (with values between 0 and 1), classification (with values between 1 and 7), scan angle rank (with values between −24 and 28), and point source ID (with values between 175 and 227).

– We use datasets `TUB1` and `FireBrigade` from the ISPRS benchmark on indoor modelling[4] [11]. `TUB1` point cloud was captured in one of the buildings of the Technische Universität Braunschweig, Germany, and `FireBrigade` was captured in the office of fire brigade in Delft, The Netherlands. These datasets do not contain values for any additional attribute, just the point coordinates. The level of clutter, defined as the amount of points belonging to elements that do not constitute the building structures, is low for `TUB1` and high for `FireBrigade`.

In all cases, LiDAR points were converted to Point Data Record Format 0 (LAS Specification 1.4). Then we created indexes for the LAZ files using the *lasindex* tool of LAStools[5]. These indexes are able to index the x and y dimensions to improve the query time. LASLib library[6] was used to execute queries on LAZ files. The k^3-lidar was configured with $k = 2$ and $l = 100$.

All the experiments were run on an isolated Intel® Core™ i7-3820 CPU @ 3.60 GHz (4 cores) with 10 MB of cache, and 64 GB of RAM. It ran Debian 9.8 *Stretch*, using gcc version 6.3.0 with −03 option.

The comparison of space is shown in the first columns of Table 2. LAZ files obtain the best results in the three cases, around 65% less than k^3-lidar, which in turn, needs around 53% less space than the uncompressed LAS.

Table 2 also shows the query times. We generated 500 random regions of different sizes. However, the LAS software failed in many queries (the reported result contains 0 points). Therefore we only considered the times of those techniques that worked properly, that is, LAZ and k^3-lidar. Our proposal is around 5 times faster than LAZ in *GetRegion* queries for `PNOA` datasets and from 16–23 times faster for `ISPRS` datasets. *FilterAttRegion* queries were only executed over `PNOA` datasets filtering by the intensity attribute, as `TUB1` and `FireBrigade` only contain the coordinates of the points, but do not include any other attribute values. For `PNOA` datasets, our proposal outperforms LAZ format, as queries are solved 5–10 times faster.

---

[4] http://www2.isprs.org/commissions/comm4/wg5/benchmark-on-indoor-modelling. html.

[5] https://github.com/LAStools/LAStools.

[6] https://github.com/LAStools/LAStools/tree/master/LASlib.

Table 1. Datasets description. We show the number of points, the minimum and maximum values for the real coordinates x, y, z, and the maximum X, Y, Z values, after the coordinates are converted (scaled and translated using an offset).

	PNOA-small	PNOA-medium	PNOA-large	TUB1	FireBridge
#points	13,265,144	25,108,130	52,627,503	32,597,694	10,406,389
min x (real)	546,000.00	544,000.00	542,000.00	−9.90	44.32
max x (real)	549,999.99	549,999.99	551,999.99	5.76	58.40
min y (real)	4,798,000.00	4,798,000.00	4,794,000.00	−24.52	23.10
max y (real)	4,801,999.99	4,803,999.99	4,805,020.45	18.30	77.67
min z (real)	−39.14	−162.43	−162.43	−1.46	1.27
max z (real)	179.58	1005.05	1005.05	1.10	12.04
max X (converted)	3,999,990	5,999,990	9,999,990	1,000,000,000	1,000,000,000
max Y (converted)	3,999,990	5,999,990	11,020,450	1,000,000,000	1,000,000,000
max Z (converted)	218,720	1,167,480	1,167,480	1,000,000,000	1,000,000,000

Table 2. Comparison of the space (MB) and average time (in milliseconds) for queries *getRegion* and *FilterAttRegion*.

Dataset	# points	Space (MB)			GetRegion (ms)		FilterAttRegion (ms)	
		LAS	LAZ	k^3-lidar	LAZ	k^3-lidar	LAZ	k^3-lidar
PNOA-small	13,265,144	254	**43**	119	1,524	**249**	1,517	**145**
PNOA-medium	25,108,130	479	**80**	225	2,521	**424**	2,655	**374**
PNOA-large	52,627,503	1004	**173**	471	6,859	**1,189**	6,283	**1,264**
TUB1	32,597,694	622	**196**	304	6,145	**383**	–	–
FireBrigade	10,406,389	199	**77**	100	1,717	**74**	–	–

6 Conclusions

In this work, we address the main drawback of the LAZ format for LiDAR data, which is its high executing times when answering to queries that retrieve a subset of points using constraints over the third dimension. LAZ is penalized by the fact that decompression is performed by blocks and the index only covers the X and Y coordinates.

We propose a new representation for LiDAR point clouds, denoted k^3-lidar, which is able to decompress random points of the cloud and, as it is based on a compact version of an octree, the k^3-tree, it can index the three dimensions. This implies significant improvements in the querying times, ranging from 5 times to more than one order of magnitude faster.

The future work will cover two main lines. First, regarding the space requirements, k^3-lidar compresses 65% less than LAZ. Therefore, the use of previous ideas from the original k^2-tree and k^3-tree, such as using different k values in different levels, or further compressing frequent submatrices, will be studied. The other line is the indexation of more dimensions, to include the attribute values stored at the points. The k^2-raster is a compact data structure designed for raster data that not only indexes the data spatially, but it also indexes the values stored at the cells of the raster. Our aim is to apply similar ideas to address the indexation of LiDAR data.

Appendix

To better understand the nature of the datasets, we show a visualization of PNOA-large in Fig. 3, and visualizations of the point clouds TUB1 and FireBrigade in Fig. 4.

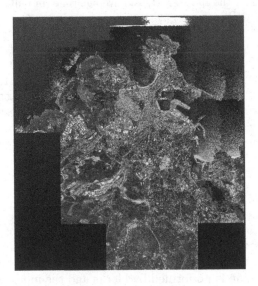

Fig. 3. Visualization of the dataset labeled as *Large*.

(a) TUB1 point cloud visualization

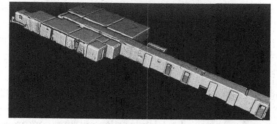

(b) TUB1 eye-dome lighting visualization

(c) FireBrigade point cloud visualization

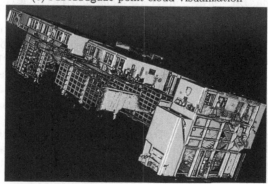

(d) FireBrigade eye-dome lighting visualization

Fig. 4. Visualization of datasets TUB1 and FireBrigade. We include the point cloud visualization and also an eye-dome lighting (EDL) visualization. EDL is a non-photorealistic, image-based shading technique designed to improve depth perception in scientific visualization images [18].

References

1. de Bernardo, G.: New data structures and algorithms for the efficient management of large spatial datasets. Ph.D. thesis, Universidade da Coruña (2014)
2. de Bernardo, G., Álvarez-García, S., Brisaboa, N.R., Navarro, G., Pedreira, O.: Compact querieable representations of raster data. In: Proceedings of the 20th SPIRE, pp. 96–108 (2013)
3. Brisaboa, N.R., Luaces, M.R., Navarro, G., Seco, D.: A new point access method based on wavelet trees. In: Heuser, C.A., Pernul, G. (eds.) ER 2009. LNCS, vol. 5833, pp. 297–306. Springer, Heidelberg (2009). https://doi.org/10.1007/978-3-642-04947-7_36
4. Brisaboa, N.R., Ladra, S., Navarro, G.: DACS: bringing direct access to variable-length codes. Inf. Process. Manage. **49**, 392–404 (2013)
5. Brisaboa, N.R., Ladra, S., Navarro, G.: Compact representation of web graphs with extended functionality. Inf. Syst. **39**, 152–174 (2014)
6. Dong, P., Chen, Q.: LiDAR Remote Sensing and Applications. CRC Press, Boca Raton (2017)
7. Grossi, R., Gupta, A., Vitter, J.S.: High-order entropy-compressed text indexes. In: Proceedings of SODA 2003. Society for Industrial and Applied Mathematics, Philadelphia, PA, Baltimore, Maryland, pp. 841–850 ,12–14 January 2003
8. Hyyppä, J., et al.: Forest inventory using laser scanning. In: Topographic Laser Ranging and Scanning, pp. 379–412. CRC Press (2018)
9. Isenburg, M.: Laszip: lossless compression of lidar data. Photogram. Eng. Remote Sens. **79**(2), 209–217 (2013)
10. Jaboyedoff, M., et al.: Use of lidar in landslide investigations: a review. Nat. Hazards **61**(1), 5–28 (2012)
11. Khoshelham, K., Vilariño, L.D., Peter, M., Kang, Z., Acharya, D.: The ISPRS benchmark on indoor modelling. Int. Arch. Photogram. Remote Sens. Spat. Inf. Sci. **42**, 367–372 (2017)
12. Klinger, A.: Pattern and Search Statistics. Academic Press (1971). https://doi.org/10.1016/B978-0-12-604550-5.50019-5
13. Ladra, S., Paramá, J.R., Silva-Coira, F.: Scalable and queryable compressed storage structure for raster data. Inf. Syst. **72**, 179–204 (2017)
14. Meagher, D.: Geometric modeling using octree encoding. Computr Graphics Image Process. **19**(2), 129–147 (1982). https://doi.org/10.1016/0146-664X(82)90104-6. http://www.sciencedirect.com/science/article/pii/0146664X82901046
15. Navarro, G.: Compact Data Structures: A Practical Approach. Cambridge University Press, Cambridge (2016)
16. Navarro, G., Nekrich, Y., Russo, L.: Space-efficient data-analysis queries on grids. Theoret. Comput. Sci. **482**, 60–72 (2013)
17. Palomer, A., Ridao, P., Youakim, D., Ribas, D., Forest, J., Petillot, Y.: 3D laser scanner for underwater manipulation. Sensors **18**(4), 1086 (2018)
18. Ribes, A., Boucheny, C.: Eye-dome lighting: a non-photorealistic shading technique. Technical report (04 2011)
19. Said, A.: Arithmetic coding. In: Lossless Compression Handbook. Elsevier (2002)
20. Samet, H.: The quadtree and related hierarchical data structures. ACM Comput. Surv. **16**, 187–260 (1984). https://doi.org/10.1145/356924.356930
21. The American Society for Photogrammetry and Remote Sensing: ASPRS LIDAR Data Exchange Format Standard. Version 1.0. Format Specification (2003)

22. The American Society for Photogrammetry and Remote Sensing: LAS Specification 1.4 - R14. Format Specification (2019)
23. Wang, R., Peethambaran, J., Chen, D.: Lidar point clouds to 3-D urban models: a review. IEEE J. Sel. Top. Appl. Earth Observations Remote Sens. **11**(2), 606–627 (2018)

Author Index